Further Elementary Analysis

Related titles in A-level mathematics:

Elementary Analysis
A. Dakin OBE, MA, B Sc and R. I. Porter MBE, MA

Third edition, metricated

With *Further Elementary Analysis* this book provides a two-year VIth form course in Pure Mathematics.

Further Mathematics
R. I. Porter MBE, MA

Second edition, metricated

Completes the course for specialists and gives adequate preparation for the 'S' papers in G.C.E.

A School Course in Vectors
R. I. Porter MBE, MA

Simple introduction to the algebra of vectors showing the application of vector methods to Geometry and Mechanics.

A Concise Course in Pure Mathematics
R. I. Porter MBE, MA

A complete treatment of the Pure Mathematics content of Advanced Level Mathematics Courses.

Further Elementary Analysis

R. I. Porter MBE, MA
Formerly Headmaster, Queen Elizabeth Grammar School, Penrith

CollinsEducational
An imprint of HarperCollins*Publishers*

Published by
Collins Educational
77-85 Fulham Palace Road
Hammersmith London W6 8JB

First published in 1951 by G. Bell & Sons Ltd
Reprinted 1952
Second edition, revised 1953, reprinted nine times
Third edition 1967, reprinted 1969
Fourth metricated edition 1970
Reprinted 1972, 1974, 1976, 1978 (twice), 1979, 1981, 1982, 1984, 1986, 1990, 1992

This edition © R. I. Porter 1986

All rights reserved. No part of this publication
may be reproduced, stored in a retrieval system,
or transmitted, in any form or by any means,
electronic, mechanical, photocopying, recording
or otherwise, without the prior permission of
Collins Educational

British Library Cataloguing in Publication Data

Porter, R. I.
 A concise course in pure mathematics.
 1. Mathematics—Examinations, questions, etc.
 I. Title
 510′.76   QA43

ISBN 0 00 322306 X

515

Typeset by Advanced Filmsetters (Glasgow) Ltd
Produced in Hong Kong

Preface

This book is a continuation of *Elementary Analysis* and is intended to complete a two-year VIth Form course in Pure Mathematics suitable for all but the mathematical specialist.

As the title suggests, the book is largely concerned with the development and applications of the calculus together with the necessary parallel developments in other branches of mathematics. In order to attain the object of a single book to cover a complete course in Pure Mathematics, some additional topics such as Choice and Chance are included.

In choosing the subject matter of this volume, due attention has been given to syllabuses suggested by the different Examining Bodies for the General Certificate of Education. The book covers completely the syllabuses for Ordinary Mathematics at Advanced level and with the exception of Pure Geometry, contains most of or all the topics included in those for Further Mathematics.

As in the previous book, the aim is to introduce pupils as quickly as possible to fresh mathematical fields and to make them acquainted with new mathematical ideas. To achieve this aim it has been necessary in most instances to dispense with formal proofs and rigid lines of approach.

Experience has proved the necessity for large numbers of examples of all types and it will be found that this requirement is very adequately met. Examples are divided into two classes, those marked [A], which are straight-forward questions of a drill type, and those marked [B], which are questions of greater length and difficulty. It is hoped that the numerous miscellaneous examples will be valuable for purposes of revision.

I am indebted to my colleague, Mr. J. Y. Rushbrooke, for the assistance he has given in reading the proofs and checking the answers.

Note on fifth edition

In this edition the opportunity has been taken to thoroughly update the text and the notation has been revised to conform with current requirements.

R.P.

Contents

1 Permutations and combinations
Elementary probability   1
The r, s principle
Arrangements of unlike
   quantities
Selections of unlike
   quantities
Conditional arrangements
   and selections

Arrangements of a number
   of quantities not all
   different
Arrangements and selections
   with repetitions

Miscellaneous examples

Probability or chance
Unit and percentage
   probability

Odds on and against an
   event happening

2 Logarithms
Logarithmic and exponential functions   20
Laws of logarithms
Change of base
Logarithmic functions
Differentiation of
   logarithmic functions

Integrals of form $\int \frac{f'(x)}{f(x)} dx$
Exponential functions
Differentiation and
   integration of exponential
   functions

Miscellaneous examples

3 Power series
Derivation and use of exponential, logarithmic and trigonometrical series   38
The exponential series
Evaluation of e
The logarithmic series
Evaluation of logarithms

Trigonometric series
Expansions of $\sin x$, $\cos x$
   and their applications

Miscellaneous examples

4 Finite algebraic series   54
Binomial theorem for
   positive integral index
Greatest term and greatest
   coefficient

Σ notation
Powers of natural numbers
Application to summation of
   series

Contents

Proof by method of mathematical induction

Miscellaneous methods of summation

Miscellaneous examples

5 Definite integration as a summation Geometrical applications 64

Definition of a definite integral as the limit of a sum.
Connection between definite and indefinite integration
Evaluation of definite integrals from the definition and by the use of indefinite integration

Approximate integration
Trapezium and Simpson's Rules
Area under a curve
Volume of revolution
Mean value and root mean square value

Miscellaneous examples

6 Further applications of definite integration 80

Centre of gravity
Moment of inertia and radius of gyration

Plane area and parallel axis theorems

7 Solution of equations 95

Simultaneous algebraic equations—linear equations in 3 unknowns: quadratic equations in 2 unknowns
Trigonometrical equations
Angles with the same sine, cosine or tangent
General solutions of equations
Equations of a quadratic form

Equations of the form $a \sin x + b \cos x = c$
Equations solvable by the method of factors
Graphical solution of equations
Approximate solution of equations
Newton's method

8 Inequalities 112

Properties of inequalities
Solution of inequalities

Modulus of a function

Miscellaneous examples

9 Equations of a curve Elementary curve tracing 120

Cartesian, parametric and polar equations of a curve

Asymptotes
Change of origin

Parametric coordinates
Polar coordinates
Connections between Polar and Cartesian coordinates
Curve tracing in Cartesian coordinates

Curve tracing in parametric coordinates
Curve tracing in polar coordinates
Sectorial areas

Miscellaneous examples

10 Analytical geometry
The straight line 137

Revision examples on the straight line
Different forms of the equation of a straight line

Equation of any straight line through the point of intersection of two given straight lines
Point dividing a line in a given ratio

Miscellaneous examples

11 The circle 149

Equation of a circle with given centre and radius
The equation
$x^2 + y^2 + 2gx + 2fy + c = 0$
Equation of the tangent at a point on a circle

Lengths of tangents
Equation of any circle passing through the points of intersection of two given circles

Miscellaneous examples

12 The parabola 162

Simplest form of the equation of a parabola
Equation of tangent at point (x_1, y_1) to parabola $y^2 = 4ax$
Parametric equations of a parabola

Parametric coordinates $(at^2, 2at)$
Tangent and normal at point $(at^2, 2at)$
Gradient form of tangent
Geometrical properties of the parabola

13 The ellipse 176

Simplest form of the equation of an ellipse
Tangent at point (x_1, y_1) to the ellipse $\dfrac{x^2}{a^2} + \dfrac{y^2}{b^2} = 1$
Parametric equations of an ellipse
Parametric coordinates $(a\cos\theta, b\sin\theta)$

The eccentric angle
Area of an ellipse
Tangent and normal at the point $(a\cos\phi, b\sin\phi)$
Gradient form of tangent
Conjugate diameters
Simple geometrical properties of the ellipse

x | Contents

14 The hyperbola — 195

Simplest form of the equation of a hyperbola
Asymptotes. Properties of a hyperbola
The rectangular hyperbola
Equation of a rectangular hyperbola with respect to its asymptotes, $xy = c^2$

Parametric equations and coordinates
Tangent and normal at the point $\left(ct, \dfrac{c}{t}\right)$

Miscellaneous examples

15 Inverse circular and hyperbolic functions — 209

Inverse circular functions and their graphs
Identities
Differentiation of inverse circular functions
The integrals $\int \dfrac{dx}{\sqrt{a^2 - x^2}}$, $\int \dfrac{dx}{a^2 + x^2}$
Hyperbolic functions and their graphs
Formulae connecting hyperbolic functions

Differentiation and integration of hyperbolic functions
Inverse hyperbolic functions and their graphs
Expression of inverse hyperbolic functions in terms of logarithmic functions
Differentiation of inverse hyperbolic functions
The integrals $\int \dfrac{dx}{\sqrt{a^2 + x^2}}$, $\int \dfrac{dx}{\sqrt{x^2 - a^2}}$ and allied types

Miscellaneous examples

16 Partial fractions and their applications — 226

Methods of expressing an algebraic fractional function as the sum of partial fractions

Use of partial fractions for the expansion and integration of algebraic fractional functions

Miscellaneous examples

17 Further methods of integration and standard types of integral — 236

Standard forms
Integration by substitution

Integration by parts
Easy reduction formulae

Miscellaneous examples

18 Differential properties of plane curves
Curvature
253

Arc length and area of surface of revolution for curves expressed in Cartesian and parametric coordinates
Surface area of a zone of a sphere
Curvature and radius of curvature of plane curves

Expressions for radius of curvature in Cartesian and parametric coordinates
Centre of curvature
Curvature at the origin, Newton's formula

Miscellaneous examples

19 Differential equations
265

Formation of differential equations
Equations of the first order and first degree
Variables separable
Homogeneous equations
Linear equations
Equations of the second order

Types $\dfrac{d^2y}{dx^2} = f(x)$, $\dfrac{d^2y}{dx^2} = f(y)$
Linear equations with constant coefficients of the forms $a\dfrac{d^2y}{dx^2} + b\dfrac{dy}{dx} + cy = 0$,
$a\dfrac{d^2y}{dx^2} + b\dfrac{dy}{dx} + cy = f(x)$

20 Complex numbers
290

Definition, notation and geometrical representation
Argand diagram
Fundamental processes
Equality, addition, subtraction, multiplication and division

The cube roots of unity
The (r, θ) form of a complex number
Modulus and argument

Miscellaneous examples

Revision papers   301

Answers   336

Index   387

1 | Permutations and combinations
Elementary probability

Arrangements and selections Before attempting to give formal definitions of the words *permutation* and *combination*, a discussion of the following problem will be of value:

Four candidates are nominated for the posts of form captain and form vice-captain and we require to know in how many ways any member of the form may cast his or her votes.

Initially, let us assume that all members of the form are instructed to vote for any two of the four candidates without indicating their choices for the separate posts.

In order to distinguish between the candidates let them be denoted by the letters A, B, C, D. Then two letters can be chosen in the following ways:

$A, B;$ $A, C;$ $A, D;$
$B, C;$ $B, D;$ $C, D.$

So there are six ways of voting for two of four candidates when no attempt is made to allocate the two different posts to the chosen candidates.

Now change the problem by assuming that all members of the form are told to vote for two candidates and to indicate which of the two is their choice for form captain. In this case there will be twice as many ways of voting as in that discussed above, for corresponding to each of the six ways of selecting two out of the four candidates, there will be two ways of choosing the form captain, as either of these two may be chosen. The possible ways of voting will be as follows, the first letter in each case being the choice for captain:

$A, B;$ $A, C;$ $A, D;$ $B, C;$ $B, D;$ $C, D;$
$B, A;$ $C, A;$ $D, A;$ $C, B;$ $D, B;$ $D, C.$

The results of this example can be expressed by saying that there are six ways of choosing two articles from four different articles when the order of choice is immaterial and twelve ways of choosing two when the order of choice is taken into account. More briefly we say that from four different articles, six *selections or combinations* of two articles, and twelve *arrangements or permutations* of two articles can be made.

The formal definitions of the terms combination and permutation are as follows:

A Combination or Selection of a given number of articles is a set or group of articles selected from those given where the order of the articles in the set or group is not taken into account.

A Permutation or Arrangement of a given number of articles is a set or group of articles chosen from those given where the order of the articles in the set or group is taken into consideration.

Thus the two sets A, B; B, A are the same combination but different permutations of the two letters A and B.

Before introducing the notation used for denoting combinations and permutations and considering some of their elementary properties, the fundamental principle which forms the basis of this subject will be discussed.

The (r, s) principle This important principle can be expressed as follows:

If one process or operation can be done in A ways and a second process or operation can be done in B ways, then the number of different ways in which both processes or operations can be performed is A × B.

Example 1 If there are 6 routes between towns A and B and 4 routes between towns B and C, how many different ways are there of travelling from A to C via B?

As any one of the 6 ways from A to B can be combined with any one of the 4 ways from B to C, it follows that there are 6×4 ways of travelling from A to C via B.

i.e. There are 24 different ways of travelling from A to C via B.

Example 2 There are 10 vacant seats in a railway compartment. In how many ways can two persons seat themselves?

The first person can occupy any of the vacant seats and consequently can choose a seat in 10 ways. There are now 9 seats available for the other person and so he can choose a seat in 9 ways.

Hence there are 10×9 ways in which the two persons can be seated.

Example 3 How many two-letter words, each consisting of one consonant and one vowel, can be made from an alphabet of twenty consonants and six vowels?

As one consonant can be chosen in 20 ways and one vowel in 6 ways, there will be 120 ways of choosing both. However, in forming a word the order of the letters must be taken into account and two different words can be formed by using one consonant and one vowel. It follows that there are 240 different two-letter words.

Extension of the r, s principle The r, s principle can be extended to cover more than two operations, for if one thing can be done in p ways, a second in q ways, a third in r ways, and so on, clearly the number of ways of performing all the operations is $p \times q \times r \times \ldots$.

Example 4 How many different ways are there of forecasting the results of eight football matches where each match can end in one of three ways?

As each of the eight results can be forecast in 3 ways, the total number of ways of forecasting all eight results is 3^8 or 6561.

Example 5 In how many ways can 3 letters be put into 3 envelopes, one in each?

For the first envelope we have a choice of 3 letters and consequently there are three ways in which it can be filled. There are now only 2 letters left and thus there are 2 ways of filling the second envelope. As there is now only one letter remaining it is only possible to fill the last envelope in one way.

Hence the total number of ways of filling all three envelopes is $3 \times 2 \times 1$, i.e. 6.

Example 6 How many different sums of money can be formed by using some or all the following coins: one fiftypence, one tenpence, one fivepence, one twopence, one penny and one twentypence?

There are two ways of disposing of each coin, it can either be taken or rejected. As there are six coins, it follows that they all can be disposed of in $2 \times 2 \times 2 \times 2 \times 2 \times 2$, i.e.

2^6 ways. However, one of these ways would involve the rejection of all the coins and is consequently not admissible.

Therefore the number of different sums of money which can be formed is $2^6 - 1$ or 63.

Example 7 How many numbers are there which consist of four digits?

The first digit may be any of the ten except 0. Thus the first digit of the number can be chosen in 9 ways. Each of the other three digits may be any of the ten and so there are 10 ways of assigning each of them.

Hence the number can be formed in $9 \times 10 \times 10 \times 10$ different ways. In other words, there are 9000 different four digit numbers.

[A] EXAMPLES 1a

1 If there are 5 routes from A to B and 3 routes from B to C, how many different ways are there of travelling from A to C via B?

2 If there are 6 steamers plying between two ports in how many ways can a man travel from one port to the other and return by a different steamer?

3 In how many ways can a boy and a girl be chosen from a group of 12 boys and 10 girls?

4 In how many ways can a consonant and a vowel be chosen from the letters of the word *volume*?

5 There are 5 candidates for a Mathematical, 6 for a Classical and 8 for a Natural Science Scholarship. In how many ways can the scholarships be awarded?

6 There are four seats in a taxi. In how many ways can 3 people seat themselves?

7 In how many ways can four travellers stay at 5 hotels so that each is at a different hotel?.

8 In how many ways can four letters be put in four envelopes, one in each?

9 How many different ways are there of forecasting the results of six football matches?

10 In how many ways can four different prizes be awarded to a form of 12 pupils, if (i) any pupil can only win one prize, (ii) there is no restriction on the number of prizes that any pupil can win?

11 How many code words of 3 letters can be formed from the 26 letters of the alphabet, repetition being allowed?

12 How many 3 digit numbers can be formed by using the digits 1, 2, 3, 4, 5, no repetition being allowed?

13 How many 5 digit numbers are there?

14 How many 4 digit numbers greater than 2000 can be formed from the digits 0, 1, 2, 3, 4, repetition being allowed?

15 In the previous question, how many of the 4 digit numbers greater than 2000 will be divisible by 5?

Arrangements of unlike quantities In order to illustrate the general principle we will consider the following examples:

Example 8 In how many ways can eight different books be arranged on a bookshelf having eight consecutive vacant places?

Imagine 8 consecutive vacant places on the bookshelf and consider in how many different ways these places can be filled. The first place can be filled by any one of the 8 books, that is, in 8 ways. There are now 7 books remaining and consequently the second

4 | Permutations and combinations

place can be filled in 7 different ways. Similarly, the third place can be filled in 6 ways and so on. Hence the total number of ways of filling the 8 vacant places is

$$8 \times 7 \times 6 \times 5 \times 4 \times 3 \times 2 \times 1,$$

or 40 320.

The number of ways of filling 8 vacant places with 8 different books is clearly the same as the number of ways of arranging the 8 books on the bookshelf or the number of ways of arranging the 8 books amongst themselves. As already defined, an arrangement of a set of articles when due regard is paid to the order of the articles is called a permutation and the result of the above example can be stated as follows:

The number of permutations of 8 unlike quantities taken all together is the product of 8 and all the integers less than it; i.e.

$$8 \times 7 \times 6 \times 5 \times 4 \times 3 \times 2 \times 1.$$

Notation The continued product of an integer n and all lesser integers is called *factorial n* and is denoted by the following notation:

$$\text{Factorial } n = n \times (n-1) \times (n-2) \ldots 3 \times 2 \times 1 = n!$$

Using this notation, the number of permutations of 8 unlike quantities taken all together can be expressed as factorial 8—i.e. 8!.

Example 9 In how many ways can 5 vacant places on a bookshelf be filled if 8 different books are available?

This example is solved in exactly the same way as Example 8. Clearly there are 8 ways of filling the first space, 7 ways of filling the second space, and so on, until for the fifth and last vacant space there will be 4 ways. Hence the total number of ways of filling all five vacant spaces is $8 \times 7 \times 6 \times 5 \times 4$, i.e. 6720.

The number of ways of filling five spaces by using any five of eight different books is the same as the number of ways of arranging groups of five articles chosen from eight different articles. As stated previously, each of these arrangements is called a permutation and consequently this example gives the following result:

The number of permutation of 8 different articles taken 5 at a time is

$$8 \times 7 \times 6 \times 5 \times 4, \text{ i.e. } \frac{8 \times 7 \times 6 \times 5 \times 4 \times 3 \times 2 \times 1}{3 \times 2 \times 1} = \frac{8!}{3!} \text{ or } \frac{8!}{(8-5)!}.$$

General results The methods used in the two previous examples lead to the following important general results:

The number of permutations or arrangements of n unlike things taken r at a time is $n(n-1)(n-2) \ldots (n-r+1)$ or $\dfrac{n!}{(n-r)!}$.

The number of permutations of n unlike things taken all together, or the number of different ways of arranging n unlike things, is

$$n(n-1)(n-2) \ldots 3.2.1 \quad \text{or} \quad n!$$

Notation The number of permutations of n things taken r at a time is denoted by the symbol $_nP_r$.

i.e.
$$_nP_r = n(n-1)(n-2) \ldots (n-r+1) = \frac{n!}{(n-r)!}.$$

Example 10 Evaluate $_5P_3$ and $_5P_5$.
$_5P_3$ = number of permutations of 5 things taken 3 at a time
   = 5.4.3 = 60.
$_5P_5$ = number of permutations of 5 things taken all together
   = 5.4.3.2.1 = 120.

Example 11 In how many ways can a number of 3 digits be formed with the figures 1, 2, 3, 4, 5, 6?
The required number of ways is the number of permutations of 6 unlike quantities taken 3 at a time.
∴ Number of ways = $_6P_3$ = 6.5.4 = 120.
The same result can, of course, be obtained by use of the r, s principle.

[A] EXAMPLES 1b

Find the number of ways of arranging:
1. 6 different coloured electric bulbs in 4 sockets.
2. 4 different pennies in a row.
3. 3 different rings on 5 hooks.
4. 6 different hats on 6 pegs.
5. The letters a, b, c, d, e taken 3 at a time.
6. 8 different flags one above the other.
7. How many permutations are there of 7 different articles taken 3 at a time?
8. How many three letter words can be formed by using 8 letters without repetition?
9. How many numbers can be formed by using (i) 4, (ii) all the digits 1, 2, 3, 4, 5, without repetition?
10. Find the values of:
    (i) 7!,   (ii) $\dfrac{10!}{9!}$,   (iii) $\dfrac{12!}{8!}$,   (iv) $\dfrac{9!}{5!4!}$,
    (v) $\dfrac{11!}{5!3!3!}$,   (vi) 6! + 5!,   (vii) 8! − 6.6!.
11. Express as factorials
    (i) 4 × 3 × 2,   (ii) 6 × 5 × 4,   (iii) 9 × 8 × 7 × 6,
    (iv) $n(n-1)(n-2)$,   (v) $2n(2n-2)(2n-4)$.
12. Find the values of
    (i) $_5P_2$,   (ii) $_6P_1$,   (iii) $_7P_4$,
    (iv) $_4P_4$,   (v) $_4P_2 \times _5P_3$,   (vi) $\dfrac{_nP_r}{_nP_{n-r}}$.
13. Write down expressions for:
    (i) $_nP_3$,   (ii) $_{n-1}P_2$,   (iii) $_nP_{n-1}$,   (iv) $_{2n}P_{2n-2}$.
14. In how many ways can a crew of 8 oarsmen be arranged in a boat?
15. How many signals can be formed by using 5 out of 8 different flags?
16. In how many ways can the letters of the word *vowels* be arranged?
17. In how many orders can 8 examinations papers be set?
18. In how many ways can 11 boys be arranged in a line if the tallest must be at the right-hand end?
19. In how many ways can 10 short stories be arranged in a book if the longest story must come first and the shortest last?

Selections of unlike quantities A selection or combination is a set of quantities selected from a given group where no attention is paid to the order of the quantities within the set.

Suppose four councillors have to be selected from six candidates and let x be the number of different ways in which the selection can be made. Each of these x ways gives a group of 4 people who can arrange themselves in 4! different ways. Consequently the number of ways of arranging groups of 4 people selected from 6 is $x \times 4!$.

But this number is the number of arrangements or permutations of 6 things taken 4 at a time, i.e. $\frac{6!}{2!}$.

Hence
$$x \cdot 4! = \frac{6!}{2!},$$

$$\therefore x = \frac{6!}{4!\,2!}.$$

The number of ways of choosing four things from a group of six unlike things is called *the number of selections or combinations of* 6 *unlike things taken* 4 *at a time*.

Consequently, the number of combinations of 6 unlike things taken 4 at a time is $\frac{6!}{4!\,2!}$.

General result By adopting the method used in the previous example, the following general result is obtained:

The number of combinations or selections of n unlike things taken r at a time is $\frac{n!}{(n-r)!\,r!}$.

Notation The number of combinations of n unlike things taken r at a time is denoted by the symbol $_nC_r$.

So, $\qquad _nC_r = \frac{n!}{(n-r)!\,r!} \quad \text{or} \quad \frac{n(n-1)(n-2)\ldots(n-r+1)}{r!}$.

e.g. $\qquad _6C_4 = \frac{6!}{2!\,4!} = 15$.

Example 12 In how many ways can 2 bats and 3 balls be selected from 3 bats and 5 balls?

2 bats can be selected from the 3 available in $_3C_2$ ways.
3 balls can be selected from the 5 available in $_5C_3$ ways.
Hence, by applying the r, s principle, it follows that the number of ways of choosing bats and balls $= {}_3C_2 \times {}_5C_3$

$$= \frac{3 \cdot 2}{2} \times \frac{5 \cdot 4 \cdot 3}{3 \cdot 2}$$

$$= 30.$$

To show (i) $r!\,_nC_r = \,_nP_r$; (ii) $_nC_r = \,_nC_{n-r}$.

(i) This result has already been derived in obtaining the value of $_nC_r$. It follows from the fact that corresponding to each combination of r things chosen from n unlike things, there are $r!$ arrangements or permutations.

Thus $\quad _nP_r$ = total number of permutations of n things taken r at a time.
$\quad\quad\quad = r! \times$ total number of combinations of n things taken r at a time.

$$\text{i.e. } _nP_r = r!\,_nC_r.$$

(ii) The result $_nC_r = \,_nC_{n-r}$ follows by either of the following methods:

(a) $_nC_r$ is the number of ways of selecting a group of r things from a given set of n unlike things. Corresponding to each group of r selected there will be a residual group of $n-r$. In other words, $_nC_r$ is merely the number of ways of dividing a set of n unlike things into two groups containing r and $n-r$ things respectively. Similarly it follows that $_nC_{n-r}$ is the number of ways of dividing a set of n unlike things into two groups containing $n-r$ and r things respectively. Obviously these two methods of division are identical and consequently $_nC_r = \,_nC_{n-r}$.

(b)
$$_nC_r = \frac{n!}{(n-r)!\,r!}.$$

Replacing r by $n-r$,

$$_nC_{n-r} = \frac{n!}{(n-\overline{n-r})!\,(n-r)!} = \frac{n!}{r!\,(n-r)!}.$$

Hence
$$_nC_r = \,_nC_{n-r}.$$

The result (ii) above is of considerable value in numerical work as can be seen from the following examples.

Example 13 Find the value of $_{21}C_{18}$.
Bearing in mind that
$$_{21}C_{18} = \frac{21.20.19\,\ldots\,.\,4}{18.17.16\,\ldots\,.\,1}$$
it will be appreciated that much time can be saved by writing
$$_{21}C_{18} = \,_{21}C_{21-18} = \,_{21}C_3 = \frac{21.20.19}{3.2} = 1330.$$

Example 14 Verify that $_{12}C_{10} + \,_{12}C_9 = \,_{13}C_{10}$.

We have
$$_{12}C_{10} = \,_{12}C_2 = \frac{12.11}{2} = 66;$$

$$_{12}C_9 = \,_{12}C_3 = \frac{12.11.10}{3.2} = 220;$$

and
$$_{13}C_{10} = \,_{13}C_3 = \frac{13.12.11}{3.2} = 286.$$

Hence
$$_{12}C_{10} + \,_{12}C_9 = \,_{13}C_{10}.$$

Permutations and combinations

The result of this example is a special case of the general result,

$$_nC_r + {_nC_{r-1}} = {_{n+1}C_r},$$

which can be verified as follows:

$$\begin{aligned} _nC_r + {_nC_{r-1}} &= \frac{n!}{(n-r)!r!} + \frac{n!}{(n-r+1)!(r-1)!} \\ &= \frac{n!}{(n+1-r)!r!}\left\{\frac{(n-r+1)!}{(n-r)!} + \frac{r!}{(r-1)!}\right\} \\ &= \frac{n!}{(n+1-r)!r!}\left\{n-r+1+r\right\} \\ &= \frac{n!(n+1)}{(n+1-r)!r!} = \frac{(n+1)!}{(n+1-r)!r!} \\ &= {_{n+1}C_r}. \end{aligned}$$

[A] **EXAMPLES 1c**

In how many ways can:

1 6 prefects be chosen from 9 boys?
2 4 clerks be chosen from 15 for promotion?
3 A committee of 3 be chosen from 10 aldermen?
4 3 books be chosen from 7 different books?
5 11 players be chosen from 15 boys?
6 A hand of 5 be chosen from 13 different cards?
7 8 matches be selected out of 10?
8 Evaluate (i) $_9C_2$, (ii) $_{10}C_7$, (iii) $_{15}C_{12}$, (iv) $\dfrac{_9C_4}{4!}$.
9 Verify that $_4C_3 + {_4C_2} = {_5C_3}$.
10 Express $_{10}C_6 + {_{10}C_5}$ as a single term and evaluate it.
11 If $_nC_{12} = {_nC_8}$, what is the value of n?
12 If $_{20}C_r = {_{20}C_{r+2}}$, find r.
13 In how many ways can 3 fivepences and 2 tenpences be selected from 5 fivepences and 3 tenpences?
14 In how many ways can 3 white and 2 black keys be simultaneously pressed down on a piano keyboard made up of 50 white and 35 black keys?
15 A bag contains 20 different pennies and 12 different fivepences. In how many ways can a group of coins each containing 2 pennies and 3 fivepences be taken from the bag?
16 Find the number of combinations of 60 things taken 58 at a time.
17 In how many ways can a committee of 5 men and 3 ladies be formed from 12 men and 8 ladies?
18 From 3 capitals, 8 consonants and 4 vowels in how many ways can a group of 6 letters consisting of 1 capital, 3 consonants and 2 vowels be chosen?
19 In how many ways can 8 different things be divided into groups of 5 and 3.
20 In how many ways can 15 people be divided into groups of 7 and 8? In how many ways can three groups containing 7, 5 and 3 people be formed?
21 In how many ways can 12 articles be divided into groups of 6, 4 and 2?

Conditional arrangements and selections

Example 15 How many numbers greater than 5000 can be formed by using some or all the digits, 7, 6, 5, 4, 3 without repetition?

The numbers formed by using all 5 digits will be greater than 5000. Also all 4 digit numbers commencing with 7, 6 or 5 will be greater than 5000.
The number of 5 digit numbers = 5! = 120.
In the 4 digit numbers, the first digit can be chosen in 3 ways. The other 3 digits are selected from 4 and can be arranged in $_4P_3$ ways.
∴ The number of 4 digit numbers = $3 \cdot {_4P_3} = 72$.
Hence the required number = 120 + 72
= 192.

Example 16 In how many ways can 4 English books and 1 Latin book be arranged on a shelf if the Latin book must always be in the middle, the selection being made from 8 English and 3 Latin books?

First consider in how many ways a selection of 4 English books and 1 Latin book can be made from the available books and then consider how these books can be arranged on the shelf.

The Latin book can be selected in $_3C_1$ or 3 ways and can be placed on the shelf in only one position, the middle.

The 4 English books can be selected in $_8C_4$ ways and arranged in the four vacant places in 4! ways. So the number of different arrangements of the English books is $_8C_4 \cdot 4!$ which is, of course, $_8P_4$.

So, the total number of arrangements = $3 \times {_8P_4} = 5040$.

Example 17 In how many ways can the letters of the word *orange* be arranged with the restriction that the three vowels must not come together?

The total number of ways of arranging the 6 letters amongst themselves

$$= 6!.$$

Now obtain the number of these arrangements in which the three vowels come together.

The three vowels can be grouped together as a single term in 3! ways and so we require to find the number of ways of arranging 4 terms amongst themselves and to combine this number, 4!, with the number of ways of arranging the three vowels amongst themselves within the single term.

It follows that the number of arrangements in which the three vowels are consecutive

$$= 4! \cdot 3!$$

Hence the total number of arrangements in which the three vowels do not come together

$$= 6! - 4! \cdot 3!$$
$$= 4!\{6 \cdot 5 - 3!\} = 576.$$

[B] EXAMPLES 1d

1 In how many ways can a crew of 8 rowers be arranged if 2 of them can only row on the bow side?

2 In how many ways can the letters of the word *article* be arranged so that the vowels occupy the even places?

3 Out of a group of 5 boys and 4 girls, in how many ways can a party of 4 be selected which includes at least 2 boys?

4 In how many ways can a party of 6 be chosen from a group of 14 so as to exclude the youngest if it includes the oldest?

5 How many 4-figure numbers greater than 7000 can be formed by using some or all the figures 0 to 9 without repetition? How many of these numbers will be divisible by 5?

6 In how many ways can 4 cards be selected from a pack of 52 cards? In how many of these ways will all the cards not be of different suits?

7 How many even numbers between 500 and 1000 can be formed with the digits 4, 5, 6, 7, 8 if no digit is repeated? How many if digits may be repeated?

8 9 girls are to be arranged in a line. In how many ways can this be done if the tallest and shortest must not come together?

9 How many different hands of 13 cards can be obtained containing at least 12 hearts?

10 In how many ways can 5 boys and 3 girls be arranged in a line if the girls are not separated?

11 There are 8 seats in a railway compartment. In how many ways can 8 people be seated if 2 must have their backs to the engine and 1 must face the engine?

12 In a football team of 11 players, the goalkeeper and centre-forward cannot play in any other position; all the other players are prepared to play in any position. How many arrangements of the team can be made?

13 Two elevens are made up of 22 players. In how many selections will 2 particular players be on opposite sides?

14 How many selections of 4 books, each containing at least 1 Latin book and 1 Greek book, can be made from 5 Latin books and 4 Greek books?

15 A detachment consists of 2 officers, 4 N.C.O's and 12 privates. In how many ways can a guard be selected to include 1 officer or N.C.O. and 5 privates? In how many of these groups will there be no officer?

16 In how many ways can 10 examination papers be arranged so that the 2 mathematical papers are not consecutive?

17 The letters of the word *nought* are to be arranged so that the *u* always follows the *o*. In how many ways can this be done?

18 A boy and his sister are included in a group of 8 boys and 6 girls. In how many ways can a tennis four (2 boys and 2 girls) be selected so as not to include both the boy and his sister?

19 In how many ways can a party of 6 or more be selected from 9 people? How many of these will contain 1 particular person?

Arrangements of a number of things not all different

Example 18 Find the number of different arrangements which can be made by using all the letters *a*, *a*, *a*, *b*, *c*, *d*.

Let the number of different arrangements be x. Suppose the three like terms are replaced by unlike terms a_1, a_2, a_3.

Then, in any one of the x arrangements, without changing the positions of b, c and d, the terms a_1, a_2, a_3 can be arranged amongst themselves in 3! ways. Thus, if the 3 like terms are replaced by 3 unlike terms, the number of different arrangements of the 6 letters will be $x \cdot 3!$.

But 6 different letters can be arranged in 6! ways.

$$\therefore x \cdot 3! = 6!,$$

$$x = \frac{6!}{3!}.$$

Arrangements of a number of things not all different | 11

Example 19 How many different arrangements can be made by using all the letters of the word *arrangement*?

Following the method used in the previous example, let the number of different arrangements be x.

Then, if the two letters a, are replaced by unlike terms a_1, a_2, the total number of arrangements will become $x \cdot 2!$.

If now the two letters r are replaced by r_1, r_2, each of the previous $x \cdot 2!$ arrangements will produce $2!$ arrangements and the total number of arrangements becomes $x \cdot 2! \cdot 2!$.

Similarly, if the two letters e are replaced by e_1, e_2 and the two letters n by n_1, n_2, the total number of arrangements becomes $x \cdot 2! \cdot 2! \cdot 2! \cdot 2!$.

But now we have 11 unlike terms which can be arranged in 11! ways.

$$\therefore x \cdot 2! \cdot 2! \cdot 2! \cdot 2! = 11!,$$

$$x = \frac{11!}{2! \, 2! \, 2! \, 2!}.$$

i.e. The number of permutations of 11 things taken all together when 2 are alike of one kind, 2 are alike of a second kind, 2 are alike of a third kind and 2 of a fourth kind is

$$\frac{11!}{2! \, 2! \, 2! \, 2!}.$$

Following the method used in this example we obtain the general result:

The number of permutations of n things taken all together when p are alike of a first kind, q alike of a second kind, r alike of a third kind and so on, is

$$\frac{n!}{p! \, q! \, r! \ldots}.$$

[A] EXAMPLES 1e

In Examples 1–5, find without assuming the general formula:

1 The number of different arrangements of the letters a, b, c, d.

2 The number of ways of arranging 3 black balls, 1 red ball and 1 white ball in a row.

3 The number of ways of arranging 2 white flags, 2 red flags and 1 blue flag on a vertical flagstaff.

4 The number of ways of arranging 3 dots and 2 dashes in a row.

5 In how many ways 3 red counters, 4 white counters and 1 black counter can be arranged in a row.

6 Find the number of arrangements that can be made out of the letters of the word *apoplexy*.

7 In how many ways can 15 balls be arranged in a line if 10 are red, 2 black, 2 yellow and 1 blue?

8 How many different signals can be given by using all the following flags, 2 red, 3 blue, 2 white?

9 Using all the digits 1, 2, 2, 3, 3, 3, 4, 5 without repetition, how many different numbers can be formed?

10 How many of the numbers in Example 9 will commence with 1?

11 Find the number of permutations of all the letters *abracadabra*.

12 In how many of the permutations in Example 11 do the *a*'s occupy the even positions?

13 Find the number of ways of arranging the letters of the word *algebra*. In how many of these will the relative positions of vowels and consonants be the same as in the original word?

14 How many numbers can be formed containing all the digits 1, 1, 2, 3, 4, 4, so that the odd digits always occupy the even places?

15 How many different arrangements can be made with the letters of the word *calculate* which start and finish with a *c*?

16 In how many different ways can 3 red balls, 4 white balls and 1 blue ball be arranged in a line if (i) the 3 red balls must come together, (ii) the 3 red balls must not all come together?

Arrangements and selections when repetitions are permitted

Example 20 Find the number of permutations of 4 things chosen out of 10 unlike things when repetitions are allowed.

In making an arrangement of 4 things we can choose any one of the 10 things to fill the first place. When the first place is filled we still have 10 things from which to choose to fill the second place since repetition is permitted. Hence the second place can be filled in 10 ways and similarly the third and fourth places can each be filled in 10 ways.

Hence, the total number of different arrangements of 4 things

$$= 10 \times 10 \times 10 \times 10$$
$$= 10^4.$$

Or the number of permutations of 10 unlike things taken 4 at a time when repetitions are allowed is 10^4.

Similarly, it follows that in general, **the number of permutations of n unlike things taken r at a time when repetitions are allowed, is n^r**.

Example 21 How many selections can be made from 12 different books when any number can be taken?

Each book can be disposed of in two ways, as we can either select it or leave it. Hence the total number of possible ways of making selections from 12 different books will be $2 \times 2 \times 2 \ldots$ (12 terms) $= 2^{12}$.

However, these 2^{12} selections include the one case when all books are rejected and this cannot be considered as a selection.

Thus, the number of selections, assuming at least one book is chosen, is $2^{12} - 1$.

If there are *n* unlike things from which selections can be made, clearly the total number of selections possible will be $2^n - 1$.

i.e. The number of ways in which a person can select some or all of a given number (n) of unlike things is $2^n - 1$.

Example 22 How many different sums of money can be made up from 3 £1 coins, 1 fiftypence, 2 tenpences, 1 twopence and 5 pennies?

As far as the £1 coins are concerned, we can take either none, one, two or three and so the £1 coins can be disposed of in 4 ways.

Similarly, the fiftypence can be disposed of in 2 ways, the tenpences in 3 ways, the twopence in 2 ways and the pennies in 6 ways.

Thus the total number of ways of disposing of all the coins is

$$4 \times 2 \times 3 \times 2 \times 6 = 288.$$

As this includes the case where none of the coins is selected, it follows that the total number of different sums of money which can be made up is $288 - 1$ or 287.

Harder miscellaneous examples | 13

[A] **EXAMPLES 1f**

If repetitions are allowed in nos. 1–4, find:
1 The number of permutations of 5 things taken 3 at a time.
2 The number of arrangements of 12 things taken 8 at a time.
3 The number of different numbers that can be formed by using all the digits, 2, 3, 4, 5.
4 The number of different 3-letter words which can be formed by using the letters a, b, c, d, e.
5 How many selections can be made from 8 different books when any number can be taken?
6 In how many ways can 3 prizes be given to a form of 30 pupils if each pupil is eligible for all the prizes?
7 In how many ways can a man invite one or more of his 4 friends to dinner?
8 In how many ways can a man travel from his home to the station once on each of three days if there are four conveyances available?
9 In how many ways can 4 articles be divided between 2 people when each person must receive at least one article?
10 In how many ways can the results of 8 football matches be forecast?
11 A combination lock consists of 4 rings each marked with 12 different letters; in how many ways can an unsuccessful attempt be made to open the lock?
12 How many different selections can be made by taking some or all the letters a, a, a, b, b, c?

Harder miscellaneous examples

Example 23 In how many ways can the letters of the word *expression* be arranged if the two s's are separated?

If there are no restrictions, the number of arrangements $= \dfrac{10!}{2!\,2!}$.

Now, consider in how many of these arrangements the two s's will come together. In this case the two letters s can be thought of as a single letter and the total number of arrangements will be $\dfrac{9!}{2!}$.

Hence, the number of arrangements in which the s's are separated

$$= \frac{10!}{2!\,2!} - \frac{9!}{2!} = 725\,760.$$

Example 24 In how many different ways can 12 examination papers be set so that no two of the three mathematical papers come together?

The 9 non-mathematical papers can be set in 9! different orders. In any one of these orders there will be 10 possible places where the first mathematical paper can be set, the beginning or end or in one of the eight intervals between the non-mathematical papers.

When the first mathematical paper has been positioned there will only be 9 possible places for the second paper and similarly there will be 8 possible places for the third paper.

Consequently, the number of ways of setting the 12 papers with no two mathematical papers consecutive is $10 \cdot 9 \cdot 8 \cdot 9!$ or $72 \cdot 10!$.

Example 25 A signaller has six flags, one red, two yellow, and three green. Find the number of different messages he can send by using any five flags.

A group of 5 must contain either 2 or 3 green flags. When 3 green flags are included, the group of 5 can be completed by taking either 2 yellow flags giving $\frac{5!}{2!\,3!}$ different arrangements, or 1 yellow and 1 red giving $\frac{5!}{3!}$ arrangements.

So when 3 green flags are used, the number of different messages is

$$\frac{5!}{2!\,3!} + \frac{5!}{3!}.$$

When 2 green flags are included the 2 yellow and 1 red must also be used. This group can be arranged in $\frac{5!}{2!\,2!}$ ways.

Therefore the total number of different messages which can be sent by using 5 flags

$$= \frac{5!}{2!\,3!} + \frac{5!}{3!} + \frac{5!}{2!\,2!}$$

$$= 60.$$

[B] MISCELLANEOUS EXAMPLES

1 In how many ways can the letters of the word *rearrangement* be arranged? In how many of these are the 2 *a*'s separated?

2 In how many ways can 4 cards be selected from a pack of 52 if at least one of the cards must be an ace?

3 Twelve different books are to be labelled, 5 white, 3 red and the rest blue. In how many different ways can this be done?

4 How many different arrangements can be made with the letters of the word *requisition*, in which the letters *q* and *r* are consecutive?

5 Find the number of ways of dividing 25 unlike quantities into 5 equal packets.

6 In how many ways can 6 boys and 4 girls be arranged in a line if the girls are not separated?

7 In how many ways can 8 people seat themselves at a round table?

8 How many odd numbers between 500 and 10,000 can be formed with the digits 1, 2, 3, 4, 5 if no digit is repeated? How many, if the digits may be repeated?

9 Find the number of permutations of all the letters of the word *adiabatic*, such that no two vowels come next to each other.

10 Prove that $_nC_r = \left(\dfrac{n+1}{r} - 1\right){}_nC_{r-1}$.

11 10 men are to play 5 sets of singles at tennis simultaneously. In how many ways can the pairs be selected?

12 In how many ways can 10 differently coloured beads be arranged on a circular wire, it being assumed that all the beads are in contact.

13 An eight-oared boat is to be manned by a crew chosen from 10 men of whom 2 can only steer. In how many ways can the boat be manned if two of the oarsmen can only row on the stroke side?

14 In how many ways can the following prizes be presented to a form of 25 pupils: first, second and third Form prizes; first and second Classical; first and second Mathematical; first French; and first Science?

15 In how many ways can 144 be expressed as a product of 2 positive integers?

16 A bag contains 3 white balls, 5 red balls and 7 green balls. In how many ways can a selection of balls be made? How many of the selections will contain at least 3 red balls?

17 On an examination paper of 10 questions, a girl obtained either 6 or 7 marks for each question. If her total mark was 63, in how many different ways could she have obtained this total?

18 In how many ways can 10 people be seated in a row of 10 chairs if 2 must be separated and another 2 must always sit together?

19 Four letters are written and four envelopes addressed. In how many ways can all the letters be placed in the wrong envelopes?

20 The results of 6 football matches have to be forecast. In how many ways is it possible to get exactly 4 results correct?

21 A committee of 6 is chosen from 8 women and 5 men so as to contain at least 3 women and 2 men. In how many different ways can this be done if 2 particular men refuse to serve on the same committee?

22 If there are 4 points on a straight line and 10 other points no three of which are collinear, what is the greatest possible number of triangles that can be formed with these points as vertices?

23 Prove that the number of ways in which 20 things can be divided into 5 packets of 4 each is to the number of ways in which they can be divided into 4 packets of 5 each as 125 is to 24.

24 Two crews of 8 are to be formed from 16 rowers of whom 6 can only row on the bow side and 4 on the stroke side. In how many ways can the crews be formed when due regard is taken to the order in the boats?

25 In how many ways can 12 people sit at a circular table so that 2 particular people are always together?

26 There are 5 different English books, 6 different French books and 4 different Latin books to be arranged on a shelf. In how many ways can this be done if the books in each language come together?

27 In how many ways can $4n$ people be divided into groups of 4? How many different foursome tennis matches can be arranged simultaneously for $4n$ people?

28 In how many ways can 9 balls, of which 4 are red, 4 white and 1 black, be arranged in a line so that no red ball is next to the black? In how many ways can the balls be arranged so that no red ball is next to a white?

29 14 equal beads are threaded on a circular wire so that consecutive beads are in contact. If 6 of the beads are blue, 4 red, 3 green and 1 yellow, find the number of different arrangements which can be made.

30 In how many ways can a lawn tennis mixed doubles be made up from 6 married couples if no husband and wife are to play in the same game?

Probability or chance

Definition If an event can happen in a ways and fail to happen in b ways where each of these $a+b$ ways is equally likely, then the **probability** or **chance** of its happening is defined as the ratio $\dfrac{a}{a+b}$ and that of its failing is $\dfrac{b}{a+b}$.

Example 26 There are 10 balls in a bag, of which 3 are red. If one ball is drawn, what is the chance of this ball being red?

Clearly, it is equally likely that any one of the 10 balls will be drawn and as there are 3 ways of drawing a red ball and 7 ways of failing, the chance of drawing a red ball is $\frac{3}{10}$ or 3 in 10.

Example 27 In a lottery there is 1 prize and 200 tickets have been sold. What is the chance of someone who holds 5 tickets winning the prize?

There are 5 ways of winning the prize and 195 ways of failing. Thus the chance of winning $= \frac{5}{200} = \frac{1}{40}$, or 1 in 40.

Example 28 What is the chance of throwing a number greater than 4 with a die whose faces are numbered from 1 to 6?

There are 6 possible ways in which the die can fall and of these 2 will be favourable to the required event, i.e. the faces numbered 5 and 6. Hence the probability or chance of throwing a number greater than 4 is $\frac{2}{6}$ or 1 in 3.

Example 29 In a competition a prize is given for correctly forecasting the results of 6 football matches. A competitor sends in 10 different forecasts, what is their chance of winning a prize?

First of all we require to know the total number of different ways of forecasting the 6 results, which is 3^6. Hence, there are 10 ways in which the matches can end in a manner favourable to the competitor and $3^6 - 10$ unfavourable ways.

Thus the chance of success is $\dfrac{10}{3^6}$ or 10 in 729.

Unit and percentage probability When an event is certain to happen, the probability or chance is unity, or *certainty is the unit of probability*.

It is readily seen that the chance of success plus the chance of failure will always be unity as an event must either happen or fail to happen. For example, if p is the chance of success, the chance of failure will be $1 - p$.

Probability or chance is sometimes expressed as a percentage of a certainty. For example, if there are 4 ways in which an event can happen, the chance of any one of these 4 arising is 1 in 4 or 25%.

Odds on and against an event happening When it is said that *the odds are three to two against an event happening*, it is meant that the ratio of the chance of the event failing to happen to the chance of the event happening is as three is to two.

Similarly, if *the odds are three to two on the event happening*, the ratio of the chance of the event happening to the chance of it failing to happen is as three is to two.

From these definitions it follows that, if the odds are three to two against an event happening, the chance of the event happening is $\dfrac{2}{3+2}$ or $\frac{2}{5}$ and the chance of it failing to happen is $\frac{3}{5}$.

In general, if the odds against an event happening are x to y then the chance of the event happening is $\dfrac{y}{x+y}$:

if the odds on an event happening are x to y then the chance of the event happening is $\dfrac{x}{x+y}$.

Example 30 There are 12 balls in a bag, 5 of which are red. If one ball is drawn, what are the odds against this being a red?

Harder examples | 17

As there are 12 balls from which the one is drawn and five of these are red, the chance of drawing a red ball is $\frac{5}{12}$.
The chance of failing to draw a red ball is $\frac{7}{12}$.
Hence, the odds against this event happening are $\dfrac{\frac{7}{12}}{\frac{5}{12}}$ or 7 to 5.

[A] **EXAMPLES 1g**

What are the chances of the events in Examples 1–6 happening?
1 Drawing a white ball from a bag containing 4 white balls and 6 others?
2 Drawing an ace out of a pack of 52 cards containing 4 aces?
3 A coin falling heads uppermost?
4 Someone holding 12 tickets winning a sweepstake for which 210 tickets have been sold?
5 Obtaining two heads when two coins are spun simultaneously?
6 Throwing a number greater than 3 with a die having faces numbered from 1 to 6?
7 If the odds against an event happening are 5 to 3, what is the percentage chance of the event happening?
8 Three successful candidates are to be drawn from 12, what are the odds against any particular candidate being successful?
9 What is the minimum number of tickets a person must hold in a lottery involving 500 tickets, in order that the odds against him holding the winning ticket are less than 3 to 1?
10 A coin is spun twice. What are the chances of obtaining (i) 2 heads, (ii) one head and one tail?
11 Out of a form of 12 boys and 8 girls, one pupil is chosen at random. What is the chance of a boy being chosen?
12 If 8 people sit on a bench, what is the chance of a particular person sitting at an end?
13 There are 4 similar balls in a bag of which 2 are red. If a pair of balls are drawn from the bag, what is the probability that they are both red?
14 A woman makes 6 different forecasts of the results of 4 football matches. What chance has she of making an all-correct forecast?
15 A card is drawn at random from a pack of 52. What are the odds against it being either a jack, queen, king or ace?

Harder examples

Example 31 Out of a form of 10 boys and 8 girls, two pupils are chosen at random. What is the chance that they are both boys?
Since there are 18 pupils altogether, two pupils can be selected in $_{18}C_2$ ways. Now, the number of ways in which two boys can be chosen is $_{10}C_2$. Thus, out of a total number of $_{18}C_2$ ways, two boys can be chosen in $_{10}C_2$ ways and hence the chance of choosing two boys is

$$\frac{_{10}C_2}{_{18}C_2} = \frac{10 \cdot 9}{18 \cdot 17} = \frac{5}{17}.$$

Example 32 From a bag containing 4 white and 6 black balls, 3 are drawn at random. What are the odds against these being all black?

18 | Permutations and combinations

The total number of ways of drawing 3 balls from a total of 10 is $_{10}C_3$; the number of ways of drawing 3 black balls is $_6C_3$.

Therefore, the chance of drawing 3 black balls $= \dfrac{_6C_3}{_{10}C_3} = \dfrac{1}{6}$.

Hence, the chance of failing to draw 3 black balls $= \frac{5}{6}$, and the odds against the event happening are 5 to 1.

Example 33 If 5 coins are spun, what is the probability of 3 falling heads uppermost? Each coin can fall in 2 ways and hence the 5 coins can fall in 2^5 ways. Three heads out of 5 can be obtained in $_5C_3$ ways and thus the chance of getting 3 heads is $\dfrac{_5C_3}{2^5}$ or $\dfrac{5}{16}$.

Example 34 Thirteen people take their places at a round table, find the odds against any two particular people sitting together.

Suppose one of the two particular people is seated. There are 12 seats remaining and so the other person can be seated in 12 ways. In 2 of these 12 ways, he would be seated next to the first person.

Hence the chance of two particular people sitting together is $\frac{2}{12}$.

The odds against the event are 10 to 2, or 5 to 1.

[B] **EXAMPLES 1h**

1 Two balls are drawn at random from a bag containing 5 red and 7 green balls. What is the chance that both are red?

2 Two dice are thrown. What are the odds against their both falling the same way up?

3 From a pack of 52 cards, 4 are drawn. Find the chance of their being all aces.

4 A man forecasts the results of 6 football matches. In how many ways can he have at least 5 results correct? What are the odds against this event happening?

5 Out of a form of 10 boys and 12 girls, 2 are chosen at random. What is the chance of their both being girls?

6 Two people are chosen at random from a group consisting of 5 married couples. What are the odds against a particular husband and wife being chosen?

7 One of two events must happen. If the chance of the one is $\frac{3}{4}$ that of the other, find the odds in favour of the second event happening.

8 12 people sit at a circular table. What is the chance of two particular people being separated?

9 10 people take their places at a round table, find the odds against two particular people sitting together.

10 What is the chance of throwing a total of 6 in a single throw with 2 dice?

11 4 coins are tossed; find the chance there should be 2 heads and 2 tails.

12 There are three events, A, B, C, one of which must happen. If the odds are 7 to 2 against A and the odds on B are $\frac{5}{8}$ the odds on C, find the odds against B and C.

13 Out of a group of 3 officers and 12 O.R.'s a party of 3 is chosen by ballot. What is the chance of the party consisting entirely of O.R.'s?

14 A has 2 tickets in a lottery containing 3 prizes and 20 blanks and B has 1 ticket in a lottery containing 1 prize and 10 blanks. Find the ratio of their chances of failure.

15 The letters of the word *dependent* are placed at random in a row. What chance is there of the 3 vowels coming together?

16 There are 10 coloured discs in a hat, of which 2 are red, 2 white, 3 green and 3 blue. If 4 discs are drawn, what are the odds against their being 2 blue, 1 red and 1 white?

17 *A* and *B* throw with 2 dice. If *A*'s total is 10, what is the chance of *B* throwing a higher number?

18 A coin is spun 3 times, what are the odds against failing to throw at least one head?

19 A bag contains 5 fiftypences and 4 fivepences. What is the chance of drawing a fivepence at the first attempt? If this event happens, what is the chance that the next draw will be a fiftypence? Deduce the chance of drawing a fivepence at the first attempt and a fiftypence at the second.

20 What is the chance of throwing a total of 9 in a single throw with 2 dice? What is the chance of a total of at least 9?

2 | Logarithms
Logarithmic and exponential functions

Definition The *logarithm* of a number to a given base, assumed positive, is the index of the power to which the base must be raised in order to equal the given number, e.g. As $8 = 2^3$, the logarithm of 8 to the base 2 is 3.

Notation The logarithm of a number N to base a is written as $\log_a N$. Thus if $N = a^x$. then $\log_a N = x$.

Example 1 Find the values of $\log_{10} 100$, $\log_2 \frac{1}{8}$, $\log_9 3$.
As $100 = 10^2$, $\log_{10} 100 = 2$.

As $\dfrac{1}{8} = \dfrac{1}{2^3} = 2^{-3}$, $\log_2 \dfrac{1}{8} = -3$.

As $3 = \sqrt{9} = 9^{\frac{1}{2}}$, $\log_9 3 = \frac{1}{2}$.

Common logarithms Logarithms to base 10 are called *Common Logarithms**. Common Logarithms are used in work involving numerical computation but, as will be seen later, are superseded by logarithms to a different base in most branches of higher mathematics.

Laws of logarithms We now prove the following important results:

(i) $\log_a MN = \log_a M + \log_a N$.

(ii) $\log_a \dfrac{M}{N} = \log_a M - \log_a N$.

(iii) $\log_a (N)^p = p \log_a N$.

(i) Let $\log_a N = x$ and $\log_a M = y$.
Then $N = a^x$ and $M = a^y$.

$\therefore MN = a^y \times a^x = a^{x+y}$, (Rule of Indices)

i.e. $\log_a MN = x + y = \log_a N + \log_a M$.

(ii) As in case (i), $N = a^x$, $M = a^y$.

$\therefore \dfrac{M}{N} = \dfrac{a^y}{a^x} = a^{y-x}$. (Rule of indices)

i.e. $\log_a \dfrac{M}{N} = y - x = \log_a M - \log_a N$.

---

* These are denoted by the symbol lg.

Change of base in logarithms | 21

(iii) As before, $N = a^x$.

$\therefore N^p = (a^x)^p = a^{px}$, (Rule of Indices)

i.e. $\log_a (N)^p = px = p \log_a N$.

Example 2 Express $\log_a \dfrac{x^3}{\sqrt{y}}$ in terms of $\log_a x$ and $\log_a y$.

$$\log_a \frac{x^3}{\sqrt{y}} = \log_a \frac{x^3}{y^{\frac{1}{2}}} = \log_a x^3 - \log_a y^{\frac{1}{2}}$$

$$= 3 \log_a x - \tfrac{1}{2} \log_a y.$$

Example 3 Simplify $\lg \tfrac{75}{16} - 2 \lg \tfrac{5}{9} + \lg \tfrac{32}{243}$.

Expression $= \lg \tfrac{75}{16} - \lg (\tfrac{5}{9})^2 + \lg \tfrac{32}{243}$

$= \lg \left(\dfrac{\tfrac{75}{16} \times \tfrac{32}{243}}{\tfrac{25}{81}} \right) = \lg 2$.

Example 4 Solve the equation $2 \cdot 4^x = 5 \cdot 3^{x-1}$.
Taking logarithms of both sides to base 10,

$x \lg 2 \cdot 4 = (x - 1) \lg 5 \cdot 3$,

$0 \cdot 3802 x = 0 \cdot 7243 (x - 1)$.

$\therefore x = \dfrac{0 \cdot 7243}{0 \cdot 3441}$

$= 2 \cdot 105$.

Change of base in logarithms To prove that $\log_a N = \dfrac{\log_b N}{\log_b a}$.

Let $\log_a N = x$, so $N = a^x$.
Taking logarithms to base b,

$\log_b N = \log_b (a^x) = x \log_b a$.

$\therefore \log_a N = x = \dfrac{\log_b N}{\log_b a}$.

Special Case. Taking $N = b$,

$\log_a b = \dfrac{\log_b b}{\log_b a} = \dfrac{1}{\log_b a}$.

Example 5 Given $\log_{10} e = 0 \cdot 4343$, find $\log_e 2$ to 3 significant figures.

$$\log_e 2 = \frac{\lg 2}{\lg e} = \frac{0 \cdot 3010}{0 \cdot 4343}$$

$= 0 \cdot 693$ to 3 sig. figs.

Example 6 Prove that $\log_a b \times \log_b c \times \log_c a = 1$.
Reduce all the logarithms to a common base, say a.

22 | **Logarithmic and exponential functions**

Then $\quad\log_a b \times \log_b c \times \log_c a = \log_a b \times \dfrac{\log_a c}{\log_a b} \times \dfrac{1}{\log_a c}.$

$$= 1.$$

Example 7 Evaluate $(0.571)^{0.24}$ to 3 significant figures.

Take $\qquad\qquad y = (0.571)^{0.24}.$

$$\lg y = 0.24 \lg 0.571 = 0.24 \times \overline{1}\cdot 7566.$$

To evaluate this product, $\overline{1}\cdot 7566$ is written as $-1 + 0.7566.$ or $-0.2434.$

$$\therefore \lg y = -(0.24 \times 0.2434)$$
$$= -0.05841 \qquad \text{using tables.}$$

This negative logarithm is now written with the decimal part positive.

$$\lg y = -1 + 0.9416$$
$$= \overline{1}\cdot 9416.$$
$$y = 0.8742.$$

i.e. $\qquad\qquad (0.571)^{0.24} = 0.874$ to 3 sig. figs.

[A] EXAMPLES 2a

1 Write down the logarithms to base 3 of

(i) 27, (ii) $\sqrt{3}$, (iii) $\dfrac{1}{3}$, (iv) $\dfrac{1}{3\sqrt{3}}$.

2 Find the logarithm of 16 to the base $\sqrt{2}$ and the logarithm of $\sqrt{2}$ to the base 16.

3 Find the values of the following logarithms:

(i) $\log_2 16$, (ii) $\log_4 64$, (iii) $\log_{64} 8$, (iv) $\log_4 0.25$, (v) $\log_a a^2$,

(vi) $\log_2 2\sqrt{2}$, (vii) $\log_4 32$, (viii) $\log_8 128$, (ix) $\log_{27} 9$, (x) $\log_{\frac{1}{4}} 32$.

4 Write down the values of:

(i) $10^{\lg 2}$, (ii) $2^{\log_2 7}$, (iii) $e^{\log_e \sin x}$, (iv) $e^{2\log_e x}$.

5 Express in terms of $\log a$, $\log b$, $\log c$, the following:

(i) $\log \dfrac{a^3 b}{c^2}$, (ii) $\log \sqrt{\dfrac{a}{b}}$, (iii) $\log \dfrac{1}{ab^3}$, (iv) $\log \sqrt{ab^2 c}$,

(v) $\log \sqrt[3]{a^{-2} b^3 c^{-1}}$, (vi) $\log(\sqrt[3]{a^2} \times \sqrt{b^3})$, (vii) $\log \dfrac{a\sqrt{b^3}}{\sqrt[3]{c^{-2} a^5}}$.

6 Reduce the following expressions to single terms:

(i) $2\log 5 - \log \tfrac{3}{7} + \tfrac{1}{2}\log \tfrac{9}{16}$,

(ii) $\tfrac{1}{2}\log 8 + 2\log \dfrac{1}{\sqrt{2}}$.

(iii) $\log(x+1) - 3\log(1-x) + 2\log x.$

(iv) $\tfrac{1}{3}\log(2x-1) + \tfrac{1}{9}\log(x+3) - \tfrac{2}{9}\log(x+1).$

(v) $\log\sqrt{x^2-1} + \tfrac{1}{2}\log\left(\dfrac{x+1}{x-1}\right).$

7 Solve the following equations, giving roots correct to 3 significant figures:
(i) $2^x = 3$, (ii) $(7 \cdot 6)^x = 5 \cdot 2$, (iii) $3^{x+1} = 4^{x-1}$
(iv) $7^x 8^{2x-1} = 6 \cdot 3$, (v) $12^{x^2} = 10^{2x}$.

8 If $a^x = cb^x$, prove that $x = \dfrac{1}{\log_c \dfrac{a}{b}}$.

9 If $a^{3-x} b^{5x} = a^{x+5} b^{3x}$, prove that $x \log \dfrac{b}{a} = \log a$.

10 Evaluate to 3 significant figures:
(i) $2 \cdot 7^{\frac{2}{3}}$, (ii) $1 \cdot 72^{-\frac{1}{2}}$, (iii) $4 \cdot 65^{\frac{4}{3}}$, (iv) $\sqrt[3]{(0 \cdot 23)^4}$,
(v) $7 \cdot 1^{1 \cdot 2}$, (vi) $\dfrac{1}{3 \cdot 86^{-2 \cdot 1}}$, (vii) $2 \cdot 34 \lg 7$, (viii) $\dfrac{\lg 17 \cdot 6}{\lg 11 \cdot 2}$.

11 Given that $\lg 2 = 0 \cdot 301030$, $\lg 3 = 0 \cdot 477121$, $\lg 7 = 0 \cdot 845098$, find the values of:
(i) $\lg 12$ (ii) $\lg 84$, (iii) $\lg 0 \cdot 128$, (iv) $\lg \sqrt{\dfrac{56}{27}}$.

12 Find the number of digits in (i) 2^{15}, (ii) 3^{12}.

13 Find, to 3 significant figures:
(i) $\log_3 4$, (ii) $\log_2 10$, (iii) $\log_{2 \cdot 5} 7 \cdot 5$, (iv) $\log_5 (2 \cdot 83)^2$.

14 Given that $\lg e = 0 \cdot 4343$, find to 3 significant figures, the logarithms of the following numbers to base e: $3 \cdot 5$, $11 \cdot 7$, $36 \cdot 4$.

[B] **EXAMPLES 2b**

Evaluate as accurately as tables permit:

1 $(0 \cdot 7)^{1 \cdot 56}$. **2** $(0 \cdot 536)^{-2 \cdot 01}$. **3** $\dfrac{\lg 0 \cdot 5}{\lg 0 \cdot 25}$.

4 $(0 \cdot 457)^{0 \cdot 263}$. **5** $0 \cdot 036 \lg 0 \cdot 817$.

6 If $e = 2 \cdot 718$, find the value of $e^{\sin x}$ when $x = \dfrac{\pi}{4}$.

7 Solve the following equations:
(i) $(0 \cdot 57)^{x+1} = (0 \cdot 83)^x$.
(ii) $(\lg e)^{2x} = \dfrac{1}{0 \cdot 722}$, where $e = 2 \cdot 718$.

8 If the relation $y = ax^n$ is satisfied by $x = 2$, $y = 10 \cdot 6$ and by $x = 3$, $y = 6 \cdot 2$, find a and n.

9 Calculate the value of $\dfrac{(4 \cdot 65)^{\frac{4}{5}} \log_e 25}{(1 \cdot 612)^3}$, where $e = 2 \cdot 718$.

10 Solve the equation $2^{2x} - 5(2^x) + 6 = 0$. (Let $2^x = y$.)

11 Given that $\sinh x = \frac{1}{2}(e^x - e^{-x})$, $\cosh x = \frac{1}{2}(e^x + e^{-x})$, where $e = 2 \cdot 718$, evaluate (i) $\sinh 2$, (ii) $\cosh 1 \cdot 5$, (iii) $\sinh 0 \cdot 5$, (iv) $\cosh 0 \cdot 2$.

12 Solve the equation $e^x - e^{-x} = 2$. (Multiply by e^x.)

13 Solve the equation $\cosh x = 2$. (See Examples 11 and 12.)
14 If $2^x 3^y = 3^x 4^y = 6$, show by taking logarithms to base 2 and eliminating $\log_2 3$, that $x^2 - 2y^2 = 2x - 3y$.
15 Given that $6^x = \frac{10}{3} - 6^{-x}$, show that $x = \pm \log_6 3$.
16 Prove that $\dfrac{1}{\log_a c^2} = \log_c \sqrt{a}$.
17 Evaluate $\log_e \sin x$ for $x = \dfrac{\pi}{6}, \dfrac{\pi}{3}$, e being equal to 2·718.
18 Evaluate $\log_e \left(\dfrac{1+t^2}{1-t^2} \right)$, where $t = \tan \dfrac{\theta}{2}$, when $\theta = \dfrac{\pi}{6}$.
19 Given that $\log_e \dfrac{T_1}{T_2} = \mu \theta$, find T_1 when $T_2 = 50$, $\theta = \dfrac{2\pi}{3}$, $\mu = 0.54$, $e = 2.718$, $\pi = 3.142$.

Logarithmic functions Logarithmic functions of a variable x consist of a function of $\log x$, e.g. $(\lg x)^2$, or of the logarithm of some function of x, e.g. $\log_e (1 + \tan x)$.

The function $\lg x$.

Suppose $\lg x = y$, so $x = 10^y$.

For all real values of y, the function 10^y is positive and so there is no value of y corresponding to a negative value of x, i.e. $\lg x$ is not defined for values of $x \leqslant 0$.

As y takes increasingly large negative values, 10^y decreases and approaches zero. Alternatively as $x \to 0, y \to -\infty$. Over its whole domain $x > 0$, $\lg x$ is an increasing function; negative for $0 < x < 1$, zero for $x = 1$, and positive for $x > 1$.

The graph of $\lg x$ is given in Fig. 1.

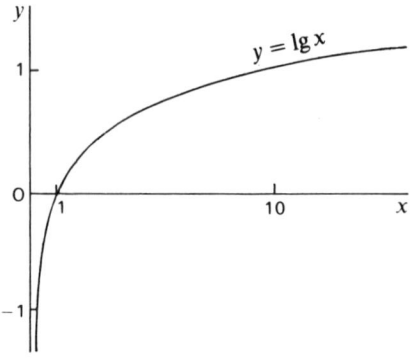

Fig. 1

Differential coefficient of lg x
Let $\qquad y = \lg x$ (x assumed positive).
Let x increase by δx and y by δy.

Then
$$y + \delta y = \lg(x + \delta x),$$
$$\delta y = \lg(x + \delta x) - \lg x$$
$$= \lg\left(\frac{x + \delta x}{x}\right)$$
$$= \lg\left(1 + \frac{\delta x}{x}\right).$$
$$\therefore \frac{\delta y}{\delta x} = \frac{1}{\delta x} \lg\left(1 + \frac{\delta x}{x}\right)$$
$$= \lg\left(1 + \frac{\delta x}{x}\right)^{\frac{1}{\delta x}}.$$

Now let $\dfrac{\delta x}{x} = t$, then $\dfrac{1}{\delta x} = \dfrac{1}{tx}$.

$$\therefore \frac{\delta y}{\delta x} = \lg(1 + t)^{\frac{1}{tx}}$$
$$= \frac{1}{x} \lg(1 + t)^{\frac{1}{t}}$$

Let $\delta x \to 0$, then $t \to 0$ and $\dfrac{\delta y}{\delta x} \to \dfrac{dy}{dx}$.

$$\therefore \frac{dy}{dx} = \frac{1}{x} \operatorname*{Lt}_{t \to 0} \lg(1 + t)^{\frac{1}{t}}$$
$$= \frac{1}{x} \lg\left\{\operatorname*{Lt}_{t \to 0}(1 + t)^{\frac{1}{t}}\right\}.$$

To complete the differentiation, it is necessary to find the limit $\operatorname*{Lt}_{t \to 0}(1 + t)^{\frac{1}{t}}$
In order to obtain a rough idea of the behaviour of the function $(1 + t)^{\frac{1}{t}}$ as t decreases and approaches zero, it has been evaluated for a series of decreasing values of t, using the Binomial Expansion.

t	$\dfrac{1}{t}$	$(1+t)^{\frac{1}{t}}$
1	1	2·000
0·5	2	2·250
0·25	4	2·441
0·10	10	2·594
0·01	100	2·706
0·001	1000	2·717
0·0001	10 000	2·718
0·00001	100 000	2·718

These tabulated results indicate that $(1+t)^{\frac{1}{t}}$ does approach a finite limit as t tends to zero, the limit having an approximate value of 2·718.
Reverting to the differentiation,

$$\frac{dy}{dx} = \frac{1}{x} \lg \left\{ \underset{t \to 0}{\text{Lt}} (1+t)^{\frac{1}{t}} \right\}.$$

The limit $\underset{t \to 0}{\text{Lt}} (1+t)^{\frac{1}{t}}$ which has an approximate value of 2·718, is denoted by the letter e.

Thus
$$\frac{dy}{dx} = \frac{1}{x} \lg e,$$

or
$$\frac{d}{dx}(\lg x) = \frac{1}{x} \lg e.$$

Replacing 10 by a, we get the general result $\frac{d}{dx}(\log_a x) = \frac{1}{x} \log_a e$.

Natural or Naperian logarithms Replacing a by e in the general result,

$$\frac{d}{dx}(\log_e x) = \frac{1}{x} \log_e e = \frac{1}{x}.$$

i.e.
$$\frac{d}{dx}(\log_e x) = \frac{1}{x}; \quad \int \frac{dx}{x} = \log_e x + c.$$

As $\log x$ does not exist if $x \leqslant 0$, these results hold only if $x > 0$.

Logarithms to base e are called *Natural* or *Naperian Logarithms* and are denoted by the symbol ln.

General differentiation of logarithmic functions The methods used to differentiate any logarithmic function are illustrated by the following example.

Example 8 Differentiate with respect to x, (i) $\ln(2x+1)$; (ii) $\ln \sin x$; (iii) $\ln \sqrt{\frac{1+x}{1-x}}$.

(i) Let
$$y = \ln(2x+1) \text{ and } u = 2x+1.$$
$$\therefore y = \ln u \text{ where } u = 2x+1.$$
$$\therefore \frac{dy}{dx} = \frac{dy}{du} \times \frac{du}{dx} = \frac{1}{u} \times 2 = \frac{2}{2x+1}.$$
i.e. $\frac{d}{dx} \ln(2x+1) = \frac{2}{2x+1}.$

(ii) $\dfrac{d}{dx} \ln (\sin x) = \dfrac{1}{\sin x} \cdot \cos x,$ using the function of a function rule
with $u = \sin x$,
$= \cot x.$

(iii) $\dfrac{d}{dx} \ln \sqrt{\dfrac{1+x}{1-x}} = \dfrac{d}{dx} \left\{ \dfrac{1}{2} \ln (1+x) - \dfrac{1}{2} \ln (1-x) \right\}$

$= \dfrac{1}{2} \dfrac{1}{1+x} - \dfrac{1}{2} \dfrac{(-1)}{1-x}$

$= \dfrac{1}{2} \left\{ \dfrac{1}{1+x} + \dfrac{1}{1-x} \right\}$

$= \dfrac{1}{1-x^2}.$

Note As illustrated in this example, if possible, simplify a logarithmic function *before* differentiation.

[A] **EXAMPLES 2c**

1 Differentiate with respect to x:

(i) $\ln 2x$, (ii) $\ln (x+6)$, (iii) $\ln (3x-2)$,
(iv) $\ln (1-x)$, (v) $\ln (x^2+x+1)$, (vi) $\ln (2-x^3)$,
(vii) $\ln \dfrac{1}{x}$, (viii) $\ln \sqrt{1+x}$, (ix) $\ln \dfrac{1+x}{1-x}$,
(x) $\ln (2x+1)^3$, (xi) $\ln \cos x$, (xii) $\ln \tan x$,
(xiii) $\ln \sin^2 x$, (xiv) $\ln \sqrt{\dfrac{1-2x}{x+1}}$, (xv) $\ln \sqrt[3]{x^3+1}$.

2 Find $\dfrac{dy}{dx}$ in the following cases:

(i) $y = \lg 3x$, (ii) $y = \log_a \sin x$, (iii) $e^y = x^2 + 1$.

3 Differentiate $x \ln x$ with respect to x. Deduce $\int \ln x \, dx$.

4 If $y = x^2 \ln x$, find $\dfrac{d^2y}{dx^2}$.

5 If $y = \dfrac{\ln x}{x}$, prove that $\dfrac{dy}{dx}$ vanishes when $x = e$.

6 Find the gradient of the curve $y = (2x^3 - 1) \ln x$ at the point where $x = 1$.

7 If $\ln s = -kt$ gives the distance travelled s m in terms of the time taken t s for a moving body, show that the velocity is directly proportional to the distance travelled.

8 The current C, flowing in a wire, is given by the equation

$$\ln (1 + C) = -kt,$$

where t is the time and k a constant. Show that the rate of decrease of current is $k(1 + C)$.

28 | Logarithmic and exponential functions

9 If $\ln y = x$, prove that $\dfrac{dy}{dx} = y$. Deduce that all derivatives of y with respect to x are equal to y.

10 By writing the equation $y = e^{2x}$ in terms of a logarithm, find $\dfrac{dy}{dx}$.

Important integral results An important class of indefinite integrals is evaluated in terms of logarithmic functions.

Consider $\dfrac{d}{dx} \ln (ax + b)$, where a and b are constants.

We have
$$\frac{d}{dx} \ln (ax + b) = \frac{a}{ax + b},$$

$$\therefore \int \frac{dx}{ax+b} = \frac{1}{a} \ln (ax+b) + c.$$

e.g.
$$\int \frac{dx}{2x+1} = \tfrac{1}{2} \ln (2x+1) + c.$$

$$\int \frac{dx}{3-x} = -\ln (3-x) + c \quad \text{if} \quad x < 3,$$

$$\text{or} \quad -\ln (x-3) + c \quad \text{if} \quad x > 3.$$

Now consider $\dfrac{d}{dx} \ln (f(x))$, where $f(x)$ is a function of x.

$$\frac{d}{dx} \ln (f(x)) = \frac{f'(x)}{f(x)} \quad \text{where } f'(x) \text{ denotes } \frac{d}{dx} (f(x)).$$

i.e.
$$\int \frac{f'(x)}{f(x)} dx = \ln (f(x)) + c.$$

e.g.
$$\int \frac{2x}{x^2+2} dx = \ln (x^2+2) + c. \quad \text{In this case } f(x) = x^2 + 2,$$
$$\text{and } f'(x) = 2x.$$

$$\int \cot x \, dx = \int \frac{\cos x}{\sin x} dx = \ln \sin x + c.$$

In words, *the integral of a fractional function of which the numerator is the differential coefficient of the denominator is the natural logarithm of the denominator. In subsequent examples involving this form of integral it will be assumed that the domain of the variable concerned is such that the logarithmic function involved is defined.*

Example 9 Integrate the following functions:

(i) $\dfrac{3}{2x-3}$; (ii) $\dfrac{x^2}{x^3+4}$; (iii) $\tan 2x$.

Important integral results

(i) $\int \dfrac{3\,dx}{2x-3} = \dfrac{3}{2}\int \dfrac{2\,dx}{2x-3} = \dfrac{3}{2}\ln(2x-3) + c.$

(ii) $\int \dfrac{x^2\,dx}{x^3+4} = \dfrac{1}{3}\int \dfrac{3x^2\,dx}{x^3+4} = \dfrac{1}{3}\ln(x^3+4) + c.$

(iii) $\int \tan 2x\,dx = \int \dfrac{\sin 2x}{\cos 2x}\,dx = -\dfrac{1}{2}\int \dfrac{(-2\sin 2x)}{\cos 2x}\,dx.$

Note. $\dfrac{d}{dx}(\cos 2x) = -2\sin 2x.$

$= -\tfrac{1}{2}\ln\cos 2x + c = \tfrac{1}{2}\ln\left(\dfrac{1}{\cos 2x}\right) + c$

$= \tfrac{1}{2}\ln\sec 2x + c.$

Example 10 Evaluate (i) $\displaystyle\int_{-2}^{-1} \dfrac{dx}{1-2x}$; (ii) $\displaystyle\int_{0}^{\pi/2} \dfrac{\cos x}{1+\sin x}\,dx.$

(i) $\displaystyle\int_{-2}^{-1}\dfrac{dx}{1-2x} = \left[-\dfrac{1}{2}\ln(1-2x)\right]_{-2}^{-1}$

$= (-\tfrac{1}{2}\ln 3) - (-\tfrac{1}{2}\ln 5)$
$= \tfrac{1}{2}\ln 5 - \tfrac{1}{2}\ln 3$
$= \tfrac{1}{2}\ln \tfrac{5}{3}.$

Usually the result can be left in this form, but if evaluation is necessary it must be remembered that the logarithm is to base e.

(ii) $\displaystyle\int_{0}^{\pi/2} \dfrac{\cos x\,dx}{1+\sin x} = \left[\ln(1+\sin x)\right]_{0}^{\pi/2}$

$= \ln(1+1) - \ln(1+0)$
$= \ln 2 - \ln 1 = \ln 2.$

Example 11 Find the area included between the curve $y(x-1) = 2$, the axis of x and the ordinates $x = 2, x = 3$.

Required area $= \displaystyle\int_{2}^{3} y\,dx$

$= \displaystyle\int_{2}^{3} \dfrac{2\,dx}{x-1} = 2\int_{2}^{3}\dfrac{dx}{x-1}$

$= 2\Big[\ln(x-1)\Big]_{2}^{3} = 2\ln 2 - 2\ln 1$

$= 2\ln 2.$

[A] EXAMPLES 2d

1 Integrate the following functions with respect to x and check the results by differentiation:

(i) $\dfrac{3}{x}.$ (ii) $\dfrac{1}{2x}.$ (iii) $1 - \dfrac{1}{x}.$ (iv) $\left(1 + \dfrac{1}{x}\right)^2.$

30 | Logarithmic and exponential functions

(v) $\dfrac{1}{x+2}$. (vi) $\dfrac{1}{2x-5}$. (vii) $\dfrac{1}{2-x}$. (viii) $\dfrac{1}{1-3x}$.

(ix) $\dfrac{2}{3(x+1)}$. (x) $\dfrac{4}{9(1-x)}$. (xi) $\dfrac{1}{q-px}$.

2 Evaluate:

(i) $\displaystyle\int \dfrac{2x\,dx}{x^2+1}$. (ii) $\displaystyle\int \dfrac{3x^2-2x}{x^3-x^2+4}\,dx$. (iii) $\displaystyle\int \dfrac{x^3}{1+x^4}\,dx$.

(iv) $\displaystyle\int \dfrac{x+1}{x^2+2x+2}\,dx$. (v) $\displaystyle\int \dfrac{x^5\,dx}{x^6+1}$. (vi) $\displaystyle\int \dfrac{\cos x\,dx}{2+\sin x}$.

(vii) $\displaystyle\int \dfrac{\sin x\,dx}{1+2\cos x}$. (viii) $\displaystyle\int \tan x\,dx$. (ix) $\displaystyle\int \cot x\,dx$.

(x) $\displaystyle\int \dfrac{x\cos x+\sin x}{x\sin x}\,dx$. (xi) $\displaystyle\int \dfrac{dx}{x\ln x}$. (xii) $\displaystyle\int \dfrac{\sec^2 x}{2+\tan x}\,dx$.

3 Find the values of:

(i) $\displaystyle\int_0^1 \dfrac{dx}{2x+1}$. (ii) $\displaystyle\int_0^{\frac{1}{2}} \dfrac{x\,dx}{1-x^2}$. (iii) $\displaystyle\int_0^{\pi/6} \tan 2x\,dx$.

(iv) $\dfrac{1}{2}\displaystyle\int_{-\frac{1}{2}}^{\frac{1}{2}} \left(\dfrac{1}{1-x}+\dfrac{1}{1+x}\right)dx$. (v) $\displaystyle\int_1^2 \dfrac{x^2+2x}{x^3+3x^2-1}\,dx$.

4 If $pv=2$, evaluate $\displaystyle\int_1^2 p\,dv$.

5 Evaluate $\displaystyle\int \dfrac{dx}{\sin x\cos x}$ by multiplying numerator and denominator by $\sec^2 x$.

6 Evaluate $\displaystyle\int \sec x\,dx$ by multiplying numerator and denominator by $(\sec x+\tan x)$.

7 Verify, by division, that $\dfrac{2x}{3x+1}=\dfrac{2}{3}-\dfrac{2}{3(3x+1)}$.

Hence, evaluate $\displaystyle\int \dfrac{2x\,dx}{3x+1}$.

8 Use the method of Example 7 to evaluate:

(i) $\displaystyle\int \dfrac{x\,dx}{x+1}$. (ii) $\displaystyle\int \dfrac{x+1}{x+2}\,dx$.

(iii) $\displaystyle\int \dfrac{3x\,dx}{1-x}$. (iv) $\displaystyle\int \dfrac{1-2x}{3x+7}\,dx$.

9 By division, show that $\dfrac{x^3}{2x-1}=\dfrac{x^2}{2}+\dfrac{x}{4}+\dfrac{1}{8}+\dfrac{1}{8(2x-1)}$, and evaluate $\displaystyle\int \dfrac{x^3\,dx}{2x-1}$.

10 Evaluate:

(i) $\displaystyle\int_1^3 \dfrac{x^2\,dx}{x+1}$. (ii) $\displaystyle\int_0^1 \dfrac{x^3\,dx}{x+2}$. (iii) $\displaystyle\int_2^3 \dfrac{x^4\,dx}{x-1}$.

Differentiation of exponential functions | 31

Exponential functions   A function of the form a^u where a is a positive constant and u a variable is called an *exponential function*. The most important function of this class is e^x, frequently spoken of as *the exponential function*.

The function e^x   For all values of x, e^x is positive.

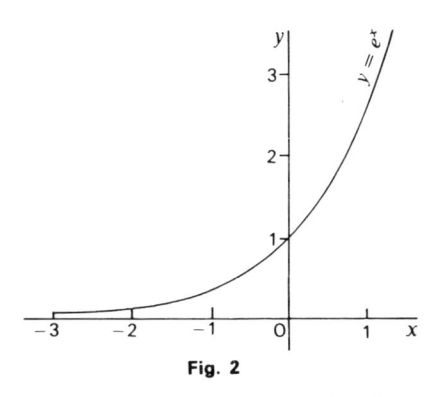

As x takes increasingly large negative values, e^x approaches zero, or, as we say mathematically, $e^x \to 0$ when $x \to -\infty$. As x increases from $-\infty$, e^x increases slowly and when $x = 0$, the function takes the value e^0 or 1. When x becomes positive, e^x increases rapidly and approaches $+\infty$ as x approaches $+\infty$.

A rough sketch graph of the function is given in Fig. 2.

Fig. 2

Example   Draw rough graphs of the functions e^{-x}, e^{2x}, $\frac{1}{2}(e^x + e^{-x})$.

Differentiation of exponential functions   Consider the general exponential function a^u, where a is a positive constant and u a function of x.
Let
$$y = a^u.$$
Take logarithms, base e, of both sides of this equation.
$$\ln y = u \ln a.$$
Differentiating with respect to x,

$$\frac{d}{dx}\ln y = \frac{d}{dx}(u \ln a) = \ln a \frac{du}{dx}.$$

But
$$\frac{d}{dx}\ln y = \frac{d}{dy}\ln y \times \frac{dy}{dx} = \frac{1}{y}\frac{dy}{dx}.$$

Hence
$$\frac{1}{y}\frac{dy}{dx} = \ln a \frac{du}{dx},$$

$$\frac{dy}{dx} = y \ln a \frac{du}{dx},$$

i.e.
$$\frac{d}{dx}(a^u) = a^u \ln a \frac{du}{dx}.$$

Taking $a = e$,

$$\frac{d}{dx}(e^u) = e^u \frac{du}{dx}.$$

32 | Logarithmic and exponential functions

Special Cases.

$$\frac{d}{dx}(e^x) = e^x; \quad \frac{d}{dx}(e^{ax}) = ae^{ax}.$$

These results give the following integrals:

$$\int e^{ax}\,dx = \frac{1}{a}e^{ax} + c; \quad \int e^u \frac{du}{dx}\,dx = e^u + c.$$

Example 12 Differentiate with respect to x: $e^{-x}, e^{\sin x}$

$$\frac{d}{dx}(e^{-x}) = e^{-x}(-1) = -e^{-x}.$$

$$\frac{d}{dx}(e^{\sin x}) = e^{\sin x}\frac{d}{dx}\sin x = \cos x\, e^{\sin x}.$$

Example 13 Integrate $e^{2x}, e^x + \frac{1}{e^x}, xe^{x^2}$.

$$\int e^{2x}\,dx = \tfrac{1}{2}e^{2x} + c.$$

$$\int \left(e^x + \frac{1}{e^x}\right)dx = \int (e^x + e^{-x})\,dx = e^x - e^{-x} + c.$$

$$\int xe^{x^2}\,dx = \tfrac{1}{2}\int e^{x^2} 2x\,dx = \tfrac{1}{2}e^{x^2} + c.$$

Example 14 Differentiate $2^{\cos x}$
Let
$$y = 2^{\cos x}.$$
Taking logarithms to base e,
$$\ln y = \cos x \ln 2.$$

Differentiate w.r. to x,
$$\frac{1}{y}\frac{dy}{dx} = -\sin x \ln 2,$$

i.e.
$$\frac{dy}{dx} = -2^{\cos x}\sin x \ln 2.$$

[A] **EXAMPLES 2e**

1 Differentiate, with respect to x:

(i) $2e^x$.   (ii) $\dfrac{1}{e^x}$.   (iii) $e^x + \dfrac{1}{e^x}$.   (iv) e^{2x}.

(v) e^{-4x}.   (vi) $e^{\frac{1}{2}x}$.   (vii) $\dfrac{1}{\sqrt{e^x}}$.   (viii) $\left(e^x - \dfrac{1}{e^x}\right)^2$.

2 Integrate the functions in Example 1 with respect to x.

3 Find $\dfrac{dy}{dx}$ if:

(i) $y = 3^x$.   (ii) $y = 10^{-x}$.   (iii) $y = e^{x^2}$.
(iv) $y = a^{2x}$.   (v) $y = e^{\tan x}$.   (vi) $ye^{3x} = 1$.

4 Evaluate:

(i) $\int_0^1 e^{-x} dx.$ (ii) $\int_0^{\frac{1}{3}} e^{6x} dx.$ (iii) $\int_0^{\frac{1}{2}} \left(1 + \frac{e^{-x}}{2}\right) dx.$ $[e = 2\cdot 718].$

5 If $I = I_0 e^{-kt}$ where I_0, k are constants, prove that $\dfrac{dI}{dt} + kI = 0.$

6 Newton's Law of Cooling can be expressed in the form

$$\frac{dT}{dt} = -k(T - T_0),$$

where T is the temperature of a hot body, T_0 a constant, the temperature of the surrounding medium and t the time. Verify that $T = T_0 + Ae^{-kt}$, A being a constant.

7 If $y = x^2 e^x$, find $\dfrac{dy}{dx}$ and show that it vanishes when $x = 0$ and $x = -2$.

8 Given $se^{-2t} = 1$, find $\dfrac{d^2 s}{dt^2}$.

9 Show that the function $e^x - e^{-x}$ increases for all values of x.

10 Evaluate: (i) $\int e^{x^2} 2x\, dx.$ (ii) $\int e^{\tan x} \sec^2 x\, dx.$ (iii) $\int \dfrac{e^{\sqrt{x}}}{\sqrt{x}} dx.$

11 Find the value of t for which the function te^{-t} has a stationary value.

12 If $x = e^{-t} \sin t$, verify that $\dfrac{d^2 x}{dt^2} + 2\dfrac{dx}{dt} + 2x = 0.$

Harder miscellaneous examples

Example 15 Differentiate $\sqrt{\dfrac{1 + x^2}{1 - x^2}}$.

The differentiation of complicated functions of this type is much simplified by the process of *logarithmic differentiation*.

Let
$$y = \sqrt{\frac{1 + x^2}{1 - x^2}}.$$

$\therefore \ln y = \tfrac{1}{2} \ln(1 + x^2) - \tfrac{1}{2} \ln(1 - x^2).$

Differentiating,
$$\frac{1}{y}\frac{dy}{dx} = \frac{1}{2}\frac{2x}{1 + x^2} - \frac{1}{2}\frac{-2x}{1 - x^2}$$

$$= \frac{2x}{(1 + x^2)(1 - x^2)}.$$

$$\therefore \frac{dy}{dx} = \sqrt{\frac{1 + x^2}{1 - x^2}} \frac{2x}{(1 + x^2)(1 - x^2)}$$

$$= \frac{2x}{(1 + x^2)^{\frac{1}{2}}(1 - x^2)^{\frac{3}{2}}}.$$

34 | Logarithmic and exponential functions

Example 16 If x is positive, prove that $\ln(1+x) > x - \dfrac{x^2}{2}$.

Let
$$f(x) = \ln(1+x) - \left(x - \dfrac{x^2}{2}\right).$$

Then
$$f'(x) = \dfrac{1}{1+x} - 1 + x = \dfrac{x^2}{1+x}.$$

If $x > 0$, $f'(x)$ is positive and so $f(x)$ is an increasing function.

Also
$$f(0) = \ln 1 = 0.$$
$$\therefore f(x) > 0 \quad \text{for} \quad x > 0.$$

i.e.
$$\ln(1+x) > x - \dfrac{x^2}{2} \quad \text{for} \quad x > 0.$$

Example 17 Show that the curve $y = xe^x$ has a minimum ordinate when $x = -1$.

$$y = xe^x.$$
$$\therefore \dfrac{dy}{dx} = e^x + xe^x = 0 \quad \text{for max. or min. points.}$$

i.e.
$$e^x(1+x) = 0.$$

As e^x does not vanish the only root is $x = -1$.

$$\dfrac{d^2y}{dx^2} = 2e^x + xe^x.$$

When $x = -1$,
$$\dfrac{d^2y}{dx^2} = e^{-1} = \dfrac{1}{e}. \quad \text{(positive)}.$$

Hence the curve has a minimum ordinate when $x = -1$.

[B] MISCELLANEOUS EXAMPLES

1 A pressure-volume curve is of the form $pv^n = c$, a constant. If $p = 90$ when $v = 4$ and $p = 40$ when $v = 6.2$, find the values of n and c.

2 Prove that $\log_a x = \log_b x \times \log_a b$.
Find the value of $\ln(0.0367)^{3.757}$ correct to 3 significant figures, taking $e = 2.718$.

3 Solve the equation $3 \cdot 4^x + 5 \cdot 2^x - 9 = 0$. [*Note.*—$4^x = 2^{2x}$.]

4 Show that the equation $3 \cdot 32x = e^x$ can be expressed as $\ln x = x - 1.2$. Plot the curve $y = \ln x$ for values of x between $\frac{1}{2}$ and 3 and hence solve the equation $3 \cdot 32x = e^x$ approximately.

5 Plot the curve $y = e^x$ for values of x between -1 and 2. Hence find approximate solutions of the equation $2e^x = 3 + 4x$.

6 Differentiate, with respect to x:

(i) $(3x^2 - x)\ln(2x+1)$. (ii) $\dfrac{x^2+2}{\ln x}$. (iii) $e^x \ln 2x$.

(iv) $e^{-2x}\cos 4x$. (v) $\ln \cos 3x$. (vi) $\ln(1+e^x)$.

(vii) $\dfrac{e^x - e^{-x}}{e^x + e^{-x}}$. (viii) $\ln\left(\dfrac{1+\sin x}{1-\cos x}\right)$. (ix) $\ln \dfrac{1-x}{\sqrt{1+x^2}}$.

(x) $\ln \dfrac{1+\sqrt{x}}{1-\sqrt{x}}$. (xi) $e^{x\sin x}$ (xii) $2^{\tan x}$

(xiii) $e^{3\ln x}$ (xiv) $\ln \sqrt{\dfrac{e^x}{1-e^{-x}}}$. (xv) $\ln \dfrac{x}{\sqrt{x^2+1}-x}$.

(xvi) $\ln \left(\dfrac{x+\sqrt{x^2+1}}{\sqrt{x^2+1}-x} \right)$.

7 Evaluate:

(i) $\displaystyle\int \dfrac{x^2+1}{1-x}\,dx$. (ii) $\displaystyle\int \dfrac{x^5\,dx}{x-1}$. (iii) $\displaystyle\int \dfrac{x+1}{x^2+2x-1}\,dx$.

(iv) $\displaystyle\int \cot \dfrac{x}{2}\,dx$. (v) $\displaystyle\int 2^x\,dx$. (vi) $\displaystyle\int \dfrac{\cos x - x\sin x}{x\cos x}\,dx$.

(vii) $\displaystyle\int \sqrt{e^x} - \dfrac{1}{\sqrt{e^x}}\,dx$. (viii) $\displaystyle\int e^{-x^2} x\,dx$.

8 If $r = \tfrac{3}{5}$, find the least integral value of n for which $\dfrac{r^n}{1-r}$ is less than 0.0001.

9 If $\dfrac{\ln p}{a} = \dfrac{\ln q}{b} = \dfrac{\ln r}{c} = \ln t$, prove that $\dfrac{q^2}{pr} = t^{2b-a-c}$.

10 If y and x are known to be connected by an equation of the form $y = ax^n$ where a and n are constants, prove that the graph of lg y against lg x will be a straight line.
Simultaneous values of x and y are given in the following table:

x	0	1·6	2·65	3·8	4·7	5·7
y	0	6	12	16·5	23·5	26·5

Show that there is an approximate relationship of the form $y = ax^n$ and determine the most suitable values of a and n.

11 N m^3 of water were measured as flowing per second over a notch when the difference of levels was x m. The following results were obtained:

x	1·2	1·4	1·6	1·8	2·0
N	6·3	9·2	12·8	17·3	22·4

Prove that x and N are connected by an equation of the form $N = ax^n$ and find the probable values of a and n.

12 Find $\dfrac{dy}{dx}$ in the following cases:

(i) $y = \dfrac{(3x+4)^6}{\sqrt{1-x^3}}$. (ii) $y = \dfrac{x^3 \sin^5 x}{\cos^3 x}$.

(iii) $y = x^{\sin x}$. (iv) $y = x^{\ln x}$.

13 If $y = \ln(x + \sqrt{x^2 + a^2})$, where a is a constant, prove that $\dfrac{dy}{dx} = \dfrac{1}{\sqrt{x^2+a^2}}$.

Hence, evaluate $\displaystyle\int_0^4 \dfrac{dx}{\sqrt{x^2+9}}$.

14 By first differentiating $\ln(x + \sqrt{x^2 - a^2})$, evaluate $\displaystyle\int_2^3 \dfrac{dx}{\sqrt{x^2-1}}$.

15 For the curve $y = e^{-x^2}$, show that $\dfrac{d^2y}{dx^2}$ vanishes when $x = \pm \dfrac{1}{\sqrt{2}}$.

16 If $y = \ln(1 + \sin x)$, prove that $\dfrac{d^3y}{dx^3} + \dfrac{d^2y}{dx^2}\dfrac{dy}{dx} = 0$.

17 If $y = e^{-2x}\cos 5x$, show that $\dfrac{dy}{dx}$ can be expressed in the form $ae^{-2x}\cos(5x + b)$ where a and b are constants. Find a and b and write down the value of $\dfrac{d^5y}{dx^5}$.

18 By writing down $\dfrac{dy}{dx}, \dfrac{d^2y}{dx^2}, \dfrac{d^3y}{dx^3}, \ldots$, find the nth differential coefficients of:
  (i) $\ln(1 + x)$.   (ii) $\ln(1 - x)$.   (iii) $\ln\left(\dfrac{1 + 2x}{1 + x}\right)$.
  (iv) xe^x.   (v) $x^2 e^x$.

19 Find the maximum and minimum points on the curve $y = xe^{-x^2}$.

20 By expressing $\dfrac{d}{dx}(e^{3x}\sin 2x)$ in the form $ae^{3x}\sin(2x + b)$ where a and b are constants, find the value of $\dfrac{d^n}{dx^n}(e^{3x}\sin 2x)$.

21 In a submarine cable, the speed of signalling is found to vary as $x^2 \ln\dfrac{1}{x}$, where x is the ratio of the radius of the core to that of the covering. If the radius of the core is $\tfrac{1}{2}$ cm, find the radius of the covering in order that the greatest possible speed is attained.

22 If $a^2 + b^2 = 7ab$, prove that $\ln\tfrac{1}{3}(a + b) = \tfrac{1}{2}(\ln a + \ln b)$.

23 Prove that $\ln(1 + x) < x - \dfrac{x^2}{2} + \dfrac{x^3}{3}$ for all positive values of x.

24 Show that the equation $\dfrac{d^2s}{dt^2} + 2n\dfrac{ds}{dt} + n^2 s = 0$ is satisfied by
$$s = (A + Bt)e^{-nt},$$
where A and B are arbitrary constants.

25 The portion of the curve $y^2 = \dfrac{x(x+1)}{x-1}$ between $x = 2$ and $x = 3$, is rotated about the x axis. Find the volume of the solid of revolution.

26 Solve the equations:
  (i) $3^{2x} - 3^{x+1} + 2 = 0$;
  (ii) $2x + y = 3$; $2^{(x-3y)} = 10$.

27 (i) Differentiate $(\ln x)^{\ln x}$.
  (ii) Prove that $\displaystyle\int_0^{\pi/4} \dfrac{dx}{9\cos^2 x - \sin^2 x} = \tfrac{1}{6}\ln 2$.   [Divide numerator and denominator by $\cos^2 x$.]

28 (i) Prove that $a^{\log b} = b^{\log a}$, whatever the base of logarithms.

(ii) Find by inspection, one root of the equation
$$2^{2x+2} - 17 \cdot 2^{x-1} + 1 = 0,$$
and determine the other.

29 If $y = a^{1/(1-\log_a x)}$ and $z = a^{1/(1-\log_a y)}$, show that $x = a^{1/(1-\log_a z)}$.

30 With the same axes, sketch the graphs of $y = e^x$ and $y = 3 - 2x - x^2$. If α is the positive root of the equation $e^x = 3 - 2x - x^2$, prove that the area in the first quadrant enclosed by the two curves and the y axis is
$$\tfrac{1}{3}(15\alpha - \alpha^3 - 6).$$

3 | Power series
Derivation and use of exponential, logarithmic and trigonometric series

Definitions A *power series* is a series of the form
$$a_0 + a_1 x + a_2 x^2 + a_3 x^3 + \ldots + a_n x^n + \ldots \infty,$$
where the coefficients $a_0, a_1, a_2, \ldots$ are constants.

If $\quad S_n = a_0 + a_1 x + a_2 x^2 + \ldots + a_n x^n,$
and if S_n approaches a definite limit S as n tends to infinity, the power series is said to be *convergent* and have a sum S.

It follows that a power series can only have a definite sum if it is convergent.

The exponential series We obtain a power series whose sum is e^x, on the assumption that such a series does exist.

Assume that $\quad e^x = a_0 + a_1 x + a_2 x^2 + a_3 x^3 + \ldots + a_n x^n + \ldots \ldots \ldots$ (i)
Letting $x = 0$, $\quad e^0 = a_0$; i.e. $a_0 = 1$.
To obtain the value of a_1, we first differentiate and then put $x = 0$.
Differentiating (i),
$e^x = a_1 + 2a_2 x + 3a_3 x^2 + \ldots \ldots$ .... (ii)
Letting $x = 0$, $\quad e^0 = a_1$; i.e. $a_1 = 1$.
Differentiating (ii),
$$e^x = 2a_2 + 3 \cdot 2a_3 x + \ldots$$
Letting $x = 0$, $\quad e^0 = 2a_2$; i.e. $a_2 = \frac{1}{2}$.
Repeating this process, we get

$$a_3 = \frac{1}{3 \cdot 2 \cdot 1} \text{ or } \frac{1}{3!}, \quad a_4 = \frac{1}{4 \cdot 3 \cdot 2 \cdot 1} \text{ or } \frac{1}{4!}, \quad a_5 = \frac{1}{5!},$$

and so on.

Hence $\quad e^x = 1 + x + \dfrac{x^2}{2!} + \dfrac{x^3}{3!} + \dfrac{x^4}{4!} + \ldots + \dfrac{x^n}{n!} + \ldots \quad \ldots$ (1)

It can be shown that the series $1 + x + \dfrac{x^2}{2!} + \dfrac{x^3}{3!} + \ldots$ is convergent for all values of x and as we have just seen, the sum must be e^x. In other words, *the above expansion of e^x as a power series is valid for all values of x.*

By replacing x by mx, the following general result is obtained:

$$e^{mx} = 1 + mx + \frac{(mx)^2}{2!} + \frac{(mx)^3}{3!} + \ldots \ldots \ldots \quad (2)$$

Further, if we denote ln a by m, that is, $a = e^m$ where $a > 0$,

then, $$a^x = e^{mx} = 1 + mx + \frac{(mx)^2}{2!} + \frac{(mx)^3}{3!} + \ldots ,$$

or, $$a^x = 1 + x \ln a + \frac{(x \ln a)^2}{2!} + \frac{(x \ln a)^3}{3!} + \ldots \quad (3)$$

Applications of the exponential series One application of the exponential series is the evaluation of the constant e to any required degree of accuracy. For, putting $x = 1$ in series (1),

$$e^1 = e = 1 + 1 + \frac{1}{2!} + \frac{1}{3!} + \frac{1}{4!} + \ldots \infty.$$

This series converges rapidly and successive terms are easily calculated by continued division, i.e.

The 3rd term is obtained by dividing the 2nd by 2.
The 4th term is obtained by dividing the 3rd by 3.
The 5th term is obtained by dividing the 4th by 4,

and so on.

Sufficient terms will be taken to obtain the value of e correct to 5 decimal places.

$$1 + 1 = 2.000\,000\,0$$

$$\frac{1}{2!} = 0.500\,000\,0$$

$$\frac{1}{3!} = 0.166\,666\,7$$

$$\frac{1}{4!} = 0.041\,666\,7$$

$$\frac{1}{5!} = 0.008\,333\,3$$

$$\frac{1}{6!} = 0.001\,388\,9$$

$$\frac{1}{7!} = 0.000\,198\,4$$

$$\frac{1}{8!} = 0.000\,024\,8$$

$$\frac{1}{9!} = 0.000\,002\,8$$

$$2.718\,281\,6$$

$\therefore e = 2.71828$ to 5 dec. places.

Example 1 Find the power series whose sum is $\left(e - \dfrac{1}{e}\right)^2$.

$$\left(e - \frac{1}{e}\right)^2 = e^2 - 2 + \frac{1}{e^2} = e^2 - 2 + e^{-2}.$$

Using the expansion $e^x = 1 + x + \dfrac{x^2}{2!} + \dfrac{x^3}{3!} + \ldots$ with $x = 2$ and -2,

$$e^2 - 2 + e^{-2} = \left(1 + 2 + \frac{2^2}{2!} + \frac{2^3}{3!} + \frac{2^4}{4!} \ldots\right) + \left(1 - 2 + \frac{2^2}{2!} - \frac{2^3}{3!} + \frac{2^4}{4!} \ldots\right) - 2,$$

$$= 2\left(\frac{2^2}{2!} + \frac{2^4}{4!} + \ldots\right).$$

Example 2 Find the coefficient of x^3 in the expansion of e^{x+3}.
Write e^{x+3} as $e^x e^3$.

So,
$$e^{x+3} = e^3 e^x = e^3 \left(1 + x + \frac{x^2}{2!} + \frac{x^3}{3!} + \ldots\right).$$

$\therefore$ The coefficient of x^3 in the expansion is $\dfrac{e^3}{3!}$.

Example 3 Find the coefficient of x^r in the expansion of the function $\dfrac{1 + x - x^2}{e^x}$.

We write $\dfrac{1 + x - x^2}{e^x}$ as $(1 + x - x^2)e^{-x}$ and expand e^{-x}.

$$\frac{1 + x - x^2}{e^x} = (1 + x - x^2)\left\{1 - x + \frac{x^2}{2!} - \frac{x^3}{3!} + \ldots + \frac{(-1)^r x^r}{r!} \ldots\right\}.$$

The sign of the term in x^r is determined by noting that it is positive for even powers of x and negative for odd powers.

$$\text{Term in } x^r = x^r \left\{\frac{(-1)^r}{r!} + \frac{(-1)^{r-1}}{(r-1)!} - \frac{(-1)^{r-2}}{(r-2)!}\right\}.$$

$\therefore$ The coefficient of $x^r = \dfrac{(-1)^r}{r!}\left\{1 - r - r(r-1)\right\}$ as $(-1)^{r-2} = (-1)^r$.

$$= \frac{(-1)^r}{r!}\left\{1 - r^2\right\}.$$

Example 4 Evaluate $\underset{x \to 0}{\mathrm{Lt}} \left\{\dfrac{e^{2x} - e^x}{x}\right\}.$

On substituting $x = 0$, the function becomes $\dfrac{1-1}{0}$ or $\dfrac{0}{0}$, an indeterminate form.
To obtain the limiting value as $x \to 0$, we expand the numerator.

$$e^{2x} - e^x = \left\{1 + 2x + \frac{(2x)^2}{2!} + \ldots\right\} - \left\{1 + x + \frac{x^2}{2!} + \ldots\right\}$$

$$\therefore \frac{e^{2x} - e^x}{x} = \frac{x + \dfrac{3x^2}{2!} + \text{higher powers of } x}{x}$$

$$= 1 + \frac{3x}{2!} + \text{higher powers of } x.$$

The logarithmic series | 41

Thus, $\underset{x\to 0}{\text{Lt}} \dfrac{e^{2x}-e^x}{x} = \underset{x\to 0}{\text{Lt}} \left\{1+\dfrac{3x}{2!}+\text{higher powers of } x\right\}$
$= 1.$

[A] EXAMPLES 3a

1 Write down infinite series whose sums are: (i) e^{-1}, (ii) $e^{\frac{1}{2}}$, (iii) e^3, (iv) $\sqrt[3]{e}$, (v) $e+\dfrac{1}{e}$, (vi) $\sqrt{e}-\dfrac{1}{\sqrt{e}}$.

2 Express as power series: (i) e^{-x}, (ii) e^{3x}, (iii) $e^{\frac{1}{x}}$, (iv) e^{-x^2}.

3 Calculate to 4 places of decimals: (i) $\dfrac{1}{e}$, (ii) $\sqrt{e}$, (iii) $\dfrac{1}{\sqrt[3]{e}}$.

4 Write down the first 4 terms in the expansions of the following functions:
   (i) $(x+1)e^x$. (ii) $(x^2-1)e^{-x}$. (iii) $\dfrac{(x+1)^2}{e^x}$. (iv) $\dfrac{e^{3x}+e^x}{e^{2x}}$.

5 Find the coefficients of x^4 in the expansions of: (i) e^{x+1}, (ii) e^{2-x}, (iii) e^{3x+2}.

6 Expand the functions $\sinh x$ and $\cosh x$ where $\sinh x = \tfrac{1}{2}(e^x - e^{-x})$, $\cosh x = \tfrac{1}{2}(e^x + e^{-x})$.

7 If x is small, prove that $\dfrac{1}{1+x} - e^{-x} = \tfrac{1}{2}x^2 - \tfrac{5}{6}x^3$ approximately.

8 If x is so small that powers above the first can be ignored, find the approximate value of $e^{3x} - 4e^x$.

9 Prove that $\dfrac{e^6 - 1}{e^3} = 2\left[3 + \dfrac{3^3}{3!} + \dfrac{3^5}{5!} + \ldots\right]$.

10 What is the coefficient of x^6 in the expansion of $(3x-2)e^x$?

11 Write down the first three terms in the expansion of the function e^{x-x^2}.

12 If $y = 10^x$, show by taking logarithms to base e that $y = e^{x \ln 10}$. Hence, write down the first 4 terms in the expansion of 10^x.

13 What is the coefficient of x^3 in the expansion of 2^x?

14 If $\ln y = \sqrt{x}$, show that $y = 1 + \sqrt{x} + \dfrac{(\sqrt{x})^2}{2!} + \ldots$.

The logarithmic series If we attempt to obtain a power series having a sum of $\ln x$, we find that it is not possible to determine values of the coefficients a_0, a_1, a_2, etc.

For, if $\quad\quad \ln x = a_0 + a_1 x + a_2 x^2 + \ldots,$

by putting $x = 0$, it follows that $a_0 = \ln 0$, which is not defined.

The significance of this result is simply that no convergent power series exists with a sum equal to $\ln x$.

This difficulty is overcome by taking the function $\ln(1+x)$.
Assume $\ln(1+x) = a_0 + a_1 x + a_2 x^2 + a_3 x^3 + a_4 x^4 + \ldots$.
Letting $x = 0$, $\ln 1 = a_0$; i.e. $a_0 = 0$.

Differentiating, $\dfrac{1}{1+x} = a_1 + 2a_2 x + 3a_3 x^2 + 4a_4 x^3 + \ldots$.

Letting $x = 0$, $1 = a_1$; i.e. $a_1 = 1$.

Differentiating again,

$$-\frac{1}{(1+x)^2} = 2a_2 + 6a_3 x + 12a_4 x^2 + \ldots$$

Letting $x = 0$, $-1 = 2a_2$; i.e. $a_2 = -\frac{1}{2}$.

Differentiating again,

$$\frac{2}{(1+x)^3} = 6a_3 + 24a_4 x + \ldots$$

Letting $x = 0$, $2 = 6a_3$; i.e. $a_3 = \frac{1}{3}$.

Continuing this process, we get

$$a_4 = -\tfrac{1}{4},\ a_5 = \tfrac{1}{5},\ a_6 = -\tfrac{1}{6},\ \text{etc.}$$

Hence, $$\ln(1+x) = x - \frac{x^2}{2} + \frac{x^3}{3} - \frac{x^4}{4} + \frac{x^5}{5} \ldots + (-1)^{n+1}\frac{x^n}{n} + \ldots \quad \ldots (4)$$

In this case, it can be shown that the infinite series is only convergent for values of x lying between -1 and $+1$ and for the value $x = 1$. In other words, $\ln(1+x)$ can only be expressed as the above power series when $-1 < x \leqslant 1$.

Applications of the logarithmic series   One of the chief practical applications of the logarithmic series is the evaluation of logarithms. The actual logarithmic series which we have obtained is of little value for this purpose, owing to the fact that it converges slowly except when x is very small. However, we can deduce other series from it by the aid of which tables of logarithms can be constructed.

Writing $-x$ for x in the series for $\ln(1+x)$, we get

$$\ln(1-x) = -x - \frac{x^2}{2} - \frac{x^2}{3} - \frac{x^4}{4} \ldots$$

$$= -\left\{x + \frac{x^2}{2} + \frac{x^3}{3} + \frac{x^4}{4} + \ldots\right\} \quad \ldots (5)$$

This result is true for $-1 < -x \leqslant 1$ or $-1 \leqslant x < 1$.
Combining series (4) and (5),

$$\ln(1+x) - \ln(1-x) = x - \frac{x^2}{2} + \frac{x^3}{3} - \frac{x^4}{4} + \ldots + \left\{x + \frac{x^2}{2} + \frac{x^3}{3} + \frac{x^4}{4} \ldots\right\}$$

$$= 2\left\{x + \frac{x^3}{3} + \frac{x^5}{5} + \ldots\right\}.$$

i.e. $$\ln\left(\frac{1+x}{1-x}\right) = 2\left\{x + \frac{x^3}{3} + \frac{x^5}{5} + \ldots\right\} \quad \ldots (6)$$

Applications of the logarithmic series | 43

This expansion holds for values of x which satisfy both the conditions,
$$-1 < x \leqslant 1 \quad \text{and} \quad -1 \leqslant x < 1.$$
Hence *the expansion holds for* $-1 < x < 1$.

For values of x less than 1, the series (6) converges rapidly and it is used as the basis of the construction of logarithm tables. For this purpose, it is convenient to modify the result in the following ways:

(i) Let $\dfrac{1+x}{1-x} = n$, so that $x = \dfrac{n-1}{n+1}$.

The restriction $-1 < x < 1$ becomes $n > 0$ for the series

$$\ln n = 2\left\{\frac{n-1}{n+1} + \frac{1}{3}\left(\frac{n-1}{n+1}\right)^3 + \frac{1}{5}\left(\frac{n-1}{n+1}\right)^5 \cdots\right\} \quad \cdots \cdots \cdots (7)$$

(ii) Put $\dfrac{1+x}{1-x} = \dfrac{n+1}{n}$, so that $x = \dfrac{1}{2n+1}$.

Thus $\quad \ln(n+1) - \ln n = 2\left\{\dfrac{1}{2n+1} + \dfrac{1}{3}\dfrac{1}{(2n+1)^3} + \dfrac{1}{5}\dfrac{1}{(2n+1)^5} + \cdots\right\} \quad \cdots \cdots (8)$

This series converges very rapidly and being convergent for $n > 0$ is of considerable value in computational work.

Example 5 Evaluate $\ln 2$ to 4 places of decimals.
Let $n = 2$ in series (7).

Then
$$\ln 2 = 2\left\{\frac{1}{3} + \frac{1}{3}\frac{1}{3^3} + \frac{1}{5}\frac{1}{3^5} + \cdots\right\}.$$

$\dfrac{1}{3} = 0.333\,333; \qquad \dfrac{1}{3} = 0.333\,333$

Divide by 9.

$\dfrac{1}{3^3} = 0.037\,037; \qquad \dfrac{1}{3}\dfrac{1}{3^3} = 0.012\,346$

Divide by 9.

$\dfrac{1}{3^5} = 0.004\,115; \qquad \dfrac{1}{5}\dfrac{1}{3^5} = 0.000\,823$

Divide by 9.

$\dfrac{1}{3^7} = 0.000\,457; \qquad \dfrac{1}{7}\dfrac{1}{3^7} = 0.000\,065$

Divide by 9.

$\dfrac{1}{3^9} = 0.000\,051; \qquad \dfrac{1}{9}\dfrac{1}{3^9} = 0.000\,006$

$\hspace{6cm}\overline{0.346\,573}$

Hence $\quad \ln 2 = 2 \times 0.346\,573 = 0.693\,146$
$\hspace{2.2cm} = 0.6931\quad$ to 4 places of decimals.

Example 6 Calculate ln 10 and lg 2 to 4 places of decimals.

If we attempted to find ln 10 directly from result (7) by letting $n = 10$, the resulting series would converge very slowly and the computation would be laborious. To overcome this difficulty, we take $n = \frac{10}{8}$ and obtain $\ln 10 - \ln 8$ or $\ln 10 - 3 \ln 2$. Substituting $n = \frac{10}{8}$ in result (7),

$$\ln 10 - 3 \ln 2 = 2\left\{\frac{1}{9} + \frac{1}{3}\frac{1}{9^3} + \frac{1}{5}\frac{1}{9^5} + \ldots\right\}.$$

Now, $\quad \dfrac{1}{9} = 0{\cdot}111\,111; \qquad \dfrac{1}{9} = 0{\cdot}111\,111$

$\dfrac{1}{9^2} = 0{\cdot}012\,346;$

$\dfrac{1}{9^3} = 0{\cdot}001\,372; \qquad \dfrac{1}{3}\dfrac{1}{9^3} = 0{\cdot}000\,457$

$\dfrac{1}{9^4} = 0{\cdot}000\,152;$

$\dfrac{1}{9^5} = 0{\cdot}000\,017; \qquad \dfrac{1}{5}\dfrac{1}{9^5} = 0{\cdot}000\,003$

$\qquad\qquad\qquad\qquad\qquad\qquad\overline{0{\cdot}111\,571}$

$\therefore \ln 10 = 3 \ln 2 + 0{\cdot}223\,142$
$\qquad\quad = 2{\cdot}079\,438 + 0{\cdot}223\,142$
$\qquad\quad = 2{\cdot}302\,580.$

i.e. $\qquad\qquad\qquad \ln 10 = 2{\cdot}3026.$

Now, $\qquad\qquad\qquad \lg 2 = \dfrac{\ln 2}{\ln 10}$

$\qquad\qquad\qquad\qquad = \dfrac{0{\cdot}6931}{2{\cdot}3026} = 0{\cdot}3010.$

i.e. $\qquad\qquad\qquad \lg 2 = 0{\cdot}3010.$

Common logarithms The method utilised in Example 6 to obtain lg 2 from ln 2 is used in the calculation of common logarithms, i.e. logarithms to base 10.

As $\qquad\qquad\qquad \lg N = \dfrac{\ln N}{\ln 10},$

$$\text{Common logarithm} = \text{Natural logarithm} \times \frac{1}{\ln 10}.$$

The constant, $\dfrac{1}{\ln 10}$, is called the *modulus* of the common system and its value to 4 decimal places is $\dfrac{1}{2{\cdot}3026}$ or $0{\cdot}4343.$

Common logarithms | 45

Example 7 Given that $2 \cdot 19722 < \ln 9 < 2 \cdot 19723$, prove that
$$2 \cdot 39789 < \ln 11 < 2 \cdot 39790.$$

The required result is obtained by finding the value of $\ln 11 - \ln 9$, or $\ln \frac{11}{9}$.
Using series (6), we equate $\frac{1+x}{1-x}$ to $\frac{11}{9}$ and obtain $x = \frac{1}{10}$.

Substituting $x = \frac{1}{10}$,

$$\ln \frac{11}{9} = 2 \left\{ \frac{1}{10} + \frac{1}{3} \frac{1}{10^3} + \frac{1}{5} \frac{1}{10^5} + \cdots \right\}.$$

Evaluating the first 3 terms of the series:

$$\frac{1}{10} = 0 \cdot 100\,000\,0; \qquad \frac{1}{10} = 0 \cdot 100\,000\,0$$

$$\frac{1}{10^3} = 0 \cdot 001\,000\,0; \qquad \frac{1}{3} \frac{1}{10^3} = 0 \cdot 000\,333\,3$$

$$\frac{1}{10^5} = 0 \cdot 000\,010\,0; \qquad \frac{1}{5} \frac{1}{10^5} = 0 \cdot 000\,002\,0$$

$$\overline{0 \cdot 100\,335\,3}$$

$$\therefore \ln 11 = \ln 9 + 2(0 \cdot 100\,335\,3)$$
$$= \ln 9 + 0 \cdot 200\,670\,6.$$

Using the given limits for $\ln 9$, we obtain the required result,
$$2 \cdot 397\,89 < \ln 11 < 2 \cdot 397\,90.$$

[A] EXAMPLES 3b

1 Calculate $\ln 1 \cdot 01$ to 5 decimal places.

2 By putting $x = \frac{1}{5}$ in the expansion of $\ln \left(\frac{1+x}{1-x} \right)$, find the value of $\ln 3 - \ln 2$, correct to 4 decimal places,

3 Calculate $\ln 13 - \ln 12$, correct to 5 decimal places. Given $\ln 2 = 0 \cdot 69315$, $\ln 3 = 1 \cdot 09861$, find the value of $\ln 13$ to 4 decimal places.

4 For what range of values of x can the function $\ln (1 + 3x)$ be expanded as a power series?

Expand the following functions as far as the third term and in each case state the values of x for which the expansions are valid:

5 $\ln (1 + 2x)$.

6 $\ln \left(1 - \frac{x}{2} \right)$.

7 $\ln \left(1 + \frac{x}{3} \right)$.

8 $\ln (1 + x)^2$

9 $\ln (1 + x)(1 - x)$.

10 $\ln \left(\frac{1-x}{1+2x} \right)$.

11 $\ln (1 + 3x)\left(1 - \frac{x}{2}\right)$.

12 $\ln (1 + x^2)$.

13 $\ln (1 + 5x + 4x^2)$.

14 $\ln\left(\dfrac{1+x^3}{1-x}\right)$.

15 $\ln\left(\dfrac{1-x^2}{1+2x}\right)$.

16 By writing $1+x+x^2$ as $\dfrac{1-x^3}{1-x}$, obtain the expansion of $\ln(1+x+x^2)$ up to and including the terms in x^6.

17 Expand $\ln(1-x+x^2)$ as far as the term in x^6.

18 Find the coefficient of x^3 in the expansion of $(1+2x)\ln(1+x)$.

19 Find the coefficient of x^4 in the expansion of $\ln(1-3x+2x^2)$ in ascending powers of x.

20 If x is small, prove that $\ln\left(\dfrac{1+x^2}{1-x^2}\right) = 2x^2 + \tfrac{2}{3}x^6$ approximately.

Trigonometric series The functions $\sin x$, $\cos x$, $\tan x$ (x in radians) can be expanded as power series by employing the method used to derive the exponential and logarithmic series.

The following results are obtained:

$$\sin x = x - \frac{x^3}{3!} + \frac{x^5}{5!} - \frac{x^7}{7!} + \ldots \qquad (9)$$

$$\cos x = 1 - \frac{x^2}{2!} + \frac{x^4}{4!} - \frac{x^6}{6!} + \ldots \qquad (10)$$

These results are true for all values of x so long as x is in radians.

In the case of $\tan x$, there is no simple law for deriving successive terms; the first three terms in the expansion are

$$\tan x = x + \frac{x^3}{3} + \frac{2x^5}{15} \ldots \qquad (11)$$

Applications of trigonometrical series. Small angles By use of the trigonometrical series, successive approximations can be made to the sine, cosine and tangent of small angles.

If x is so small that all powers above the first can be ignored,

$$\sin x \approx x: \qquad \cos x \approx 1: \qquad \tan x \approx x.$$

If all powers of x above the second are ignored,

$$\sin x \approx x: \qquad \cos x \approx 1 - \frac{x^2}{2}: \qquad \tan x \approx x.$$

If all powers of x above the third are ignored,

$$\sin x \approx x - \frac{x^3}{6}: \qquad \cos x \approx 1 - \frac{x^2}{2}: \qquad \tan x \approx x + \frac{x^3}{3}.$$

Example 8 Assuming the result $\sin\theta \approx \theta$ if θ is small, prove geometrically that an approximate value of $\cos\theta$ is $1 - \theta^2/2$.

Approximate solution of equations | 47

In Fig. 3, $\triangle ABC$ is isosceles with $AB = AC = 1$ unit and $\hat{A} = \theta$, a small angle measured in radians. AD, the altitude from A, bisects BC at D.

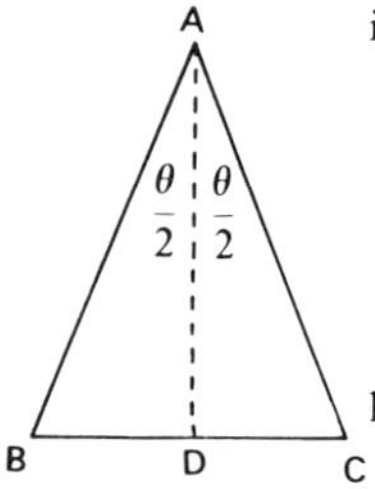

In $\triangle ABC$, $\quad a^2 = b^2 + c^2 - 2bc \cos A$.

i.e. $\quad BC^2 = 2 - 2\cos\theta$. .... (i)

But $\quad BD = AB \sin\dfrac{\theta}{2};\quad$ i.e. $BD = \sin\dfrac{\theta}{2}$.

As θ is small, we can take $\sin\dfrac{\theta}{2} = \dfrac{\theta}{2}$,

hence, $\quad BD = \dfrac{\theta}{2}$.

$\therefore BC^2 = 4BD^2 = \theta^2$ .... (ii)

Fig. 3

From (i) and (ii), $\quad \theta^2 = 2 - 2\cos\theta$.

i.e. $\quad \cos\theta = 1 - \dfrac{\theta^2}{2}$ approximately.

Approximate solution of equations The trigonometric series can be used to obtain approximate values of the roots of certain types of equations.

Example 9 Obtain an approximate solution of the equation $\dfrac{\sin\theta}{\theta} = \dfrac{99}{100}$.

As $\dfrac{\sin\theta}{\theta}$ is approximately equal to 1, θ must be a small angle. So, ignoring powers of θ above the third,

$$\dfrac{\sin\theta}{\theta} = \dfrac{\theta - \dfrac{\theta^3}{6}}{\theta} = 1 - \dfrac{\theta^2}{6}.$$

$\therefore \quad \dfrac{\theta^2}{6} = \dfrac{1}{100}$,

i.e. $\quad \theta = \dfrac{\sqrt{6}}{10}$ rads. $= 14°\ 2'$ approximately.

Example 10 Find an approximate solution of the equation $\cos\theta = 2\theta$, if θ is positive.

Take $\quad \cos\theta = 1 - \dfrac{\theta^2}{2}$ approximately.

Then $\quad 1 - \dfrac{\theta^2}{2} = 2\theta,\quad$ or $\quad \theta^2 + 4\theta - 2 = 0$.

$\therefore \theta = -2 + \sqrt{6}\quad$ (taking the positive root)

$= 0{\cdot}45$ rad. approx.

i.e. $\quad$ Approximate value of $\theta = 25.8°$.

Tables show that this is a close approximation even though the result used is normally applicable only for small angles.

Power series

Expansions of sines and cosines of multiple angles From series (9) and (10), replacing x by mx,

$$\sin mx = mx - \frac{(mx)^3}{3!} + \frac{(mx)^5}{5!} - \ldots,$$

$$\cos mx = 1 - \frac{(mx)^2}{2!} + \frac{(mx)^4}{4!} - \ldots.$$

Example 11 Express $\sin^2 x$ as a power series.
$\sin^2 x = \frac{1}{2}(1 - \cos 2x)$

$$= \frac{1}{2} - \frac{1}{2}\left(1 - \frac{(2x)^2}{2!} + \frac{(2x)^4}{4!} - \frac{(2x)^6}{6!} \ldots\right) = \frac{1}{2}\left(\frac{(2x)^2}{2!} - \frac{(2x)^4}{4!} + \frac{(2x)^6}{6!} - \ldots\right).$$

Example 12 Express $\cos\left(\frac{\pi}{4} + x\right) \sin x$ as a power series.

$$\cos\left(\frac{\pi}{4} + x\right) \sin x = \sin x \left(\frac{\cos x}{\sqrt{2}} - \frac{\sin x}{\sqrt{2}}\right)$$

$$= \frac{1}{2\sqrt{2}}(\sin 2x + \cos 2x - 1)$$

$$= \frac{1}{2\sqrt{2}}\left\{\left(2x - \frac{(2x)^3}{3!} + \frac{(2x)^5}{5!} \ldots\right) + \left(1 - \frac{(2x)^2}{2!} + \frac{(2x)^4}{4!} \ldots\right) - 1\right\}$$

$$= \frac{1}{2\sqrt{2}}\left(2x - \frac{(2x)^2}{2!} - \frac{(2x)^3}{3!} + \frac{(2x)^4}{4!} + \frac{(2x)^5}{5!} - \ldots\right).$$

Example 13 Evaluate $\underset{\theta \to 0}{\text{Lt}} \frac{3\sin\theta - \sin 3\theta}{\theta(\cos\theta - \cos 3\theta)}$.

$$\underset{\theta \to 0}{\text{Lt}} \frac{3\sin\theta - \sin 3\theta}{\theta(\cos\theta - \cos 3\theta)} = \underset{\theta \to 0}{\text{Lt}} \frac{3\left(\theta - \frac{\theta^3}{6} + \ldots\right) - \left(3\theta - \frac{(3\theta)^3}{6} + \ldots\right)}{\theta\left\{1 - \frac{\theta^2}{2} + \ldots - \left(1 - \frac{(3\theta)^2}{2} + \ldots\right)\right\}}$$

$$= \underset{\theta \to 0}{\text{Lt}} \frac{4\theta^3 + \ldots}{\theta(4\theta^2 + \ldots)} = \underset{\theta \to 0}{\text{Lt}} \frac{1 + \text{terms in } \theta}{1 + \text{terms in } \theta} \quad \text{(dividing by } 4\theta^3\text{)}$$

$$= 1.$$

[B] EXAMPLES 3c

Given that $\pi = 3\cdot141593$, find the values of the following expressions, correct to 6 places of decimals:

1 $\sin 15''$.
2 $\cos 1'$.
3 $\tan 45''$.
4 $\sin 1°$.
5 $\sin 89° 30'$.
6 $\cot 89° 45'$.

7 Prove that $\tan x - \sin x \approx \frac{x^3}{2}$ if x is small. Use four-figure tables to assess the approximate error when $x = 0\cdot3$ radians.

8 Calculate $\sin^2 1° 30'$ from the series, correct to 4 places of decimals.

9 If $\tan\theta = 1\cdot002\theta$, find an approximate positive value of θ.

10 Given that $1014\sin\theta = 1013\theta$, prove that θ is approximately $0\cdot077$ radians.

Harder miscellaneous examples | 49

11 Obtain an approximate solution of the equation $\cos\theta = 5\theta$.
12 Expand $\cos^2 x$, x in radians, as far as the term in x^6.
13 Verify that the general term of the expansion of $\sin x$ can be expressed as
$(-1)^{n+1}\dfrac{x^{2n-1}}{(2n-1)!}$ and find the corresponding term in the expansion of $\cos x$.
14 Using the identity $4\sin^3 x = 3\sin x - \sin 3x$, obtain the first 3 terms in the expansion of $\sin^3 x$ as a power series in x.
15 Express $\sin\left(\dfrac{\pi}{4}+x\right)\cos x$ as a power series in x.
16 Show that $8\sin\dfrac{\theta}{2} - \sin\theta = 3\theta$, if θ is so small that powers of θ greater than the 4th can be ignored. Using this result to express θ in terms of $\sin\theta$ and $\sin\dfrac{\theta}{2}$, obtain an approximate solution of the equation $26\sin\theta = 25\theta$.

Sum to infinity the series in Examples 17–20.

17 $\dfrac{1}{1!} - \dfrac{1}{3!} + \dfrac{1}{5!} - \cdots$ **18** $1 - \dfrac{1}{2^2\,2!} + \dfrac{1}{2^4\,4!} - \cdots$

19 $\dfrac{2^3}{3!} - \dfrac{2^5}{5!} + \dfrac{2^7}{7!} - \cdots$ **20** $\dfrac{1}{2!} - \dfrac{\pi^2}{4^2\,4!} + \dfrac{\pi^4}{4^4\,6!} - \cdots$

Find the limits of the following fractions as $\theta \to 0$:

21 $\dfrac{\theta - \sin\theta}{\theta^3}$.   **22** $\dfrac{\theta^2}{1 - \cos 2\theta}$.   **23** $\dfrac{\sin\theta}{\sin 3\theta}$.   **24** $\dfrac{\tan 2\theta - 2\sin\theta}{\theta^3}$.

Harder miscellaneous examples

Example 14 If x is so small that powers above the third can be neglected, prove that
$$\ln\left(\dfrac{1+e^x}{2}\right) = \tfrac{1}{2}x + \tfrac{1}{8}x^2.$$

We have,
$$\dfrac{1+e^x}{2} = \dfrac{1}{2}\left\{1 + 1 + x + \dfrac{x^2}{2!} + \dfrac{x^3}{3!}\right\}, \text{ as higher powers can be neglected.}$$

$$= 1 + \dfrac{x}{2} + \dfrac{x^2}{4} + \dfrac{x^3}{12},$$

$$= 1 + \dfrac{x}{2}\left(1 + \dfrac{x}{2} + \dfrac{x^2}{6}\right).$$

$$\therefore \ln\dfrac{1+e^x}{2} = \ln\left\{1 + \dfrac{x}{2}\left(1 + \dfrac{x}{2} + \dfrac{x^2}{6}\right)\right\}$$

$$= \dfrac{x}{2}\left(1 + \dfrac{x}{2} + \dfrac{x^2}{6}\right) - \dfrac{\left\{\dfrac{x}{2}\left(1 + \dfrac{x}{2} + \dfrac{x^2}{6}\right)\right\}^2}{2} + \dfrac{\left\{\dfrac{x}{2}\left(1 + \dfrac{x}{2} + \dfrac{x^2}{6}\right)\right\}^3}{3}$$

$$= \dfrac{x}{2} + x^2\left(\dfrac{1}{4} - \dfrac{1}{8}\right) + x^3\left(\dfrac{1}{12} - \dfrac{1}{8} + \dfrac{1}{24}\right)$$

$$= \dfrac{x}{2} + \dfrac{x^2}{8}, \text{ if } x^4 \text{ and higher powers are ignored.}$$

Example 15 Prove that $\dfrac{1+3\cos\theta}{3+\cos\theta} = 1 - \dfrac{\theta^2}{4} - \dfrac{\theta^4}{96}$ if θ^6 is negligible.

$$\dfrac{1+3\cos\theta}{3+\cos\theta} = (1+3\cos\theta)(3+\cos\theta)^{-1}$$

$$= \left(1+3-\dfrac{3\theta^2}{2}+\dfrac{\theta^4}{8}\right)\left(3+1-\dfrac{\theta^2}{2}+\dfrac{\theta^4}{24}\right)^{-1} \quad \text{ignoring higher powers of } \theta.$$

$$= \left(4-\dfrac{3\theta^2}{2}+\dfrac{\theta^4}{8}\right)\dfrac{1}{4}\left\{1-\dfrac{\theta^2}{8}\left(1-\dfrac{\theta^2}{12}\right)\right\}^{-1}.$$

As θ is small we can assume that $\dfrac{\theta^2}{8}\left(1-\dfrac{\theta^2}{12}\right)$ is less than 1 and expand the second bracket by using the Binomial Expansion.

$$\therefore \dfrac{1+3\cos\theta}{3+\cos\theta} = \dfrac{1}{4}\left(4-\dfrac{3\theta^2}{2}+\dfrac{\theta^4}{8}\right)\left\{1+\dfrac{\theta^2}{8}\left(1-\dfrac{\theta^2}{12}\right)+\dfrac{\theta^4}{64}\left(1-\dfrac{\theta^2}{12}\right)^2\right\}$$

$$= \dfrac{1}{4}\left(4-\dfrac{3\theta^2}{2}+\dfrac{\theta^4}{8}\right)\left(1+\dfrac{\theta^2}{8}+\dfrac{\theta^4}{192}\right)$$

$$= \dfrac{1}{4}\left\{4-\theta^2+\theta^4\left(\dfrac{1}{48}-\dfrac{3}{16}+\dfrac{1}{8}\right)\right\}$$

$$= 1-\dfrac{\theta^2}{4}-\dfrac{\theta^4}{96}.$$

Example 16 Sum to infinity the following series:

(i) $1 - \dfrac{1}{2!} + \dfrac{1}{3!} - \dfrac{1}{4!} + \ldots$;

(ii) $\dfrac{2^2}{1!} + \dfrac{3^2}{2!} + \dfrac{4^2}{3!} + \ldots$;

(i) The form of the series suggests it is something to do with e^{-1}.

$$e^{-1} = 1 - 1 + \dfrac{1}{2!} - \dfrac{1}{3!} + \ldots$$

$$= \dfrac{1}{2!} - \dfrac{1}{3!} + \dfrac{1}{4!} - \ldots$$

$\therefore$ The sum of the series is $1 - e^{-1}$ or $1 - \dfrac{1}{e}$.

(ii) In all but the simplest of cases, the summation of an infinite series usually involves the splitting up of the general (nth) term of the series into the sum or difference of simpler terms.

In this case, the nth term $= \dfrac{(n+1)^2}{n!}$.

This expression is reduced to a series of terms with constant numerators of the form $\dfrac{A}{(n-a)!}$, where a is an integer.

We have $\dfrac{(n+1)^2}{n!} = \dfrac{n^2+2n+1}{n!} = \dfrac{n}{(n-1)!} + \dfrac{2}{(n-1)!} + \dfrac{1}{n!}$

$= \dfrac{(n-1)+1}{(n-1)!} + \dfrac{2}{(n-1)!} + \dfrac{1}{n!}$

$= \dfrac{1}{(n-2)!} + \dfrac{3}{(n-1)!} + \dfrac{1}{n!}.$

Clearly this relationship only holds for $n > 2$, otherwise $(n-2)!$ is meaningless.

Let S_1, S_2, S_3 represent the sums of the series whose nth terms are $\dfrac{1}{(n-2)!}, \dfrac{1}{(n-1)!}, \dfrac{1}{n!}$ for n taken from 3 to infinity.

$\therefore$ Sum of given series $= \dfrac{2^2}{1!} + \dfrac{3^2}{2!} + S_1 + 3S_2 + S_3$

$= 4 + \dfrac{9}{2} + \left(\dfrac{1}{1!} + \dfrac{1}{2!} + \dfrac{1}{3!} + \ldots\right)$

$+ 3\left(\dfrac{1}{2!} + \dfrac{1}{3!} + \ldots\right) + \left(\dfrac{1}{3!} + \dfrac{1}{4!} + \ldots\right).$

Remembering that $e = 1 + \dfrac{1}{1!} + \dfrac{1}{2!} + \dfrac{1}{3!} + \ldots,$

the sum of the given series $= 4 + \tfrac{9}{2} + (e-1) + 3(e-2) + (e - 2 - \tfrac{1}{2})$

$= 4 + \tfrac{9}{2} + 5e - 9\tfrac{1}{2}$

$= 5e - 1.$

[B] MISCELLANEOUS EXAMPLES

1 Find the coefficient of x^6 in the expansion of $(1-x)e^{2-x}$.

2 Expand in powers of x as far as the term in x^3, $\ln\dfrac{1+2x}{1-2x} - \ln\left(\dfrac{1+x}{1-x}\right)^2$.
For what values of x is the expansion valid?

3 Find the coefficient of x^8 in the expansion of $\cos^3 x$ as a power series in x.

4 By substituting $x = \dfrac{1}{10}$ in the series for $\ln\dfrac{1+x}{1-x}$, find the value of $\ln 11$ correct to six places of decimals, assuming that $\ln 3 = 1\cdot 0986123$.

5 Prove that $x\sqrt{1+x} + \ln(1-x) \approx -\tfrac{11}{24}x^3 - \tfrac{3}{16}x^4$ if x is small.

6 Expand $\dfrac{e^x - e^{5x}}{e^{3x}}$ in ascending powers of x.

Find the first 4 terms of the following expansions; in each case give the range of values of x for which the expansion is valid:

7 $e^x \sin x$.   8 $e^x \cos x$.   9 $e^x \ln(1+x)$.   10 $\dfrac{\ln(1-x)}{e^x}$.

11 $\dfrac{\ln(1+x)}{1-x}$.   12 $\dfrac{\sin x}{\sqrt{1+x}}$.   13 $\dfrac{1}{e^{2x}(1-x^2)}$.

14 Find the value of $\ln\tfrac{128}{125}$ correct to 6 decimal places. Hence, assuming $\ln 2 = 0\cdot 6931472$, calculate the value of $\ln 10$ to 6 decimal places.

15 Show that $\ln(n+a) - \ln(n-a) = 2\left(\dfrac{a}{n} + \dfrac{a^3}{3n^3} + \dfrac{a^5}{5n^5} + \ldots\right)$, if $-n < a < n$.

16 By writing $1 + x + x^2 = \dfrac{1-x^3}{1-x}$, find the coefficient of x^{3n} in the expansion of $\ln(1 + x + x^2)$.

17 Obtain the coefficient of x^{4n} in the expansion of $\ln\dfrac{1}{1+x+x^2+x^3}$.

$\left[\text{Hint: } 1 + x + x^2 + x^3 = \dfrac{1-x^4}{1-x}\right].$

18 Writing $\tan\theta = \dfrac{\sin\theta}{\cos\theta}$, obtain the first three terms in the expansion of $\tan\theta$ as a power series.

19 Prove that $\dfrac{3\sin\theta}{2+\cos\theta}$ differs from θ by approximately $\dfrac{\theta^5}{180}$, if θ is small.

20 If $y = x - \dfrac{x^2}{2} + \dfrac{x^3}{3} - \dfrac{x^4}{4} + \ldots,$

show that $x = y + \dfrac{y^2}{2!} + \dfrac{y^3}{3!} + \ldots$

21 If $\ln\tfrac{10}{9} = a$, $\ln\tfrac{25}{24} = b$, $\ln\tfrac{81}{80} = c$, prove that $\ln 2 = 7a - 2b + 3c$, $\ln 3 = 11a - 3b + 5c$, $\ln 5 = 16a - 4b + 7c$.
Calculate the values of a, b, c and deduce the values of $\ln 2$, $\ln 3$, $\ln 5$ to 5 places of decimals.

22 If $e^y(1-x) = 1 + 2x$, express y as a function of x. Expand this function as a series in ascending powers of x giving the terms as far as that in x^4 and state for what values of x the expansion is valid.

23 Find the coefficients of x, x^2 and x^3 in the expansion of $e^{ax} + b\ln(1+x) - \sqrt{1+2x}$ in ascending powers of x. Show that a, b can be determined so that the coefficients of x, x^2 and x^3 in this expansion are all zero. Find the values of a, b.

24 If $n > 1$, prove that

$$2\ln n - \ln(n+1) - \ln(n-1) = \dfrac{1}{n^2} + \dfrac{1}{2n^4} + \dfrac{1}{3n^6} + \ldots$$

25 Prove that $\theta\cot\theta \approx 1 - \tfrac{1}{3}\theta^2 - \tfrac{1}{45}\theta^4$, if θ is small.

26 Find the coefficient of x^n in the expansion of $\ln\left(1 + \dfrac{1}{2-x}\right)$ as a power series in x, assuming x is positive and < 2.

27 Prove that $\dfrac{\ln 10 - 3\ln 2}{2} = \dfrac{1}{9} + \dfrac{1}{3 \cdot 9^3} + \dfrac{1}{5 \cdot 9^5} + \ldots$

Given $\ln 2 = 0.69315$, deduce the approximate value of $\ln 10$.
Evaluate the following limits:

28 $\underset{x \to 0}{\text{Lt}} \dfrac{6\sin x - \sin 6x}{\cos x - \cos 6x}.$

29 $\underset{x \to 0}{\text{Lt}} \dfrac{x\ln(1+x)}{1-\cos x}.$

30 $\underset{x \to 0}{\text{Lt}} \dfrac{e^x - 1 + \ln(1-x)}{\sin^2 x}.$

31 $\underset{\theta \to 0}{\text{Lt}} \dfrac{\cos m\theta - \cos n\theta}{\theta(\sin n\theta - \sin m\theta)}.$

32 $\underset{x \to 0}{\text{Lt}} \dfrac{1 - \cos(1 - \cos x)}{x^4}.$

33 $\underset{\theta \to 0}{\text{Lt}} \dfrac{\ln(1+\sin\theta) - \theta}{\theta^2}.$

34 Write down the expansion of $\ln n$ in ascending powers of $\left(\dfrac{n-1}{n+1}\right)$. If S_5 denotes the sum of the first five terms in the expansion for $\ln 3$ (i.e. the series obtained by putting $n=3$), prove that $S_5 < \ln 3 < S_5 + \dfrac{1}{33 \cdot 2^8}$.

35 Prove the following results:

(i) $\dfrac{1}{1!} + \dfrac{1+2}{2!} + \dfrac{1+2+2^2}{3!} + \ldots = e^2 - e.$

(ii) $\dfrac{1}{2 \cdot 2} + \dfrac{1}{3 \cdot 2^2} + \dfrac{1}{4 \cdot 2^3} + \ldots = 2\ln 2 - 1.$

(iii) $4\left\{\dfrac{1}{1!} + \dfrac{1}{5!} + \dfrac{1}{9!} + \ldots\right\} = e - e^{-1} + 2\sin 1.$

36 Sum to infinity the following series:

(i) $\dfrac{\pi^2}{3!} - \dfrac{\pi^4}{5!} + \dfrac{\pi^6}{7!} - \ldots$

(ii) $\dfrac{1}{2} - \dfrac{1}{4} + \dfrac{1}{6} - \dfrac{1}{8} + \ldots$

(iii) $\dfrac{1}{1 \cdot 2} - \dfrac{1}{2 \cdot 3} + \dfrac{1}{3 \cdot 4} - \ldots$

(iv) $1 + \dfrac{2^2}{1!} + \dfrac{3^2}{2!} + \ldots$

4 | Finite algebraic series

The Binomial Theorem The Binomial Expansion in the form
$$(1+x)^n = 1 + nx + \frac{n(n-1)}{2!}x^2 + \frac{n(n-1)(n-2)}{3!}x^3 + \ldots$$
has been introduced in the earlier volume. It will be recalled that in the case where n is not a positive integer, the series is infinite and the expansion is only true if x is numerically less than unity.

The case when *n is a positive integer* will be discussed further. When this is so, the coefficients of the second and successive terms can be expressed more briefly as $_nC_1$, $_nC_2$, $_nC_3$, etc.,

i.e. $(1+x)^n = 1 + {_nC_1}x + {_nC_2}x^2 + \ldots + {_nC_r}x^r + \ldots + x^n$,

or, more generally,
$$(a+x)^n = a^n + {_nC_1}a^{n-1}x + {_nC_2}a^{n-2}x^2 + \ldots + {_nC_r}a^{n-r}x^r + \ldots + x^n.$$

Example 1 Find the term independent of x in the expansion of $\left(\frac{3x^2}{2} - \frac{1}{3x}\right)^9$.

$$\text{Expression} = \left(\frac{3x^2}{2}\right)^9 \left(1 - \frac{2}{9x^3}\right)^9.$$

Thus the term required will be the term in $\frac{1}{x^{18}}$ in the expansion of $\left(1 - \frac{2}{9x^3}\right)^9$, i.e. the 7th term.

$$\text{7th term} = {_9C_6}\left(-\frac{2}{9x^3}\right)^6$$

$$= \frac{9.8.7}{3!} \cdot \frac{2^6}{9^6 x^{18}}. \quad ({_9C_6} = {_9C_3}.)$$

$$\therefore \text{Required term} = \frac{(3x^2)^9}{2^9} \cdot \frac{9.8.7}{3!} \cdot \frac{2^6}{9^6 x^{18}}$$

$$= \frac{7}{18}.$$

Greatest term or coefficient

Example 2 Find the numerically greatest terms in the expansion of $(3-2x)^{15}$ when $x = \frac{5}{2}$.

It is simpler to write the expression as $3^{15}\left(1 - \frac{2x}{3}\right)^{15}$. Let the rth and $(r+1)$th terms in the expansion of $\left(1 - \frac{2x}{3}\right)^{15}$ be u_r, u_{r+1} respectively.

Then
$$\frac{u_{r+1}}{u_r} = \frac{{}_{15}C_r}{{}_{15}C_{r-1}}\left(-\frac{2x}{3}\right)$$
$$= \frac{15-r+1}{r}\left(\frac{5}{3}\right) \quad \text{numerically when } x = \tfrac{5}{2}.$$

$\therefore u_{r+1}$ is numerically greater than u_r, so long as

$$\frac{16-r}{r}\cdot\frac{5}{3} > 1.$$

i.e. $\qquad 80 > 8r.$

The greatest integral value of r satisfying this condition is 9. Consequently terms of the series increase in numerical value up the 10th term. In this example the terms do not immediately decrease for values of r greater than 9, for when $r = 10$,

$$u_{r+1} = u_r.$$

Hence there are two numerically greatest terms, the 10th and 11th.
Numerical value of greatest terms $= {}_{15}C_{10}3^5(5)^{10}$
$$= 3003.3^5.5^{10}.$$

Example 3 Find the greatest coefficient in the expansion of $(2+3x)^7$.

$$(2+3x)^7 = 2^7\left(1+\frac{3x}{2}\right)^7.$$

Let v_{r+1}, v_r be the coefficients of the $(r+1)$th, rth terms respectively in the expansion of $\left(1+\dfrac{3x}{2}\right)^7$.

$$\frac{v_{r+1}}{v_r} = \frac{7-r+1}{r}\cdot\frac{3}{2}.$$

$\therefore v_{r+1} > v_r$, so long as

$$\frac{8-r}{r}\cdot\frac{3}{2} > 1.$$

i.e. $\qquad 24 > 5r.$

The greatest integral value of r satisfying this inequality is 4 and so the 5th term has the greatest coefficient.
Greatest coefficient in given expansion $= {}_7C_4\, 2^3\, 3^4 = 22\,680.$

Proof of the Binomial Theorem for a positive integral index The method of proof used is known as *mathematical induction*. We start by assuming that the Binomial Theorem is true for the index n,

i.e. $(a+x)^n = a^n + {}_nC_1 a^{n-1}x + {}_nC_2 a^{n-2}x^2 + \ldots$
$$+ {}_nC_{r-1} a^{n-r+1}x^{r-1} + {}_nC_r a^{n-r}x^r + \ldots + x^n.$$

Multiplying by $(a+x)$,

$(a+x)^{n+1} = a^{n+1} + a^n x(1 + {}_nC_1) + a^{n-1}x^2({}_nC_1 + {}_nC_2) + \ldots$
$$+ a^{n+1-r}x^r({}_nC_{r-1} + {}_nC_r) + \ldots + x^{n+1}.$$

56 | Finite algebraic series

But, $\qquad 1 + {}_nC_1 = 1 + n = {}_{n+1}C_1;$

$\qquad\qquad {}_nC_1 + {}_nC_2 = {}_{n+1}C_2;$

and generally, $\quad {}_nC_{r-1} + {}_nC_r = {}_{n+1}C_r.$ (See p. 8.)

$\therefore (a+x)^{n+1} = a^{n+1} + {}_{n+1}C_1 a^n x + {}_{n+1}C_2 a^{n-1} x^2 + \ldots + {}_{n+1}C_r a^{n+1-r} x^r$
$\qquad\qquad\qquad + \ldots + x^{n+1},$

i.e. the Binomial Theorem for index $(n+1)$.

Thus, *if* the Binomial Theorem is true for index n, *then* it is also true for index $(n+1)$.

Clearly the result is true for $n = 2$, as $(a+x)^2$ does equal $a^2 + {}_2C_1 ax + x^2$. It follows then that the result is true for $n = 3$ and hence for $n = 4$, and so on for all positive integral values of n.

The method of mathematical induction is of value in the proof of *stated* mathematical formulae.

[B] EXAMPLES 4a

1. Find the term independent of a in $\left(\dfrac{3}{2}a^2 - \dfrac{1}{3a}\right)^9$.

2. What is the coefficient of x^7 in the expansion of $(1+2x)^4(1-2x)^6$?

3. Find the terms in a^5 and a^6 in $\left(3a^2 - \dfrac{1}{a}\right)^6 \left(a + \dfrac{1}{a}\right)^4$.

4. Find the coefficients of x^3 and x^5 in the expansion of $(1+x+x^2)^5$.

5. What are the terms in x^3 and x^{10} in $(1+x)^5(1-x+x^2)^4$?

6. Expand $(1-2x)^4(1+2x+4x^2)^3$.

7. Find the term independent of x in the expansion of $\left(x + \dfrac{1}{x}\right)^6 \left(x - \dfrac{1}{x}\right)^8$.

8. Find the value of r if the coefficients of x^r and x^{r+1} in $(3x+2)^{19}$ are equal.

9. If $x = \frac{1}{4}$, find the ratio of the 8th and the 7th terms in $(1+2x)^{15}$.

10. Find which are the greatest terms in the following expansions

    (i) $(1+3x)^{18}$ when $x = \frac{1}{4}$;   (ii) $\left(1 + \dfrac{x}{2}\right)^{12}$ when $x = \frac{1}{2}$;

    (iii) $(4+x)^8$ when $x = 3$;   (iv) $(x+y)^n$ when $n = 14$, $x = 2$, $y = \frac{1}{3}$.

11. Find the numerically greatest terms in the expansions:

    (i) $(2-x)^{12}$ when $x = \frac{2}{3}$;   (ii) $(3a-b)^n$ when $a = 1$, $n = 16$, $b = \frac{1}{2}$.

12. Find which terms have the greatest coefficients in:

    (i) $(1+x)^{10}$.   (ii) $(2+x)^{11}$.   (iii) $(1+x)^{2n+1}$.

13. In the following expansions, show there are two greatest terms and find their values:

    (i) $(3+2x)^{15}$ when $x = \frac{5}{2}$;   (ii) $(a+x)^n$ when $a = \frac{1}{2}$, $x = \frac{1}{3}$, $n = 9$.

14. Find the coefficients of x^2 and x^3 in $(2+2x+x^2)^n$.

15. Expand $(1-x+x^2)^n$ in ascending powers of x as far as the term in x^3.

16 Show that the coefficients of all odd powers of x in
$$(1-x+x^2)^4(1+x+x^2)^4$$
are zero and find the coefficient of x^6.

Summation of series A series of terms is completely defined if the rth term, u_r, is given as a function of r.
The sum of the first n terms of the series
$$u_1, u_2, u_3, \ldots u_r, \ldots u_n,$$
is denoted by $\sum_{r=1}^{r=n} u_r$ or more briefly, $\sum_1^n u_r$.

Example 4 Use the Σ notation to express the sums of the following series:

(i) $1 + \dfrac{1}{2} + \dfrac{1}{3} + \ldots + \dfrac{1}{n}$;

(ii) $1.2 + 2.3 + 3.4 + \ldots 2n$ terms.

(i) $\qquad\qquad$ Sum of series $= \sum_1^n \dfrac{1}{r}$.

(ii) $\qquad\qquad$ rth term of series, $u_r = r(r+1)$.

$$\text{Sum of } 2n \text{ terms} = \sum_1^{2n} r(r+1).$$

Sums of powers of the first n natural numbers
Series $1 + 2 + 3 + 4 + \ldots + n$.
This series is an A.P. with common difference 1.

$$\therefore \text{Sum} = \frac{n}{2}(2 + \overline{n-1} \cdot 1)$$

$$= \frac{n(n+1)}{2}.$$

i.e. $\qquad\qquad \sum_1^n r = \dfrac{n(n+1)}{2}.$

Series $1^2 + 2^2 + 3^2 + 4^2 + \ldots + n^2$.
This series is summed by using the identity,
$$(n+1)^3 - n^3 = 3n^2 + 3n + 1.$$
Replacing n by $(n-1)$,
$$n^3 - (n-1)^3 = 3(n-1)^2 + 3(n-1) + 1.$$
Similarly, $\quad (n-1)^3 - (n-2)^3 = 3(n-2)^2 + 3(n-2) + 1.$
$$\cdots\cdots\cdots\cdots\cdots\cdots\cdots$$
$$3^3 - 2^3 = 3.2^2 + 3.2 + 1.$$
$$2^3 - 1^3 = 3.1^2 + 3.1 + 1.$$

Adding,
$$(n+1)^3 - 1^3 = 3\sum_{1}^{n} r^2 + 3\sum_{1}^{n} r + n.$$
$$\therefore 3\sum_{1}^{n} r^2 = n^3 + 3n^2 + 2n - \frac{3n(n+1)}{2}$$
$$= \tfrac{1}{2}n(2n^2 + 3n + 1)$$
$$= \tfrac{1}{2}n(n+1)(2n+1).$$

i.e.
$$\sum_{1}^{n} \mathbf{r^2} = \tfrac{1}{6}\mathbf{n(n+1)(2n+1)}.$$

Series $1^3 + 2^3 + 3^3 + 4^3 + \ldots + n^3$.

This series is summed by using the previous method and starting with the identity,
$$(n+1)^4 - n^4 = 4n^3 + 6n^2 + 4n + 1.$$

We find that,
$$\sum_{1}^{n} \mathbf{r^3} = \left\{\frac{\mathbf{n(n+1)}}{\mathbf{2}}\right\}^2 = \left(\sum_{1}^{n} \mathbf{r}\right)^2.$$

Example 5 Sum the series $1^2 + 3^2 + 5^2 + \ldots + 29^2$.

Sum $= 1^2 + 2^2 + 3^2 + \ldots + 30^2 - (2^2 + 4^2 + 6^2 + \ldots + 30^2)$
$$= \sum_{1}^{30} r^2 - 4\sum_{1}^{15} r^2$$
$$= \frac{30 \cdot 31 \cdot 61}{6} - 4 \cdot \frac{15 \cdot 16 \cdot 31}{6}$$
$$= 4495.$$

Example 6 Find the sum of n terms of the series
$$1.1 + 2.3 + 3.5 + \ldots.$$
rth term of series $= r(2r-1)$.
$$\therefore \text{Sum to } n \text{ terms} = \sum_{1}^{n} r(2r-1)$$
$$= 2\sum_{1}^{n} r^2 - \sum_{1}^{n} r$$
$$= 2 \cdot \tfrac{1}{6}n(n+1)(2n+1) - \frac{n(n+1)}{2}$$
$$= \tfrac{1}{6}n(n+1)\{2(2n+1) - 3\}$$
$$= \tfrac{1}{6}n(n+1)(4n-1).$$

[A] EXAMPLES 4b

What expressions are represented by the symbols in nos. 1–9?

1 $\sum_{r=1}^{r=3} r.$ **2** $\sum_{r=1}^{r=4} 2r^2.$ **3** $\sum_{r=1}^{r=8} r(r+1).$

4 $\sum_{r=1}^{r=20} r^3.$ 5 $\sum_{r=1}^{r=10} \frac{1}{r}.$ 6 $\sum_{r=1}^{r=15} \frac{1}{r(r+2)}.$

7 $\sum_{r=1}^{r=n} r^4.$ 8 $\sum_{r=1}^{r=n} (r+1)(r+2).$ 9 $\sum_{r=1}^{r=n} \frac{1}{r!}.$

10 What is the 10th term of the series $\sum_{r=1}^{r=n} \frac{1}{r(r+3)}$?

11 What is the 7th term of the series $\sum_{1}^{n} \frac{r+2}{r!}$?

12 How many terms are there in the series $\sum_{1}^{2n} (r^2+r+1)$? What is the $(n+1)$th term?

For the series in Examples 13–19, write down the rth terms and express the sums in the Σ notation:

13 $1+4+7+10+\ldots$ 30 terms.
14 $1^2+3^2+5^2+7^2+\ldots$ 20 terms.
15 $1+\frac{1}{2}+\frac{1}{3}+\frac{1}{4}+\ldots n$ terms.
16 $\frac{1}{1^2}+\frac{1}{3^2}+\frac{1}{5^2}+\frac{1}{7^2}+\ldots n$ terms.
17 $3.4+4.5+5.6+6.7+\ldots n$ terms.
18 $\frac{1}{2.3}+\frac{1}{3.4}+\frac{1}{4.5}+\ldots 2n$ terms.
19 $\frac{2}{1.2.3}+\frac{3}{3.4.5}+\frac{4}{5.6.7}+\ldots n$ terms.

20 What is the 6th term of the series $2\sum_{1}^{n} (-1)^{r+1} \frac{1}{r(r+1)}$?

21 Express the series $1-\frac{1}{3}+\frac{1}{5}-\frac{1}{7}+\ldots n$ terms in the Σ notation.

What are the values of the series in nos. 22–32?

22 $\sum_{1}^{3} \frac{1}{r}.$ 23 $\sum_{1}^{4} \log r.$ 24 $\sum_{3}^{5} \frac{1}{r^2}.$ 25 $\sum_{1}^{20} r^2.$

26 $\sum_{1}^{8} r^3.$ 27 $\sum_{1}^{12} 3r.$ 28 $\sum_{1}^{24} \frac{r^2}{2}.$ 29 $\sum_{1}^{15} (2r+r^2).$

30 $\sum_{1}^{14} r(1+r^2).$ 31 $\sum_{1}^{11} 1.$ 32 $\sum_{1}^{21} (1+r+r^2).$

33 Find the sum of all the integers between 30 and 70.
34 Sum the series $10^2+12^2+14^2+\ldots+20^2$.
35 What is the sum of the cubes of the integers from 11 to 20 inclusive?
36 Find the sum of the squares of the first 20 odd numbers.
37 Find the sums of the series:
   (i) $2^2+4^2+6^2+\ldots+(2n)^2$.
   (ii) $1^2+3^2+5^2+\ldots+(2n-1)^2$.

38 The nth term of a series is $n^2(n+2)$. Find the sum of the series (i) to n terms, (ii) to $2n$ terms.

60 | Finite algebraic series

39 If the nth term of a series can be written in the form $n(n-1)+1$, find the sum to n terms.

Miscellaneous series

Example 7 Sum to n terms, the series $1 + 3x^2 + 5x^4 + \ldots$

The coefficients $1, 3, 5, \ldots$ are successive terms in an A.P., first term 1, common difference 2. The terms in x are successive terms in a G.P., common ratio x^2.

$$n\text{th term} = (1 + \overline{n-1}.2)(x^2)^{n-1} = (2n-1)x^{2n-2}.$$

Let
$$S_n = 1 + 3x^2 + 5x^4 + 7x^6 + \ldots + (2n-3)x^{2n-4} + (2n-1)x^{2n-2}.$$
$$\therefore x^2 S_n = x^2 + 3x^4 + 5x^6 + \ldots\ldots\ldots + (2n-3)x^{2n-2} + (2n-1)x^{2n}.$$

Subtracting,
$$S_n(1-x^2) = 1 + 2x^2 + 2x^4 + 2x^6 + \ldots + 2x^{2n-2} - (2n-1)x^{2n}$$
$$= 1 + 2x^2(1 + x^2 + x^4 + \ldots + x^{2n-4}) - (2n-1)x^{2n}$$
$$= 1 + 2x^2 \frac{(1-(x^2)^{n-1})}{1-x^2} - (2n-1)x^{2n} \quad \text{(Summing the G.P.)}$$

i.e.
$$S_n = \frac{1-(2n-1)x^{2n}}{1-x^2} + \frac{2x^2(1-x^{2n-2})}{(1-x^2)^2}.$$

The sum S_n of n terms of a series of this type where each term is the product of successive terms of an A.P. and a G.P., can always be found by obtaining an expression for $S_n(1-r)$ where r is the common ratio of the G.P.

Example 8 Find the sum of the series,

$$\frac{1}{2.3} + \frac{1}{3.4} + \frac{1}{4.5} + \ldots \frac{1}{(n+1)(n+2)}.$$

$$n\text{th term} = \frac{1}{(n+1)(n+2)} = \frac{1}{n+1} - \frac{1}{n+2}.$$

Writing $(n-1)$ for n, $\quad (n-1)\text{th term} = \dfrac{1}{n} - \dfrac{1}{n+1}.$

Similarly, $\quad (n-2)\text{th term} = \dfrac{1}{n-1} - \dfrac{1}{n},$

$\quad (n-3)\text{th term} = \dfrac{1}{n-2} - \dfrac{1}{n-1}.$

$\ldots\ldots\ldots\ldots\ldots\ldots$

3rd term $= \frac{1}{4} - \frac{1}{5}.$
2nd term $= \frac{1}{3} - \frac{1}{4}.$
1st term $= \frac{1}{2} - \frac{1}{3}.$

On addition, terms on the R.H.S. go out in pairs and the sum is

$$\frac{1}{2} - \frac{1}{n+2}, \quad \text{or} \quad \frac{n}{2(n+2)},$$

i.e. $$\text{Sum of series} = \frac{n}{2(n+2)}.$$

Example 9 Prove by induction, that
$$\frac{1}{1.4} + \frac{1}{4.7} + \ldots + \frac{1}{(3n-2)(3n+1)} = \frac{n}{3n+1}.$$

Assume that
$$\frac{1}{1.4} + \frac{1}{4.7} + \ldots + \frac{1}{(3n-2)(3n+1)} = \frac{n}{3n+1}.$$

Adding the $(n+1)$th term to both sides,

$$\text{sum to } (n+1) \text{ terms} = \frac{n}{3n+1} + \frac{1}{(3n+1)(3n+4)}$$

$$= \frac{3n^2 + 4n + 1}{(3n+1)(3n+4)}$$

$$= \frac{(3n+1)(n+1)}{(3n+1)(3n+4)}$$

$$= \frac{n+1}{3n+4}.$$

i.e. if the sum to n terms is $\frac{n}{3n+1}$, the sum to $(n+1)$ terms is $\frac{n+1}{3(n+1)+1}$.

So if the result is true for n terms, it is also true for $(n+1)$ terms.

But, $$\text{sum of 2 terms} = \frac{1}{1.4} + \frac{1}{4.7} = \frac{2}{7}$$

$$= \frac{n}{3n+1} \text{ with } n = 2.$$

Hence, the result is true for $n = 2$ and therefore it is also true for $n = 3$, and as it is true for $n = 3$ it must also be true for $n = 4$ and so on for all integral powers of n.

[B] EXAMPLES 4c

Find the sums to n terms of the series whose rth terms are given in Examples 1–9:

1. $4r - 3$.
2. $3r^2 - r$.
3. $r^3 + 3r$.
4. $(r+1)(2r+1)$.
5. $r^2(2r+3)$.
6. $1 - r$.
7. $3^r - 2^r$.
8. $\frac{1}{r} - \frac{1}{r+1}$.
9. $\frac{1}{r} - \frac{1}{r+2}$.
10. The sum of n terms of a series is $2n + 3n^2$, find:
    (i) the first term.   (ii) the second term.   (iii) the nth term.
11. Sum the series $1 + 2a + 3a^2 + \ldots$, to n terms.
12. Show that the rth term of the series
$$\frac{1}{4.5} + \frac{1}{5.6} + \frac{1}{6.7} + \ldots$$
can be expressed as $\frac{1}{r+3} - \frac{1}{r+4}$. Deduce the sum to n terms.

62 | Finite algebraic series

13 Sum the series $1 + 4(\frac{1}{2}) + 7(\frac{1}{2})^2 + \ldots 28(\frac{1}{2})^9$.

14 Find the sum to n terms of the series, $2 + 4(\frac{1}{5}) + 6(\frac{1}{5})^2 + \ldots$, and deduce the sum to infinity.

15 Obtain the sum to n terms of the series,
$$\frac{1}{3.5} + \frac{1}{5.7} + \frac{1}{7.9} + \ldots$$

16 If $\sum_{1}^{n} u_r = 3n^2 + 4n$, what is the value of $\sum_{1}^{n-1} u_r$? Deduce the value of u_n and find $\sum_{n+1}^{2n} u_r$.

Prove by induction, the results in Examples 17–20:

17 $1.2 + 2.3 + 3.4 + \ldots n$ terms $= \frac{1}{3}n(n+1)(n+2)$.

18 $\frac{1}{1.2} + \frac{1}{2.3} + \frac{1}{3.4} + \ldots n$ terms $= \frac{n}{n+1}$.

19 $1.1! + 2.2! + 3.3! + \ldots n$ terms $= (n+1)! - 1$.

20 $2.5 + 5.8 + 8.11 + \ldots n$ terms $= n(3n^2 + 6n + 1)$.

[B] MISCELLANEOUS EXAMPLES

1 Find the coefficients of x^8 and x^{12} in the expansion of $(1 - 2x)^4(x + x^3)^7$.

2 Prove that the $(n+1)$th term of a G.P., first term a and third term b is equal to the $(2n+1)$th term of a G.P., first term a and fifth term b.

3 Find the term independent of a in $\left(a - \frac{1}{a^2}\right)^{12}$.

4 Sum to n terms and to infinity, the series $1 + \frac{4}{5} + \frac{7}{5^2} + \frac{10}{5^3} + \ldots$.

5 Sum to n terms, $1.5 + 3.7 + 5.9 + \ldots$.

6 If the coefficients of the first three terms in the expansion of $\left(1 + \frac{x}{2}\right)^n$ are in A.P., find the 4th term.

7 Find the first three terms in the expansions of $(1 + x)^{10}$ and $(1 + 2x)^7$. If $x = \frac{4}{3}$ show that the fifth term of the first is equal to the fourth term of the second and if x has this value, find the numerically greatest term in the expansion of $(1 + 2x)^7$.

8 The first two terms of a G.P. are a and b ($b < a$). If the sum of the first n terms is equal to the sum to infinity of the remaining terms, prove $a^n = 2b^n$.

9 Sum to n terms, $\frac{1}{1.4} + \frac{1}{4.7} + \frac{1}{7.10} + \ldots$.

10 Find the greatest coefficient in the expansion of $(3 + 7x)^{10}$.

11 Sum to n terms each of the series:

(i) $1^2.2 + 2^2.3 + 3^2.4 + \ldots$.

(ii) $1 + 3x + 5x^2 + 7x^3 + \ldots$.

12 Show that the nth term of the series
$$3 + 4x + 6x^2 + 10x^3 + 18x^4 + 34x^5 + \ldots$$
is $2x^{n-1} + (2x)^{n-1}$ and hence find the sum to n terms.

13 Prove by induction,

$$\frac{1}{3.4.5}+\frac{2}{4.5.6}+\frac{3}{5.6.7}+\ldots n \text{ terms} = \frac{1}{6}-\frac{1}{n+3}+\frac{2}{(n+3)(n+4)}.$$

14 If $(1+ax+bx^2)^{10} = 1 - 30x + 410x^2 + \ldots$, find the values of a and b.

15 A series is obtained by taking the logarithms of each term of a G.P. with first term a and common ratio r. Prove that the sum of n terms of this series is $\log a^n r^{\frac{1}{2}n(n-1)}$. Simplify this expression when $a = \sqrt{r}$ and the logarithms are to base r.

16 Find the numerically greatest term in $(3-2x)^{11}$ when $x = \frac{3}{5}$.

17 Prove that the sum of the series

$$(2n-1) + 2(2n-3) + 3(2n-5) + \ldots n \text{ terms} = \tfrac{1}{6}n(n+1)(2n+1).$$

18 Find an A.P. with first term unity such that the second, tenth and thirty-fourth terms form a geometric series.

19 Find the term independent of x in $\left(\dfrac{4x^2}{3} - \dfrac{3}{2x}\right)^9$.

20 By squaring the result $1 + 2 + 3 + 4 + \ldots + n = \dfrac{n(n+1)}{2}$, find the sum of the products of the first n natural numbers taken two at a time and show that it is equal to half the difference between the sum of their cubes and the sum of their squares.

21 If $(1+x+x^2)^n = a_0 + a_1 x + a_2 x^2 + \ldots + a_{2n} x^{2n}$, find the values of:

(i) $a_0 + a_1 + a_2 + \ldots + a_{2n}$;

(ii) $a_0 + a_2 + a_4 + \ldots + a_{2n}$.

22 Show that the nth term of the series $1 + (1+2) + (1+2+2^2) + \ldots$, is $2^n - 1$. Hence find the sum of n terms.

23 If $S_n = 3 + 8a + 15a^2 + 24a^3 + \ldots + n(n+2)a^{n-1}$, obtain the value of $S_n(1-a)^2$ and deduce the value of S_n. If $a = \frac{1}{3}$, what is the limit of S_n as $n \to \infty$?

24 Show that in the expansion of $(1+x)^n$ where x is positive, the terms will increase until a term containing x^r is reached where r is the least integer $\geqslant \dfrac{nx-1}{x+1}$. Calculate the greatest term in the expansion of $(1+2x)^{12}$ where $x = \frac{1}{4}$.

5 | Definite integration as a summation Geometrical applications

Integration as a method of summation In *Elementary Analysis* Chapter 9, we introduced the idea of a definite integral in discussing the area under a curve. Having first defined the process of *indefinite integration* as the reverse of differentiation—i.e. if

$$\frac{dy}{dx} = \phi(x),$$

then y is the indefinite integral of $\phi(x)$ w.r. to x, or in the notation used

$$y = \int \phi(x)\,dx + c,$$

Fig. 4

c being any constant, it was shown that A, the area included between the curve $y = f(x)$, the x axis and the ordinates $x = a$, $x = b$, was given by the result

$$A = \left\{\text{Value of } \int f(x)dx \text{ when } x = b\right\} - \left\{\text{Value of } \int f(x)dx \text{ when } x = a\right\}.$$

For brevity, this result was written as

$$A = \int_a^b y\,dx,$$

the integral $\int_a^b y\,dx$ being called a *definite integral*.

We will now approach the idea of a definite integral from a totally different standpoint and treat definite integration as a process of summation, the method used in the original development of the subject.

Determination of an area by summation Suppose we require to find the area $ABML$ bounded by the curve $y = f(x)$. Let the coordinates of L and M be a and b. Now let LM be divided into $(n+1)$ equal small parts, LQ_1, Q_1Q_2, Q_2Q_3, ... Q_nM, each of length h. (Fig. 5).

Then $\qquad b - a = (n+1)h.$

The successive ordinates $LA, Q_1P_1, Q_2P_2, \ldots Q_nP_n$ are of lengths $f(a), f(a+h)$, $f(a+2h), \ldots f(a+nh)$.

By completing the rectangles as shown in Fig. 5, it is seen that the area

Determination of an area by summation | 65

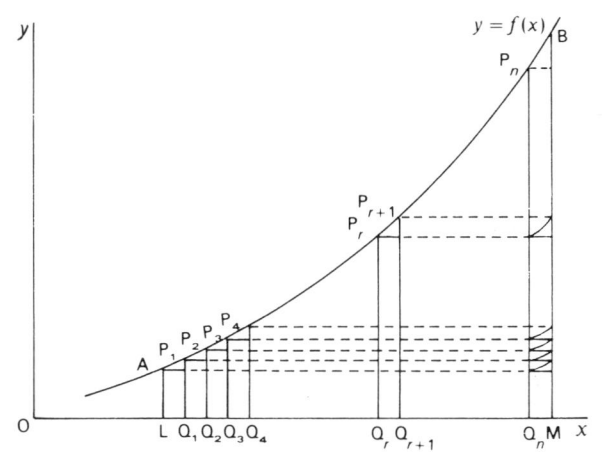

Fig. 5

required $\approx hf(a) + hf(a+h) + hf(a+2h) + \ldots + hf(a+nh)$. This sum can be written as

$$\sum_{r=0}^{r=n} hf(a+rh),$$

where $hf(a+rh)$ is the general term, $(r+1)$th, of the series and the summation extends for all terms from $r = 0$ to $r = n$.

Thus, $\qquad$ area $\approx \sum_{r=0}^{r=n} hf(a+rh)$.

The approximation is due to the fact that allowance has not been made for the small triangles (actually curvilinear triangles as one side is curved), but if we imagine each of the small triangles moved parallel to Ox into a corresponding position in the largest strip ($Q_n M B P_n$ in the case illustrated), it is readily seen that their sum is less than the area of this strip.

Consequently, if we allow n to increase, i.e. take more and more strips, the width h of a strip will decrease so also will the area of each strip and the sum of the areas of the small triangles.

It follows then that we get a closer and closer approximation to the true area if we allow n to get larger and larger.

In other words, the actual area is given by the limit of the above sum as $n \to \infty$ or as $h \to 0$.

$$\therefore \text{Area} = \operatorname*{Lt}_{h \to 0} \sum_{r=0}^{r=n} hf(a+rh).$$

Regarding $(a+rh)$ as the variable x, (the x coordinate of the point Q_r) and the infinitesimal increment as δx, the limits of the summation become $x = a$ (corresponding to $r = 0$) and $x = b - \delta x$ (corresponding to $r = n$).

$$\therefore \text{Area} = \underset{\delta x \to 0}{\text{Lt}} \sum_{x=a}^{x=b-\delta x} f(x)\delta x$$

$$= \underset{\delta x \to 0}{\text{Lt}} \sum_{x=a}^{x=b} f(x)\delta x.$$

Notation This limit of a sum is written as $\int_a^b f(x)\,dx$, and is called *the definite integral of $f(x)$ with respect to x between the limits $x = a$ and $x = b$*. The reason for the use of the elongated S as the sign for integration is now apparent.

Relationship between definite and indefinite integration It has been shown (*Elementary Analysis*, Ch. 9), that the area included between a curve $y = f(x)$, the x axis and the ordinates $x = a, x = b$, is given by the result,

Area under curve = Value of indefinite integral of $f(x)$ when $x = b$
 − Value of indefinite integral of $f(x)$ when $x = a$;

where the indefinite integral of $f(x)$ is defined as that function of x which when differentiated will give $f(x)$.

Also, we have just seen that the same area is given by the definite integral, $\int_a^b f(x)\,dx$ where this integral is defined as the limit of a sum. Consequently, the definite integral

$$\int_a^b f(x)\,dx = \text{Value of the indefinite integral of } f(x) \text{ when } x = b$$

− Value of the indefinite integral when $x = a$.

In anticipation of this result we have already introduced the notation $\int f(x)\,dx$ for an indefinite integral. Using this notation,

$$\int_a^b f(x)\,dx = \text{Value of } \int f(x)\,dx \text{ when } x = b - \text{Value of } \int f(x)\,dx \text{ when } x = a,$$

or $$\left[\int f(x)\,dx \right]_a^b.$$

This result, now proved to be true on the basis of entirely different definitions of definite and indefinite integration was previously used as the definition of a definite integral, a step which of course precludes the use of integration as a means of summation.

Evaluation of definite integrals from the definition Whilst it is apparent that the simple relationship between definite and indefinite integration

Evaluation of definite integrals | 67

will always lead to a quick evaluation of a definite integral when the indefinite integration can be performed, it is of interest to show how simple definite integrals can be evaluated from the definition. Furthermore, we will use the definition of a definite integral as the basis for approximate methods of evaluation in cases where the corresponding indefinite integration cannot be performed.

Example 1 Evaluate $\int_a^b e^x \, dx$ from the definition.

We have,

$$\int_a^b e^x \, dx = \underset{h \to 0}{\text{Lt}} \, h \left\{ e^a + e^{a+h} + e^{a+2h} + \ldots + e^{a+nh} \right\}$$

where $b - a = (n+1)h$.

But $e^a + e^{a+h} + e^{a+2h} + \ldots + e^{a+nh} = e^a(1 + e^h + e^{2h} + \ldots + e^{nh})$,

$$= e^a \frac{(1 - e^{\overline{n+1}h})}{1 - e^h}$$

the series being a G.P., with common ratio e^h.

Hence, $\int_a^b e^x \, dx = \underset{h \to 0}{\text{Lt}} \, h \frac{e^a - e^{a+\overline{n+1}h}}{1 - e^h}$

$$= \underset{h \to 0}{\text{Lt}} \, (e^a - e^b) \frac{h}{1 - e^h} \quad \text{as } a + \overline{n+1}h = b,$$

$$= (e^b - e^a) \underset{h \to 0}{\text{Lt}} \left(\frac{h}{h + \frac{h^2}{2!} + \frac{h^3}{3!} \ldots} \right) \quad \text{as } e^h = 1 + h + \frac{h^2}{2!} + \ldots,$$

$$= (e^b - e^a) \underset{h \to 0}{\text{Lt}} \left(\frac{1}{1 + \frac{h}{2!} + \ldots} \right)$$

$$= e^b - e^a.$$

Evaluation of definite integrals The common methods of evaluation of definite integrals are
 (i) the direct method using the relationship between definite and indefinite integrals, and
 (ii) approximate methods dependent on the representation of a definite integral as an area.

Example 2 Evaluate

(i) $\int_1^2 \left(x + \frac{1}{x} \right)^2 dx;$ (ii) $\int_0^{\frac{\pi}{4}} \cos 2x \, dx;$ (iii) $\int_0^1 \frac{dx}{x+2};$ (iv) $\int_0^{\frac{1}{2}} e^{2x} \, dx.$

(i) $\int_1^2 \left(x + \frac{1}{x}\right)^2 dx = \int_1^2 \left(x^2 + 2 + \frac{1}{x^2}\right) dx$

$= \left[\frac{x^3}{3} + 2x - \frac{1}{x}\right]_1^2$

$= (\frac{8}{3} + 4 - \frac{1}{2}) - (\frac{1}{3} + 2 - 1)$

$= 4\frac{5}{6}.$

(ii) $\int_0^{\frac{\pi}{4}} \cos 2x \, dx = \left[\frac{\sin 2x}{2}\right]_0^{\frac{\pi}{4}} = \frac{\sin \frac{\pi}{2}}{2} - \frac{\sin 0}{2}$

$= \frac{1}{2}.$

(iii) $\int_0^1 \frac{dx}{x+2} = \left[\ln(x+2)\right]_0^1 = \ln 3 - \ln 2$

$= \ln \frac{3}{2}.$

(iv) $\int_0^{\frac{1}{2}} e^{2x} dx = \left[\frac{e^{2x}}{2}\right]_0^{\frac{1}{2}} = \frac{e}{2} - \frac{e^0}{2}$

$= \frac{1}{2}(e - 1).$

[A] EXAMPLES 5a

Evaluate the following definite integrals:

1 $\int_0^1 x^{10} dx.$

2 $\int_{-1}^1 (x^3 - 2x) dx.$

3 $\int_4^9 \frac{dx}{\sqrt{x}}.$

4 $\int_0^4 (1 + \sqrt{x})^2 dx.$

5 $\int_1^2 \frac{dx}{x+3}.$

6 $\int_0^{\frac{\pi}{4}} \sin 2x \, dx.$

7 $\int_{-2}^{-1} 3x(1 - x^2) dx.$

8 $\int_0^1 (e^x + e^{-x}) dx.$

9 $\int_1^3 \frac{dx}{2x - 1}.$

10 $\int_1^2 \left(2x - \frac{1}{2x}\right)^2 dx.$

11 $\int_0^{\frac{\pi}{4}} (x + \sin x) dx.$

12 $\int_{-1}^1 e^{-2x} dx.$

13 $\int_0^2 \frac{x \, dx}{x^2 + 1}.$

14 $\int_{-\frac{\pi}{4}}^{\frac{\pi}{4}} \sin x \cos x \, dx.$

15 $\int_0^{\frac{\pi}{4}} \sec x \tan x \, dx.$

16 $\int_1^2 \frac{1 + x + x^2}{x^3} dx.$

17 $\int_{-2}^{-1} \frac{dx}{e^x}.$

18 $\int_{-1}^0 \frac{4 \, dx}{1 - 2x}.$

19 $\int_0^1 (2x + 1)^4 dx.$

20 $\int_2^4 \frac{1}{\sqrt{e^x}} dx.$

21 $\int_1^2 2^x dx.$

Approximate integration Various methods have been devised for the approximate evaluation of a definite integral when the corresponding indefinite integral is difficult or impossible to find. The methods use the relationship between an area under a curve and a definite integral.

Estimation of area | 69

Method 1 Estimation of area The most elementary method of determining the approximate value of a definite integral, $\int_a^b f(x)\,dx$, is to draw the curve $y = f(x)$ for x between a and b and to estimate the area under the curve by "counting squares". This is a laborious process and it is simpler to obtain an estimate of this area by other means, one of which is to express the required area approximately as the sum of a series of trapeziums. For example, in Fig. 6, the area under the curve between $x = a$ and $x = b$ has been split up into 5 parts by taking 4 equally spaced ordinates between $x = a$ and $x = b$.

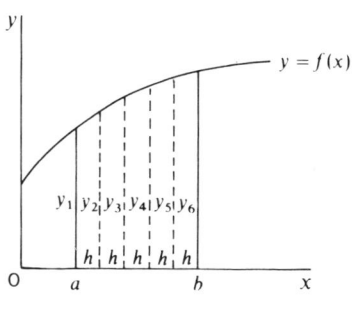

Fig. 6

Represent the ordinates as y_1, y_2, y_3, y_4, y_5, y_6 where $y_1 = f(a)$, $y_2 = f(a+h)$, ... $y_6 = f(b)$ and $b - a = 5h$.

Then, taking each piece of area as approximately equal to a trapezium, we obtain the result,

$$\text{area under curve} = \int_a^b f(x)\,dx \approx \frac{h}{2}(y_1 + y_2) + \frac{h}{2}(y_2 + y_3) + \frac{h}{2}(y_3 + y_4)$$

$$+ \frac{h}{2}(y_4 + y_5) + \frac{h}{2}(y_5 + y_6)$$

$$\approx h\left(\frac{y_1 + y_6}{2} + y_2 + y_3 + y_4 + y_5\right),$$

i.e. $\int_a^b f(x)\,dx \approx \{\text{Mean of first and last ordinates added to the sum of the intervening ordinates}\} \times \text{the common interval}.$

Clearly, the result will hold when the required area is divided into any number of parts. Furthermore, the greater the number of parts, i.e. the greater the number of intervening ordinates, and the closer will be the approximation.

Example 3 Obtain an approximate value of $\int_1^2 \frac{dx}{1+x}$ by the trapezium method.
Divide the required area into 10 parts by ordinates, $x = 1, 1\cdot1, 1\cdot2, \ldots 2\cdot0$. The common interval between the ordinates is $0\cdot1$.

Definite integration

Denoting the lengths of the ordinates by $y_1, y_2, \ldots, y_{11}$, where $y_1 = \dfrac{1}{1+1}$.

$y_2 = \dfrac{1}{1+1\cdot1}, \ldots y_{11} = \dfrac{1}{1+2}$, we have

$$\int_1^2 \dfrac{dx}{1+x} \approx \text{Common interval} \times \left\{ \dfrac{y_1 + y_{11}}{2} + y_2 + y_3 + \ldots + y_{10} \right\}$$

$$\approx 0.1 \times \left\{ \dfrac{5}{12} + \dfrac{1}{2\cdot1} + \dfrac{1}{2\cdot2} + \ldots + \dfrac{1}{2\cdot9} \right\}$$

$$\approx 0.1\{0.417 + 0.476 + 0.454 + 0.435 + 0.417 + 0.400 + 0.385 \\ + 0.370 + 0.357 + 0.345\}$$

$$\approx 0.1(4.066).$$

i.e.
$$\int_1^2 \dfrac{dx}{1+x} \approx 0.4066.$$

The accuracy of this result can be checked as the integral is readily evaluated as ln 1·5 or 0·4055.

Method 2 Simpson's Rule The basis of this method of approximate integration is the following result.

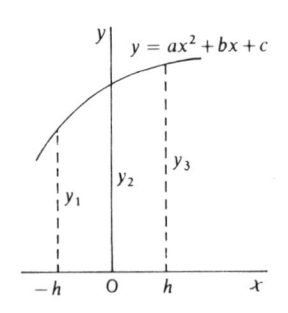

Consider the curve $y = ax^2 + bx + c$, where a, b, c are constants.

Area under curve between $x = -h$ and $x = h$

$$= \int_{-h}^{h} (ax^2 + bx + c)\,dx = \left[\dfrac{ax^3}{3} + \dfrac{bx^2}{2} + cx \right]_{-h}^{h}$$

$$= \left(\dfrac{ah^3}{3} + \dfrac{bh^2}{2} + ch \right) - \left(-\dfrac{ah^3}{3} + \dfrac{bh^2}{2} - ch \right)$$

$$= \dfrac{2ah^3}{3} + 2ch.$$

Fig. 7

Now, the expression

$$\dfrac{h}{3}(y_1 + y_3 + 4y_2) = \dfrac{h}{3}\{(ah^2 - bh + c) + 4c + (ah^2 + bh + c)\}$$

$$= \dfrac{h}{3}\left\{ 2ah^2 + 6c \right\} = \dfrac{2ah^3}{3} + 2ch.$$

$\therefore$ Area under curve $= \dfrac{h}{3}(y_1 + 4y_2 + y_3).$

This result can be applied to give approximate values of the areas under curves whose equations are not of the form considered.

Example 4 Find an approximate value of $\int_0^{\frac{\pi}{3}} \tan x \, dx$.

Imagine the area split into two parts by the ordinate $x = \frac{\pi}{6}$. The areas of the separate parts are assessed by using the result obtained above.

Area between $x = 0$ and $x = \frac{\pi}{6}$

$$\approx \frac{1}{3}\frac{\pi}{12}\left(\tan 0 + \tan \frac{\pi}{6} + 4 \tan \frac{\pi}{12}\right).$$

Area between $x = \frac{\pi}{6}$ and $x = \frac{\pi}{3}$

$$\approx \frac{1}{3}\frac{\pi}{12}\left(\tan \frac{\pi}{6} + \tan \frac{\pi}{3} + 4 \tan \frac{\pi}{4}\right).$$

∴ Total area

$$\approx \frac{\pi}{36}\left(\tan 0 + \tan \frac{\pi}{3} + 2 \tan \frac{\pi}{6} + 4 \tan \frac{\pi}{12} + 4 \tan \frac{\pi}{4}\right)$$

$$\approx \frac{\pi}{36}(0 + 1\cdot 732 + 1\cdot 155 + 1\cdot 072 + 4\cdot 000)$$

$$\approx 0\cdot 695.$$

i.e. $\int_0^{\frac{\pi}{3}} \tan x \, dx \approx 0\cdot 695.$

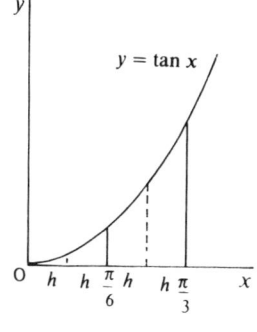

Fig. 8

The accuracy of this result can be checked as $\int \tan x \, dx$ is readily evaluated as $\ln \sec x$. The true value of the integral is $0\cdot 693$ to 3 sig. figs. A closer approximation is obtained if the given area is split up into more parts.

[B] **EXAMPLES 5b**

In Examples 1–4, use the trapezium rule to obtain approximate values of the integrals:

1 $\int_1^3 \frac{dx}{x+3}$ —take ordinates at intervals of $0\cdot 2$.

2 $\int_0^1 \frac{dx}{x^2+1}$ —take ordinates at intervals of $0\cdot 1$.

3 $\int_1^{10} \lg x \, dx$ —take ordinates at intervals of 1.

4 $\int_1^2 2^x \, dx$ —take ordinates at intervals of $0\cdot 2$.

Use Simpson's Rule to evaluate approximately the integrals in Examples 5–10

5 $\int_0^1 \frac{dx}{\sqrt{4-x^2}}$ —divide the interval into 2 equal parts.

72 | **Definite integration**

6    $\int_0^2 \sqrt{x^2+1}\,dx$—divide the interval into 4 equal parts.

7    $\int_0^{\pi/3} \sec x\,dx$—divide the interval into 2 equal parts.

8    $\int_0^{\pi} \sqrt{\sin x}\,dx$—divide the interval into 4 equal parts.

9    $\int_1^4 \dfrac{dx}{1+x^3}$—divide the interval into 3 equal parts.

10    $\int_0^1 e^{-x^2}\,dx$—divide the interval into 2 equal parts.

11   The work done by a force P newtons in moving through a distance $s = a$ to $s = b$ m., is given by $\int_a^b P\,ds$ joules. Find the work done by the force in moving from $s = 0$ to $s = 60$ m., if P is given in terms of s by the following table:

P (newtons)	0	17	27	33	37	39	39·5
s(m)	0	10	20	30	40	50	60

12   The net force on a train of mass W kg is a function $F(v)$ of the speed v km h^{-1}. The time t s to reach a speed v km h^{-1}. is given by the formula, $t = \int_0^v \dfrac{W\,dv}{60^2\, F(v)}$. If $W = 4 \times 10^5$ kg and $F(v)$ is given by the table below, find the approximate time taken to reach a speed of 30 km h^{-1}.:

v	0	5	10	15	20	25	30
$F(v)$	8·93	8·67	8·36	7·99	7·56	7·07	6·52

13   Use Simpson's Rule to find the value of $\int_0^2 x^3\,dx$. By evaluating the integral exactly, show that Simpson's rule gives the correct result.

14   Verify that Simpson's rule gives the exact value of the integral
$\int_{-1}^1 (a+bx+cx^2+dx^3)\,dx$. [Take $x = \pm 1$ as the first and last ordinates and $x = 0$ as the middle ordinate.]

Applications of definite integration   Definite integration has many applications both to geometrical problems and to problems arising in applied mathematics. First, we will reconsider very briefly the application of definite integration to the evaluation of areas and volumes.

Area under a curve   We have the result, that the area contained between the curve $y = f(x)$, the x axis and the ordinates $x = a, x = b$ is given by

$$\text{Area} = \int_a^b y\,dx \quad \text{or} \quad \int_a^b f(x)\,dx.$$

Sign of area

Interchanging y and x, it follows that the area contained between the curve $x = \phi(y)$, the y axis and the abscissae $y = p$, $y = q$ is

$$\int_p^q x\, dy \quad \text{or} \quad \int_p^q \phi(y)\, dy.$$

Sign of area The sign of $\int_a^b y\, dx$ is fixed by the sign of y. Thus, an area below the x axis will appear as a negative area. Consequently, if the curve $y = f(x)$ cuts the x axis between a and b at the point $x = c$, then the *numerical* value of the area under the curve from $x = a$ to $x = b$, is

$$\int_a^c y\, dx - \int_c^b y\, dx,$$

the latter integral being negative.

Note. $\int_a^b y\, dx$ gives the *difference* between the two areas A_1 and A_2 (Fig. 9).

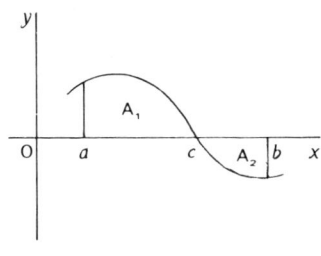

Fig. 9

Example 5 Find the area included between the curve $y = x + \sin 2x$, the x axis and the ordinates $x = 0$, $x = \dfrac{\pi}{2}$.

In the domain $x = 0$ to $x = \dfrac{\pi}{2}$, y is positive and the whole area is above the x axis.

$$\text{Area} = \int_0^{\frac{\pi}{2}} (x + \sin 2x)\, dx = \left[\frac{x^2}{2} - \frac{\cos 2x}{2} \right]_0^{\frac{\pi}{2}}$$

$$= \left(\frac{\pi^2}{8} - \frac{\cos \pi}{2} \right) - \left(0 - \frac{\cos 0}{2} \right)$$

$$= \frac{\pi^2}{8} + 1.$$

Example 6 Find the area included between the curves $y^2 = 4ax$ and $x^2 = 4ay$. The point of intersection A of the curves is given by the equations

$$y^2 = 4ax. \quad \ldots\ldots\ldots\ldots (i)$$

$$x^2 = 4ay. \quad \ldots\ldots\ldots\ldots (ii)$$

Substituting $y = \dfrac{x^2}{4a}$, in (i),

$$\frac{x^4}{16a^2} = 4ax,$$

or $\quad x^4 - 64a^3 x = 0,$

i.e. $\quad x(x^3 - 64a^3) = 0.$

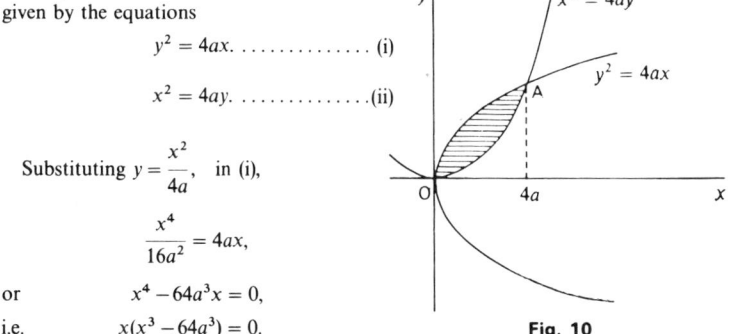

Fig. 10

74 | Definite integration

Giving $x = 0$ (the origin) and $x^3 = 64a^3$,
i.e. $x = 4a$.

Thus A is the point $(4a, 4a)$.

$$\text{Required area} \int_0^{4a} \sqrt{4ax}\, dx - \int_0^{4a} \frac{x^2}{4a}\, dx$$

$$= 2\sqrt{a} \int_0^{4a} x^{\frac{1}{2}}\, dx - \frac{1}{4a} \int_0^{4a} x^2\, dx$$

$$= 2\sqrt{a} \left[\frac{2x^{\frac{3}{2}}}{3} \right]_0^{4a} - \frac{1}{4a} \left[\frac{x^3}{3} \right]_0^{4a}$$

$$= 2\sqrt{a} \left(\frac{16}{3} a^{\frac{3}{2}} \right) - \frac{1}{4a} \cdot \frac{64a^3}{3}$$

$$= a^2 \left(\frac{32}{3} - \frac{16}{3} \right) = \frac{16a^2}{3}.$$

Volume of revolution It has been shown in *Elementary Analysis* that the volume swept out by the rotation of the area under the curve $y = f(x)$ between $x = a$ and b, about the x axis is given by

$$\text{Volume of Revolution} = \pi \int_a^b y^2\, dx.$$

We will now prove this result by using the definition of a definite integral as a summation.

Referring to the diagram, Fig. 11, suppose the volume of revolution is divided up into thin discs by planes drawn perpendicular to the axis of revolution.

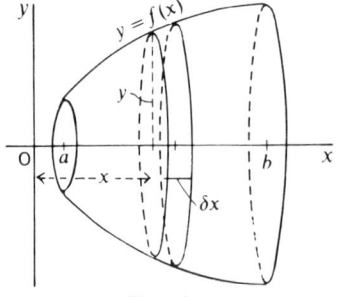

Fig. 11

Consider a typical disc, distant x from the origin, and of thickness δx.

The disc is approximately a thin cylinder and its volume is approximately equal to $\pi y^2\, \delta x$ where y, the radius of the disc, is connected to x by the relation $y = f(x)$. The whole volume of revolution can be expressed as approximately equal to the volumes of all such discs between $x = a$ and $x = b$.

i.e. $$\text{Volume of Revolution} \approx \sum_{x=a}^{x=b} \pi y^2\, \delta x.$$

Clearly, the approximation gets closer as we take thinner discs and, of course, more of them. In the limit, as $\delta x \to 0$, the sum is equal to the volume.

$$\therefore \text{Volume} = \underset{\delta x \to 0}{\text{Lt}} \sum_{x=a}^{x=b} \pi y^2\, \delta x.$$

i.e. $$\textbf{Volume of Revolution} = \int_a^b \pi y^2\, dx.$$

Similarly, if the portion of the curve $x = \phi(y)$ between $y = p$ and $y = q$ is rotated about the y axis, the volume of revolution will be $\int_p^q \pi x^2 \, dy$.

Mean value

The mean value of a function f(x) in the interval $a \leqslant x \leqslant b$ is defined as

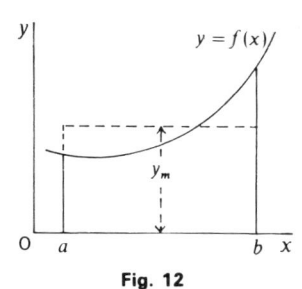

$$\frac{1}{b-a} \int_a^b f(x) \, dx.$$

Geometrically, the mean value of the function $f(x)$ or y, is the altitude of the rectangle on base $(b - a)$ whose area is equal to the area under the curve from $x = a$ to $x = b$.

In applying the idea of a mean value it is essential to have a clear understanding as to what is the independent variable with regard to which the mean is being obtained. This point is illustrated by the following example.

Fig. 12

Example 7 For a particle falling freely from rest with acceleration g, find the mean value of the velocity with respect to time t and with respect to distance travelled, s, for a period starting from the rest position.

Mean value of velocity w.r. to $t = \dfrac{1}{t_2 - t_1} \int_{t_1}^{t_2} v \, dt$.

As we are considering a period starting from the rest position, $t_1 = 0$ and we can write t_2 as t_1.

∴ Mean value of velocity in interval $t = 0$ to $t_1 = \dfrac{1}{t_1} \int_0^{t_1} gt \, dt = \tfrac{1}{2} g t_1$.

i.e. Mean velocity with respect to time in interval $t = 0$ to t_1 is equal to $\tfrac{1}{2}$(velocity at $t = t_1$).

Mean velocity w.r. to distance $= \dfrac{1}{s_1} \int_0^{s_1} v \, ds$ where s_1 is the distance travelled in time t_1,

$$= \frac{1}{s_1} \int_0^{s_1} \sqrt{2gs} \, ds \quad \text{as } v^2 = 2gs,$$

$$= \tfrac{2}{3} \sqrt{2gs_1} = \tfrac{2}{3} \text{(velocity at } t = t_1\text{)}.$$

Thus, mean velocity w.r. to time $= \tfrac{3}{4}$ (mean velocity w.r. to distance).

Geometrically, the different results arise from the fact that in the first case we plot v against t and in the second case v against s. As the curves are different, the mean values are also different.

Root mean square value Consider the case of alternating current. If C is the instantaneous value of the current and t the time we can write C in the form, $C = A \cos \omega t$ where A and ω are constants.

76 | Definite integration

If we wish to define the magnitude of an alternating current, account must be taken not only of A, the peak current, but also of the period, $\dfrac{2\pi}{\omega}$. The mean value would not be a suitable measure as it would be zero over an interval of time equal to one period owing to the fact that in this interval, the area above the axis would be equal to the area below the axis. To overcome this difficulty we introduce a measure called *the root mean square* or *R.M.S. value*, defined as follows:

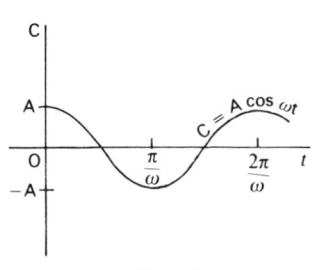

Fig. 13

R.M.S. value of C in the time interval t_1 to $t_2 = \sqrt{\dfrac{1}{t_2 - t_1} \int_{t_1}^{t_2} C^2 \, dt}$.

In the case of periodic or alternating functions, it is usual to consider the R.M.S. value over an interval of a complete period.

$$\therefore \text{(R.M.S. current)}^2 = \dfrac{1}{\tfrac{2\pi}{\omega}} \int_0^{\frac{2\pi}{\omega}} A^2 \cos^2 \omega t \, dt$$

$$= \dfrac{A^2 \omega}{2\pi} \int_0^{\frac{2\pi}{\omega}} \dfrac{\cos 2\omega t + 1}{2} \, dt$$

$$= \dfrac{A^2 \omega}{2\pi} \left[\dfrac{\sin 2\omega t}{4\omega} + \dfrac{t}{2} \right]_0^{\frac{2\pi}{\omega}}$$

$$= \dfrac{A^2 \omega}{2\pi} \left(\dfrac{\sin 4\pi}{4\omega} + \dfrac{2\pi}{2\omega} \right)$$

$$= \dfrac{A^2}{2}.$$

i.e. R.M.S. current $= \dfrac{A}{\sqrt{2}} = 0.7071$ (Peak Current).

In general, **the R.M.S. value of a function f(x) for x = a to b**

$$= \sqrt{\dfrac{1}{b-a} \int_a^b \{f(x)\}^2 \, dx}.$$

[A] EXAMPLES 5c

Find the areas in Examples 1–7:
1. Between the curve $y = 1 + \cos 2x$, the x axis and ordinates $x = 0$, $x = \pi$.
2. Contained between the curve $y = 5(1 + 2x - 3x^2)$ and the x axis.
3. Between the curve $y^2 x = 8$, the y axis and the abscissae $y = 1$, $y = 2$.

4 Included between the curve $y^2 = 8x$ and the double ordinate $x = 2$.
5 Between the curve $y(x+1) = 4$, the x axis and the ordinates $x = 2$, $x = 4$.
6 Between the curve $y = \ln x$, the y axis and the abscissa $y = 1$, $y = 3$.
7 Bounded by the x axis and the portion of the curve $y = x^2(x-1)(x-4)$ which lies below it.

Obtain the volumes of revolution about the x axis of the areas in Examples 8–14:

8 Included between the curve $y = x - \dfrac{1}{x}$, the x axis and ordinates $x = -2$, $x = -1$.

9 Included between the curve $y = e^{2x}$, the x axis and ordinates $x = -1$, $x = 1$.
10 The area contained between the curve $xy = 4$, the x axis and $x = 2$, $x = 3$.
11 The loop of the curve $y^2 = x(x-4)^2$ [i.e. the portion of the curve between $x = 0$, $x = 4$].
12 Included between the curve $y = \sin x - \cos x$, and the axes.

13 Between the curve $y = \dfrac{1}{x-1}$, the x axis and $x = 4$, $x = 6$.

14 Between the curve $y = \tan x$, the line $y = 1$ and the y axis.
15 Find the area of the segment cut off from the curve $y = x(2-x)$ by the line $y = x$. Find the volume swept out by the rotation of this segment about the x axis.
16 What is the mean value of the function $3t(t^2 + 1)$ over the interval $t = 0$ to 4?

17 Find the mean value of the function $2\sin 2t$ over the interval $t = \dfrac{\pi}{8}$ to $\dfrac{\pi}{4}$.

18 Find the R.M.S. value of the function $\dfrac{1}{x}$ in the interval $x = -2$ to -1.

19 What area is included between the curves $y^2 = x$ and $x^2 = 8y$? In what ratio is this area divided by the line $x = 2y$?
20 Find the ratio of the mean value and R.M.S. value of the function $(1 + \sin x)$ in the interval $0 \leqslant x \leqslant \tfrac{1}{2}\pi$.
21 A force P is given in terms of the distance s by the equation
$$P = s(1+s)^2.$$
Find the mean force for values of s between 0 and 10.

22 The velocity v of a body moving with simple harmonic motion is given by the equation $v = a\omega \sin \omega t$, where a and ω are constants. Find the mean velocity and the R.M.S. velocity for $t = 0$ to $\dfrac{\pi}{2\omega}$.

23 Find the mean height of the curve $y = 8x - x^3$ between the origin and the point where it meets the positive part of the x axis and show that this average height is about 65% of the greatest height.
24 What is the ratio of the mean and R.M.S. values of the function $e^x + e^{-x}$ in the interval $0 \leqslant x \leqslant 2$?
25 Prove that the curves $y^2 = 4x$, $x^2 = 4y$ divide into three equal areas the square bounded by the axes and the lines $x = 4$, $y = 4$.

[B] MISCELLANEOUS EXAMPLES

Evaluate the definite integrals in Examples 1–12:

Definite integration

1. $\displaystyle\int_{1}^{6} \frac{dx}{\sqrt{x+3}}.$
2. $\displaystyle\int_{0}^{\pi/3} \sec^2 x(1+\sin x)dx.$
3. $\displaystyle\int_{0}^{\pi/3} \sin x \cos^5 x\,dx.$

4. $\displaystyle\int_{0}^{\pi/4} \sin 3x \cos x\,dx.$
5. $\displaystyle\int_{0}^{\pi/12} \cos 4x \cos 2x\,dx.$
6. $\displaystyle\int_{0}^{2\pi/3} \sin x \sin \frac{x}{2}dx.$

7. $\displaystyle\int_{1}^{2} \frac{x\,dx}{x+3}.$
8. $\displaystyle\int_{0}^{\pi/4} \tan x(1+\sec^2 x)dx.$
9. $\displaystyle\int_{0}^{2} \sqrt{4-x^2}\,x\,dx.$

10. $\displaystyle\int_{-1}^{0} \frac{x^3\,dx}{1-x}.$
11. $\displaystyle\int_{0}^{\pi/4} e^{\tan x} \sec^2 x\,dx.$
12. $\displaystyle\int_{0}^{\pi} \frac{\sin x}{\cos^3 x}dx.$

13 A cup is obtained by rotating the parabola $y^2 = 4ax$ about the x axis. If the diameter of the top of the cup is 7 cm and it holds exactly 49π cm^3 find its depth.

14 Make a rough sketch of the curve $y^2 = x(x-4)^2$ and find the area of the loop.

15 A sphere radius r is divided into two segments by a plane at distance c from the centre, c being less than r; prove that the volumes of the segments are in the ratio
$$\frac{2r^3 + 3r^2 c - c^3}{2r^3 - 3r^2 c + c^3}.$$

16 A sphere of radius a has a hollow concentric spherical cavity of radius $\dfrac{2a}{3}$. Show that any plane whose distance from the centre is $\dfrac{19a}{45}$ divides the sphere into two portions whose volumes are in the ratio $3:1$.

17 The curve $y = 1 + \sin x$ is rotated about the axis of x. Prove that the volume contained between the surface of revolution so formed and the planes $x = 0$, $x = \pi$ is
$$\pi\left(4 + \frac{3\pi}{2}\right).$$

18 Find the area included between the two curves $2y^2 - 3y = x - 1$ and $y^2 - 2y = x - 3$.

19 A point is such that its velocity v at time t is given by $v = u + ft$, where u and f are constants. Prove that the R.M.S. velocity over a period of time $t = t_0$ to t_1 is equal to $\sqrt{\frac{1}{3}(v_0^2 + v_0 v_1 + v_1^2)}$, where v_0, v_1 are the velocities at times t_0 and t_1.

20 Find the points in which the circle $x^2 + y^2 = 9a^2$ intersects the parabola $y^2 = 8ax$. The smaller of the two areas into which the circle is divided is rotated about Ox. Prove that the volume formed is $\dfrac{40\pi a^3}{3}$.

21 A variable force P newtons is given in terms of the time t s and distance s m by the following table:

P	160	640	1120	1600	2080
t	0	1	2	3	4
s	0	5	24	63	128

Use Simpson's rule to evaluate the integrals $\displaystyle\int_0^4 P\,dt$ and $\displaystyle\int_0^{128} P\,ds$ and deduce the mean forces in the period with respect to t and with respect to s.

22 Make a rough sketch of the curve $y = \dfrac{1}{1+x^2}$. Find the area included between the curve and the line $y = \frac{1}{2}$ and the volume obtained by rotating this area about the y axis.

Miscellaneous examples | 79

23 A mound in the form of a segment of a sphere is formed on a horizontal plane. The radius of the base of the mound is 153 m and the height of the highest point above the plane is 16 m. Prove that the volume of the mound is approximately $5 \cdot 905 \times 10^5$ m^3.

24 The power consumed by an electric current I at any instant is given by I^2R, where R is a constant. Prove that if $I = I_0 \sin \omega t$, where I_0 and ω are constants, the mean power consumed over a period, i.e. from $\omega t = 0$ to $\omega t = 2\pi$, is $\frac{1}{2} I_0^2 R$ and obtain the R.M.S. power over the same period.

25 The density ρ at a point P of a rod AB of length $2a$ is given by $\rho = \rho_0 \left(1 + \dfrac{x}{a}\right)$, where x is the distance of P from A. Prove that the mass of the rod is $4a\rho_0$ and deduce the mean density.

26 The surface density σ of the charge at a point on an electrified circular disc of radius a is given by $\sigma = \dfrac{a\sigma_0}{\sqrt{a^2 - r^2}}$, where σ_0 is a constant and r is the distance of the point from the centre of the disc. Prove that the total charge on the disc is $2\pi a\sigma_0 \displaystyle\int_0^a \dfrac{r\,dr}{\sqrt{a^2 - r^2}}$ and obtain the mean surface density.

6 | Further applications of definite integration

Centre of gravity It is readily shown by taking moments about the axes, that the coordinates of the centre of gravity of a system of particles, masses $m_1, m_2, \ldots$, located at points $(x_1, y_1), (x_2, y_2) \ldots$, are given by the formulae,

$$\bar{x} = \frac{m_1 x_1 + m_2 x_2 + m_3 x_3 + \ldots}{m_1 + m_2 + m_3 + \ldots} = \frac{\sum mx}{\sum m};$$

$$\bar{y} = \frac{m_1 y_1 + m_2 y_2 + m_3 y_3 + \ldots}{m_1 + m_2 + m_3 + \ldots} = \frac{\sum my}{\sum m}.$$

We will now apply these results to the determination of the centres of gravity of continuous bodies which can be considered as made up of an infinite number of infinitesimally small particles.

Example 1 Find the C.G. of a uniform plane area bounded by the curve $y = x^2$, the x axis and the ordinates $x = 1, x = 3$.

Imagine the area to be divided up into a large number of thin strips parallel to the y axis. Consider a typical strip or element, distance x from the origin and of width δx. If y is the length of the strip, we note that $y = x^2$.

This small element is approximately rectangular in shape and of area $y\delta x$. As the area is assumed of uniform density σ, it can be replaced approximately by a mass $\sigma y \delta x$ at the centre of the rectangle, i.e. the point $\left(x, \dfrac{y}{2}\right)$.

Hence, the whole area can be replaced by a series of particles and its C.G. can be found by using the formulae quoted above, the result being approximate until we go to the limit as $\delta x \to 0$.

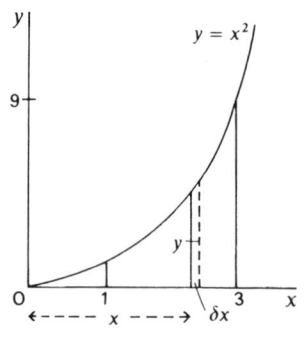

Fig. 14

We have
$$\bar{x} = \frac{\sum mx}{\sum m}.$$

In this case
$$\bar{x} \approx \frac{\sum_{x=1}^{x=3} \sigma y \delta x \, x}{\sum_{x=1}^{x=3} \sigma y \delta x}$$

Centre of gravity | 81

$$\therefore \bar{x} = \frac{\int_1^3 \sigma xy\,dx}{\int_1^3 \sigma y\,dx} = \frac{\sigma \int_1^3 x^3\,dx}{\sigma \int_1^3 x^2\,dx} \quad \text{as } \sigma \text{ is constant}$$

$$= \frac{\left[\dfrac{x^4}{4}\right]_1^3}{\left[\dfrac{x^3}{3}\right]_1^3} = \frac{20}{8\frac{2}{3}} = \frac{30}{13}.$$

To find $\bar{y}$, we use the result $\bar{y} = \dfrac{\sum my}{\sum m}$, noting that *the y coordinate of the mass* $\sigma y\,\delta x$ is $\dfrac{y}{2}$.

Hence,
$$\bar{y} = \frac{\int_1^3 \dfrac{\sigma y^2}{2}\,dx}{\int_1^3 \sigma y\,dx} = \frac{\frac{1}{2}\int_1^3 x^4\,dx}{8\frac{2}{3}}$$

$$= \frac{\dfrac{1}{2}\left[\dfrac{x^5}{5}\right]_1^3}{8\frac{2}{3}} = \frac{\dfrac{242}{10}}{\dfrac{26}{3}} = \frac{363}{130}.$$

Therefore, the coordinates of the C.G. of the plane area are $(\frac{30}{13}, \frac{363}{130})$.

In general, the C.G. of a uniform plane area bounded by the curve y = f(x), the x axis and ordinates x = a, x = b, is given by

$$\bar{x} = \frac{\int_a^b xy\,dx}{\int_a^b y\,dx} \;;\; \bar{y} = \frac{\int_a^b \dfrac{y^2}{2}\,dx}{\int_a^b y\,dx}.$$

Example 2 Find the C.G. of a plane area bounded by the parabola $y^2 = 4ax$ and the double ordinate $x = h$.

As the curve $y^2 = 4ax$ is symmetrical about the x axis, the C.G. of the area will lie on the x axis, or $\bar{y} = 0$.

Consider the area to be split up into strips by lines parallel to the y axis.

Let a typical strip, distance x from the origin, be of width δx and length $2y$; y and x will be connected by the relation $y^2 = 4ax$.

Replacing the element which is approximately rectangular, by a particle mass $2y\sigma\,\delta x$ at the point $(x, 0)$, we have

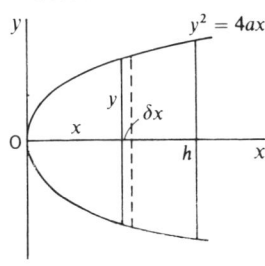

Fig. 15

$$\bar{x} \approx \frac{\sum_0^h 2y\sigma\,\delta x \cdot x}{\sum_0^h 2y\sigma\,\delta x},$$

82 | Further applications of definite integration

or,
$$\bar{x} = \frac{\int_0^h 2\sigma xy\, dx}{\int_0^h 2\sigma y\, dx}$$

$$= \frac{\int_0^h x\sqrt{4ax}\, dx}{\int_0^h \sqrt{4ax}\, dx} = \frac{\int_0^h x^{\frac{3}{2}}\, dx}{\int_0^h x^{\frac{1}{2}}\, dx}$$

$$= \left[\frac{2}{5}x^{\frac{5}{2}}\right]_0^h \bigg/ \left[\frac{2}{3}x^{\frac{3}{2}}\right]_0^h = \frac{3}{5}h.$$

i.e. The C.G. of the area is the point $\left(\dfrac{3h}{5}, 0\right)$.

Example 3 Find the C.G. of a uniform wire in the shape of circular arc, radius a, subtending an angle 2α at the centre (α in radians).

Let the x axis be the axis of symmetry.

Consider the arc to be divided up into small elements by radial lines drawn through the centre O.

Then, measuring θ from the x axis with the usual sign convention, we have $\delta s = a\, \delta\theta$.

Replacing the element δs by a particle at its centre,

$$\bar{x} = \frac{\sum mx}{\sum m} = \frac{\int_{-\alpha}^{\alpha} ax\, d\theta}{\int_{-\alpha}^{\alpha} a\, d\theta}.$$

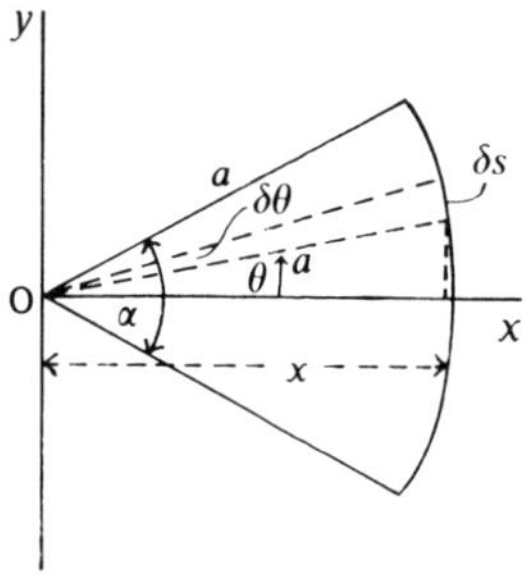

Fig. 16

But $x = a\cos\theta$, from the diagram.

$$\therefore \bar{x} = \frac{a^2 \int_{-\alpha}^{\alpha} \cos\theta\, d\theta}{2a\alpha} = \frac{a}{2\alpha}\{\sin\alpha - \sin(-\alpha)\}$$

$$= \frac{2a\sin\alpha}{2\alpha} \text{ as } \sin(-\alpha) = -\sin\alpha,$$

$$= \frac{a\sin\alpha}{\alpha}.$$

By symmetry, $\bar{y} = 0$.

Hence, the C.G. of the wire is on the axis of symmetry, distant $\dfrac{a\sin\alpha}{\alpha}$ from the origin.

This result can be used to deduce the C.G. of a semicircle. For, taking $\alpha = \dfrac{\pi}{2}$, we see that, for a semicircular wire, $\bar{x} = \dfrac{2a}{\pi}$.

Centre of gravity | 83

In the case of a semicircular area, consider the area to be split up into a large number of triangular elements by radial lines drawn through the centre. As the distance of the C.G. of a triangle from the vertex is two-thirds of the height, it follows that each triangle can be replaced by an equivalent particle distant $\frac{2a}{3}$ from O. In other words, the whole semicircle can be replaced by an equivalent arc, radius $\frac{2a}{3}$. Consequently, as the C.G. of the arc is given by,

$$\bar{x} = \frac{2\,\text{Radius}}{\pi},$$

Fig. 17   for the semicircle, $\bar{x} = \frac{4a}{3\pi}.$

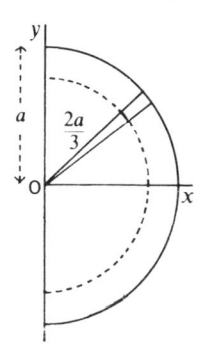

This result can also be obtained by direct integration.

Example 4 Find the C.G. of a uniform hemisphere, radius a, density ρ.
Let the hemisphere be divided into thin discs by planes perpendicular to Ox.
Then, for a typical disc distance x from O, thickness δx,

mass, $\delta m \approx \rho \pi y^2 \, \delta x.$

Replacing the disc by a particle, mass δm, at its centre, we have

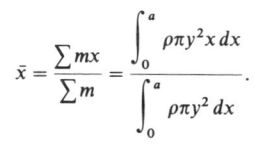

$$\bar{x} = \frac{\sum mx}{\sum m} = \frac{\int_0^a \rho \pi y^2 x \, dx}{\int_0^a \rho \pi y^2 \, dx}.$$

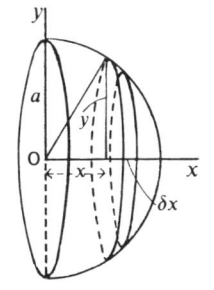

To evaluate the integrals, we require y in terms of x.
By Pythagoras, $y^2 = a^2 - x^2.$

Fig. 18

$$\therefore \bar{x} = \frac{\int_0^a x(a^2 - x^2)\,dx}{\int_0^a (a^2 - x^2)\,dx} = \frac{\left[\dfrac{a^2 x^2}{2} - \dfrac{x^4}{4}\right]_0^a}{\left[a^2 x - \dfrac{x^3}{3}\right]_0^a}$$

$$= \frac{\dfrac{a^4}{4}}{\dfrac{2}{3}a^3} = \frac{3a}{8}.$$

i.e. The C.G. of the hemisphere is on the radius of symmetry and distant $\frac{3a}{8}$ from the base.

Further applications of definite integration

[A] EXAMPLES 6a

1 The diagram in Fig. 19 represents a rectangular plate of density ρ. Taking axes as shown and dividing the area into elements of which a typical one is indicated, show that the x coordinate of the C.G. is given by

$$\bar{x} = \int_0^a \rho x \, dx \bigg/ \int_0^a \rho \, dx.$$

Hence, evaluate $\bar{x}$ when (i) the density of the plate is constant, (ii) the density of the plate is given by

$$\rho = \rho_0 \left(1 + \frac{x}{a}\right) \text{ where } \rho_0 \text{ is constant.}$$

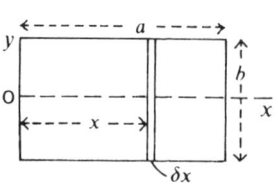

Fig. 19

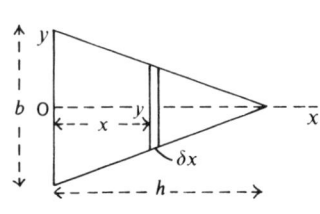

2 The diagram in Fig. 20 represents an isosceles triangle of height h, base b and density ρ. Taking elements parallel to Oy of which a typical one of length y, width δx is shown, prove the following results:

(i) $y = \dfrac{b}{h}(h - x)$.

Fig. 20

(ii) $\bar{x} = \int_0^h \rho x (h - x) \, dx \bigg/ \int_0^h \rho (h - x) \, dx.$

Evaluate $\bar{x}$ for the cases $\rho = \rho_0$ and

$$\rho = \rho_0 \left(1 + \frac{x}{h}\right), \text{ where } \rho_0 \text{ is constant.}$$

3 Fig. 21 represents a uniform plane area bounded by the curve $y^2 = x^3$ and the double ordinate $x = 2$. Taking strips as indicated of length $2y$ and width δx, show that $\bar{x} = \int_0^2 x^{\frac{5}{2}} \, dx \bigg/ \int_0^2 x^{\frac{3}{2}} \, dx$, and evaluate it.

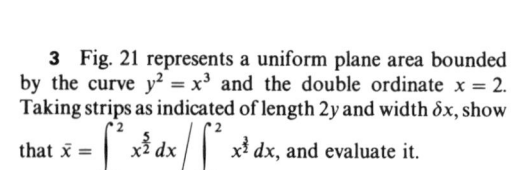

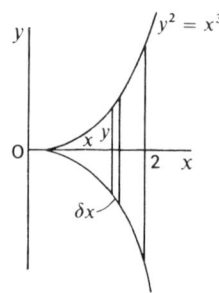

Fig. 21

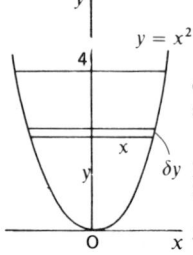

4 Fig. 22 represents a uniform plane area bounded by the curve $y = x^2$ and the line $y = 4$. Taking strips parallel to Ox as shown, find the position of the centre of mass of the area.

5 What is the y coordinate of the C.G. of the part of the plane area in the previous question to the right of the y axis? Show that the x coordinate of this area is given by

$$\bar{x} = \tfrac{1}{2} \int_0^4 x^2 \, dy \bigg/ \int_0^4 x \, dy \text{ and find its value.}$$

Fig. 22

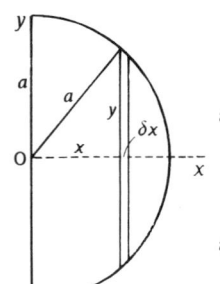

6 The diagram in Fig. 23 represents a uniform semicircular area of radius a. Prove that

$$\bar{x} = 2\int_0^a \sqrt{a^2 - x^2}\, x\, dx \bigg/ \text{Area of semicircle},$$

and deduce that $\quad \bar{x} = \dfrac{4a}{3\pi}.$

Fig. 23

[B] EXAMPLES 6b

Find the centres of gravity of the uniform areas in Examples 1–6:
1. Area bounded by $y = x^3$, $y = 0$ and $x = 2$.
2. Area bounded by $y = x(x+1)$, $y = 0$ and $x = 3$.
3. Area bounded by $y^2 = 4x$ and the double ordinate $x = 2$.
4. Area bounded by $y = x(3 - x)$ and $y = 0$.
5. Area bounded by $y = \dfrac{1}{x}$, $y = 1$ and $y = 4$.
6. Area bounded by the loop of the curve $y^2 = x(1 - x)^2$ between $x = 0$ and 1.

Find the centres of mass of the uniform solids formed by the revolution of the areas in Examples 7–10 about the x-axis:

7. Area bounded by $y^2 = x$ and $x = 2$.
8. Area bounded by $y = x^2$, $y = 0$ and $x = 3$.
9. Area bounded by $y^2 = x^3$ and $x = 4$.
10. Area bounded by $xy = 4$, $y = 0$, $x = 1$ and $x = 3$.
11. Find the area of the segment of the curve $y = x(4 - x)$ cut off by the straight line $y = x$. What are the coordinates of the mass centre of this area?
12. Find the centre of gravity of the area bounded by the curves $y = x^2$ and $x = y^2$.
13. The portion of the curve $y = x^2 - x^3$ between $x = 0$ and $x = 1$ is rotated about the x axis; find the centre of gravity of the volume obtained.
14. Prove that the centre of mass of a uniform right circular cone, base radius r, height h, is on the axis of the cone and distant $\dfrac{3h}{4}$ from the vertex.
15. A uniform solid is made up of a circular cylinder with a conical end. If the C.G. of the solid is on the axis of symmetry and distant $\dfrac{3h}{4}$ from the base of the cylinder, find the height of the cone in terms of h, the height of the cylinder.
16. Find the C.G. of a uniform pyramid whose base is a square, side 6 cm and whose height is 8 cm, the vertex being vertically above the centre of the square. [Divide the volume into elements by means of planes drawn parallel to the base.]
17. Show by integration that the centre of area of any triangle is a third the way along a median from the base.
18. Find the C.G. of a frustrum of a cone with circular ends having radii 8 cm and 4 cm and 6 cm apart.

86 | Further applications of definite integration

19 A cup forming part of a sphere of radius 6 cm contains water to a depth of 2 cm, find the depth below the surface of the centre of mass of the water.

20 The area contained between the curve $y = x(1 + x)$ and the line $y = 4x$ is rotated about Ox. Find the coordinates of the C.G. of the volume of revolution.

21 The density of a rod AB, length $2a$, is given by $\rho = \rho_0 \left(2 - \sqrt{\dfrac{x}{a}}\right)$, where x is the distance from A. Find the mass of the rod and the distance of its centre of mass from A.

22 Find the C.G. of a closed uniform wire in the form of a semicircle, radius a.

23 Find the C.G. of a uniform thin hemispherical shell of radius r.

24 A hemispherical vessel has internal and external radii a and b. If the C.G. of the vessel lies on the inner surface, find the relationship between a and b.

25 The density at any point of a circular cone of semi-vertical angle α and height h is proportional to the distance of the point from the base of the cone and the greatest density is ρ_0. Find the C.G. of the cone.

26 A hollow vessel made of thin metal sheeting is in the shape of a circular cone with a hemispherical base lying outside the cone. If the height of the cone is half its base radius r, find the C.G. of the vessel.

27 A thin hemispherical vessel of radius 6 cm is filled with water to a depth of 3 cm. If the mass of the vessel is 100 g and the density of water 1 g cm^{-3}, find the C.G. of vessel and water.

Moments of inertia

Definitions *The moment of inertia* (M.I.) *of a body about an axis* is equal to the sum of the products of each element of mass of the body and the square of its perpendicular distance from the axis.

If we consider a body made up of elements of mass $m_1, m_2, \ldots$ at distances $x_1, x_2, \ldots$ from a given axis, then

$$\text{M.I. of body about the given axis} = m_1 x_1^2 + m_2 x_2^2 + \ldots$$
$$= \sum m x^2.$$

In the case of continuous bodies made up of an infinite number of infinitesimally small elements of mass the summation is replaced by its limit, a definite integral.

The radius of gyration of a body about an axis is defined by the relationship,

$$\text{Mass (Radius of Gyration)}^2 = \text{Moment of Inertia,}$$

i.e.
$$Mk^2 = \text{Moment of Inertia}$$
$$k^2 = \frac{\text{Moment of Inertia}}{M},$$

k being the radius of gyration.

Example 5 Find the M.I. of a uniform rod, mass M, length $2a$, about an axis through one end perpendicular to the rod.

Moments of inertia | 87

Let the line density of the rod be σ. Consider the rod to be divided up into a large number of small elements. If a typical element at distance x from A is of length δx, then the M.I. of the rod about axis $Ay \approx \sum_{0}^{2a} \sigma \delta x \, x^2$, i.e. the sum of the products of all elements of mass and the square of their distances from the axis.

Fig. 24

In the limit, as $\delta x \to 0$, we have

$$\text{M.I. about } Ay = \underset{\delta x \to 0}{\text{Lt}} \sum_{0}^{2a} \sigma x^2 \, \delta x$$

$$= \int_0^{2a} \sigma x^2 \, dx = \sigma \int_0^{2a} x^2 \, dx \quad \text{as } \sigma \text{ is constant}$$

$$= \frac{8a^3 \sigma}{3}.$$

The M.I. is usually expressed in terms of the mass and substituting $M = 2a\sigma$, we have the result

M.I. of uniform rod, length 2a, about a perpendicular axis through one end $= \dfrac{4}{3} Ma^2$.

By an identical method, it is easily shown that **the M.I. of the rod about a perpendicular axis through its centre $= \dfrac{1}{3} Ma^2$.**

The method used in this example is applicable to the case of a non-uniform rod where the line density σ is given as a function of x say $f(x)$.

The M.I. about $Ay = \displaystyle\int_0^{2a} x^2 f(x) \, dx$, which can be evaluated when $f(x)$ is known.

Example 6 Find the radius of gyration of a uniform triangular lamina about one side.

Let the side AB of $\triangle ABC$ be of length c, and let the height be h.
Then if the uniform surface density is σ,

$$M = \tfrac{1}{2} ch\sigma.$$

Consider the triangle to be divided up into thin strips parallel to the axis AB. Letting the typical strip at distance x from the axis have length y and width δx, we have,

$$\text{M.I. about } AB = \underset{\delta x \to 0}{\text{Lt}} \sum_{0}^{h} \sigma y \, \delta x \, x^2$$

$$= \sigma \int_0^h x^2 y \, dx \quad \text{as } \sigma \text{ is constant.}$$

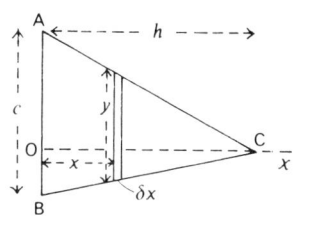

Fig. 25

To evaluate the integral, we obtain y as a function of x.
By similar triangles,

$$\frac{y}{c} = \frac{h-x}{h};$$

$$y = \frac{c(h-x)}{h}.$$

88 | Further applications of definite integration

$$\therefore \text{M.I. about } AB = c\sigma \int_0^h \left(x^2 - \frac{x^3}{h} \right) dx$$

$$= c\sigma \left(\frac{h^3}{3} - \frac{h^3}{4} \right) = \frac{1}{12} ch^3 \sigma.$$

Expressing the result in terms of M where $M = \frac{1}{2}ch\sigma$,

M.I. of $\triangle ABC$ about $AB = \frac{1}{6}Mh^2$.

$$\therefore (\text{Radius of gyration})^2 = \frac{\text{M.I.}}{M} = \frac{1}{6}h^2.$$

i.e. Radius of gyration about side $AB = \dfrac{h}{\sqrt{6}}$.

Example 7 A plane lamina is enclosed between the curve $y = \dfrac{1}{x}$, the x axis and the ordinates $x = 1$, $x = 3$. Find the M.I. about the axes Oy and Ox, if the surface density σ varies according to the relation $\sigma = kx$ where k is constant.

First consider the M.I. about Oy.
Split the area up into strips parallel to Oy.

$$\text{M.I. about } Oy = \underset{\delta x \to 0}{\text{Lt}} \sum_1^3 \sigma y \, \delta x \, x^2$$

$$= \int_1^3 \sigma y x^2 \, dx.$$

But $y = \dfrac{1}{x}$ as (x, y) are the coordinates of a point P on the curve and $\sigma = kx$.

$$\therefore \text{M.I. about } Oy = \int_1^3 kx \frac{1}{x} x^2 \, dx = k \int_1^3 x^2 \, dx$$

$$= \frac{26k}{3}.$$

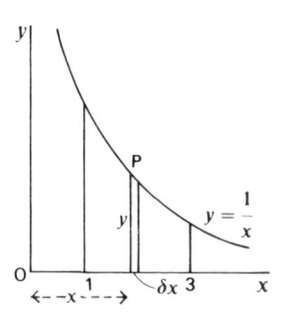

Fig. 26

As the result is purely numerical there is no need to express it in terms of the mass of the lamina.

Now, consider the M.I. about Ox.

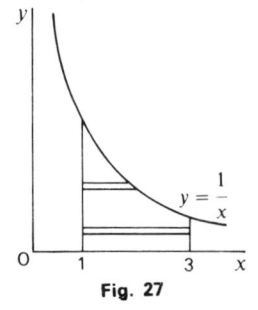

Fig. 27

We might be tempted to try splitting the area into strips parallel to Ox, but the result could not be obtained in this manner for two reasons:

(i) Owing to the nature of the area the variable used to denote the lengths of the strips would be discontinuous; some of the strips would be of constant length and the remainder would be variable (see Fig. 27).

(ii) In this example, the surface density is not constant but varies with the distance from the y axis. This necessitates the use of strips parallel to the y axis to ensure that the surface density is constant at all points of a strip.

Moments of inertia | 89

So, in considering the M.I. about Ox, we take strips parallel to Oy.

$$\text{Mass of typical strip} \approx \sigma y\, \delta x \approx kxy\, \delta x.$$

In this case, all elements of the strip are not at the same distance from the axis under consideration. To overcome this difficulty we treat each strip as a uniform rod and use the result of Example 5.

$$\text{M.I. of strip about } Ox \approx \tfrac{4}{3}\left(kxy\,\delta x\right)\left(\frac{y}{2}\right)^2$$

$$\approx \tfrac{1}{3}kxy^3\,\delta x.$$

Summing for all strips between $x = 1$ and $x = 3$,

$$\text{M.I. of area about } Ox = \tfrac{1}{3}k\int_1^3 xy^3\,dx$$

$$= \tfrac{1}{3}k\int_1^3 \frac{dx}{x^2}$$

$$= \frac{2k}{9}.$$

Example 8 Find the M.I. of a uniform circular disc radius a, about a perpendicular axis through the centre.

Let σ be the constant surface density.

Split the area up into elements by means of a series of concentric circles.

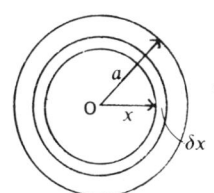

For a strip, width δx, at distance x from O,

$$\text{mass} \approx 2\pi x\,\delta x\,\sigma.$$

∴ M.I. about perpendicular axis through O

$$= \underset{\delta x \to 0}{\text{Lt}} \sum_0^a 2\pi x\,\delta x\,\sigma x^2$$

$$= 2\pi\sigma \int_0^a x^3\,dx = \frac{\pi\sigma a^4}{2}.$$

Fig. 28

Mass of disc $M = \pi a^2 \sigma$.

∴ **M.I. of circular disc about perpendicular axis through centre** $= \tfrac{1}{2}Ma^2$.

General theorems Two important theorems on moments of inertia are now given without proofs. The latter are given in most books on Applied Mathematics.

Theorem 1 *The M.I. of a plane lamina about an axis Oz perpendicular to its plane is equal to the sum of the M.I.s. about any two perpendicular axes Ox, Oy in the plane of the lamina.*

e.g. the M.I. of a uniform rectangular lamina about a perpendicular axis Oz through one vertex

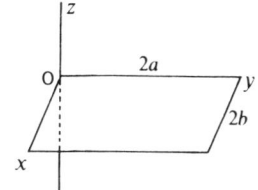

$$= \text{M.I. about } Ox + \text{M.I. about } Oy$$
$$= \tfrac{4}{3}Ma^2 + \tfrac{4}{3}Mb^2$$
$$= \tfrac{4}{3}M(a^2 + b^2).$$

Fig. 29

Further, using the result in the case of a uniform circular lamina, we have (Fig. 30),

M.I. about Oz
$$= \text{M.I. about } Ox + \text{M.I. about } Oy.$$
But, M.I. about $Oz = \tfrac{1}{2}Ma^2$,
and, by symmetry, the M.I. about diameter Ox
$$= \text{M.I. about diameter } Oy.$$
Hence,

M.I. of circular disc about a diameter
$$= \tfrac{1}{4}\mathbf{Ma^2}.$$

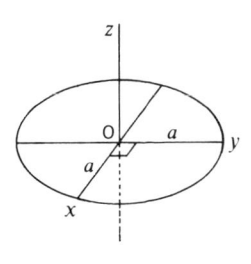

Fig. 30

Theorem 2 The parallel axis theorem *The M.I. of a body about a given axis AB is equal to the M.I. about a parallel axis $A'B'$ through the centre of gravity of the body plus the product of the mass of the body and the square of the distance between the axes AB, $A'B'$.*

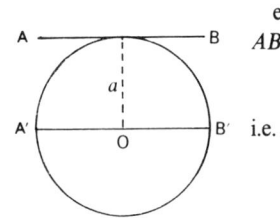

e.g. the M.I. of a circular disc about a tangent
$AB = $ M.I. of the disc about the diameter $A'B'$
+ Mass times square of distance between AB and $A'B'$.

i.e. M.I. about tangent AB
$$= \frac{Ma^2}{4} + Ma^2 = \tfrac{5}{4}Ma^2.$$

Fig. 31

Solid of revolution The method of finding the M.I. of a solid of revolution is illustrated by the following example:

Example 9 A uniform solid is formed by the rotation of the part of the curve $y^2 = 4ax$ bounded by the ordinate $x = h$, about the x axis. Find the radii of gyration of the solid about the axes Ox and Oy.

Divide the solid into thin circular discs by planes drawn perpendicular to Ox.

Mass of typical disc $\approx \rho \pi y^2 \, \delta x$,

where ρ is the density.

In finding the M.I. of the whole solid about an axis, we first write down the M.I. of the disc about that axis and then sum the result for all discs as x varies from O to h.

M.I. of disc about $Ox = \dfrac{\text{Mass} \times (\text{Radius})^2}{2}$

$\approx \dfrac{\rho \pi y^2 \, \delta x \, y^2}{2}.$

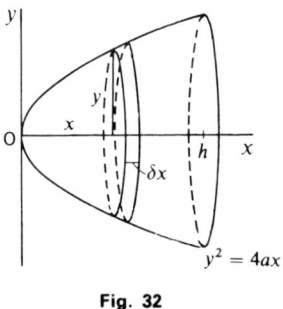

Fig. 32

∴ M.I. of solid about $Ox \approx \sum_0^h \dfrac{\rho \pi y^4 \, \delta x}{2}$

$$= \dfrac{\pi \rho}{2} \int_0^h y^4 \, dx \quad \text{where } y^2 = 4ax,$$

$$= \dfrac{\pi \rho}{2} \int_0^h 16 a^2 x^2 \, dx$$

$$= 8a^2 \pi \rho \dfrac{h^3}{3}.$$

To obtain the radius of gyration, we require the mass of the solid, M.

$$M = \pi \rho \int_0^h y^2 \, dx = 4a\pi\rho \int_0^h x \, dx = 2a\pi h^2 \rho.$$

Hence, radius of gyration about $Ox = \sqrt{\dfrac{\tfrac{8}{3} a^2 h^3 \pi \rho}{2 a h^2 \pi \rho}}$

$$= \sqrt{\tfrac{4}{3} ah}.$$

In order to find the M.I. of the circular disc about the axis Oy, we use the parallel axis theorem, the parallel axis through the C.G. of the disc being a diameter.

M.I. of disc about $Oy \approx \rho \pi y^2 \, \delta x \dfrac{y^2}{4} + \rho \pi y^2 \, \delta x \, x^2$

$$\approx \rho \pi y^2 \, \delta x \left(\dfrac{y^2}{4} + x^2 \right).$$

Hence, summing for all discs between $x = 0$ and $x = h$,

M.I. of solid of revolution about $Oy = \rho \pi \int_0^h \left(\dfrac{y^4}{4} + x^2 y^2 \right) dx$

$$= \rho \pi \int_0^h (4a^2 x^2 + 4a x^3) \, dx$$

$$= 4a\rho\pi \left(\dfrac{ah^3}{3} + \dfrac{h^4}{4} \right)$$

$$= 2M \left(\dfrac{ah}{3} + \dfrac{h^2}{4} \right).$$

i.e. Radius of gyration about $Oy = \sqrt{\dfrac{h}{6}(4a + 3h)}.$

[A] EXAMPLES 6c

1 Using the elements of area shown in Fig. 33, prove that the M.I. of a uniform rectangular area $ABCD$, $AD = 2a$, $AB = 2b$, about the side AB is $\tfrac{4}{3} Ma^2$. Deduce the

M.I. about an axis through the centre of the area parallel to AB. What are the corresponding results if side AD is substituted for AB?

2 Use the Parallel Axis theorem to write down the M.I. of the rectangular area in Example 1 about an axis parallel to AB and distant $\dfrac{a}{2}$ from the centre.

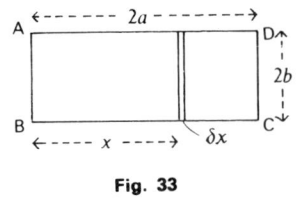

Fig. 33

3 Use the results of Example 1 and the Plane Area theorem to deduce the M.I. of a uniform square plate of side $2a$ about a perpendicular axis through the centre.

4 In Fig. 34, ABC is a triangle with $AB = c$ and height h. Find y, the length of a typical element, in terms of x, h and c. Deduce that the M.I. of the uniform triangular area ABC about an axis $A'B'$ through C parallel to side AB, is $\tfrac{1}{2}Mh^2$.

Use the standard results obtained in the text and the two theorems in the following examples.

5 Find the M.I. of a uniform rod, mass M, length $2a$, about a perpendicular axis, distance b from the centre of the rod.

6 Find the radii of gyration of a uniform circular disc, radius 20 cm, about (i) a tangent, (ii) an axis perpendicular to the disc passing through a point on the circumference.

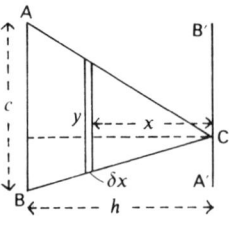

Fig. 34

7 A plane lamina is in the form of an isosceles right-angled triangle with equal sides each 30 cm long and mass 1 kg. Find the M.I. about (i) one of the equal sides, (ii) a perpendicular axis through the point of intersection of the equal sides.

8 What is the M.I. of a uniform circular wire, radius a, mass M, about an axis perpendicular to the plane through its centre? Deduce the M.I. about a diameter.

9 A uniform circular hoop has radius 1 m and mass 6 kg. Find its M.I. and radius of gyration about an axis tangential to it.

10 A flywheel consists of a uniform circular disc, radius 2 m, mass 100 kg, to which is attached a uniform circular rim, radius 2 m, mass 50 kg. Find (i) the M.I., (ii) the radius of gyration of the flywheel about the perpendicular axis through its centre.

11 A uniform rod AB, length 6 m, mass 5 kg, has a particle of mass 1 kg attached at the end B. What is the M.I. of this particle about a perpendicular axis through A? Deduce the radius of gyration of the rod and particle about this axis.

12 A uniform lamina is of the shape indicated in Fig. 35. If the mass of the lamina is 8 kg, find its M.I. about (i) axis Ox, (ii) axis Oy, (iii) the perpendicular axis through O.

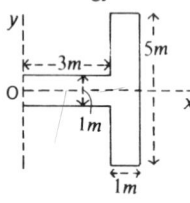

Fig. 35

13 Using the result for the M.I. of a triangle about one side, find the M.I. of a uniform square lamina side 4 m, mass 8 kg about a diagonal.

14 Fig. 36 represents a uniform circular cylinder, mass M, radius a. Show that the M.I. of the circular element about axis Ox is $\dfrac{\rho\pi a^4\,\delta x}{2}$, where ρ is the density and deduce that the M.I. of the cylinder about its axis is $\tfrac{1}{2}Ma^2$.

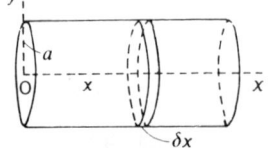

Fig. 36

[B] EXAMPLES 6d

Find the M.I.'s of the following areas about the x and y axes, assuming unit density:

1 The area bounded by the curve $y = 2x^2$, $y = 0$, $x = 1$, $x = 3$.

2 The area bounded by the curve $y^2 = x$ and $x = 2$.

3 The area bounded by the curve $x^2 = y^3$ and $y = 1$.

4 Find the radius of gyration of a uniform rectangular lamina with sides 3 m and 4 m about a diagonal.

5 A circular plate, radius 4 m, mass 40 kg, has a circular hole diameter 2 m punched in it, the circumference of the second circle passing through the centre of the first. Find the M.I. of the resulting figure about an axis perpendicular to the plate and passing through the centre of the original circle.

6 A clock pendulum consists of a uniform rod AB, length 1 m, mass 1 kg, to the end B of which is attached a flat circular disc, radius 10 cm, mass 3 kg, the end of the rod being fixed to a point on the circumference of the disc. Find the radius of gyration of the pendulum about an axis through A perpendicular to rod and disc.

7 Prove that the M.I. of a sphere about a diameter is $\frac{2}{5}Ma^2$, where a is the radius.

8 Find the radius of gyration of a uniform hollow sphere of internal and external radii, 6 cm and 8 cm, about a diameter.

9 Prove that the M.I. of a uniform right circular cone, mass M, base radius a about its axis of symmetry is $\dfrac{3Ma^2}{10}$.

10 Find the M.I. about the line $y = 0$ of the uniform solid, mass M, obtained by rotating the area enclosed by the curve $y^2 = x^3$ and the line $x = a$ about the x axis.

11 A straight rod AB, of length $2a$, has a line density $\rho = \rho_0 \left(1 + \sqrt{\dfrac{x}{a}}\right)$ where ρ_0 is constant and x is the distance from A. Find the radius of gyration of the rod about a perpendicular axis through A.

12 Prove that the M.I. of a uniform cubical block, mass M, side $2a$, about an axis through the centre of one face perpendicular to two other faces is $\frac{5}{3}Ma^2$.

13 Find the M.I. of a uniform circular cylinder radius a, height h, mass M about a diameter of one of the plane ends.

14 Find the M.I. of a uniform right circular cone, radius a, height h, mass M, about a diameter of the base.

15 The area bounded by the curve $y^2 = hx$ and the double ordinate $x = h$ is rotated about Ox. Find the radius of gyration of the solid of revolution about axis Oy.

16 Find the M.I. about a diameter of a uniform hollow sphere of internal and external radii b and a. By allowing $b \to a$, deduce the radius of gyration of a uniform spherical shell of radius a about a diameter and also about a tangent.

17 Find the M.I. of a uniform solid hemisphere radius a about a diameter of the base. Deduce the M.I. about an axis through the centre of gravity parallel to the base.

18 The density of a circular plate, radius a, varies according to the relationship $\rho = \rho_0 \left(2 - \dfrac{x}{a}\right)$, where x is the distance from the centre. Find the radii of gyration about a perpendicular axis through the centre and a diameter.

19 The density of a sphere radius a at a distance x from the centre is $\rho_0 \left(1 + \dfrac{cx}{a}\right)$ where ρ_0 and c are constants. Write down the M.I. of a spherical shell, radius x,

thickness δx, about a diameter and hence deduce that the square of the radius of gyration of the sphere about a diameter is $\dfrac{4(6+5c)}{15(4+3c)}a^2$.

20 Find the M.I. of a uniform rod AB, length $2a$, mass M, about an axis through A making an angle α with the rod.

7 | Solution of equations

Algebraic equations The methods of solution of the following types of simultaneous equations will be illustrated by worked examples:

(i) Linear equations in three unknowns;
(ii) Quadratic equations in two unknowns.

Linear equations The method of solution of simultaneous linear equations in three unknowns is an extension of that used to solve linear equations in two unknowns.

Example 1 Solve the equations,

$$2x - y + z = 4,$$
$$4x + 3y - 2z = -6,$$
$$6x + y - 3z = -2.$$

Number the equations (i), (ii) and (iii) respectively.
Adding (i) and (iii),

$$8x - 2z = 2,$$

or, $\qquad 4x - z = 1.\ \ldots\ldots\ldots\ldots\ldots\ldots\ldots\ldots$(iv)

Now take a second pair of the equations (i), (ii), (iii), and again eliminate y. Multiplying (i) by 3 and adding to (ii),

$$10x + z = 6.\ \ldots\ldots\ldots\ldots\ldots\ldots\ldots\ (v)$$

Adding (iv) and (v), $\qquad 14x = 7.$
i.e. $\qquad x = \frac{1}{2}.$
By substitution, $\qquad y = -2.$
$\qquad z = 1.$

Quadratic equations Two types of simultaneous quadratic equations will be considered,

(i) where one equation is solvable for one unknown in terms of the other,
(ii) where the equations are homogeneous.

Example 2 Find the points of intersection of the straight line $2x + 3y = 1$ and the curve $3x^2 + 7xy + 4y^2 = 0$.

The coordinates of the common points are the roots of the equations

$$3x^2 + 7xy + 4y^2 = 0,\ \ldots\ldots\ldots\ldots\ldots\ldots\ldots\ldots\ldots$(i)$$
$$2x + 3y = 1.\ \ldots\ldots\ldots\ldots\ldots\ldots\ldots\ldots\ldots$(ii)$$

From (ii), $\qquad x = \dfrac{1 - 3y}{2}.$

Substituting in (i),

$$3\left(\frac{1-3y}{2}\right)^2 + 7y\left(\frac{1-3y}{2}\right) + 4y^2 = 0,$$

$$3\left(\frac{1-6y+9y^2}{4}\right) + 7y\left(\frac{1-3y}{2}\right) + 4y^2 = 0,$$

giving
$$y^2 - 4y + 3 = 0,$$
$$(y-1)(y-3) = 0,$$
$$y = 1, \quad 3.$$

Hence,
$$x = -1, \quad -4.$$

Therefore, the straight line and the curve intersect at the points $(-1, 1)$ $(-4, 3)$.

Example 3 Solve the equations

$$\frac{x^2}{y} + \frac{y^2}{x} = 9, \quad x + y = 6.$$

Multiplying the first equation by xy,

$$x^3 + y^3 = 9xy.$$

This cubic equation reduces to a quadratic because

$$x^3 + y^3 = (x+y)(x^2 - xy + y^2) = 6(x^2 - xy + y^2),$$

using the second equation.

$$\therefore 6(x^2 - xy + y^2) = 9xy$$

or
$$2x^2 - 5xy + 2y^2 = 0; \dots\dots\dots\dots\dots\dots\dots\dots \text{(i)}$$

and
$$x + y = 6 \dots\dots\dots\dots\dots\dots\dots\dots\dots\text{(ii)}$$

Substituting $x = 6 - y$ in (i),

$$2(6-y)^2 - 5y(6-y) + 2y^2 = 0,$$
$$9y^2 - 54y + 72 = 0,$$
$$y^2 - 6y + 8 = 0,$$
$$(y-4)(y-2) = 0.$$

hence
$$\left.\begin{array}{l} y = 2, 4. \\ x = 4, 2. \end{array}\right\}$$

Example 4 Solve the equations:

$$x(x + 2y) = 4x - y, \dots\dots\dots\dots\dots\dots\dots\dots \text{(i)}$$

$$y(2x - y) = 6x - 5y \dots\dots\dots\dots\dots\dots\dots\dots \text{(ii)}$$

From (i),
$$2xy + y = 4x - x^2,$$

$$y = \frac{x(4-x)}{2x+1} \dots\dots\dots\dots\dots\dots\dots\dots\dots \text{(iii)}$$

Substituting in (ii),

$$\frac{x(4-x)}{2x+1}\left\{\frac{2x(2x+1) - x(4-x)}{2x+1}\right\} = 6x - \frac{5x(4-x)}{2x+1},$$

$$x(4-x)(5x^2 - 2x) = 6x(4x^2 + 4x + 1) - 5x(4-x)(2x+1).$$

Expanding and rearranging,
$$5x^4 + 12x^3 - 3x^2 - 14x = 0,$$
$$x(5x^3 + 12x^2 - 3x - 14) = 0.$$
One root is $x = 0$ and the other roots are the solutions of
$$5x^3 + 12x^2 - 3x - 14 = 0. \quad\quad\quad\quad\quad\text{(iv)}$$
The general solution of cubic equations is outside the scope of this book but in simple cases it is possible to spot an obvious integral solution.

Clearly (iv) is satisfied by $x = 1$.

So $(x-1)$ is a factor of the cubic function and long division gives the remaining factor, $5x^2 + 17x + 14$.
$$\therefore (x-1)(5x^2 + 17x + 14) = 0,$$
$$(x-1)(5x+7)(x+2) = 0.$$
i.e. $\quad\quad\quad\quad x = 1, \ -\tfrac{7}{5}, \ -2 \text{ and } 0.$

By substitution, y is obtained and the complete solution is

$x = 0,$	$x = 1,$	$x = -2,$	$x = -\tfrac{7}{5},$
$y = 0.$	$y = 1.$	$y = 4.$	$y = \tfrac{21}{5}.$

Homogeneous quadratic equations A homogeneous quadratic equation is one in which all the variable terms are of the second degree.

e.g. $\quad\quad\quad 3x^2 - 2xy = 6 \quad$ is a homogeneous equation,

$\quad\quad\quad\quad 4x^2 - y^2 = 3x - 2 \quad$ is not a homogeneous equation.

Homogeneous equations can be solved by substituting $y = mx$.

Example 5 Solve the equations
$$2x^2 - y^2 = 7,$$
$$2y^2 - xy = 4.$$
Substitute $y = mx$, then
$$2x^2 - m^2x^2 = 7, \quad\quad\quad\quad\quad\text{(i)}$$
$$2m^2x^2 - mx^2 = 4. \quad\quad\quad\quad\quad\text{(ii)}$$
Dividing,
$$\frac{x^2(2-m^2)}{x^2(2m^2-m)} = \frac{7}{4},$$
$$4(2-m^2) = 7(2m^2 - m),$$
$$18m^2 - 7m - 8 = 0,$$
$$(9m-8)(2m+1) = 0,$$
$$\therefore m = \tfrac{8}{9}, \ -\tfrac{1}{2}.$$
Substituting in (i),
when $m = \tfrac{8}{9}$, $\quad\quad\quad\quad x^2(2 - \tfrac{64}{81}) = 7,$
$$x^2 = \tfrac{81}{14}.$$
$$x = \pm \frac{9}{\sqrt{14}}.$$

98 | Solution of equations

Similarly, the value $m = -\frac{1}{2}$, gives $\quad x = \pm 2$.

The relationship $y = mx$ gives the values of y and we get the roots

$$x = 2, \quad x = -2, \quad x = \frac{9}{\sqrt{14}}, \quad x = -\frac{9}{\sqrt{14}},$$

$$y = -1. \quad y = 1. \quad y = \frac{8}{\sqrt{14}}. \quad y = -\frac{8}{\sqrt{14}}.$$

[A] EXAMPLES 7a

Solve the following equations:

1 $5x - y = 3$, $\quad$ **2** $y - 2x = 0$, $\qquad\qquad$ **3** $4x - 3y = 1$,
$\quad\;\; y^2 - 6x^2 = 25$. $\quad\;\; 3x^2 - xy + y^2 - 4x + 3y = 7$. $\quad\;\; 12xy + 13y^2 = 25$.

4 $2x + 3y = 5$, $\quad$ **5** $p + 2q - r = 3$, $\quad$ **6** $2x - 5y - z = -4$,
$\quad\;\; y - 2z = 1$, $\qquad\;\; 3p - q + 2r = 0$, $\qquad\;\; x + 3y + 2z = 5$,
$\quad\;\; 4z + x = 2$. $\qquad\;\; 2p + 3q + r = 7$. $\qquad\;\; 3x - y - z = -10$.

7 $2xy - x^2 = 24$, $\quad$ **8** $x^2 + 2xy = 3$, $\quad$ **9** $xy + y^2 - 4 = 0$,
$\quad\;\; 2y^2 - xy = 30$. $\qquad\;\; y^2 - xy = 4$. $\qquad\;\; x^2 - xy + 2y^2 - 8 = 0$.

10 $\dfrac{1}{x} + \dfrac{1}{y} + \dfrac{1}{z} = 10$, $\;\; \dfrac{2}{x} - \dfrac{1}{y} + \dfrac{1}{z} = 2$, $\;\; \dfrac{3}{x} + \dfrac{2}{y} - \dfrac{3}{z} = 7$. $\quad$ **11** $2x^4 - 3x^2 + 1 = 0$.

12 $5m^2 + \dfrac{3}{m^2} = 16$. $\quad$ **13** $(x + 2y)(2x + y) - 20 = y^2 + 4x(x + y) - 16 = 0$.

14 $\dfrac{1}{x} - \dfrac{1}{y} = \dfrac{1}{x + 1} - \dfrac{1}{y + 2} = \dfrac{1}{12}$.

[B] EXAMPLES 7b

Solve the equations 1–8:

1 $2x^2 - xy - 3y^2 + x + 4y = 9$, $\qquad$ **2** $x^2 + y^2 = 2(a^2 + b^2)$,
$\quad\;\; 2x - 3y = 5$. $\qquad\qquad\qquad\qquad\quad\;\; ax + by = a^2 + b^2$.

3 $3x^2 - xy - y^2 = 9$, $\qquad\qquad\qquad$ **4** $x + y = 7 + \sqrt{xy}$,
$\quad\;\; 5x^2 - xy = 18$. $\qquad\qquad\qquad\qquad\;\; x^2 + y^2 + xy = 133$.

5 $x^3 + 8y^3 = 117$, $\qquad\qquad\qquad\qquad$ **6** $\dfrac{x^2}{y} + \dfrac{y^2}{x} = \dfrac{9}{2}$,
$\quad\;\; x + 2y = 3$. $\qquad\qquad\qquad\qquad\qquad\;\; x + y = 3$.

7 $x(2x - y) = x + 2y$, $\qquad\qquad\qquad$ **8** $x(3x - y) = y$,
$\quad\;\; 2x^2 - 5xy - 3y^2 = x - 3y$. $\qquad\qquad 8x(x - 1) + y(y + 4) = 6xy$.

9 By writing $y = x^2 + 6x$, solve the equation $x^2 + 6x + 2\sqrt{x^2 + 6x} = 24$.

10 Obtain one root of the following equations by inspection and hence complete their solutions:

(i) $x(ax + 1) - x(a^2 + ab) - a - b = 0$; (ii) $(1 - a^2)(x + a) - 2a(1 - x^2) = 0$.

11 Use the equations $3x - y + 4z = 3$, $2x + 3y = 4$ to express y and z in terms of x and hence solve these equations simultaneously with the equation $x^2 - xy + yz = 7$.

12 By dividing by x^2 and writing $z = x - \dfrac{1}{x}$, solve the equation

$$6x^4 - 25x^3 - 12x^2 + 25x + 6 = 0.$$

Trigonometrical equations A trigonometrical equation differs from an algebraic equation in that it has an infinite number of roots. For instance, the roots of the equation $\sin\theta = 1$ are $90°, 450°, 810°$, etc., or, in radians, $\frac{\pi}{2}, \frac{5\pi}{2}, \frac{9\pi}{2}$, etc.

No general rules can be given for the solution of trigonometrical equations. Certain types listed below occur very frequently; the methods of solving these types will be illustrated by means of worked examples.

Type (i)—Equations of a quadratic form.
Type (ii)—Equations of the form $a\sin x + b\cos x = c$, where a, b and c are constants.
Type (iii)—Equations solvable by the method of factors.

Before discussing the solutions of these types of equations, it is of value to obtain expressions for all angles which have the same sine, cosine or tangent.

Angles with the same sine Suppose $\sin\theta = k$, where $-1 \leqslant k \leqslant 1$.
There will always be one solution between $-\frac{\pi}{2}$ and $+\frac{\pi}{2}$.

Let $\theta = \alpha$ represent this solution.

Figs. 37 a, b illustrate the graphical representation of α in the cases when k is positive and negative. In both cases, angles defined by the two positions, OP, OP'

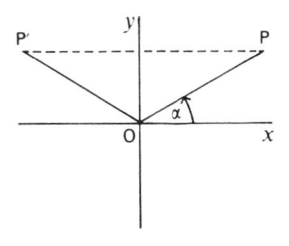

Fig. 37a (k positive) 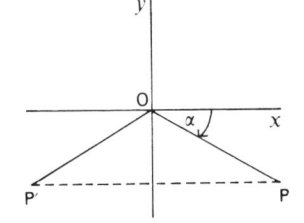   **Fig. 37b** (k negative)

of the rotating radius vector will have the same sine as α, i.e. k. Thus the solutions of the equation are

$$\theta = \alpha,\ \pi - \alpha,\ 2\pi + \alpha,\ 3\pi - \alpha,\ 4\pi + \alpha,\ \text{etc.}$$
$$-\pi - \alpha,\ -2\pi + \alpha,\ -3\pi - \alpha,\ -4\pi + \alpha,\ \text{etc.}$$

i.e. $\qquad\theta = n\pi + (-1)^n\alpha,\ \textbf{where}\ n\ \textbf{is any integer.}$

Angles with the same cosine Suppose $\cos\theta = k$ where $-1 \leqslant k \leqslant 1$. There will always be one solution of this equation between 0 and π.

Proceeding as before, let this value of θ be α.

Using Figs. 38 a, b which illustrate the graphical representation of α in cases where k is positive and negative and noting that all angles defined by the two

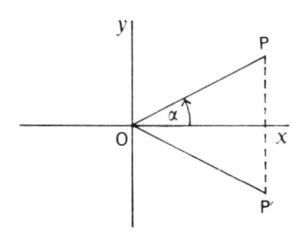

100 | Solution of equations

Fig. 38a (*k* positive) 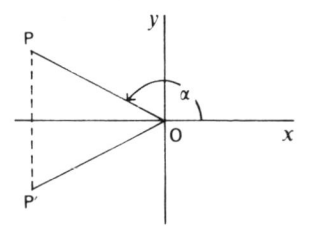   **Fig. 38b** (*k* negative)

positions OP, OP' of the radius vector have the same cosine as α, we have

$$\theta = \alpha, \quad 2\pi - \alpha, \quad 2\pi + \alpha, \quad 4\pi - \alpha, \quad 4\pi + \alpha, \text{ etc.}$$
$$-\alpha, \quad -2\pi + \alpha, \quad -2\pi - \alpha, \quad -4\pi + \alpha, \quad -4\pi - \alpha, \text{ etc.}$$

i.e. $\quad \theta = 2n\pi \pm \alpha$, where n is any integer.

Angles with the same tangent Suppose $\tan\theta = k$, where $-\infty < k < \infty$. Let $\theta = \alpha$ be the solution of this equation between $-\dfrac{\pi}{2}$ and $\dfrac{\pi}{2}$.

Using Figs. 39 *a, b* which illustrate the graphical representation of α in cases where k is positive and negative and noting that all angles defined by the two

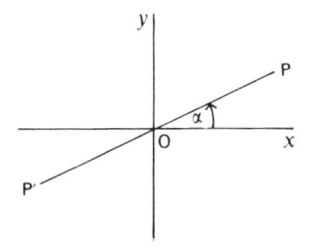

Fig. 39a (*k* positive) 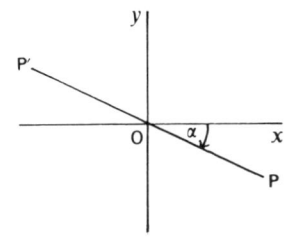   **Fig. 39b** (*k* negative)

positions OP, OP' of the radius vector have the same tangent as α, we have

$$\theta = \alpha, \alpha + \pi, \alpha + 2\pi, \alpha + 3\pi, \alpha + 4\pi, \text{ etc.}$$
$$\alpha - \pi, \alpha - 2\pi, \alpha - 3\pi, \alpha - 4\pi, \text{ etc.}$$

i.e. $\quad \theta = n\pi + \alpha$, where n is any integer.

Example 6 Write down the general solutions of the following equations:

(i) $\sin\theta = \tfrac{1}{2}$; (ii) $\cos\theta = -\dfrac{\sqrt{3}}{2}$; (iii) $\tan\theta = -1$.

(i) $\sin\theta = \tfrac{1}{2}$. The value of θ between $-\dfrac{\pi}{2}$ and $+\dfrac{\pi}{2}$ is $\dfrac{\pi}{6}$, i.e. $\alpha = \dfrac{\pi}{6}$.

∴ General solution is $\theta = n\pi + (-1)^n \dfrac{\pi}{6}$.

(ii) $\cos\theta = -\dfrac{\sqrt{3}}{2}$. The value of θ between 0 and π is $\dfrac{5\pi}{6}$, i.e. $\alpha = \dfrac{5\pi}{6}$.

$\therefore$ General solution is $\theta = 2n\pi \pm \dfrac{5\pi}{6}$.

(iii) $\tan\theta = -1$. The value of θ between $-\dfrac{\pi}{2}$ and $+\dfrac{\pi}{2}$ is $-\dfrac{\pi}{4}$, i.e. $\alpha = -\dfrac{\pi}{4}$.

$\therefore$ General solution is $\theta = n\pi - \dfrac{\pi}{4}$.

Example 7 Find the general solution of the equation $2\cos\dfrac{\theta}{2} = 1$.

Here $\cos\dfrac{\theta}{2} = \dfrac{1}{2}$ and hence, the first solution, α, is $\dfrac{\pi}{3}$. General solution is

$$\dfrac{\theta}{2} = 2n\pi \pm \dfrac{\pi}{3}.$$

i.e. $$\theta = 4n\pi \pm \dfrac{2\pi}{3}.$$

Note It is a *common error* to write down the first value of θ, $\dfrac{2\pi}{3}$, and then give the general solution as $\theta = 2n\pi \pm \dfrac{2\pi}{3}$.

Example 8 Find all values of x between 0 and 2π for which $\tan 4x = \tan x$.

As the angles $4x$ and x have the same tangent,

$$4x = n\pi + x,$$

where n is any integer.

i.e. $$3x = n\pi,$$

$$x = \dfrac{n\pi}{3}.$$

This is the general solution of the equation and to obtain the particular solutions between 0 and 2π, we take $n = 0, 1, 2, 3, 4, 5, 6$, and obtain

$$x = 0, \dfrac{\pi}{3}, \dfrac{2\pi}{3}, \pi, \dfrac{4\pi}{3}, \dfrac{5\pi}{3}, 2\pi.$$

[A] EXAMPLES 7c

Write down the general solutions of the following equations:

1. $\sin\theta = \dfrac{\sqrt{3}}{2}$.
2. $\cos\theta = \tfrac{1}{2}$.
3. $\tan\theta = 1$.
4. $\sin\theta = -\dfrac{\sqrt{2}}{2}$.
5. $\cos\theta = -\dfrac{\sqrt{2}}{2}$.
6. $\tan\theta = -\sqrt{3}$.
7. $\operatorname{cosec}\theta = -2$.
8. $\sec\theta = \sqrt{2}$.
9. $\cot\theta = -\sqrt{3}$.
10. $\sin\dfrac{\theta}{2} = 0{\cdot}3827$.
11. $\cos 2\theta = \dfrac{\sqrt{3}}{2}$.
12. $\cos\dfrac{2\theta}{3} = 0{\cdot}9659$.

13 $\tan 3\theta = -1$. **14** $\cot \dfrac{\theta}{2} = 0.4142$. **15** $\sin^2 \theta = 1$.

16 $2\cos^2 \dfrac{\theta}{2} = 1$. **17** $3\cot^2 \theta - 1 = 0$.

18 $(2\sin \theta - 1)(2\sin \theta + 1) = 0$. **19** $\tan \dfrac{\theta}{2}\left(\tan \dfrac{\theta}{2} + 1\right) = 0$.

20 $(3\sin \theta - 2)(2\sin \theta + 1) = 0$. **21** $\left(4\tan \dfrac{\theta}{2} - 3\right)\left(3\tan \dfrac{\theta}{2} + 1\right) = 0$.

Find the general solutions of the following equations and deduce those between 0 and 2π:

22 $\sin 2\theta = \sin \theta$. **23** $\cos 3\theta = \cos 2\theta$. **24** $\tan 3\theta = \tan \theta$.

25 $\sin 5\theta - \sin 3\theta = 0$. **26** $\sec 4\theta = \sec \theta$. **27** $\tan \theta - \tan \dfrac{\theta}{2} = 0$.

28 $\cot \dfrac{5\theta}{2} = \cot \dfrac{\theta}{2}$. **29** $\sin 4\theta = \cos \theta \left\{\text{write } \cos \theta = \sin\left(\dfrac{\pi}{2} - \theta\right)\right\}$.

30 $\cos 3\theta = \sin 2\theta$. **31** $\tan 3\theta = \cot \theta$. **32** $\tan \dfrac{3\theta}{2} - \cot \dfrac{\theta}{2} = 0$.

Find the solutions of the following equations lying between $\pm \pi$:

33 $\sin 3\theta = \sin \dfrac{\pi}{10}$. **34** $\cos \dfrac{\theta}{2} = \sin \dfrac{\pi}{5}$. **35** $\tan 2\theta = \tan \dfrac{2\pi}{3}$.

36 $\sec 2\theta = \operatorname{cosec} \dfrac{\pi}{3}$. **37** $\sin 7\theta = \sin 3\theta$. **38** $\tan 5\theta = \cot 2\theta$.

Equations of a quadratic form These are trigonometrical equations which can be reduced to quadratic equations in a single ratio. In dealing with this type of equation it is essential to be familiar with the following fundamental identities:

$$\sin^2 \theta + \cos^2 \theta = 1.$$
$$\sec^2 \theta = 1 + \tan^2 \theta.$$
$$\operatorname{cosec}^2 \theta = 1 + \cot^2 \theta.$$
$$\cos 2\theta = 2\cos^2 \theta - 1 = 1 - 2\sin^2 \theta.$$

Example 9 Find the general solution of the equation $6\sin^2 x + 5\cos x = 7$, where x is measured in degrees.

Using the relation $\sin^2 x = 1 - \cos^2 x$, this equation can be transformed into a quadratic equation in $\cos x$.

$$6\sin^2 x + 5\cos x = 7.$$
$$\therefore \ 6(1 - \cos^2 x) + 5\cos x = 7.$$
$$6\cos^2 x - 5\cos x + 1 = 0.$$

Factorizing, $\qquad (3\cos x - 1)(2\cos x - 1) = 0.$

i.e. $\qquad \cos x = \tfrac{1}{3}$ and $\cos x = \tfrac{1}{2}$.

When $\cos x = \tfrac{1}{3}$, the first value of $x = 70° \, 32'$.

General solution is $x = 360° n \pm 70° \, 32'$.

When $x = \frac{1}{2}$, the first value of x is $60°$.

General solution is $x = 360°n \pm 60°$.

i.e. the general solution is
$$x = 360°n \pm 70° 32'$$
and $$x = 360°n \pm 60°$$

Equations of the type $a \sin x + b \cos x = c$ where a, b, c are constants

A method of solving this type of equation by expressing

$$a \sin x + b \cos x$$

as a single sine or cosine has been discussed in Chapter 14 of the previous volume. We will revise this method and also introduce an alternative method of solution.

Example 10 Solve the equation $3 \sin x + 2 \cos x = 1$ for values of x between $0°$ and $360°$.

Let $\qquad 3 \sin x + 2 \cos x \equiv R \cos(x - \alpha)$ where α is acute.

i.e. $\qquad 3 \sin x + 2 \cos x \equiv R \cos \alpha \cos x + R \sin \alpha \sin x$.

As this relationship is true for all values of x, the coefficients of $\sin x$ and $\cos x$ must be identical.

Hence, $\qquad R \cos \alpha = 2$ and $R \sin \alpha = 3$.

Squaring and adding, $\qquad R^2 = 13; \qquad R = \sqrt{13}$.

Dividing, $\qquad \tan \alpha = \frac{3}{2}; \qquad \alpha = 56° 18'$.

$$\therefore \sqrt{13} \cos(x - \alpha) = 1,$$

$$\cos(x - \alpha) = \frac{1}{\sqrt{13}} = 0.2773.$$

The first value of the angle $(x - \alpha)$ is $73° 54'$.

The general solution is $x - \alpha = 360°n \pm 73° 54'$.

i.e. $\qquad x = 360°n \pm 73° 54' + 56° 18'$.

The solutions between $0°$ and $360°$ are obtained by taking $n = 0$ and $n = 1$, giving
$$x = 73° 54' + 56° 18' = 130° 12'$$
$$x = 360° - 73° 54' + 56° 18' = 342° 24'$$

Note The expression $3 \sin x + 2 \cos x$ could equally well have been expressed in the form $R \sin(x + \beta)$.

Now consider an alternative method of solving this type of equation.

Example 11 Find the general solution of the equation $20 \cos x - 25 \sin x = 1$.

Let $\tan \frac{x}{2} = t$. Then $\tan x = \frac{2t}{1 - t^2} \left(\text{using } \tan 2A = \frac{2 \tan A}{1 - \tan^2 A} \right)$.

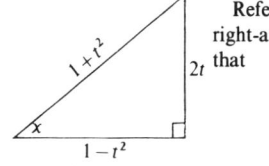

Referring to Fig. 40 and noting that the hypotenuse of a right-angled triangle with sides $2t$ and $1 - t^2$ is $1 + t^2$, it follows that
$$\sin x = \frac{2t}{1 + t^2}, \qquad \cos x = \frac{1 - t^2}{1 + t^2}.$$

Fig. 40

104 | **Solution of equations**

These results are of considerable importance and should be remembered.
Substituting for sin x and cos x in terms of t, the given equation becomes

$$20\frac{(1-t^2)}{1+t^2} - 25\frac{2t}{1+t^2} = 1.$$

i.e. $\qquad 20(1-t^2) - 50t = 1 + t^2,$

$\qquad\qquad 0 = 21t^2 + 50t - 19.$

Factorizing, $\qquad (7t+19)(3t-1) = 0,$

$\qquad\qquad t = -\frac{19}{7} = -2\cdot7143.$

and $\qquad\qquad t = \frac{1}{3} = 0\cdot3333.$

So $\qquad \tan\frac{x}{2} = -2\cdot7143;\quad$ i.e. $\frac{x}{2} = n\pi - 1\cdot218$ rad.,

and $\qquad \tan\frac{x}{2} = 0\cdot3333;\quad$ i.e. $\frac{x}{2} = n\pi + 0\cdot3217$ rad.

∴ The general solution of the equation is $\left.\begin{array}{l} x = 2n\pi - 2\cdot436 \\ x = 2n\pi + 0\cdot6434 \end{array}\right\}$ rad.

There is little to choose between the alternative methods of solution from the point of view of simplicity when the constants a, b, c are integers. However, when all or some of these constants are not integral, the first method of solution will usually be simpler.

Equations solvable by the method of factors

Example 12 Solve the equation $\cos x - \cos 4x = \cos 2x - \cos 3x$ for values of x lying between $-\pi$ and π.

$$\cos x - \cos 4x = \cos 2x - \cos 3x.$$

Rearranging, $\qquad \cos x + \cos 3x = \cos 2x + \cos 4x.$

Factorizing, $\qquad 2\cos\frac{3x+x}{2}\cos\frac{3x-x}{2} = 2\cos\frac{4x+2x}{2}\cos\frac{4x-2x}{2},$

$$\cos 2x \cos x = \cos 3x \cos x.$$

∴ $\cos x\{\cos 2x - \cos 3x\} = 0.$

i.e. $\qquad \cos x\left\{2\sin\frac{3x+2x}{2}\sin\frac{3x-2x}{2}\right\} = 0,$

$$\cos x \sin\frac{5x}{2}\sin\frac{x}{2} = 0.$$

∴ $\cos x = 0,\ \sin\frac{5x}{2} = 0$ or $\sin\frac{x}{2} = 0.$

$\cos x = 0$ gives $x = 2n\pi \pm \frac{\pi}{2}.$

$\sin\dfrac{5x}{2} = 0$ gives $\dfrac{5x}{2} = n\pi;\ x = \dfrac{2n\pi}{5}.$

$\sin\dfrac{x}{2} = 0$ gives $\dfrac{x}{2} = n\pi;\ x = 2n\pi.$

Hence, the roots between $-\pi$ and π are

$$x = -\dfrac{4\pi}{5},\ -\dfrac{\pi}{2},\ -\dfrac{2\pi}{5},\ 0,\ \dfrac{2\pi}{5},\ \dfrac{\pi}{2},\ \dfrac{4\pi}{5}.$$

Example 13 Find the general solution of the equation $\sin 4\theta - \sin 2\theta = \cos 3\theta$, where θ is in radians.

$$\sin 4\theta - \sin 2\theta = \cos 3\theta.$$

Factorizing, $\quad 2\cos\dfrac{4\theta + 2\theta}{2} \sin\dfrac{4\theta - 2\theta}{2} = \cos 3\theta,$

$$2\cos 3\theta \sin\theta = \cos 3\theta,$$

$$\cos 3\theta (2\sin\theta - 1) = 0.$$

$\therefore\ \cos 3\theta = 0\ $ or $\ \sin\theta = \tfrac{1}{2}.$

$\cos 3\theta = 0$ gives $3\theta = 2n\pi \pm \dfrac{\pi}{2};\quad \theta = \dfrac{2n\pi}{3} \pm \dfrac{\pi}{6}.$

$\sin\theta = \tfrac{1}{2}$ gives $\theta = n\pi + (-1)^n \dfrac{\pi}{6}.$

The general solution is $\quad \theta = \dfrac{2n\pi}{3} \pm \dfrac{\pi}{6}.$

$\left.\begin{array}{l} \theta = \dfrac{2n\pi}{3} \pm \dfrac{\pi}{6}. \\[6pt] \theta = n\pi + (-1)^n\dfrac{\pi}{6}. \end{array}\right\}$ rad.

[B] EXAMPLES 7d

Solve the following equations, giving all roots between $\pm 180°$:

1. $15\sin^2\theta + 2\cos\theta = 14.$
2. $3\sec^2\theta - 5\tan\theta = 5.$
3. $\tan\theta + 2\cot\theta = 3.$
4. $3\cos x - 4\sin x = 4.$
5. $3\sin 2\theta = 2\cos\theta.$
6. $\sin 3x + \sin x = \sin 2x.$
7. $\tan 2\theta = 3\tan\theta.$
8. $12\sin x + 5\cos x = 7.$
9. $4\cot\theta + \operatorname{cosec}\theta = 3.$
10. $3\sin x = 2\sin(60° - x).$
11. $\tan\theta + \tan(\theta + 45°) = 1.$
12. $2\cos 2x = \cos x - \sin x.$

Express in radians, the general solutions of the following equations:

13. $\cos 2\theta = \cos\theta - 1.$
14. $\tan\theta = \operatorname{cosec} 2\theta.$
15. $\sin 7\theta - \sin\theta = \sin 3\theta.$
16. $6\cos\theta - 3\sin\theta + 4 = 0.$
17. $2\sin^2\theta + \sin^2 2\theta = 2.$
18. $4\sin\theta \sin 3\theta = 1.$
19. $\tan 3\theta = \cot\theta.$
20. $\cos\theta + \cos 2\theta + \cos 3\theta = 0.$
21. $\sin\theta + \sqrt{3}\cos\theta = 1.$
22. $\sec\theta = \sec 2\theta.$
23. $4\sin 2\theta - 3\cos 2\theta = 3.$
24. $\cos 2\theta + \sin\theta = 0.8.$
25. $\sin 5\theta + \cos 3\theta = 0.$
26. $\cos 3\theta + 2\cos 5\theta + \cos 7\theta = 0.$
27. $3\sin 2\theta = 1 + 2\cos 2\theta.$

106 | Solution of equations

28 Prove that $\sin 3\theta = 3\sin\theta - 4\sin^3\theta$. By replacing $\sin\theta$ by x and solving the equation $\sin 3\theta = \frac{1}{2}$ for θ between $0°$ and $360°$, find the roots of the equation $3x - 4x^3 = \frac{1}{2}$.

29 Solve the equation $\cos 3\theta = 0.4$ for θ between $0°$ and $360°$, and hence find the roots of the equation $4x^3 - 3x = 0.4$.

30 Solve the equation $\sin 3\theta = \cos 2\theta$ for θ between 0 and 2π. By expressing $\sin 3\theta$ and $\cos 2\theta$ in terms of $\sin\theta$, deduce the roots of the equation $4x^3 - 2x^2 - 3x + 1 = 0$.

31 Use the equation $\cos 3\theta = \cos 2\theta$ to prove that $\cos\dfrac{2\pi}{5} = \frac{1}{4}(\sqrt{5} - 1)$ and $\cos\dfrac{\pi}{5} = \frac{1}{4}(\sqrt{5} + 1)$.

Graphical solution of equations The approximate solution of equations by a graphical method has been discussed in the previous volume. The method can be briefly summarised as follows:

Suppose we have an equation $f(x) = 0$, where $f(x)$ is any function of the variable x.

The approximate solutions of the equation can be obtained by plotting the graph of the function $f(x)$ and reading off the values of x where the curve cuts the x axis.

Usually it is more convenient to split $f(x)$ into a difference of two simpler functions, i.e. $p(x) - q(x)$ and write the equation as $p(x) = q(x)$.

The graphical solution is readily obtained by plotting graphs of the two functions $p(x)$ and $q(x)$ and noting the values of x at the points of intersection.

For example, suppose approximate solutions of the equation

$$2x^3 - 3x + 4 = 0$$

are required.

The simplest solution is obtained by writing the equation in the form $2x^3 = 3x - 4$ and finding the points of intersection of the graphs $y = 2x^3$ and $y = 3x - 4$.

Example 14 Find the acute angle which satisfies the equation $2 - \dfrac{x}{30} = \tan x$, where x is measured in degrees, giving the result correct to within $5'$.

An approximate value of the required solution of the equation

$$2 - \frac{x}{30} = \tan x$$

is obtained by plotting the two graphs $y = 2 - \dfrac{x}{30}$ and $y = \tan x$ on the same diagram for x between $0°$ and $60°$ (Fig. 41).

We see that $x = 37°$ is an approximation to the root of the equation.

To obtain a closer approximation, we proceed as follows:

When $x = 37°$,

$$2 - \frac{x}{30} = 0.767$$

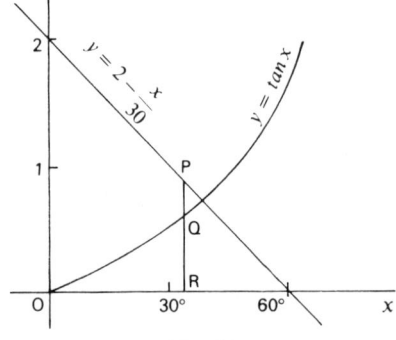

Fig. 41

and $\tan x = 0.754$;

i.e. $$\left(2 - \frac{x}{30}\right) > \tan x.$$

So in Fig. 41, if $OR = 37°$, $PR > RQ$.
Thus the required root is greater than $37°$.

When $x = 38°$, $2 - \frac{x}{30} = 0.733$ and $\tan x = 0.767$;

i.e. $$\left(2 - \frac{x}{30}\right) < \tan x.$$

Consequently, $x = 38°$ lies to the right of the point of intersection of the graphs. Thus the required root is less than $38°$.
From the above it is seen that $37°$ < Required root < $38°$.
Now consider values of x between $37°$ and $38°$. To aid in choosing a value close to the actual root, note that the difference between $2 - \frac{x}{30}$ and $\tan x$ is considerably smaller when $x = 37°$ than when $x = 38°$. Hence, the root is nearer $37°$ than $38°$ and the value $x = 37°\,20'$ will be taken.

When $x = 37°\,20'$, $2 - \frac{x}{30} = 0.756$ and $\tan x = 0.762$;

i.e. $$2 - \frac{x}{30} < \tan x.$$

Thus the required root is less than $37°\,20'$.

When $x = 37°\,10'$, $2 - \frac{x}{30} = 0.761$ and $\tan x = 0.758$;

i.e. $$2 - \frac{x}{30} > \tan x.$$

Thus the required root is greater than $37°\,10'$.

i.e. $37°\,10'$ < Required root < $37°\,20'$.

So, to within 5 min., the required root = $37°\,15'$.
Repeated application of this procedure would give the root to any required degree of accuracy.

[B] EXAMPLES 7e

1 Draw the graph of $y = x^3$ for values of x between -3 and $+3$. Hence obtain approximate solutions of the equations:
    (i) $x^3 + 2 = 0$.   (ii) $x^3 = 2x + 5$.   (iii) $2x^3 - 15x + 9 = 0$.

2 Plot the graph of the function e^x for values of x between -4 and $+2$. Obtain the approximate solutions in this range of the equation $2e^x = 2x + 3$.

3 Plot the graph of $y = \lg x$ for values of x between 0.1 and 5 and deduce approximate solutions of the equation $6 \lg x = 2x - 3$.

4 Plot the graph of $y = \cos 2\theta$ for values of θ between 0 and $\frac{\pi}{4}$. Use the graph to find approximate solutions of the equations:
    (i) $\cos 2\theta = \theta$,   (ii) $2\cos^2 \theta = \theta + 0.8$,
θ being measured in radians.

5 By drawing the graphs of $y = 2^x$ and $y = \dfrac{4}{x}$ for values of x between 0·5 and 2, find the approximate solution of the equation $x2^x = 4$.

6 Given that $\cosh x = \tfrac{1}{2}(e^x + e^{-x})$, by drawing the graph of the function for values of x between -2 and 2, find approximate solutions of the following equations:

(i) $\cosh x = 3$,   (ii) $2\cosh x = 3 - 2x$.

7 Show that the equation $2x^3 + x - 5 = 0$ has only one real root and find its value correct to 2 places of decimals.

8 Find from graphical considerations, the number of real roots of the equation $\sin x = 1 - \dfrac{x}{2\pi}$.

9 Solve graphically the equation $x^2 = 4(1 - \sin x)$ for positive values of x, x being measured in radians.

10 Solve the equation $\cos x = \dfrac{x}{80}$, where x is positive and measured in degrees, giving the root accurate to 6'.

11 Show graphically that the equation $\tfrac{1}{2}\ln(1+x) = x - 2$ has two roots and determine the positive root correct to 2 significant figures.

12 Show that the equation $x = 2\sin x$, x measured in radians, has one positive root other than zero and find it correct to 2 decimal places.

13 Find the acute angle θ in radians correct to 2 decimal places, for which
$$\theta \tan \theta = 1.$$

14 Find the positive root of the equation $x^3 + 2x^2 - 5x - 7 = 0$, correct to 1 place of decimals.

15 Solve graphically, the equation $e^x - e^{-x} = 1$.

Approximate solution of equations. Newton's method

Newton's method for the approximate solution of an equation is dependent on some method of obtaining rough approximations to the values of the roots (e.g. a graphical method).

Example 15 Verify that $x = -0·5$ is an approximate root of the equation $x^3 - 4x - 2 = 0$ and obtain the value of the root correct to two decimal places.

A rough sketch of the graphs of $y = x^3$, $y = 4x + 2$, shows that the equation $f(x) = x^3 - 4x - 2 = 0$ has a root between $x = -1$ and $x = 0$. As $f(-0·5) = -0·125$, a small quantity, the root is approximately $x = -0·5$.

Suppose the accurate root is $x = -0·5 + h$ where h is small. Then
$$(h - 0·5)^3 - 4(h - 0·5) - 2 = 0.$$

As h is small, ignore powers above the first.

$$\therefore \quad -0·125 + 0·75h - 4h + 2 - 2 = 0 \quad \text{approximately.}$$

$$h = -\dfrac{0·125}{3·25} = -0·038.$$

So a closer approximation to the root is
$$x = -0·5 - 0·038 = -0·54 \text{ to 2 dec. places.}$$

If a closer approximation was required, the above method could be repeated using $-0·54$ in place of $-0·5$ as the approximate root.

General result Suppose $x = a$ is an approximate root of the equation
$$f(x) = 0.$$
Let the accurate root be $x = a + h$, where h is small.
$$\therefore f(a + h) = 0.$$
The value of the differential coefficient of $f(x)$ at the point $x = a$ is given by
$$f'(a) = \operatorname*{Lt}_{h \to 0} \left\{ \frac{f(a+h) - f(a)}{h} \right\}$$
$$\approx \frac{f(a+h) - f(a)}{h}$$
if h is small.

i.e.
$$h \approx \frac{f(a+h) - f(a)}{f'(a)}$$

Consequently, as $f(a+h) = 0$
$$h \approx -\frac{f(a)}{f'(a)}.$$

∴ **A closer approximation to the root is**
$$\mathbf{x = a - \frac{f(a)}{f'(a)}}$$

Example 16 Show that the equation $e^x = 2 - x$ has only one real root and find its value correct to three decimal places.

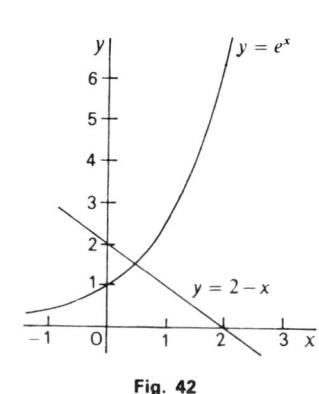

A sketch of the graphs of $y = e^x$ and $y = 2 - x$ (Fig. 42) shows that there is only one common point and so the equation $e^x = 2 - x$ has only one root which is approximately $x = 0.5$.

Use Newton's method, with
$$f(x) = e^x - 2 + x;$$
$$f'(x) = e^x + 1.$$
Taking $a = 0.5$,
$$f(a) = e^{0.5} - 2 + 0.5 = 0.149;$$
$$f'(a) = e^{0.5} + 1 = 2.649.$$
∴ A closer approximation to the root is
$$x = a - \frac{f(a)}{f'(a)} = 0.5 - \frac{0.149}{2.649}$$
$$= 0.444.$$

Fig. 42

To obtain a closer approximation, we take $a = 0.444$.
$$\therefore f(a) = 0.003; \quad f'(a) = 2.559.$$
i.e.
$$x = 0.444 - \frac{0.003}{2.559}$$
$$= 0.443 \quad \text{correct to 3 places of decimals.}$$

Example 17 Show that the equation $2 \sin x = x$ (x in radians) has a root between $x = 1$ and $x = 2$. Find the root correct to three significant figures.

Let $\qquad f(x) = 2 \sin x - x$.

Putting $x = 1$, $\quad f(1) = 2 \sin 1 - 1 = 2(0.8415) - 1 = 0.683$.

Putting $x = 2$, $\quad f(2) = 2 \sin 2 - 2 = 2(0.9092) - 2 = -0.182$.

As $f(x)$ changes sign between $x = 1$ and $x = 2$, $f(x) = 0$ for some value of x between 1 and 2.

Clearly the root is closer to $x = 2$ than to $x = 1$, but to illustrate a case where Newton's method fails, we will attempt to obtain the root by using $a = 1$.

$$f(1) = 0.683;$$
$$f'(1) = 2 \cos 1 - 1 = 0.08.$$

Newton's method then gives a closer approximation to the root as

$$x = 1 - \frac{0.683}{0.08} = -7.5.$$

Clearly this is absurd.

The failure of the method is due to the fact that the approximate root $x = 1$ is close to a root of $f'(x) = 0$. Newton's method will always fail under those circumstances.

This difficulty does not arise when a is taken as 2.

A closer approximation is

$$x = 2 - \frac{f(2)}{f'(2)}$$

$$= 2 - \frac{-0.182}{-1.833}$$

$$= 1.9$$

Repeating the method with $a = 1.9$, we find that a closer approximation is

$$x = 1.90 \text{ to 3 sig. figs.}$$

[B] EXAMPLES 7f

1 By means of rough graphs show that the equation $x^3 - 2x^2 - 1 = 0$ has only one real root. Verify that the root lies between 2·20 and 2·25.

2 Show that the equation $x^3 - 3x^2 - 4x + 4 = 0$ has 3 real roots which lie between -2 and -1, 0 and 1, 3 and 4 respectively.

3 Prove that the equation $2x^3 + x^2 = 1$ has a root between 0 and 1 and no other real roots.

4 Show graphically that the equation $\tan x° = 1 - \frac{x°}{180}$ has a root between 0° and 90°. Verify that the root lies between 38° and 39°.

5 Prove that the equation $x^3 - 4x - 1 = 0$ has 3 real roots.

6 Show graphically that the equation $e^x - \frac{1}{x} = 0$ has only one root and verify that it lies between 0·5 and 0·6.

7 Find graphically the range of values of m for which the equation $x^3 = mx - 1$ has 3 real roots.

8 Show that the equation $\sin \frac{\theta}{2} = \frac{2}{\theta}$ has two roots between 0 and 2π and verify that one root is approximately 2·23.

9 Find to 2 places of decimals, the root of the equation $x^3 + 3x = 7$, close to 1·5.

10 Show that the equation $x^2 e^x = 1$ has a root approximately equal to 0·7. Use Newton's method to obtain the root correct to 3 places of decimals.

11 Show that $x = 3$ is an approximate root of the equation $3 \ln x = 6 - x$ and determine the root correct to 2 decimal places.

12 Show that the equation $\cos 2\theta = \theta$ has a root close to $\theta = 0·5$ radians and determine this root correct to 2 decimal places.

13 Find to three places of decimals, the positive root of the equation $x^3 + 2x^2 - 5x - 7 = 0$.

14 The equation $x^3 - 2x - 5 = 0$ has a root close to 2. By taking $x = 2 + h$ and neglecting powers of h above the second prove that $6h^2 + 10h - 1 = 0$ and hence deduce the value of the root correct to 2 decimal places.

15 Find the root of the equation $x^3 - 9x + 14 = 0$ correct to 2 places of decimals.

16 Show that the equation $10x^3 - 17x^2 + x + 6 = 0$ has one root between -1 and 0 and determine its value.

17 Show by Newton's method that approximate roots of the equation $(x-2)(x+1) = \varepsilon x$, where ε is small, are $-1 + \dfrac{\varepsilon}{3}$ and $2 + \dfrac{2\varepsilon}{3}$.

18 Prove that an approximate root of the equation $\sin x = \tfrac{1}{2} - \varepsilon x^2$, where ε is small, is $\dfrac{\pi}{6}\left(1 - \dfrac{\pi\varepsilon}{3\sqrt{3}}\right)$.

8 | Inequalities

Properties of inequalities

(i) *An inequality will still hold after each side has been increased, diminished, multiplied or divided by the same POSITIVE quantity.*

For clearly, if
$$x > y,$$
then
$$x + 2 > y + 2,$$
$$x - 2 > y - 2,$$
$$2x > 2y,$$
and
$$\frac{x}{2} > \frac{y}{2}.$$

(ii) *In an inequality any term may be transposed from one side to the other if its sign is changed.*

This result follows from property (i).

e.g. If
$$3 - 4x > 2x,$$
adding $4x$ to both sides,
$$3 > 2x + 4x,$$
i.e.
$$\tfrac{1}{2} > x.$$

(iii) *Both sides of an inequality can be multiplied or divided by the same NEGATIVE number so long as the inequality sign is REVERSED.*

Clearly the inequality $2 > 1$, on multiplying both sides by -1, must be written
$$-2 < -1.$$
e.g. If
$$-2x < 4,$$
then
$$2x > -4,$$
$$x > -2.$$

Solution of inequalities The methods of solution of inequalities correspond very closely to those used for the solution of equalities or equations and can be classified as

(i) Analytical methods;
(ii) Graphical or semi-graphical methods.

Example 1 For what values of x is $\frac{2}{3}x - 3 > 4$?
Multiplying both sides by 3,
$$2x - 9 > 12,$$
$$2x > 9 + 12.$$
i.e. $\qquad x > \frac{21}{2}.$

Example 2 For what values of x is $2x^2 > x + 3$?
Transposing all the terms to the L.H.S.,
$$2x^2 - x - 3 > 0.$$
Factorising the L.H.S.,
$$(2x - 3)(x + 1) > 0.$$
The product of two factors is positive, if both factors have like signs.
Both factors are positive if $x > \frac{3}{2}$,
and both factors are negative if $\quad x < -1$.
Hence the given inequality holds for values of x greater than $\frac{3}{2}$ or less than -1.

Example 3 Show that $2x^2 + 2 > 3x$ for all real values of x.
We have $\qquad 2x^2 - 3x + 2 > 0.$

The function on the L.H.S. will not factorise and we complete the square, as in the case of a quadratic equation.
Dividing by 2, $\qquad x^2 - \frac{3}{2}x + 1 > 0,$
$$(x - \tfrac{3}{4})^2 + 1 - \tfrac{9}{16} > 0,$$
$$(x - \tfrac{3}{4})^2 + \tfrac{7}{16} > 0.$$
The term $(x - \frac{3}{4})^2$ is positive for all real values of x and consequently the inequality is satisfied by all such values.

Example 4 Find the range of values of x for which $5 - 3x$ is less than $\dfrac{2}{x}$.

This example is best solved by a semi-graphical method. First find the values of x for which
$$5 - 3x = \frac{2}{x}.$$
i.e. the roots of
$$3x^2 - 5x + 2 = 0.$$
$$(3x - 2)(x - 1) = 0.$$
$$\therefore x = \tfrac{2}{3} \text{ and } 1.$$

Now consider sketch graphs of $y = 5 - 3x$ and $y = \dfrac{2}{x}$ (Fig. 43).

The points of intersection, A and B, have x coordinates, $\frac{2}{3}$ and 1 respectively. As $5 - 3x < \dfrac{2}{x}$ when the straight line is below the curve, it follows that the inequality is satisfied for values of x between 0 and $\frac{2}{3}$ and values of x greater than 1.

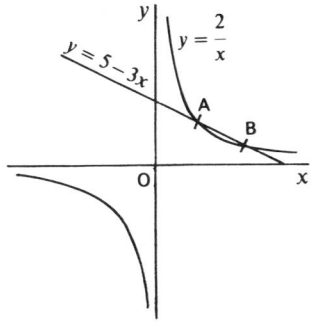

Fig. 43

Modulus of a function The modulus of a function $f(x)$ is denoted by $|f(x)|$. $|f(x)|$ is defined as a function which has the same *numerical* value as $f(x)$ for all values of x.

e.g. When $x = 2$, $\qquad 4 - 3x = -2$

and $\qquad\qquad\qquad |4 - 3x| = 2.$

The modulus notation is frequently used to express inequalities in more concise forms.

e.g. the statement, x is numerically less than 1, can be written

$$|x| < 1,$$

instead of $\qquad\qquad -1 < x < 1.$

Example 5 For what values of x is $|2x - 3| > 5$?

$2x - 3$ is numerically greater than 5.

$\therefore\ 2x - 3 > 5\ $ and $\ 2x - 3 < -5.$

i.e. $\qquad\qquad x > 4\ $ and $\ x < -1.$

Example 6 Show that if $x^2 - 2x < 6$ then $|x - 1| < \sqrt{7}$.

$$x^2 - 2x < 6.$$

Completing the square, $\qquad (x - 1)^2 < 7.$

Hence, $(x - 1)$ must be numerically less than $\sqrt{7}$, or, $|x - 1| < \sqrt{7}.$

[A] EXAMPLES 8a

For what values of x are the inequalities in nos. 1–18 satisfied?

1 $2x + 1 > 5.$ **2** $4x - 3 > x + 1.$ **3** $2(1 - 2x) < x.$
4 $-2x < 6.$ **5** $\frac{2}{3}(x - 1) > \frac{1}{2}(1 - 2x) + 1.$ **6** $x(x - 1) < 0.$
7 $(2x + 1)(x - 2) < 0.$ **8** $(1 - 3x)(x + 1) > 0.$ **9** $2x > x^2 - 3.$
10 $2x^2 - 3x > 2.$ **11** $\dfrac{x + 1}{x} > 1.$ **12** $x^2 > x + 6.$
13 $(x + 8)(x - 3) < 3x.$ **14** $(2x - 1)(x - 2) > 5.$ **15** $(x + 2)^2 + 1 > 0.$
16 $x^2 + x + 1 > 0.$ **17** $2x^2 + 7 > 4x.$ **18** $x^3 > 4x.$

19 If $x = 2$, find the values of:

(i) $|3x - 1|$; (ii) $|1 - x^2|$; (iii) $\left|\dfrac{1 + x}{1 - x}\right|$; (iv) $|4x - 2| - |1 - 3x|.$

20 Sketch the graph of $|\cos x|$ for values of x between 0 and π.

Express in the modulus form the inequalities in nos. 21–30:

21 x is numerically less than 3.   **22** $-4 < x < 4.$
23 $-1 < 2x - 1 < 1.$ **24** $2 < x < 6.$
25 $-4 < x < 0.$ **26** $-3 < 2x < 5.$
27 $(x + 1)^2 > 4.$ **28** $(x - 2)^2 < 3.$
29 $(x - 2)^2 - 3 > 0.$ **30** $(x + \frac{5}{2})^2 - \frac{7}{4} < 0.$

31 Find the range of values of x for which:

(i) $|x + 3| < 2$; (ii) $|4x - 1| > 3$; (iii) $|1 - 3x| > 7$; (iv) $|5 - 2x| < 3.$

32 For what values of θ between -2π and 2π is $|\cos \theta| > \frac{1}{2}$?

Miscellaneous examples

Example 7 For what values of m has the equation $4x^2 + 8x - 8 = m(4x - 3)$ no real roots?

The equation can be written
$$4x^2 + x(8 - 4m) - 8 + 3m = 0.$$

Roots are imaginary if $b^2 < 4ac$.

i.e. $\qquad (8 - 4m)^2 < 4 \cdot 4(3m - 8).$

Dividing by 16, $\qquad (2 - m)^2 < 3m - 8,$

$\qquad m^2 - 7m + 12 < 0.$

Factorizing, $\qquad (m - 3)(m - 4) < 0.$

i.e. $\qquad 3 < m < 4.$

Example 8 For what values of x is the function $\dfrac{(2x - 1)(x + 2)}{(2x + 1)(x - 2)}$ positive?

Consider the signs of the numerator and denominator independently.
$(2x - 1)(x + 2)$ is positive when $x < -2$ and $x > \frac{1}{2}$ and negative when
$$-2 < x < \tfrac{1}{2}.$$
$(2x + 1)(x - 2)$ is positive when $x < -\frac{1}{2}$ and $x > 2$ and negative when
$$-\tfrac{1}{2} < x < 2.$$

The function is positive when numerator and denominator have like signs, i.e.

when $\qquad x < -2, x > 2$
and $\qquad -\frac{1}{2} < x < \frac{1}{2}.$ $\qquad$ See Fig. 44 (a) and (b).

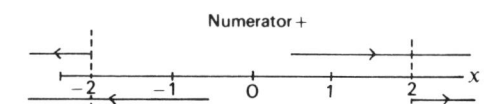

Numerator +

$\qquad\qquad$ −2 $\qquad$ −1 $\qquad$ 0 $\qquad$ 1 $\qquad$ 2 $\qquad\longrightarrow x$

Denominator +

Fig. 44a

Numerator −

$\qquad\qquad$ −2 $\qquad$ −1 $\qquad$ 0 $\qquad$ 1 $\qquad$ 2 $\qquad\longrightarrow x$

Denominator −

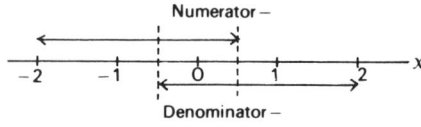

Fig. 44b

Example 9 If x is real and $y = \dfrac{x^2 + 1}{x^2 + x + 1}$, prove that $|y - \tfrac{4}{3}| \leqslant \tfrac{2}{3}$.

$$y = \frac{x^2 + 1}{x^2 + x + 1}.$$

Rearranging as a quadratic equation in x,
$$x^2(1 - y) - xy + (1 - y) = 0.$$

As x is real, the roots of this equation are real and
$$b^2 \geqslant 4ac.$$

116 | Inequalities

$$\therefore (-y)^2 \geq 4(1-y)(1-y),$$
$$y^2 \geq 4 - 8y + 4y^2,$$
$$0 \geq 3y^2 - 8y + 4.$$

Dividing by 3 and completing the square,

$$0 \geq (y - \tfrac{4}{3})^2 + \tfrac{4}{3} - \tfrac{16}{9},$$
$$\tfrac{4}{9} \geq (y - \tfrac{4}{3})^2.$$

i.e. $|y - \tfrac{4}{3}| \leq \tfrac{2}{3}.$

Example 10 Show that the sum of any real positive quantity and its reciprocal is never less than 2.

Let x be any real positive quantity.

We need to prove $$x + \frac{1}{x} > 2.$$

i.e. $$x + \frac{1}{x} - 2 > 0,$$

or $$\frac{x^2 - 2x + 1}{x} > 0.$$

This result follows as

$$\frac{x^2 - 2x + 1}{x} = \frac{(x-1)^2}{x},$$

which is positive when x is positive.

[B] EXAMPLES 8b

For what values of x are the functions in Examples 1–12 positive?

1 $\dfrac{(x-1)(x-2)}{x}.$ **2** $\dfrac{(2x+3)(x-4)}{x-1}.$ **3** $\dfrac{(1-x)(1+2x)}{x+1}.$

4 $\dfrac{x^2 - x - 2}{2x - 3}.$ **5** $\dfrac{3x^2 - 8x - 3}{4 - x}.$ **6** $\dfrac{(x+2)(x-2)}{(x+1)(x-1)}.$

7 $\dfrac{x(2x-1)}{(x+1)(x-2)}.$ **8** $\dfrac{1-2x}{x(3x+2)}.$ **9** $\dfrac{4 - 3x - x^2}{x^2 + x + 1}.$

10 $x(2x-3)(3x+5).$ **11** $(1-x)(2-x)(5+2x).$ **12** $\dfrac{x^2(x+1)}{x-1}.$

Find the ranges of values of x which satisfy the inequalities in Examples 13–18:

13 $x^2 - x > 6(x - 1).$ **14** $x + 1 < \dfrac{6}{x}.$ **15** $\dfrac{x}{2} > \dfrac{18}{x}.$

16 $x + 2 > \dfrac{30}{x+1}.$ **17** $3x + 8 < \dfrac{3}{x}.$ **18** $x + \dfrac{1}{4x} < 1.$

Find the ranges of values of x between $\pm \pi$ for which the following inequalities hold:

19 $|\sin 2x| > \tfrac{1}{2},$ **20** $\sin x > 2\cos x.$ **21** $\sin x < \sqrt{3}\cos x.$

22 $\sin 2x > \sqrt{2}\sin x.$ **23** $2\cos 2x < 3\cos x - 1.$

24 For what values of a has the equation $3x^2 + 5x - 4 = a(2x - 1)$ real roots?

25 Find the condition that the equation $x^2 + 2(k+2)x + 9k = 0$ has no real roots.

26 Prove that the roots of the equation $x^2 - 2ax + a^2 - b^2 - c^2 = 0$ are real.

27 Prove that $x^3 + 3x + 14 > 0$ for $x > -2$.

28 If $2x^2 + 6 > 7x$, prove that $|4x - 7| > 1$.

Prove the results of Examples 29–32, if x is real:

29 $-\dfrac{1}{11} \leqslant \dfrac{x}{x^2 - 5x + 9} \leqslant 1.$

30 $\dfrac{1}{3} \leqslant \dfrac{x^2 - x + 1}{x^2 + x + 1} \leqslant 3.$

31 $-\dfrac{1}{13} \leqslant \dfrac{x+2}{2x^2 + 3x + 6} \leqslant \dfrac{1}{3}.$

32 $5 \geqslant \dfrac{x^2 + 34x - 71}{x^2 + 2x - 7} \geqslant 9.$

33 If $y = \dfrac{x^2 + 2x - 11}{2(x - 3)}$ and x is real, show that $|y - 4| \geqslant 2$.

34 Use the inequality $(x - y)^2 > 0$ to prove the result $x^2 + y^2 > 2xy$, x and y being unequal. Deduce that the arithmetic mean of two unequal positive numbers is greater than their geometric mean.

35 By writing $a^2 + b^2 + c^2 = \tfrac{1}{2}\{(a^2 + b^2) + (b^2 + c^2) + (c^2 + a^2)\}$, deduce that
$$a^2 + b^2 + c^2 \geqslant bc + ca + ab.$$

36 Find the range of values of x for which $\dfrac{x^2 - 4x + 3}{x^2 + 1} < 1$.

[B] MISCELLANEOUS EXAMPLES

1 Solve the equations: (i) $3x^2 + y^2 = 3, \quad 2x^2 - xy + 3y^2 = 8.$
(ii) $3x + 2y + 4z = 19,$
$2x - y + z = 3,$
$6x + 7y - z = 17.$

2 Find between what integers the roots of the following equation lie.
$$x^3 - 3x^2 - 4x + 11 = 0$$

3 Find all values of θ between $\pm 360°$ for which $2\cos 2\theta = \cos \theta + 1$.

4 What is the general solution of the equation $\tan 2x + \cot 3x = 0$?

5 Determine the sign of the function $2x^2 - 6x + 7$.

6 Solve the equation $\sin \theta = \cos(\theta + 45°)$, θ being between $0°$ and $360°$.

7 Solve graphically the equation $xe^{\frac{x}{2}} = 1$.

8 Find to 3 places of decimals, the positive root of the equation
$$x^3 = 2x + 5.$$

9 Prove that the roots of the following equations are real:
(i) $(x - a)(x - b) = h^2$.
(ii) $(a - b + c)x^2 + 4(a - b)x + (a - b - c) = 0$.

10 Find the range of values of x for which $\dfrac{1}{x^2} > 8 + \dfrac{2}{x}$.

11 Solve the equation $\sin 4x + \cos 3x = 0$.

12 Find graphically the number of roots of the equation
$$2x = 3\pi(1 - \cos x).$$

13 Solve (i) $x - \dfrac{1}{y} = 3,$ (ii) $3x(x-y) = y+5,$
$y - \dfrac{1}{z} = -2,$ $3y(y-x) = 5y-x.$
$z - \dfrac{1}{x} = \dfrac{1}{2}.$

14 Apply Newton's method twice to find to 3 decimal places the root of the equation $x^4 - 12x + 7 = 0$ near 2.

15 Find values of θ between $\pm\pi$ for which
 (i) $\cos 4\theta + \cos 2\theta + \cos \theta = 0;$
 (ii) $3\sin 2\theta + 4\cos 2\theta = 2\cdot 5.$

16 For what values of x is $x^3 + 1 \geqslant x^2 + x$?

17 By putting $y = mx$, solve the equations $x^3 + \tfrac{7}{2}xy^2 = y^3 + \tfrac{7}{2}yx^2 = 1.$

18 Prove graphically that the least positive root of $\dfrac{x}{2\pi} = \sec x$ is 2π and that there is another root between 2π and $\dfrac{5\pi}{2}$. Determine the value of this root to 2 significant figures.

19 Find the values of x for which the following inequalities are true:
 (i) $\dfrac{x^2 + 2x - 19}{x - 4} > 4;$ (ii) $\dfrac{x^2}{(x-1)(x-2)} > 1.$

20 Solve the equations: (i) $\dfrac{(1-x)^2}{2-x^2} = \dfrac{(1-a)^2}{2-a^2};$
 (ii) $x + 2y + 3z = 7,$
 $3x + y + 2z = 2,$
 $x^2 + y^2 + z^2 = 6.$

21 Prove that $\dfrac{2}{27} \leqslant \dfrac{x^2 - 2x + 2}{x^2 + 3x + 9} \leqslant 2.$

22 Draw the graph of the function $x^2(1+x)$ and prove that the equation $x^3 + x^2 = 2x + 1$ has 3 real roots. Find the values of the roots correct to 2 places of decimals.

23 Solve the following equations for values of x and y between $0°$ and $360°$:
 (i) $\sin(x+y) = \tfrac{1}{2},$ (ii) $\sin x + \sin y = \dfrac{\sqrt{3}}{2},$
 $\cos(x-y) = -\dfrac{\sqrt{3}}{2}.$ $\cos x + \cos y = \tfrac{1}{2}.$

24 Solve graphically the equation $x = \cos^2 x.$ $\left[\text{Use } \cos^2 x = \dfrac{\cos 2x + 1}{2}\right].$

25 Prove that the equation $\dfrac{x}{2} = \sin^2 x$ has 3 and only 3 real roots.

26 Show that $x^3 - 2 > x^2 + x$ for $x > 2.$

Miscellaneous examples | 119

27 Find the values of x for which $x^2 + 17 < 7x + \dfrac{11}{x}$.

28 Use the result $\cos 3\theta = 4\cos^3 \theta - 3\cos \theta$ to solve the equation
$$8x^3 - 6x + 1 = 0.$$

29 Show that the approximate roots of the equation $x^2 - 1 = \varepsilon x^3$ where ε is small, are $1 + \dfrac{\varepsilon}{2}$ and $-1 + \dfrac{\varepsilon}{2}$.

30 Find an approximate solution of the equation $\tan 2x = 1 + \varepsilon x$ when second and higher powers of ε can be neglected.

31 Using the result $a^2 + b^2 \geqslant 2ab$, deduce that $a^2 - ab + b^2 \geqslant ab$. Hence show if a, b, c are positive and unequal
   (i) $a^3 + b^3 > ab(a+b)$;
   (ii) $2(a^3 + b^3 + c^3) > bc(b+c) + ca(c+a) + ab(a+b)$.

32 For what ranges of values of x are the following functions negative:
   (i) $(x-1)(x-2)$;   (ii) $\dfrac{(x-1)(x-3)}{(x-2)(x-4)}$?

33 Draw the graph $y = \tfrac{3}{2}\sin 2x - \tfrac{1}{4}\sin 4x$ between $x = 0$ and $180°$. Find to the nearest degree, the acute angles for which $y = 1$ and show that the equation you have solved is
$$1 = \sin 2x (1 + \sin^2 x).$$

34 Show that $(x^2 - 2x - 3)/(2x^2 + 2x + 1)$ must lie between -4 and $+1$ if x is real, and find the range of values of x for which the expression is negative.

35 Establish the identity:
$$a^2 + b^2 + c^2 - bc - ca - ab \equiv \tfrac{1}{2}\{(b-c)^2 + (c-a)^2 + (a-b)^2\}$$
and deduce that
   (i) $a^2 + b^2 + c^2 > bc + ca + ab$;   (ii) $a^3 + b^3 + c^3 > 3abc$,
where a, b, c are positive and unequal.

36 Find graphically the approximate value of θ between 0 and $\dfrac{\pi}{2}$ which satisfies the equation $\tan \theta = 1 + \sin \theta$.

By substituting for $\tan \theta$ and $\sin \theta$ in terms of $\tan \dfrac{\theta}{2}$, deduce an approximate value for the positive root of the equation $x^4 + 4x^3 - 1 = 0$ and use Newton's method to obtain a closer approximation.

37 Find general solutions of the equations:
   (i) $\sin x + \cos x \cos 2x = \cos 2x \cos 3x$;
   (ii) $(1 - \tan x)(1 + \sin 2x) = 1 + \tan x$.

38 Draw the graph of $\lg x$ from $x = 0\cdot 2$ to $x = 5\cdot 0$. Show from your graph that if $m > 0$, the equation $mx = \lg x$ has two, one or no roots, according to the value of m. Find the approximate value of m for which the equation has one root and obtain the value of this root correct to two places of decimals.

39 Solve the equations: $x^2 + 3xy - y^2 = 3(x+y)$,
$$xy + y^2 = 5y - x.$$

40 Find the general solution of the equation
$$\tan(\pi \cos x) = \cot(\pi \sin x).$$

9 | Equations of a curve
Elementary curve tracing

Equations of a curve Fig. 45(a) represents a circle, radius 1 unit. The origin O is on the circumference and rectangular axes Ox, Oy are chosen as shown.

Let $P(x, y)$ be any point on the circle.
From $\triangle PCQ$, by Pythagoras,

$$CQ^2 + QP^2 = CP^2.$$

i.e. $(x-1)^2 + y^2 = 1.$

This is *the Cartesian equation* of the circle.

Referring again to Fig. 45(a), let angle $PCQ = \phi$.

Then

$$PQ = \sin\phi \quad \text{and} \quad CQ = \cos\phi.$$

Hence
$$\left. \begin{array}{l} x = 1 + \cos\phi \\ y = \sin\phi \end{array} \right\}$$

These equations are called *parametric equations, ϕ being a parameter*. The coordinates of any point P on the circle are $(1 + \cos\phi, \sin\phi)$, *the parametric coordinates of P*.

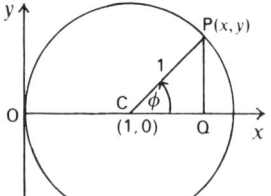

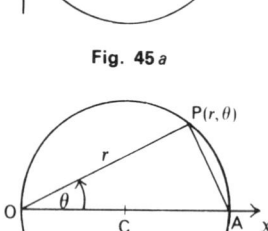

Fig. 45a

Fig. 45b

Corresponding to every point on the curve there is one value of the parameter ϕ.

In Fig. 45(b), the position of the point P on the circle is fixed by denoting the distance OP by r and the angle AOP by θ. These values (r, θ) are called *the polar coordinates* of P.

The connection between r and θ is obtained by using $\triangle OPA$.

$$OP = OA\cos\theta.$$

i.e. $r = 2\cos\theta.$

This is *the polar equation* of the circle.

Parametric coordinates Suppose the Cartesian coordinates of any point on a curve are $(f(t), g(t))$ where $f(t), g(t)$ are given functions of a variable t.

Then the curve can be considered as the locus of the point (x, y) where

$$x = f(t); \ y = g(t).$$

t is called a parameter and the equations are called the parametric equations of the curve.

$(f(t), g(t))$ are the parametric coordinates of any point on the curve; any particular point is determined by one particular value of t.

Parametric coordinates and parametric equations frequently lead to considerable simplification in the derivation of the properties of plane curves.

Example 1 Find the Cartesian equation of the locus of the point $\left(t+\dfrac{1}{t}, t-\dfrac{1}{t}\right)$

$$x = t + \frac{1}{t}, \quad y = t - \frac{1}{t}.$$

Adding, $\quad x+y = 2t;$ i.e. $\quad t = \tfrac{1}{2}(x+y).$

Substituting for t,

$$x = \tfrac{1}{2}(x+y) + \frac{1}{\tfrac{1}{2}(x+y)},$$

giving, $\quad x^2 - y^2 = 4.$

Example 2 Find the coordinates of the points of intersection of the curve $x = 2t^2 - 1$, $y = 3(t+1)$ and the straight line $3x - 4y = 3$.

Substituting for x and y in the equation of the line,

$$3(2t^2 - 1) - 12(t+1) = 3,$$
$$6t^2 - 12t - 18 = 0,$$
$$6(t-3)(t+1) = 0.$$

i.e. $\quad t = 3, -1.$

The parameters of the points on the curve at its intersections with the straight line are $t = 3, t = -1$.

∴ Coordinates of points of intersection are $(17, 12)$ and $(1, 0)$.

Example 3 Find the gradient of the locus $(2t^3 + 1, 3t^2 - 1)$ at the point parameter m. Write down the equation of the normal at this point.

We have $\quad x = 2t^3 + 1; \quad y = 3t^2 - 1.$

$$\therefore \frac{dx}{dt} = 6t^2; \quad \frac{dy}{dt} = 6t.$$

$$\frac{dy}{dx} = \frac{dy}{dt} \times \frac{dt}{dx} = \frac{6t}{6t^2} = \frac{1}{t}.$$

∴ Gradient of locus at point where $t = m$ is $\dfrac{1}{m}$.

Gradient of normal $= -m.$

Equation of normal is

$$y - (3m^2 - 1) = -m(x - \overline{2m^3 + 1}) \quad [y - k = m(x - h)].$$

i.e. $\quad y + mx = 2m^4 + 3m^2 + m - 1.$

[A] EXAMPLES 9a

1 Find the coordinates of the points on the locus $(3 \sin \phi, 2 \cos \phi)$ where ϕ has values $0, \dfrac{\pi}{2}, \dfrac{\pi}{4}, \dfrac{2\pi}{3}, -\dfrac{\pi}{3}.$

Equations of a curve

2 Find the length of the chord joining the points on the curve
$$x = 1 + \sin\theta, \quad y = \cos\theta,$$
for which θ has values 0 and $\dfrac{\pi}{2}$.

3 Find the gradient of the chord joining the points $t = -1$, $t = 2$ on the curve $x = 4t^2$, $y = 8t$.

4 Find the gradient of the curve $x = 2t - t^3$, $y = 1 - t^2$ at the point parameter 2.

5 Obtain the coordinates of the points on the locus $(2\sin^3 t, 3\cos^3 t)$ where $t = \dfrac{\pi}{6}, \dfrac{\pi}{4}, \dfrac{\pi}{3}, -\dfrac{\pi}{2}$.

6 Show that the locus of the point $(2\cos\theta, 2\sin\theta)$ is a circle, centre the origin, radius 2. Sketch the locus and mark the points where $\theta = \pm\dfrac{\pi}{4}$.

7 Find the point of intersection of the locus $x = 3t + 2$, $y = 1 - t$ and the straight line $y + x = 2$.

8 Find the coordinates of the points where the curve $x = 2 - 3t + t^2$, $y = (3 + t)^2$ meets the y axis. Show that the x axis is a tangent to the curve.

9 Show that the locus of the point $\left(\dfrac{t+2}{2t+1}, \dfrac{2-5t}{2t+1}\right)$ as t varies is a straight line.

10 Prove that the straight line $y = x + 2$ is a tangent to the locus $(2t^2, 4t)$.

Find the Cartesian equations of the curves with the following parametric equations:

11 $x = \dfrac{2-3t}{1+t}$, $y = \dfrac{3+2t}{1+t}$. **12** $x = 2 - m$, $y = m^3 + 4$.

13 $x = 4t$, $y = \dfrac{4}{t}$. **14** $x = 2\cos\phi$, $y = 3\sin\phi$.

15 $x = 2 - 5\cos\phi$, $y = 1 - 3\sin\phi$. (Use $\sin^2\phi + \cos^2\phi = 1$.)

Find the Cartesian equations of the following loci:

16 $(3t^2, 6t)$. **17** $(2t^2 - t^3, 1 + t)$. **18** $\left(t - \dfrac{1}{t}, t + \dfrac{1}{t}\right)$.

19 $(3 + 2\cos\theta, 2\sin\theta)$. **20** $(\sin\theta, \cos 2\theta)$. **21** $(\sin^3\theta, \cos^3\theta)$.

22 Find the equation of the normal to the curve $x = t^2$, $y = 2t$ at the point parameter t.

23 Find the equation of the tangent to the curve $xy = c^2$ at the point $\left(ct, \dfrac{c}{t}\right)$.

24 Prove that the equation of the chord joining the points $t = t_1$, $t = t_2$ on the locus $(at^2, 2at)$ is $y(t_1 + t_2) = 2x + 2at_1 t_2$.

25 Find the values of t for which points on the locus $\left(ct, \dfrac{c}{t}\right)$ lie on the straight line $x - 2y = c$.

Connection between polar and Cartesian coordinats | 123

Polar coordinates Referring to Fig. 46, it is seen that the position of a point P in a plane is fixed if the distance OP, r, and the angle θ are known.

(r, θ) *are the polar coordinates of P.*

O is called the origin or pole and OA the initial line.

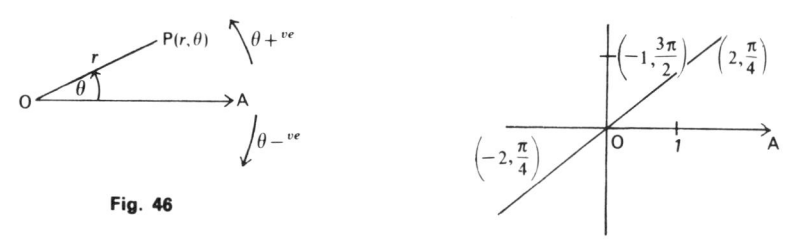

Fig. 46

Fig. 47

Sign Conventions: θ is positive when measured in a counterclockwise direction and negative when measured in a clockwise direction.

When the θ of a point has been fixed, a positive value of r is measured in the outward direction OP and a negative value in the opposite direction.

e.g. the points $\left(2, \dfrac{\pi}{4}\right), \left(-2, \dfrac{\pi}{4}\right), \left(-1, \dfrac{3\pi}{2}\right)$ are shown in Fig. 47.

Example 4 Find the distance between the points $\left(2, \dfrac{\pi}{4}\right)$ and $\left(3, \dfrac{7\pi}{12}\right)$.

In $\triangle POQ$ (Fig. 48),

$$\widehat{POQ} = \frac{7\pi}{12} - \frac{\pi}{4} = \frac{\pi}{3}.$$

By the cosine rule,

$$PQ^2 = 9 + 4 - 12\cos\frac{\pi}{3},$$
$$= 7.$$
$$PQ = \sqrt{7}.$$

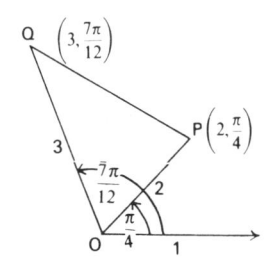

Fig. 48

Connection between polar and Cartesian coordinates

Taking the initial line as the positive direction of the x axis, it follows from Fig. 49, that if P is the point (x, y) or (r, θ),

$$x = r\cos\theta;\ y = r\sin\theta.$$

or

$$r = \sqrt{x^2 + y^2};\ \tan\theta = \frac{y}{x}.$$

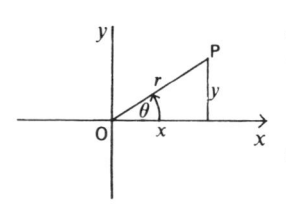

Fig. 49

Consequently, a curve given in Cartesian coordinates can be expressed in polar coordinates and vice versa.

Example 5 Find the polar equation of the curve $x^3 + y^3 = 3xy$.
Substituting $x = r\cos\theta$, $y = r\sin\theta$,
$$r^3\cos^3\theta + r^3\sin^3\theta = 3r^2\sin\theta\cos\theta,$$
$$r = \frac{3\sin\theta\cos\theta}{\cos^3\theta + \sin^3\theta}.$$

[A] EXAMPLES 9b

1 Show on a diagram the points with polar coordinates:

$(2, 0)$, $(1, \pi)$, $\left(3, \frac{\pi}{4}\right)$, $\left(1, \frac{3\pi}{2}\right)$, $\left(\frac{3}{2}, -\frac{\pi}{3}\right)$, $\left(-2, \frac{\pi}{6}\right)$, $\left(1, \frac{5\pi}{3}\right)$,

$\left(-1, \frac{5\pi}{6}\right)$, $\left(-2, -\frac{\pi}{2}\right)$.

2 Find the squares of the distances between the following pairs of points:

(i) $(2, 0)$, $\left(1, \frac{\pi}{4}\right)$. (ii) $\left(3, \frac{\pi}{3}\right)$, $\left(1, \frac{2\pi}{3}\right)$. (iii) $\left(2, \frac{5\pi}{4}\right)$, $\left(4, -\frac{\pi}{3}\right)$.

3 If O is the origin and P, Q the points $\left(3, \frac{3\pi}{4}\right)$, $\left(1, -\frac{3\pi}{4}\right)$, find angle POQ.

4 Find the areas of $\triangle OPQ$ where P and Q are given by:

(i) $P\left(5, \frac{\pi}{4}\right)$, $Q\left(2, \frac{\pi}{2}\right)$; (ii) $P\left(1, \frac{\pi}{6}\right)$, $Q\left(2, -\frac{\pi}{3}\right)$;

(iii) $P\left(4, \frac{5\pi}{6}\right)$, $Q\left(1, \frac{2\pi}{3}\right)$; (iv) $P(r_1, \theta_1)$, $Q(r_2, \theta_2)$.

5 Plot the points $P\left(2, \frac{\pi}{6}\right)$, $Q\left(5, \frac{\pi}{4}\right)$, $R\left(3, \frac{\pi}{3}\right)$. Find the areas of triangles OPQ, OQR, ORP and deduce that of triangle PQR.

6 Draw a straight line perpendicular to the initial line at distance 2 from the pole. Denoting the polar coordinates of any point on the line as (r, θ), show that $r\cos\theta = 2$ (the polar equation of the line).

7 Assuming a common origin and the x axis as the initial line, find the (x, y) coordinates of the points with polar coordinates:

$\left(4, \frac{\pi}{3}\right)$, $\left(2, \frac{\pi}{2}\right)$, $\left(1, \frac{3\pi}{4}\right)$, $\left(5, -\frac{\pi}{4}\right)$, $\left(6, -\frac{7\pi}{6}\right)$.

8 With the same assumptions as in Example 7, find the polar coordinates of the points with Cartesian coordinates:

$(3, 4)$, $(-3, 4)$, $(5, -12)$, $(-1, 1)$, $(-6, -8)$, $(1, -\sqrt{3})$.

9 Find the polar equations of the curves:

(i) $x^2 + y^2 = 4$. (ii) $x^2 + (y-2)^2 = 4$. (iii) $xy = c^2$.

(iv) $\dfrac{x^2}{4} + \dfrac{y^2}{2} = 1$. (v) $x^2 - y^2 = a^2$. (vi) $(x^2 + y^2)^2 = a^2(x^2 - y^2)$.

Curve tracing. Cartesian equation | 125

10 Find the Cartesian equations of the curves:

(i) $r = 3$.   (ii) $\theta = \dfrac{\pi}{4}$.   (iii) $r\cos\theta = 2$.

(iv) $r^2 = a^2 \sin\theta$.   (v) $\dfrac{2}{r} = 1 + \cos\theta$.   (vi) $r = a\cos^2\theta$.

Curve tracing. Cartesian equation

Asymptotes Consider the curve $y = \dfrac{1}{x-1}$.

As x approaches the value 1, y approaches either $+\infty$ or $-\infty$, according as x is greater than or less than 1. We say that the line $x = 1$ touches the curve at infinity.

The line $x = 1$ is called *an asymptote* of the curve.

Asymptotes parallel to the axes can be found by solving the equation of the curve for y and for x.

Example 6 Find the asymptotes parallel to the axes of the curve $y(x^2 - 1) = x$.

Solving for y,
$$y = \frac{x}{(x+1)(x-1)}.$$

Hence, $x = -1$, $x = 1$ are asymptotes.

Solving for x, $x^2 y - x - y = 0$,
$$x = \frac{1 \pm \sqrt{1 + 4y^2}}{2y}.$$

Hence, $y = 0$ is an asymptote.

The asymptotes parallel to the axes are $x = -1$, $x = 1$, $y = 0$.

General rule *It will be found that asymptotes parallel to the axes can be obtained by equating to zero the coefficients of the highest powers of x and y.*

e.g. For the curve $xy^3 + x^3 y = 1$, the asymptotes parallel to the axes are $x = 0$ (equating coefficient of y^3 to zero), and $y = 0$ (equating coefficient of x^3 to zero).

Change of origin The equation of a curve can often be simplified by a change of origin.

e.g. If the origin is moved to the point $(2, 0)$, the equation $y^2 = 4(x - 2)$ becomes $y^2 = 4x$.

Example 7 Find the equation of the curve $x^2 + y^2 - 2x + 4y - 4 = 0$, when the origin is moved to the point $(1, -2)$.

Let P, any point on the curve, have coordinates (x, y) with respect to the old axes and (X, Y) with respect to the new axes.

Then, in Fig. 50,
$$PQ = x, \quad PQ' = X, \quad PR = y, \quad PR' = Y.$$
Clearly, $\quad x = X+1; \quad x-1 = X,$
and $\quad\quad\quad y = Y-2; \quad y+2 = Y.$

Substituting for x, y in the given equation,
$$(X+1)^2 + (Y-2)^2 - 2(X+1) + 4(Y-2) - 4 = 0,$$
$$X^2 + Y^2 = 9.$$

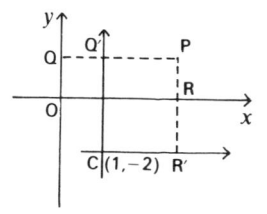

Fig. 50

i.e. When the point $(1, -2)$ is taken as origin, the equation of the curve becomes
$$x^2 + y^2 = 9.$$

Example 8 The equation of a curve is $\dfrac{(x-2)^2}{4} + \dfrac{(y+3)^2}{2} = 1$. Choose a suitable change of origin to simplify the equation.

If (X, Y) be the coordinates of a point relative to the new origin, arrange that $\quad\quad\quad x-2 = X \quad \text{and} \quad y+3 = Y.$
i.e. $\quad\quad\quad\quad x = X+2, \quad y = Y-3.$

Consequently the origin must be moved to the point $(2, -3)$.

In general, if the origin is moved to the point (h, k) the equation of the curve $f(x, y) = 0$ becomes $f(x+h, y+k) = 0$.

Systematic curve tracing

1. *Inspect the equation to detect any symmetry* using the rules:
 (i) If no odd powers of y appear the curve is symmetrical about the x axis.
 (ii) If no odd powers of x appear the curve is symmetrical about the y axis.
2. *Look out for an obvious change of origin* which will lead to simplification.
 e.g. The curve $(y-1)^2 = 4(x+2)^3$ reduces to $y^2 = 4x^3$, a curve symmetrical about the new x axis, when the origin is moved to the point $(-2, 1)$.
3. *Determine any asymptotes parallel to the axes.*
4. *Determine any obvious points on the curve*, e.g. the points where the curve meets the axes.
 e.g. The curve $y^2 = (x+1)(x+4)$ meets the x axis at the points $(-1, 0), (-4, 0)$ and the y axis at the points $(0, 2), (0, -2)$.
5. *Find $\dfrac{dy}{dx}$. Determine maximum and minimum points and any other points where $\dfrac{dy}{dx} = 0$ or $\to \pm\infty$*, i.e. where the tangent to the curve is parallel to or perpendicular to the x axis.
 Also find the gradients of the curve at the points determined in (4).
6. *Determine any limitations on the possible domain of x and range of y.* This can usually be done by solving for y and/or x.
 e.g. Curve $\quad\quad\quad\quad x^2 y^2 = a^2(x^2 - y^2).$

Solving for y, $\quad\quad\quad\quad y^2 = \dfrac{a^2 x^2}{x^2 + a^2}.$

Systematic curve tracing | 127

As $\dfrac{a^2x^2}{x^2+a^2}$ is positive for all real values of x, all values of x are permissible.

Solving for x, $\qquad x^2 = \dfrac{a^2y^2}{a^2-y^2}.$

For real values of x, $\dfrac{a^2y^2}{a^2-y^2}$ must be positive and consequently

$$y^2 < a^2 \quad \text{or} \quad |y| < a.$$

i.e. the whole curve is included between the asymptotes $y = \pm a$.

Example 9 Trace the curve $y = \dfrac{x(x+1)}{x-1}$.

1 There is no symmetry.

2 There is no suitable change of origin.

3 Equating to zero the highest power of y (i.e. y), $x = 1$ is an asymptote.
The term in the highest power of x is $1x^2$ and so there is no asymptote parallel to the x axis.

4 The curve meets the x axis at the points $(0, 0)$ and $(-1, 0)$, and the y axis only at $(0, 0)$.

5
$$\frac{dy}{dx} = \frac{x^2 - 2x - 1}{(x-1)^2}.$$

$\therefore \dfrac{dy}{dx} = 0$ when $x^2 - 2x - 1 = 0.$

$$x = 1 \pm \sqrt{2}.$$
$$y = 3 \pm \sqrt{8}.$$
$$\frac{d^2y}{dx^2} = \frac{(2x-2)(x-1)^2 - 2(x-1)(x^2-2x-1)}{(x-1)^4} = \frac{4}{(x-1)^3}.$$

Consequently $(1+\sqrt{2}, 3+\sqrt{8})$ is a minimum point and $(1-\sqrt{2}, 3-\sqrt{8})$ is a maximum point.

Further the gradients at $(0, 0)$ and $(-1, 0)$ are -1 and $\frac{1}{2}$ respectively.

6 Clearly y exists for all values of x excepting $x = 1$.
Solving for x,
$$x^2 + x(1-y) + y = 0.$$
This equation has real roots only if
$$(1-y)^2 \geq 4y,$$
$$y^2 - 6y + 1 \geq 0,$$
$$(y-3)^2 \geq 8.$$
i.e. $\qquad |y-3| \geq \sqrt{8}.$

Consequently, the curve does not exist for values of y between $3+\sqrt{8}$ and $3-\sqrt{8}$.

A sketch of the curve is given in Fig. 51.

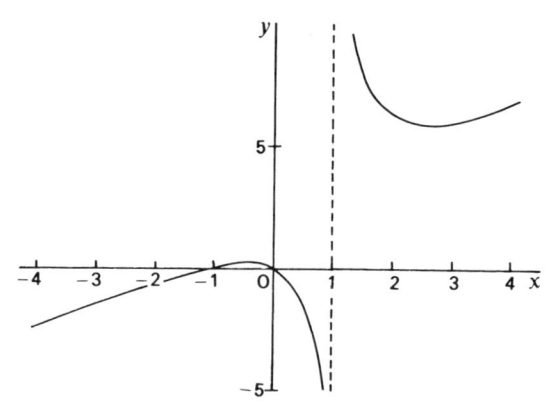

Fig. 51

Example 10 Trace the curve $y = \dfrac{x^2 + 2x + 3}{x^2 + 3x + 2}$.

  1 There is no symmetry.
  2 There is no suitable change of origin.
  3 The asymptotes parallel to Oy are given by
$$x^2 + 3x + 2 = 0;$$
i.e. $\qquad\qquad x = -2 \quad \text{and} \quad x = -1.$

The asymptote parallel to Ox is given by
$$y - 1 = 0$$
(equating to zero the coefficient of x^2).
i.e. $\qquad\qquad y = 1.$

On substituting $y = 1$ in the equation of the curve in addition to an infinite root, there is a root $x = 1$.
i.e. The curve cuts the asymptote $y = 1$ at the point $(1, 1)$.

  4 The curve does not meet the x axis and meets the y axis only when
$$y = \tfrac{3}{2}.$$

  5 $\qquad \dfrac{dy}{dx} = \dfrac{(2x+2)(x^2+3x+2) - (2x+3)(x^2+2x+3)}{(x^2+3x+2)^2}$

$\qquad\qquad = \dfrac{x^2 - 2x - 5}{(x^2+3x+2)^2}.$

$\therefore \dfrac{dy}{dx} = 0 \quad \text{when} \quad x = 1 \pm \sqrt{6}.$

In passing through the value $x = 1 - \sqrt{6}$, $\dfrac{dy}{dx}$ changes from $+$ to $-$;

In passing through the value $x = 1 + \sqrt{6}$, $\dfrac{dy}{dx}$ changes from $-$ to $+$.

Curve tracing. Parametric coordinates | 129

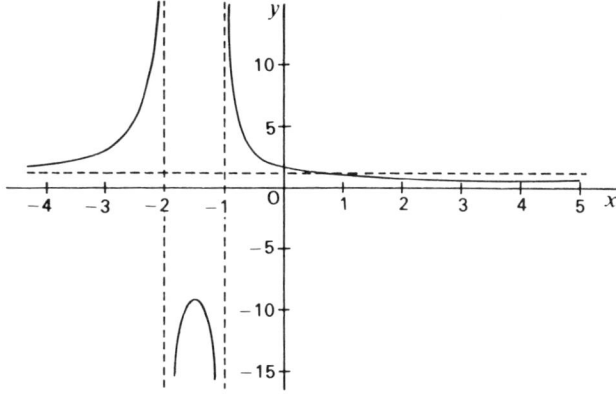

Fig. 52

∴ $(1 - \sqrt{6}, -4 - 2\sqrt{6})$ is a maximum point, and $(1 + \sqrt{6}, 2\sqrt{6} - 4)$ is a minimum point.

6 Solving for x,
$$x^2(y-1) + x(3y-2) + 2y - 3 = 0.$$
For real roots,
$$(3y-2)^2 \geqslant 4(y-1)(2y-3),$$
$$y^2 + 8y \geqslant 8;$$
$$(y+4)^2 \geqslant 24.$$
i.e.
$$|y+4| \geqslant 2\sqrt{6}.$$

Hence the curve does not exist for values of y between $2\sqrt{6} - 4$ and $-4 - 2\sqrt{6}$. A sketch of the curve is given in Fig. 52.

Curve tracing. Parametric coordinates There are no fixed rules of procedure in the case of curve tracing when the curve is given by parametric equations.

If possible the Cartesian equation should be obtained and used.

Example 11 Trace the curve $x = 3t^2 + 1$, $y = 2t^3 - 1$, where t is a parameter.
$$\frac{x-1}{3} = t^2 \quad \text{and} \quad \frac{y+1}{2} = t^3.$$

Dividing,
$$t = \frac{3}{2}\left(\frac{y+1}{x-1}\right).$$

Substituting for t,
$$x - 1 = 3\left\{\frac{3}{2}\left(\frac{y+1}{x-1}\right)\right\}^2,$$

$4(x-1)^3 = 27(y+1)^2$, the Cartesian equation.

1 Changing the origin to $(1, -1)$, the equation becomes $4x^3 = 27y^2$.

130 | Equations of a curve

2 There is symmetry about the new axis of x.
3 There are no asymptotes.
4 The curve only meets the new axes at the origin.
5 Differentiating,
$$12x^2 = 54y\frac{dy}{dx}.$$
$$\frac{dy}{dx} = \pm\frac{2x^2}{9 \cdot \frac{2x^{\frac{3}{2}}}{\sqrt{27}}} = \pm\frac{\sqrt{3}}{3}x^{\frac{1}{2}}.$$

So $\dfrac{dy}{dx} = 0$ when $x = 0$ (i.e. at the new origin).

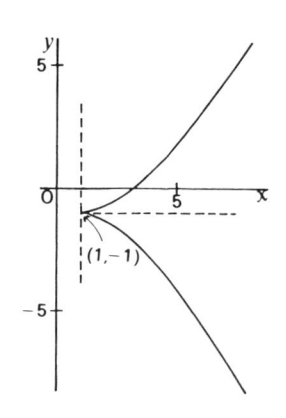

6 As y^2 must be positive, x must be positive.
So the curve only exists if $x > 0$ (referred to new origin).

Fig. 53

A sketch of the curve is given in Fig. 53.

Example 12 Sketch the curve $x = a\,(\theta - \sin\theta)$, $y = a\,(1 - \cos\theta)$ for values of θ between 0 and 2π. Find the area included between this part of the curve and the x axis.

To assist in sketching the curve, we note the following facts:
1 y is always positive and $\leqslant 2a$.
2 $y = 0$ when $\theta = 0, 2\pi$.

3
$$\frac{dy}{dx} = \frac{dy}{d\theta} \times \frac{d\theta}{dx} = \frac{a\sin\theta}{a(1 - \cos\theta)}$$

$$= \frac{2a\sin\dfrac{\theta}{2}\cos\dfrac{\theta}{2}}{2a\sin^2\dfrac{\theta}{2}} = \cot\dfrac{\theta}{2}.$$

$\therefore \dfrac{dy}{dx} \to \infty$ as $\theta \to 0$ from above and $\to -\infty$ as $\theta \to 2\pi$ from below.

Also $\qquad\dfrac{dy}{dx} = 0$ when $\theta = \pi$.

Clearly y has a maximum value $2a$ at this point.

4 The following table of values is obtained:

θ	0	$\dfrac{\pi}{4}$	$\dfrac{\pi}{2}$	$\dfrac{3\pi}{4}$	π	$\dfrac{3\pi}{2}$	2π
x	0	$0{\cdot}08a$	$0{\cdot}57a$	$1{\cdot}65a$	$3{\cdot}14a$	$4{\cdot}71a$	$6{\cdot}28a$
y	0	$0{\cdot}29a$	a	$1{\cdot}71a$	$2a$	a	0

A sketch of the curve is given in Fig. 54.
The area enclosed between the curve and the x axis is $\displaystyle\int_0^{2a\pi} y\,dx$.

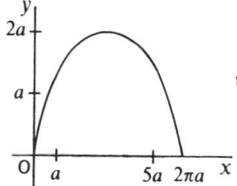

This integral is expressed in terms of the variable θ using the parametric equations of the curve:

$$y = a(1 - \cos\theta),$$
$$\frac{dx}{d\theta} = a(1 - \cos\theta).$$

Fig. 54

Noting that $\int y\,dx = \int y\frac{dx}{d\theta}\,d\theta = \int a^2(1-\cos\theta)^2\,d\theta$, and the limits for θ are 0 and 2π, it follows that

$$\text{the area} = a^2 \int_0^{2\pi} (1 - \cos\theta)^2\,d\theta$$

$$= a^2 \int_0^{2\pi} \left(1 - 2\cos\theta + \frac{\cos 2\theta + 1}{2}\right)d\theta$$

$$= a^2 \left[\frac{3\theta}{2} - 2\sin\theta + \frac{\sin 2\theta}{4}\right]_0^{2\pi}$$

$$= 3\pi a^2.$$

[B] EXAMPLES 9c

1 Sketch the curve $y^2 = x$. Use the result to sketch the following curves
  (i) $y^2 = 4x$.   (ii) $y^2 = 4(x-2)$.   (iii) $(y+1)^2 = x - 3$.
  (iv) $y^2 - 2y = x$.   (v) $y^2 + 3y + 2x + 1 = 0$.

2 Sketch the curves:
$$y = \frac{1}{x}, \quad y = \frac{1}{x+2}, \quad x(y+2) = 1, \quad y = \frac{x}{x+1}.$$

3 Sketch the curves:
  (i) $x = 3\cos\theta$, $y = 3\sin\theta$.   (ii) $x = 2 + 3\cos\theta$, $y = 3\sin\theta - 1$.

4 Trace the curves:
  (i) $x = 2t^2$, $y = 4t$.   (ii) $x = 3 + 2t^2$, $y = -4t$.

Sketch the following curves:

5 $y = x^2(x-1)$.

6 $y = x(2x-1)(x+1)$.

7 $y = \dfrac{x^2}{x+1}$.

8 $y = x - \dfrac{1}{x}$.

9 $y^2 = 4x(1-x)$.

10 $y^2 = x(x+1)$.

11 $y^2(x+1) = 2$.

12 $4(y-1) + x^3 = 0$.

13 $y^2 x = x - 2$.

14 $y = \dfrac{x-1}{x(x+1)}$.

15 $y = \dfrac{x^3}{1-x}$.

16 $y^2 = x^2(4 - x^2)$.

132 | **Equations of a curve**

17 $y^2 = x + \dfrac{1}{x}$.

18 $y^2 = \dfrac{1}{x(x-1)}$.

19 $y = x^2(a-x)$.

20 $y = xe^{-x}$.

21 $x^2 y^2 = a^2(x^2 - y^2)$.

22 $y(a^2 + x^2) = x^3$.

23 Prove that $-\tfrac{1}{2} \leqslant \dfrac{x}{x^2+1} \leqslant \tfrac{1}{2}$. Sketch the graph of the function.

24 Show that the function $\dfrac{2x^2 + x + 2}{2x^2 - x + 2}$ can only take values between $\tfrac{3}{5}$ and $\tfrac{5}{3}$ and sketch its graph.

25 Show that $\dfrac{3x^2 - 3}{6x - 10}$ has a minimum value 3 and a maximum value $\tfrac{1}{3}$. Explain the apparent paradox by means of a rough graph.

26 Show that the function $\dfrac{2x - 1}{2x^2 - 4x + 1}$ is capable of all values and sketch its graph.

Sketch the graphs of the following functions:

27 $\dfrac{(2x+3)(x-6)}{(x+1)(x-2)}$.

28 $\dfrac{1}{(x-2)(x-4)}$.

29 $\dfrac{x^2 - x}{x^2 + x + 1}$.

30 Sketch the curve $x = 2 + 3\cos\theta$, $y = 1 + 2\sin\theta$ and find its area.

31 Find the area of the loop of the curve $4y^2 = x^2(4-x)$.

32 Find the volume of the solid produced by the rotation of the loop of the curve $y^2 = \dfrac{x^2(a+x)}{a-x}$ about the x axis.

33 Trace the curve $x = \cos^3 t$, $y = \sin^3 t$ and find its area.

34 Sketch the curves $y^2 = 4(x-2)$, $y^2 = 2x$, and find the area enclosed between them.

35 Find the area of the curve $x = a\sin 2\theta$, $y = 2a\cos\theta$.

36 The curve $x = a\cos\theta$, $y = b\sin\theta$, is rotated about the x axis, find the volume of revolution.

Curve tracing. Polar coordinates

1. *Look for symmetry.*
   (i) If r is a function of $\cos\theta$ only, there is symmetry about the initial line.
   (ii) If r is a function of $\sin\theta$ only, there is symmetry about the line $\theta = \dfrac{\pi}{2}$.
   (iii) If only even powers of r appear, there is symmetry about the origin.

2. *Determine any limits to the possible values of r and θ.*
   e.g. For the curve $r^2 = a^2 \cos\theta$, it is clear that $|r| \leqslant a$ and also $\cos\theta$ must be positive or zero, restricting θ to the domain $\pm\dfrac{\pi}{2}$.

3. *Form a table of values with* $\theta = 0$, $\pm\dfrac{\pi}{6}$, $\pm\dfrac{\pi}{4}$, $\pm\dfrac{\pi}{3}$, $\pm\dfrac{\pi}{2}$ ... $\pm\pi$.

Example 13 Trace the curve $r = a \cos 2\theta$.

1 As $\cos 2\theta = \cos(-2\theta)$ there is symmetry about the initial line. Also as $\cos 2\theta = 1 - 2\sin^2\theta$, there is symmetry about $\theta = \dfrac{\pi}{2}$. Consequently we need only take values of θ between 0 and $\dfrac{\pi}{2}$.

2 $|r| \leqslant a$.

3 The following table is obtained:

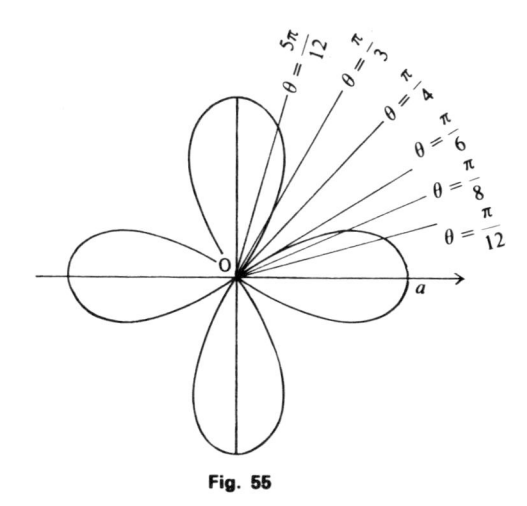

θ	0	$\dfrac{\pi}{12}$	$\dfrac{\pi}{8}$	$\dfrac{\pi}{6}$	$\dfrac{\pi}{4}$	$\dfrac{\pi}{3}$	$\dfrac{3\pi}{8}$	$\dfrac{5\pi}{12}$	$\dfrac{\pi}{2}$
r	0	$\dfrac{\sqrt{3}}{2}a$	$\dfrac{\sqrt{2}}{2}a$	$\dfrac{a}{2}$	0	$\dfrac{-a}{2}$	$\dfrac{-\sqrt{2}}{2}a$	$\dfrac{-\sqrt{3}}{2}a$	$-a$

Fig. 55

Example 14 Trace the curve $r^2 = a^2 \sin 2\theta$.

1 There is symmetry about the line $2\theta = \dfrac{\pi}{2}$, i.e. $\theta = \dfrac{\pi}{4}$, and also about the origin.

2 $\sin 2\theta$ must be positive or zero and hence $0 \leqslant 2\theta \leqslant \pi$.

3 The following table is obtained:

θ	0	$\dfrac{\pi}{12}$	$\dfrac{\pi}{8}$	$\dfrac{\pi}{6}$	$\dfrac{\pi}{4}$
r	0	$\pm 0{\cdot}71a$	$\pm 0{\cdot}84a$	$\pm 0{\cdot}93a$	$\pm a$

(Fig. 56)

134 | Equations of a curve

Sectorial areas To find the sectorial area bounded by the curve $r = f(\theta)$ and the radii vectors OA, OB. Imagine the area split up into small elements of which POQ is typical (Fig. 57).

$$\delta A \approx \tfrac{1}{2} r(r + \delta r) \sin \delta\theta$$
$$\approx \tfrac{1}{2} r^2 \delta\theta.$$

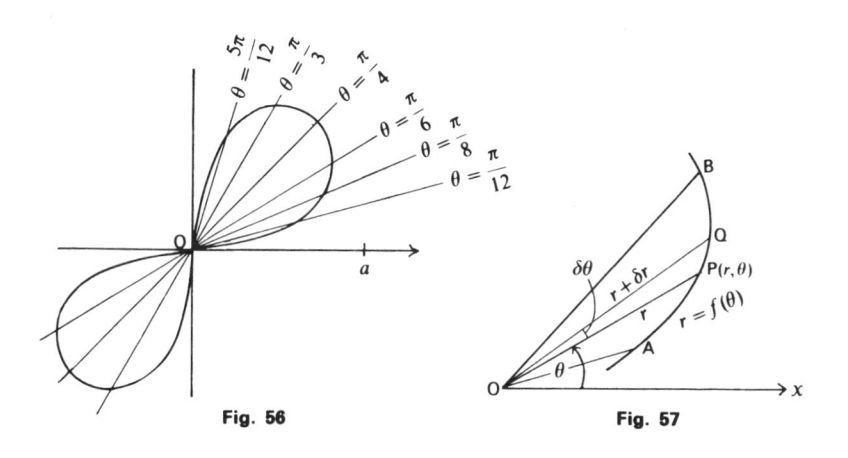

Fig. 56

Fig. 57

$\therefore$ **Sectorial Area** $\approx \underset{\delta\theta \to 0}{\text{Lt}} \sum_{\theta=\alpha}^{\theta=\beta} \tfrac{1}{2} r^2 \delta\theta$ where $\widehat{AOx} = \alpha, \widehat{BOx} = \beta.$

$$= \frac{1}{2} \int_\alpha^\beta r^2 \, d\theta.$$

Example 15 Find the area of one loop of the curve $r = a \cos 2\theta.$

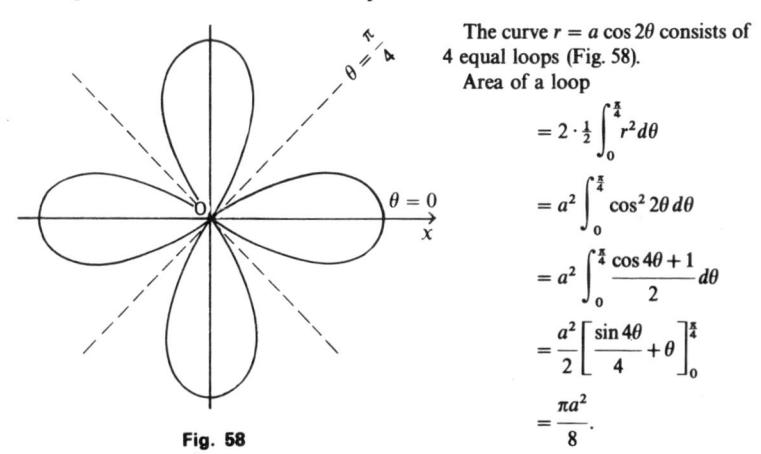

The curve $r = a \cos 2\theta$ consists of 4 equal loops (Fig. 58).
Area of a loop

$$= 2 \cdot \tfrac{1}{2} \int_0^{\frac{\pi}{4}} r^2 d\theta$$

$$= a^2 \int_0^{\frac{\pi}{4}} \cos^2 2\theta \, d\theta$$

$$= a^2 \int_0^{\frac{\pi}{4}} \frac{\cos 4\theta + 1}{2} d\theta$$

$$= \frac{a^2}{2} \left[\frac{\sin 4\theta}{4} + \theta \right]_0^{\frac{\pi}{4}}$$

$$= \frac{\pi a^2}{8}.$$

Fig. 58

[B] EXAMPLES 9d

Trace the following curves:

1 $r = a$.
2 $r = a\cos\theta$.
3 $r = a\sin\theta$.
4 $r = a\sin 2\theta$.
5 $r = a\cos\dfrac{\theta}{2}$,
6 $r^2 = a^2\cos 2\theta$.
7 $r = a(1+\cos\theta)$.
8 $r = a(1-\cos\theta)$.
9 $r = a\cos^2\theta$.
10 $\dfrac{a}{r} = 1 + \cos\theta$.
11 $r\theta = a$.
12 $r = a\cos 3\theta$.

13 Find the area bounded by the spiral $r = a\theta$ and the radii vectors $\theta = \dfrac{\pi}{4}$, $\theta = \dfrac{\pi}{2}$.

14 Find the area of the sector of the curve $r = a(2+\cos\theta)$ bounded by the lines $\theta = 0$, $\theta = \dfrac{\pi}{2}$.

15 In what ratio is the area bounded by the curve $r = a(1+\sin\theta)$ and the lines $\theta = 0$, $\theta = \dfrac{\pi}{2}$, divided by the line $\theta = \dfrac{\pi}{4}$?

16 Show that the curve $r = 2a\cos\theta$ is a circle, radius a. Hence obtain the formula for the area of a circle

17 Find the area enclosed by the curve $r = a(1-\cos\theta)$.

18 Sketch the curve $r = a\sin 2\theta$ and find the area of one loop.

19 Sketch the curve $r^2 = a^2\cos 2\theta$ and find its total area.

20 Obtain sketches of the following curves by using their polar equations:
(i) $(x^2+y^2)^2 = 2a^2xy$.
(ii) $(x^2+y^2)^3 = a^2x^4$.

[B] MISCELLANEOUS EXAMPLES

1 If $y = \dfrac{x}{1+x+x^2}$, find the maximum and minimum values of y and sketch the curve.

2 Plot the curve $y = x\sin x$ for values of x between 0 and 2π. Calculate the areas of the two loops formed by the curve and the axis of x.

3 The curve $y = 1 + \sin x$ is rotated about the x axis. Prove that the volume of revolution between $x = 0$ and $x = \pi$ is $\pi\left(4 + \dfrac{3\pi}{2}\right)$.

4 Trace the curve $x^4 - x^2 + y^2 = 0$ and prove that its area is $\tfrac{4}{3}$.

5 Sketch the curve $y = \dfrac{1-x}{1+x^2}$ and state the possible values of y.

6 Sketch the curve $2y^2 = x(1-x^2)$ and find the volume of the solid formed by rotating the loop about the x axis.

7 Plot the curve $y^2 = \dfrac{a(x-a)(x-4a)}{x-5a}$ for values of x between 0 and $8a$.

8 Trace the curve $r = a(1 + 2\cos\theta)$ and prove that the area of the inner loop is $0.5435a^2$.

9 Prove that $\dfrac{2x^2 - 14x + 11}{2x^2 - 2x + 5}$ lies between -1 and 3. Draw a graph of the function.

10 Sketch the curve $y^2 = x(x-1)^2$ and find the distance of the centre of mass of the loop from the origin.

11 Find the area of the curve $x = a\cos^3 t$, $y = b\sin^3 t$.

12 Trace the curve $xy^2 = a^2(x-a)$ and find the equations of the tangents to the curve which pass through the origin.

13 Find the area of the loop of the curve $r = a\theta\cos\theta$ between $\theta = 0$ and $\theta = \dfrac{\pi}{2}$.

14 Prove that the function $\dfrac{x-a}{x^2-2x+a}$ is capable of all values if $0 < a < 1$. Sketch the graphs of the function when $a = \tfrac{2}{3}$ and $a = 2$.

15 Sketch the curve $y = \dfrac{x^2-1}{x^2-4}$. Find the equation of the tangent at $(1, 0)$ and find where it cuts the curve again.

16 The cycloid has equations $x = a(\theta + \sin\theta)$, $y = a(1 + \cos\theta)$. Find the area of the undulation of the curve between $\theta = \pm\pi$.

17 Find the area of a loop of the curve $r = a(\cos 3\theta + \sin 3\theta)$.

18 Trace the curve $x = a\cos^3 t$, $y = a\sin^3 t$. Find the equation of the tangent at any point parameter t and prove that the length of the tangent intercepted between the axes is constant.

19 Find the area enclosed by the curve $x = 2\cos t$, $y = \cos 2t$ and the x axis.

20 Trace the curve $x = \dfrac{3m}{1+m^3}$, $y = \dfrac{3m^2}{1+m^3}$ for values of m between 0 and $+\infty$. Show that x is a maximum when $2m^3 = 1$.

21 Trace the curve $r = a(3 + 2\cos\theta)$ and show that its area is $11\pi a^2$.

22 Find the turning points on the curve $y^2 = \dfrac{x}{1+x^2}$. Sketch the curve.

23 Find the turning points on the curve $y = 2\sin x - \sin 2x$, for x between 0 and 2π, and trace this portion of the curve.

24 Prove that the area of a loop of the curve $x = a\sin 2t$, $y = a\sin t$ is $\dfrac{4a^2}{3}$.

25 Trace the curve $x = t - t^3$, $y = 1 - t^4$ and prove that it forms a loop of area $\tfrac{16}{35}$. Find the coordinates of the centre of gravity of this loop.

26 Calculate the area between the curves $y = xe^{-x}$, $y = xe^x$ and the line $x = 1$.

27 Find the equation of the tangent at the point $(4am^2, 8am^3)$ on the curve $x^3 = ay^2$ and prove that it meets the curve again at the point $(am^2, -am^3)$. If $9m^2 = 2$, show that the tangent is also a normal to the curve.

28 For the curve $x = (1+t)^2$, $y = t(1+t)^2$, find the turning values of y and sketch the curve. Show that the normal to the curve at the origin intersects the curve again and find the point of intersection.

29 Find the Cartesian equation of the curve $x = \sin^2 t$, $y = \sin^3 t \cos t$, and sketch the curve. Show that the line $x = \tfrac{1}{2}$ divides the area enclosed by the curve in the ratio $3\pi - 4$ to $3\pi + 4$.

30 Sketch the curve $x = a\left(\dfrac{1}{t} - t\right)$, $y = a(1-t^2)$.

Find the area enclosed between the curve and the line $y = a$ and show that the volume of revolution of this area about the x axis is $64\pi a^3/15$.

10 | Analytical geometry
The straight line

The straight line The following important results related to the straight line have already been obtained:

(i) *Any equation of the first degree represents a straight line and conversely the equation of a straight line is always of the first degree.*

(ii) *The equation $y = mx + c$ represents a straight line of gradient m, meeting the y axis at distance c from the origin.*

(iii) *The equation of the straight line passing through the point (h, k) and having gradient m is $y - k = m(x - h)$.*

(iv) *The angle θ between two straight lines gradients m_1 and m_2 is given by*
$$\tan\theta = \frac{m_1 - m_2}{1 + m_1 m_2}.$$

The lines are parallel if $m_1 = m_2$ and perpendicular if $m_1 m_2 = -1$.

(v) *The length of the perpendicular from the point (h, k) to the straight line $ax + by + c = 0$ is $\pm \dfrac{ah + bk + c}{\sqrt{a^2 + b^2}}$.*

The sign is chosen in order to make the perpendicular from the origin positive.

(vi) *The equations of the bisectors of the angles between two straight lines $ax + by + c = 0$, $a'x + b'y + c' = 0$ are given by*
$$\frac{ax + by + c}{\sqrt{a^2 + b^2}} = \pm \frac{a'x + b'y + c'}{\sqrt{a'^2 + b'^2}}.$$

Example 1 PN, the perpendicular from P(3, 4) to the line $2x + 3y = 1$ is produced to Q such that NQ = PN. Find the coordinates of Q.

Gradient of PQ $= -1/-\frac{2}{3} = \frac{3}{2}$.
Equation of PQ is
$$y - 4 = \tfrac{3}{2}(x - 3),$$
i.e. $\quad 3x - 2y = 1$.

Solving this equation simultaneously with the equation $2x + 3y = 1$, we obtain the coordinates of N, $(\tfrac{5}{13}, \tfrac{1}{13})$.

If the coordinates of Q are (α, β), then the coordinates of the mid-point of PQ,
i.e. N, are $\dfrac{\alpha + 3}{2}, \dfrac{\beta + 4}{2}$.

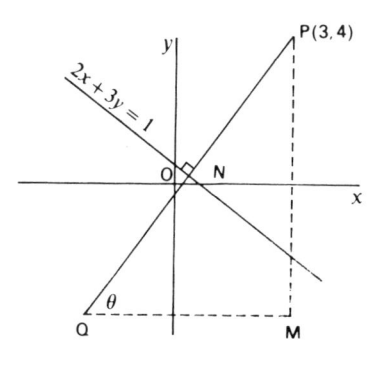

Fig. 59

138 | Analytical geometry. The straight line

$$\therefore \frac{\alpha+3}{2} = \frac{5}{13}; \quad \alpha = -2\tfrac{3}{13}.$$

$$\frac{\beta+4}{2} = \frac{1}{13}; \quad \beta = -3\tfrac{11}{13}.$$

Coordinates of Q are $(-2\tfrac{3}{13}, -3\tfrac{11}{13})$.

Example 2 Determine whether the points $(3, -2)$, $(-1, 7)$ are on the same or opposite sides of the line $2x - 5y = 13$.

Length of perpendicular from the origin to the given line $= \pm \dfrac{-13}{\sqrt{29}}$.

To make this positive, the negative sign is chosen.

Length of perpendicular from $(3, -2) = -\dfrac{6+10-13}{\sqrt{29}} = -\dfrac{3}{\sqrt{29}}$.

Length of perpendicular from $(-1, 7) = -\dfrac{-2-35-13}{\sqrt{29}} = +\dfrac{50}{\sqrt{29}}$.

Hence $(-1, 7)$ is on the same, and $(3, -2)$ the opposite side, of the line to the origin. Consequently the points are on opposite sides of the line.

Example 3 ABCD is a square; A is the point $(0, -2)$ and C the point $(5, 1)$, AC being a diagonal. Find the coordinates of B and D.

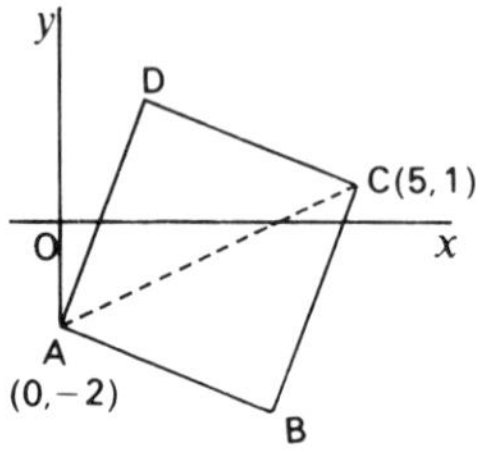

AD, AB each make angles of $45°$ with AC.

Gradient of $AC = \dfrac{1-(-2)}{5-0} = \dfrac{3}{5}$.

So, if m_1, m_2 are the gradients of AD and AB,

$$\tan 45° = \frac{m_1 - \tfrac{3}{5}}{1 + \dfrac{3m_1}{5}},$$

and

$$\tan 45° = \frac{\tfrac{3}{5} - m_2}{1 + \dfrac{3m_2}{5}}.$$

Fig. 60

i.e.
$$1 = \frac{5m_1 - 3}{5 + 3m_1}; \quad m_1 = 4.$$

$$1 = \frac{3 - 5m_2}{5 + 3m_2}; \quad m_2 = -\tfrac{1}{4}.$$

Gradients of AB and $DC = -\tfrac{1}{4}$.
Gradients of AD and $BC = 4$.
Equation of AB is $y + 2 = -\tfrac{1}{4}x$ or $4y + x + 8 = 0$.
Equation of BC is $y - 1 = 4(x - 5)$ or $y - 4x + 19 = 0$.
Solving these equations, we find B is the point $(4, -3)$.
Similarly, using the equations of AD and DC, we find D is the point $(1, 2)$.

[A] EXAMPLES 10a

1 Find the equations of the straight lines through $(2, -1)$ parallel and perpendicular to the line $2y - 5x = 4$.

2 Find the acute angle between the two lines $2y - x = 3$, $3y + 4x = 5$.

Various forms of the equation of a straight line | 139

3 The coordinates of two points A, B are $(3, -1), (-2, 2)$. Find the equation of the perpendicular bisector of AB.

4 Find the distance of the point $(-2, 1)$ from the line $2y - x - 7 = 0$.

5 The vertices of a triangle are $A(0, 0), B(-2, 1), C(1, 4)$. Find the length and the equation of the median through B.

6 Find the equations of the lines bisecting the angles between the lines $y = 3x$, $y = x + 2$. Verify that the bisectors are perpendicular.

7 Prove that the quadrilateral with vertices $(2, 1), (2, 3), (5, 6), (5, 4)$ is a parallelogram.

8 Find the equations of the straight lines drawn through the point $(1, -2)$, making angles of $45°$ with the x axis.

9 Find the coordinates of the foot of the perpendicular from the point $(-1, 1)$ to the line $2y + 4x = 7$.

10 The vertices of a triangle have coordinates $A(3, 1), B(1, 5), C(5, 3)$. Prove that the triangle is isosceles and find its area.

11 Prove that the lines $2y - x - 5 = 0$, $2y + x - 5 = 0$, $y + 2x - 5 = 0$ are all tangents to a circle whose centre is at the origin. What is the radius of the circle?

12 The coordinates of A, B, C, three vertices of a rectangle $ABCD$, are $(1, 3), (-1, 1)$, $(2, -2)$. Find the coordinates of D and the area of the rectangle.

13 Find the equation of the interior bisector of the angle A of $\triangle ABC$ whose sides BC, CA, AB have equations $y = 0, y - x = 0, y = 3x - 4$.

14 Find the coordinates of the point which lies midway between the origin and the line $2y - 5x = 6$.

15 The two straight lines $3y + ax = 5$, $12y - ax = 1$ are perpendicular. Find the values of a.

16 The coordinates of A, B, C, the vertices of triangle ABC, are $(1, 1), (2, 5), (4, -1)$. Find (i) angle ABC, (ii) the length of the altitude through A.

17 What are the coordinates of the reflection of the point $(1, 1)$ in the line $\frac{x}{3} + \frac{y}{4} = 1$?

18 Find the equations of the two lines drawn through the point $(-2, -1)$ which are inclined at $45°$ to the line $y - 2x = 3$.

19 Prove that the lines $2x - 3y = 4, 6y = 4x + 3$ are parallel and find the distance between them.

20 Are the points $(1, -2), (-2, 1)$ on the same or opposite sides of the line $3x - 5y = 2$?

Various forms of the equation of a straight line

The gradient form Any equation of the first degree can be expressed in the form
$$y = mx + c.$$
m is the gradient and c the intercept the line makes on the y axis.

The intercept form Consider the equation.
$$\frac{x}{a} + \frac{y}{b} = 1.$$

140 | Analytical geometry. The straight line

It is of the first degree and consequently represents a straight line.
Also when $y = 0$, $x = a$; when $x = 0$, $y = b$.
∴ a, b are the intercepts made by the line on the x and y axes respectively.

The perpendicular form   Let the perpendicular OM from the origin to the straight line AB be p, always considered positive. Let α be the angle OM makes with the positive direction of the x axis, considered positive in a counterclockwise direction.

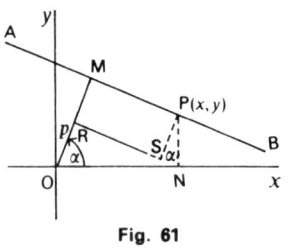

Let $P(x, y)$ be any point on the line.
Then, from Fig. 61,

$$OM = p = OR + RM$$
$$= OR + SP$$
$$= ON \cos \alpha + PN \sin \alpha$$
$$= x \cos \alpha + y \sin \alpha.$$

i.e. **$x \cos \alpha + y \sin \alpha = p$.**

Fig. 61

To express the equation of a straight line in the perpendicular form
Take the straight line $5x - 12y + 39 = 0$.
We require to express this equation in the form

$$x \cos \alpha + y \sin \alpha = p,$$

where p is positive.

Rearranging the given equation,

$$-5x + 12y = 39.$$

Dividing by $\sqrt{(-5)^2 + (12)^2}$, i.e. 13,

$$-\tfrac{5}{13}x + \tfrac{12}{13}y = 3.$$

As $(-\tfrac{5}{13})^2 + (\tfrac{12}{13})^2 = 1$, we can take
$-\tfrac{5}{13} = \cos \alpha$ and $\tfrac{12}{13} = \sin \alpha$.

∴ $x \cos \alpha + y \sin \alpha = 3$,
where $\cos \alpha = -\tfrac{5}{13}$,
              $\sin \alpha = \tfrac{12}{13}$,
              $\alpha = 112° \, 36'$.

The result is illustrated in Fig. 62.

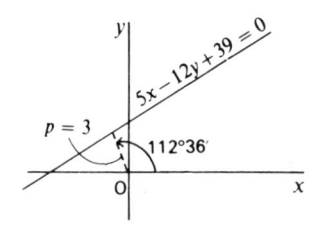

Fig. 62

Length of perpendicular from the point (h, k) to the straight line $x \cos \alpha + y \sin \alpha = p$.

Let P be the point (h, k) (Fig. 63).
Take a parallel line through P to the given line. The length of the perpendicular from the origin to this line is $p + p'$.
∴   Equation of parallel line is

$$x \cos \alpha + y \sin \alpha = p + p'.$$

Fig. 63

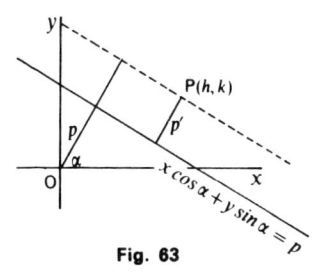

But (h, k) lies on this line,

$$\therefore \quad h\cos\alpha + k\sin\alpha = p + p',$$
$$p' = h\cos\alpha + k\sin\alpha - p.$$

Thus, *the perpendicular distance of a given point from the straight line* $x\cos\alpha + y\sin\alpha = p$ *is obtained by substituting the coordinates of the point in the expression*

$$\mathbf{x\cos\alpha + y\sin\alpha - p.}$$

Example 4 Find the distance between the parallel lines $ax + by + c = 0$, $ax + by + d = 0$.

Length of perpendicular from the origin on to line $ax + by + c = 0$

$$= \frac{c}{\sqrt{a^2 + b^2}}.$$

Length of perpendicular from the origin on to line $ax + by + d = 0$

$$= \frac{d}{\sqrt{a^2 + b^2}}.$$

$\therefore$ Distance between the parallel lines $= \dfrac{c}{\sqrt{a^2+b^2}} \sim \dfrac{d}{\sqrt{a^2+b^2}}$

$$= \frac{c \sim d}{\sqrt{a^2 + b^2}}.$$

Equation of a straight line passing through the point of intersection of two given straight lines Let the equations of the two given straight lines be

$$a_1 x + b_1 y + c_1 = 0,$$
$$a_2 x + b_2 y + c_2 = 0.$$

Consider the equation

$$a_1 x + b_1 y + c_1 + \lambda(a_2 x + b_2 y + c_2) = 0,$$

where λ is independent of x and y, and may take any constant value.

For any value of λ, this equation, being of the first degree, represents a straight line. Furthermore, the values of x and y which satisfy the two given equations (i.e.

the coordinates of the point of intersection of the given lines) clearly satisfy the third equation. In other words, the third equation represents a straight line passing through the point of intersection of the two given straight lines.

Thus, *all straight lines passing through the point of intersection of the two lines* $a_1x+b_1y+c_1 = 0$, $a_2x+b_2y+c_2 = 0$, *are given by the equation*

$$a_1x+b_1y+c_1 + \lambda(a_2x+b_2y+c_2) = 0,$$

where λ is any constant.

Example 5 Find the equation of the straight line joining the point of intersection of the lines $4x - y = 7$, $2x + 3y = 1$ to the origin.

Any line through the intersection of the given lines is of the form

$$4x - y - 7 + \lambda(2x + 3y - 1) = 0.$$

If this line passes through the origin, the equation is satisfied when $x = y = 0$.

$$\therefore \quad -7 - \lambda = 0; \quad \lambda = -7.$$

Hence, the required equation is

$$4x - y - 7 - 7(2x + 3y - 1) = 0$$

or,
$$-10x - 22y = 0$$
$$5x + 11y = 0.$$

Example 6 Show that, for all values of m, the line

$$x(5m+1) + y(2m-3) + 2 - 3m = 0$$

passes through the point of intersection of two fixed lines.

The equation $\quad x(5m+1) + y(2m-3) + 2 - 3m = 0$

can be written $\quad x - 3y + 2 + m(5x + 2y - 3) = 0.$

This is the equation of a straight line passing through the point of intersection of the lines $x - 3y + 2 = 0$, $5x + 2y - 3 = 0$.

To find the coordinates of the points which divide the line joining two given points in a given ratio.

Internal division In Fig. 64(a), P divides AB internally in the ratio $m:n$.
External division In Fig. 64(b), P divides AB externally in the ratio $m:n$.

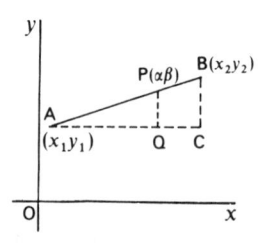

Fig. 64 a

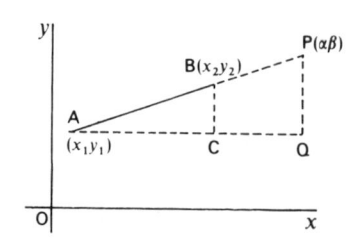

Fig. 64 b

Internal and external division | 143

In both cases take A, B as the points (x_1, y_1), (x_2, y_2) and let P be the point (α, β).

Internal division	*External division*
By similar triangles,	
$\dfrac{AP}{PB} = \dfrac{AQ}{QC} = \dfrac{m}{n}.$	$\dfrac{AP}{PB} = \dfrac{AQ}{QC} = \dfrac{m}{n}.$
But $AQ = \alpha - x_1$, $QC = x_2 - \alpha$.	But $AQ = \alpha - x_1$, $QC = \alpha - x_2$.
$\therefore \dfrac{\alpha - x_1}{x_2 - \alpha} = \dfrac{m}{n},$	$\dfrac{\alpha - x_1}{\alpha - x_2} = \dfrac{m}{n},$
$n\alpha - nx_1 = mx_2 - m\alpha$	$n\alpha - nx_1 = m\alpha - mx_2$
$\alpha(m+n) = nx_1 + mx_2$	$\alpha(n-m) = nx_1 - mx_2$
$\alpha = \dfrac{nx_1 + mx_2}{m+n}.$	$\alpha = \dfrac{nx_1 - mx_2}{n-m}.$
Similarly,	Similarly,
$\beta = \dfrac{ny_1 + my_2}{m+n}.$	$\beta = \dfrac{ny_1 - my_2}{n-m}.$

It is useful to notice that the results for external division can be obtained from those for internal division by merely changing the sign of *either* m or n.

Example 7 A, B are the points $(-1, 2)$, $(3, 1)$. Find the coordinates of the points P and Q which divide AB internally and externally in the ratio $3:5$.

Let P be the point (α_1, β_1).

Then $\alpha_1 = \dfrac{nx_1 + mx_2}{m+n}$ where $x_1 = -1$, $x_2 = 3$
$\hspace{10em} m = 3$, $n = 5$

$= \dfrac{-5+9}{8} = \dfrac{1}{2}.$

$\beta_1 = \dfrac{ny_1 + my_2}{m+n}$ where $y_1 = 2$, $y_2 = 1$.

$= \dfrac{10+3}{8} = \dfrac{13}{8}.$

i.e. P is the point $(\tfrac{1}{2}, \tfrac{13}{8})$.

Let Q be the point (α_2, β_2).

Then $\alpha_2 = \dfrac{nx_1 + mx_2}{m+n}$ where $x_1 = -1$, $x_2 = 3$
$\hspace{10em} m = -3$, $n = 5$

$= \dfrac{-5-9}{-3+5} = -7.$

$$\beta_2 = \frac{ny_1 + my_2}{m+n} \qquad \text{where } y_1 = 2, \quad y_2 = 1$$

$$= \frac{10-3}{-3+5} = \frac{7}{2}.$$

i.e. Q is the point $(-7, \frac{7}{2})$.

Example 8 Find the coordinates of the centre of gravity of the $\triangle ABC$ with vertices $A(2, 0)$, $B(-3, 2)$, $C(0, 4)$.

The centre of gravity G of the triangle is at the point of intersection of the medians, i.e. if CZ is a median, G lies one third the way up the median from C or G divides ZC in the ratio $1:2$.

Coordinates of Z, the midpoint of AB, are $(-\frac{1}{2}, 1)$.

Let G be the point (α, β).

Then $\qquad \alpha = \dfrac{nx_1 + mx_2}{m+n}$

where $\qquad m = 1, \qquad n = 2$
$\qquad\qquad x_1 = -\frac{1}{2}, \quad x_2 = 0$

$$\alpha = \frac{-1+0}{3} = -\frac{1}{3}.$$

Similarly, $\qquad \beta = \dfrac{2+4}{3} = 2.$

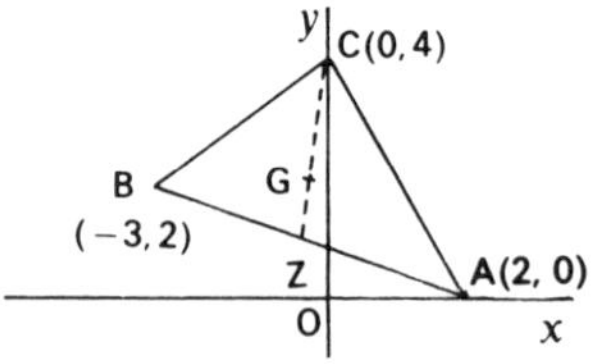

Fig. 65

Example 9 Find the ratio in which the line $4x - y = 3$ divides the line joining the points $(2, -1)$, $(-3, 2)$.

Any point P on the line joining $(2, -1)$, $(-3, 2)$ can be written as $\left(\dfrac{2-3\lambda}{1+\lambda}, \dfrac{-1+2\lambda}{1+\lambda} \right)$, as these are the coordinates of the point which divides the given line in the ratio $\lambda : 1$.

Now P lies on the line $4x - y = 3$, if

$$4\left(\frac{2-3\lambda}{1+\lambda}\right) - \left(\frac{-1+2\lambda}{1+\lambda}\right) = 3,$$

$$8 - 12\lambda + 1 - 2\lambda = 3 + 3\lambda,$$

$$17\lambda = 6,$$

$$\lambda = \tfrac{6}{17}.$$

Hence, the line $4x - y = 3$ divides the line joining $(2, -1)$, $(-3, 2)$ internally in the ratio $6:17$.

[A] EXAMPLES 10b

1 Find the equation of the line joining the points $(2, 3)$, $(-3, 1)$ in (i) the gradient form, (ii) the intercept form.

2 Sketch the following lines:

(i) $\dfrac{x}{3} + \dfrac{y}{1} = 1$.

(ii) $x \cos 20° + y \sin 20° = 2$.

(iii) $5x - 3y = 6$.

(iv) $y = x \tan 35° - 2$.

3 Express the following equations in the perpendicular form:
   (i) $x \cos 60° - y \sin 60° = 4$. (ii) $x \cos 40° + y \sin 40° = -2$.
   (iii) $3x + 4y = 10$. (iv) $4x - 3y + 1 = 0$.
   (v) $5x - 12y + 26 = 0$. (vi) $\sqrt{2}x - y = \sqrt{3}$.

4 Write down the lengths of the perpendiculars from the point (2, 3) to the lines:
   (i) $x \cos 45° + y \sin 45° = 1$. (ii) $x \cos 60° + y \sin 60° = 3$.
   (iii) $5x + 12y - 13 = 0$. (iv) $x \sin \alpha + y \cos \alpha = 1$.

5 Determine whether the points $(-4, 8)$, $(-5, -8)$ are on the same or opposite sides of the line $3x - 4y + 20 = 0$.

6 Prove that the lines $7x + 2y = 5$, $6x + 3y = 5$, $5x + 4y = 5$ are concurrent.

7 Find the equation of the line joining the point $(1, -1)$ to the common point of the lines $2x - y = 6$, $x - 3y = 2$.

8 Find the coordinates of the points which divide the line joining $(-3, 0)$, $(2, 1)$ internally and externally in the ratio $4:5$.

9 Find the equation of the line through the intersection of the lines $y - x = 5$, $2x + y = 7$, which is parallel to $y + x = 0$.

10 Find the ratio in which the line joining the points $(0, 1)$, $(3, 2)$ is divided by the line $x + y = 2$.

11 Find the equation of the line drawn through the intersection of the lines $3x - 4y = 10$, $2y - 3x + 8 = 0$, which is perpendicular to the line $4x - 5y = 3$.

12 P is the point $(3, 4)$ and O is the origin. The line through P perpendicular to OP is drawn, find its equation in the form $x \cos \theta + y \sin \theta = p$, giving θ to the nearest minute.

13 The lengths of the perpendiculars from P on to $3x + 4y + 19 = 0$ and $3x - 4y + 13 = 0$ are in the ratio $2:3$. Show that P must lie on one or other of the lines $15x + 4y + 83 = 0$ or $3x + 20y + 31 = 0$.

14 Find the ratio in which the line joining $(-1, 0)$, $(2, 2)$ is divided by the line joining the origin to the point $(-1, 4)$.

15 Find the equation of the straight line through the intersection of $2x - 5y = 1$, $x + 3y = 2$, perpendicular to the former line.

16 A triangle is formed by the lines $y = 0$, $y - x = 0$, $2y - 5x + 6 = 0$. Find the coordinates of its centre of gravity.

17 Show that for all values of m the straight line
$$x(2m - 3) + y(3 - m) + 1 - 2m = 0$$
passes through the point of intersection of two fixed lines. For what values of m does the given line bisect the angles between the two fixed lines?

18 Find the equation of the straight line drawn through the point $(2, 0)$ which divides the line joining $(-2, -1)$, $(4, 3)$ internally in the ratio $2:3$.

19 If the gradients m_1, m_2 of two lines are the roots of the quadratic equation $3m^2 - 6m - 2 = 0$, find the tangents of the angles between the lines.

20 A triangle is formed by the lines $(AB)x - 2y = 0$, $(AC)x + y = 3$, $(BC)2y + x = 5$. Find the equation of the altitudes through A and B and hence find the coordinates of their point of intersection. Verify that the third altitude also passes through the point, the orthocentre of the triangle.

21 If P, Q divide the line joining $(4, 1)$, $(0, 3)$ internally and externally in the ratio $3:5$, find the angle POQ, where O is the origin.

22 The equations of two sides AB, AD of a parallelogram are $x - y = 2$,

$2x + 3y = 4$; the coordinates of vertex C are (4, 5). Find the equation of diagonal AC and the angles CAB, CAD.

23 Two lines $4x - 8y = 11$, $7x + 3y = 22$ intersect at P. Find the equation of the straight line joining P to the origin.

[B] MISCELLANEOUS EXAMPLES

1 The coordinates of the vertices of a triangle are $(2, 4)$, $(-1, 8)$, $(-1, -2)$. Find (i) the equations of the sides of the triangle, (ii) its angles to the nearest minute.

2 Find the equations of the two lines through the point $(2, -3)$ which make angles of $45°$ with the line $2x - y = 2$.

3 Obtain the equations of the straight lines which pass through the point $(5, 6)$ and are parallel to the lines $2x + 3y - 8 = 0$, $5x - 4y - 3 = 0$. Find the area of the parallelogram formed by the four lines.

4 Find the equations of the six bisectors of the angles between the lines $x + 7y - 3 = 0$, $17x - 7y + 3 = 0$, $x - y + 1 = 0$ and prove that three of these lines pass through the point $(1, 1)$.

5 A straight line AB is produced to C so that $BC = 2AB$. If the coordinates of A and B are $(-1, 1)$, $(3, -1)$ respectively, find the coordinates of C. Find also the equation of the perpendicular bisector of AB.

6 Prove that the lines joining the point $(1, -2)$ to the points $(-3, 0)$, $(5, 6)$ are perpendicular. Calculate the fourth vertex of the rectangle of which these points are three vertices.

7 The coordinates of A, B, C are $(-3, -1)$, $(11, 13)$, $(-1, -3)$ respectively. Find the coordinates of the point of intersection of the medians of triangle ABC.

8 Obtain the equation of the bisector of that angle formed by the lines $2x - y = 3$, $3x + 4y = 8$, which contains the origin.

9 Find the values of the angles of the triangle formed by the lines $y - 2x + 1 = 0$, $y + 3x - 19 = 0$, $x - 3y + 7 = 0$. Also find the coordinates of the circumcentre of the triangle.

10 Show that the feet of the perpendiculars from the point $(6, 4)$ to the sides of the triangle whose vertices are $(5, 1)$, $(-3, 1)$, $(1, 9)$ lie on a straight line.

11 Without drawing a figure, determine whether the points $(1, 2)$, $(4, -7)$ lie in the same, adjacent, or opposite angles formed by the lines $2y - 5x = 3$, $3y + x = 11$.

12 Find the area of the triangle with sides $2y + x = 0$, $3y + 2x + 4 = 0$, $2y - 3x + 8 = 0$.

13 The equations of four lines EAB, BCF, CDE, FDA are

$$3x - 2y + 1 = 0,\ 4x - y + 2 = 0,\ 2x + y + 2 = 0$$

and $$2(3x - 2y + 1) - 3(4x - y + 2) - (2x + y + 2) = 0.$$

Obtain, without finding the coordinates of B and D, the equation of the straight line BD.

14 Find the coordinates of the incentre of the triangle formed by the lines $y - x = 0$, $y + 2x = 3$, $2y - 4x = 7$.

15 Prove that the area of the triangle formed by the three lines $y = x$, $y = 2x$, $3x + 4y = 10$ is $\frac{50}{77}$.

Find also the coordinates of the centre of the circumscribed circle of the triangle and verify the result by careful drawing.

16 Find the equation of the straight line through the point $(-2, 2)$ which divides the line joining the points $(1, 5)$, $(4, -1)$ internally in the ratio $3:2$.

Find the ratio in which the line $2x - 3y + 1 = 0$ divides the line joining $(2, 1)$, $(4, -1)$.

17 Find the area of the convex quadrilateral whose vertices are the points $(2, -1)$, $(5, 0)$, $(4, 6)$, $(0, 3)$.

18 PN, the perpendicular from $P(3, 4)$ on the line $2x + 3y = 1$, is produced to Q so that $2NQ = 3PN$. Find the coordinates of Q.

19 Find the coordinates of the point on the line $3x - 4y = 14$ which is equidistant from the points $(2, 6)$, $(-6, 2)$.

20 The coordinates of A, C opposite vertices of a square $ABCD$, are $(-1, 1)$, $(4, -3)$. Find the coordinates of B and D.

21 Find the area of the figure bounded by the lines $4y - 3x = 13$, $y + 2x = 6$, $y - x + 6 = 0$, $y + x + 2 = 0$. What are the coordinates of the point of intersection of the diagonals of the figure?

22 Show that the equation of the perpendicular bisector of the line joining the points (a, b), (c, d) is $(a - c)x + (b - d)y = \frac{1}{2}(a^2 + b^2 - c^2 - d^2)$.

Find the coordinates of the centre of the circle which passes through the points $(1, 2)$, $(-1, -1)$, $(2, -2)$.

23 Two adjacent sides of a parallelogram have equations $4x + 5y = 18$, $7x + 2y = 0$ and one diagonal is given by the equation $11x + 7y = 9$. Find the coordinates of the vertices of the parallelogram.

24 Find λ so that the lines with equations

$$8x + 15y + \lambda = 0$$

and

$$(221 + 16\lambda)x + (30\lambda - 1428)y + \lambda^2 - 2023 = 0,$$

are perpendicular.

25 Write down the general equation of a straight line through the point (a, b). A line is drawn through $(1, 2)$ to cut the axes Ox, Oy at H, K respectively and parallelogram $OHKL$ is completed. Find the coordinates of L in terms of m the gradient of the line and deduce that the locus of L as m varies is the curve $2x - y = xy$.

26 The triangle ABC has vertices $(0, h)$, $(-k, 0)$, $(k, 0)$. Find the equations of its medians and determine their point of concurrence.

27 If the line $p \equiv ax + by + c = 0$ bisects at right angles the line joining the points $P_1(x_1, y_1)$, $P_2(x_2, y_2)$, find x_2, y_2 in terms of a, b, c, x_1, y_1.

Prove that, if P_2 lies on the line $p' \equiv a'x + b'y + c' = 0$, then P_1 lies on the line $2(aa' + bb')p - (a^2 + b^2)p' = 0$.

28 The vertices A, B, C of a triangle are $(0, 0)$, (x_1, y_1), (x_2, y_2) respectively. Show that the tangent of angle B is

$$\pm (x_1 y_2 - x_2 y_1)/\{x_1(x_1 - x_2) + y_1(y_1 - y_2)\}.$$

29 Prove that the coordinates of the image of the point (x', y') in the line $lx + my + n = 0$ are given by

$$\frac{x - x'}{l} = \frac{y - y'}{m} = -\frac{2(lx' + my' + n)}{l^2 + m^2}.$$

30 Show that the points (x_1, y_1), (x_2, y_2) are the ends of one diagonal of a parallelogram and (x_3, y_3), (x_4, y_4) those of the other, provided that $x_1 + x_2 = x_3 + x_4$ and $y_1 + y_2 = y_3 + y_4$. Show also that the parallelogram is a rectangle if $x_1 x_2 + y_1 y_2 = x_3 x_4 + y_3 y_4$.

31 Prove that the portions of the axes Ox, Oy intercepted between the bisectors of the angles between the lines $ax + by + c = 0$, $a'x + b'y + c' = 0$ will be equal in length if $c'(a \pm b) = c(a' \pm b')$.

32 Through a given point F on the diagonal BD of square $ABCD$, lines are drawn parallel to the sides meeting AB in G, DA in H; show that the lines BH, CF and DG are concurrent.

33 Without drawing a figure, determine whether the point $(7, 11)$ is inside or outside the triangle formed by the lines $4x + 3y = 60$, $y = x + 8$, $y = 2x - 20$.

11 | The circle

Definition The circle is defined as the locus of a point moving at a given distance, the radius, from a fixed point, the centre.

Equation of a circle of given radius and centre Let $C(a, b)$ be the centre of the circle and r the radius.

Let $P(x, y)$ be any point on the circle. In Fig. 66, $PQ = y - b$, $CQ = x - a$.
By Pythagoras,
$$PC^2 = PQ^2 + CQ^2,$$
i.e. $\quad r^2 = (y - b)^2 + (x - a)^2.$

The equation of the circle is
$$(x - a)^2 + (y - b)^2 = r^2.$$

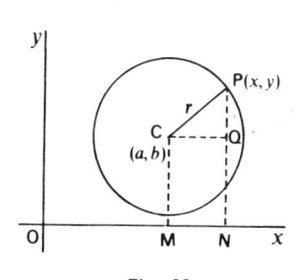

Fig. 66

Special case If the origin is the centre of the circle, the equation reduces to
$$x^2 + y^2 = r^2.$$

Example 1 Find the equation of the circle, centre $(2, -1)$, radius 4.
Equation is $\quad (x - 2)^2 + (y + 1)^2 = 4^2$
or $\quad x^2 + y^2 - 4x + 2y - 11 = 0.$

Example 2 Find the centre and radius of the circle whose equation is
$$x^2 + y^2 - 6x + 4y + 4 = 0.$$
The equation $\quad x^2 + y^2 - 6x + 4y + 4 = 0$
can be written
$$x^2 - 6x + (\tfrac{6}{2})^2 + y^2 + 4y + (\tfrac{4}{2})^2 = -4 + 9 + 4,$$
i.e. $\quad (x - 3)^2 + (y + 2)^2 = 9.$

Comparing this equation with the standard form, it is seen that the circle has centre $(3, -2)$ and radius 3.

Example 3 Find the equation of the circle, centre $(1, 2)$, which touches the line $3x - 4y + 10 = 0$.

As the line is a tangent, the radius of the circle is the perpendicular distance of the point $(1, 2)$ from the line $3x - 4y + 10 = 0$.

i.e. $\quad$ Radius $= \dfrac{3 \cdot 1 - 4 \cdot 2 + 10}{\sqrt{3^2 + 4^2}}$

$\quad = \tfrac{5}{5} = 1.$

∴ Equation of circle is $(x - 1)^2 + (y - 2)^2 = 1^2$
or $\quad x^2 + y^2 - 2x - 4y + 4 = 0.$

Example 4 Prove that the circle $x^2 + y^2 + 10(x + y) + 25 = 0$ touches the x and y axes and find the points of contact.

The equation can be written

$$x^2 + 10x + (5)^2 + y^2 + 10y + (5)^2 = -25 + 25 + 25$$

or $$(x+5)^2 + (y+5)^2 = 25.$$

∴ The circle has centre $(-5, -5)$ and radius 5.

As the distances of the centre from the two axes are each equal to the radius, the circle touches both axes.

Clearly the points of contact are $(-5, 0)$ and $(0, -5)$.

[A] **EXAMPLES 11a**

Find the centres and radii of the following circles:

1 $x^2 + y^2 - 2x - 2y - 2 = 0$.
2 $x^2 + y^2 - 6x + 4y + 3 = 0$.
3 $x^2 + y^2 - 8x = 0$.
4 $x^2 + y^2 + 10y - 25 = 0$.
5 $2x^2 + 2y^2 - 8x + 8y + 3 = 0$.
6 $3(x^2 + y^2) - 6x + 9y + 5 = 0$.

Find the equations of the following circles:

7 Centre $(0, 0)$, radius 5.
8 Centre $(1, 2)$, radius 3.
9 Centre $(-2, 3)$, radius 7.
10 Centre $(3, -7)$, radius 6.
11 Centre $(4, 3)$, touching the x axis.
12 Centre $(-2, -3)$, touching the y axis.
13 Centre (a, a), radius $\sqrt{2a}$.
14 Centre $(a, a+2)$, radius $a+1$.
15 Centre $(1, 1)$, touching the line $2x - y + 4 = 0$
16 Centre $(3, -2)$, touching the line $x + y - 3 = 0$.
17 Centre $(0, 5)$, touching the line $3y - 4x + 5 = 0$.
18 Find the equation of the diameter of the circle $x^2 + y^2 - 4x + 2y = 0$ which passes through the origin.
19 Show that the line $3x - 2y + 13 = 0$ touches the circle

$$x^2 + y^2 - 12x + 8y - 65 = 0.$$

20 Prove that the line $4y + 3x = 75$ touches the circle

$$x^2 + y^2 - 12x - 16y + 75 = 0$$

at the point $(9, 12)$.

21 Find the equation of the diameter of the circle

$$x^2 + y^2 - 6x + 2y - 10 = 0,$$

one extremity of which is the point $(1, 3)$. Also find the coordinates of the other extremity.

22 Show that the line $2x - 3y + 26 = 0$ is a tangent to the circle

$$x^2 + y^2 - 4x + 6y - 104 = 0,$$

and find the equation of the diameter through the point of contact.

23 Find the equation of the tangent to the circle $x^2 + y^2 = 25$ at the point $(3, -4)$.

24 A circle passes through the origin and through the points $(4, 0)$, $(2, 3)$. Find the coordinates of its centre and the equation of the tangent at the origin.

25 Prove that the circles

$$x^2 + y^2 - 4x + 8y - 30 = 0, \qquad x^2 + y^2 + 8x - 16y + 30 = 0.$$

are equal in area. Find the length of their common chord.

General equation of the circle | 151

26 Find the equation of the tangent to the circle $x^2 + y^2 = 10x$ at the point (2, 4).

27 Find the length of the tangents which can be drawn from the point (5, 6) to the circle $x^2 + y^2 - 2x - 4y = 4$.

28 Find the equation of the normal to the circle $x^2 + y^2 - 8x + 2y = 9$ which passes through the origin.

29 Show that the circles $x^2 + y^2 = 9$, $x^2 + y^2 - 6x - 8y + 9 = 0$ intersect at right angles.

30 Show that the circles $x^2 + y^2 - 2x - 4y - 4 = 0$, $x^2 + y^2 - 8x - 12y + 48 = 0$ touch externally and find the point of contact.

31 Find the equation of the circle centre (1, −2), which touches the circle $x^2 + y^2 - 2x - 15 = 0$ internally.

32 Prove that the circle $x^2 + y^2 + 6x - 8y = 0$ lies entirely inside the circle $x^2 + y^2 + 4x - 4y - 53 = 0$.

33 P is any point (x, y) on the circle with diameter AB, where A, B are the points (1, 1), (2, 3) respectively. By writing down the gradients of AP, PB, show that the equation of the circle can be written as

$$(x-1)(x-2) + (y-1)(y-3) = 0.$$

34 Use the method of example 33 to prove that the equation of the circle having points (x_1, y_1), (x_2, y_2) as the ends of a diameter is

$$(x-x_1)(x-x_2) + (y-y_1)(y-y_2) = 0.$$

General equation of the circle. The equation of the circle, centre (a, b), radius r, is

$$(x-a)^2 + (y-b)^2 = r^2,$$

or, $$x^2 + y^2 - 2ax - 2by + a^2 + b^2 - r^2 = 0.$$

Conversely, the equation $x^2 + y^2 + 2gx + 2fy + c = 0$ for values of the constants g, f and c satisfying the condition $g^2 + f^2 - c > 0$ represents a circle.

This equation is called the *general equation* of the circle and it is important to note that in this general equation,

(i) *the coefficients of x^2 and y^2 are equal,*

(ii) *there is no term in xy.*

To find the centre and radius of the circle $x^2 + y^2 + 2gx + 2fy + c = 0$
Completing the squares of the terms in x and y,

$$x^2 + y^2 + 2gx + 2fy + c = (x+g)^2 + (y+f)^2 + c - g^2 - f^2.$$

So the equation of the circle is

$$(x+g)^2 + (y+f)^2 = g^2 + f^2 - c.$$

Hence the centre is the point $(-g, -f)$ and the radius is $\sqrt{g^2 + f^2 - c}$.

Example 5 Find the centre and radius of the circle $2x^2 + 2y^2 - 7x + 4y - 3 = 0$.

First divide throughout by 2 to make the coefficients of x^2 and y^2 unity,

$$x^2 + y^2 - \tfrac{7}{2}x + 2y - \tfrac{3}{2} = 0.$$

Comparing this equation with the general equation, it follows that
$$g = -\tfrac{7}{4}, \quad f = 1, \quad c = -\tfrac{3}{2}.$$
Hence the centre is $(\tfrac{7}{4}, -1)$ and the radius $\sqrt{(\tfrac{7}{4})^2 + 1^2 + \tfrac{3}{2}}$ or $\dfrac{\sqrt{89}}{4}$.

Equation of the tangent at any point (x_1, y_1) on the circle $x^2 + y^2 = r^2$

$$x^2 + y^2 = r^2.$$

Differentiating w.r. to x, $\quad 2x + 2y\dfrac{dy}{dx} = 0,$

i.e. $\qquad\qquad\qquad\qquad \dfrac{dy}{dx} = -\dfrac{x}{y}.$

So the gradient of the tangent at the point $(x_1, y_1) = -\dfrac{x_1}{y_1}$.

Equation of tangent at (x_1, y_1) is
$$y - y_1 = -\dfrac{x_1}{y_1}(x - x_1),$$
or, $\qquad\qquad\qquad yy_1 + xx_1 = x_1^2 + y_1^2.$

Since the point (x_1, y_1) lies on the circle $x^2 + y^2 = r^2$,
$$x_1^2 + y_1^2 = r^2.$$

∴ The equation of the tangent at (x_1, y_1) is
$$\mathbf{xx_1 + yy_1 = r^2}.$$

Example 6 Write down the equations of the tangent and normal at the point $(5, -12)$ on the circle $x^2 + y^2 = 169$.

Equation of tangent at (x_1, y_1) is $xx_1 + yy_1 = 169$.
∴ Equation of tangent at $(5, -12)$ is $5x - 12y = 169$.
Gradient of this tangent is $\tfrac{5}{12}$ and hence the gradient of the normal at $(5, -12)$ is $-\tfrac{12}{5}$.
Equation of normal at $(5, -12)$ is $y + 12 = -\tfrac{12}{5}(x - 5)$
or, $\qquad\qquad\qquad\qquad 12x + 5y = 0.$

Equation of the tangent at any point (x_1, y_1) on the circle, $x^2 + y^2 + 2gx + 2fy + c = 0.$

$$x^2 + y^2 + 2gx + 2fy + c = 0.$$

Differentiating w.r. to x, $\quad 2x + 2y\dfrac{dy}{dx} + 2g + 2f\dfrac{dy}{dx} = 0.$

$$\dfrac{dy}{dx} = -\dfrac{(x + g)}{(y + f)}.$$

Length of the tangents from a given point to a circle | 153

So, gradient of tangent at the point $(x_1, y_1) = -\dfrac{(x_1 + g)}{(y_1 + f)}$.

The equation of the tangent at (x_1, y_1) is

$$y - y_1 = -\dfrac{(x_1 + g)}{(y_1 + f)}(x - x_1),$$

or, $\quad yy_1 + yf - y_1^2 - y_1 f = -xx_1 + x_1^2 - gx + gx_1$

$$xx_1 + yy_1 + gx + fy = x_1^2 + y_1^2 + gx_1 + fy_1.$$

Adding $gx_1 + fy_1 + c$ to each side and noting that, as (x_1, y_1) lies on the circle,

$$x_1^2 + y_1^2 + 2gx_1 + 2fy_1 + c = 0,$$

$$\mathbf{xx_1 + yy_1 + g(x + x_1) + f(y + y_1) + c = 0.}$$

[Note that in the equation of the circle, x^2 is replaced by xx_1, y^2 by yy_1, $2gx$ by $g(x + x_1)$ and $2fy$ by $f(y + y_1)$.]

Example 7 Find the equation of the tangent at the point $(0, 2)$ on the circle

$$x^2 + y^2 + 8x - 6y + 8 = 0.$$

Use the general result, $xx_1 + yy_1 + g(x + x_1) + f(y + y_1) + c = 0$, where $x_1 = 0$, $y_1 = 2$, $g = 4$, $f = -3$, $c = 8$.

Required equation is $\quad 2y + 4(x + 0) - 3(y + 2) + 8 = 0$,

i.e. $\quad\quad\quad\quad\quad\quad\quad\quad 4x - y + 2 = 0.$

Alternatively, this result can be obtained from first principles.

To find the length of the tangents from a given point to a circle

Example 8 Find the length of the tangents from the point $P(5, 7)$ to the circle

$$2x^2 + 2y^2 + 8x - 5y = 0.$$

First obtain the centre and radius of the circle.

Equation is $\quad\quad\quad x^2 + y^2 + 4x - \tfrac{5}{2}y = 0.$

$\therefore$ Centre O is the point $(-2, \tfrac{5}{4})$; radius $= \sqrt{4 + \tfrac{25}{16}} = \dfrac{\sqrt{89}}{4}$.

If T is the point of contact of one of the tangents, it follows by Pythagoras that,

(Length of tangent)^2 = (Distance between point and centre)^2 − (Radius)^2,

i.e. $\quad\quad\quad PT^2 = OP^2 - (\text{radius})^2$

$\quad\quad\quad\quad\quad\quad = (5 - (-2))^2 + (7 - \tfrac{5}{4})^2 - \tfrac{89}{16}$

$\quad\quad\quad\quad\quad\quad = 49 + \tfrac{529}{16} - \tfrac{89}{16}$

$\quad\quad\quad\quad\quad\quad = \tfrac{1224}{16}.$

$\therefore \quad PT = \dfrac{\sqrt{1224}}{4} = \dfrac{\sqrt{306}}{2} = 8\cdot 75$ units.

Length of tangents from P to circle = $8\cdot 75$ units.

The circle

General Case. To find the length of the tangents from the point (h, k) to the circle $x^2 + y^2 + 2gx + 2fy + c = 0$

Centre of circle, O, is $(-g, -f)$; radius is $\sqrt{g^2 + f^2 - c}$.
∴ Square of length of tangents from (h, k)

$$= \{(h+g)^2 + (k+f)^2\} - (g^2 + f^2 - c)$$
$$= h^2 + k^2 + 2gh + 2fk + c.$$

Thus, the *square of the length of the tangent from a given point (h, k) to the circle $x^2 + y^2 + 2gx + 2fy + c = 0$ is obtained by substituting $x = h$ and $y = k$ in the equation of the circle.*

E.g., taking Example 8, equation of circle is

$$x^2 + y^2 + 4x - \frac{5y}{2} = 0.$$

Note The coefficients of x^2 and y^2 must be unity before the general result can be applied.

Substituting $x = h = 5$, $y = k = 7$,

$$(\text{Tangent})^2 = 25 + 49 + 20 - \tfrac{35}{2}$$
$$= \tfrac{306}{4}.$$

i.e. Length of tangent $= \dfrac{\sqrt{306}}{2}$ as before.

[A] EXAMPLES 11b

Write down the centres and radii of the following circles:

1. $x^2 + y^2 - 5x + y - 3 = 0$.
2. $x^2 + y^2 + 8y = 0$.
3. $2x^2 + 2y^2 - 4x + 6y - 5 = 0$.
4. $3x^2 + 3y^2 - 2x + 3y + 1 = 0$.
5. $(x-2)(x-4) + (y-3)(y-5) = 0$.
6. $x(x+3) + y(y-4) = 0$.
7. $(x+y+3)^2 + (x-y-3)^2 = 36$.

8 Find the equation of the circle centre $(-1, 2)$ which passes through the point $(2, 5)$.

9 Find the coordinates of the points in which the circle, centre $(2, -3)$, radius 4, cuts the x axis.

10 Find the length of the chord cut off on the y axis by the circle
$$x^2 + y^2 - 2x + 2y - 3 = 0.$$

Find the equations of the tangents and normals to the following circles at the specified points:

11. $x^2 + y^2 = 5$; $(-2, 1)$.
12. $x^2 + y^2 = 25$; $(3, -4)$.
13. $4(x^2 + y^2) = 9$; $(0, \tfrac{3}{2})$.
14. $x^2 + y^2 + 3x - 2y = 0$; $(0, 0)$.
15. $x^2 + y^2 - 6x + 3y - 5 = 0$; $(-1, -2)$.
16. $2(x^2 + y^2) - x + 4y = 15$; $(3, 0)$.
17. $x^2 + y^2 - 6x + 4y + 3 = 0$; $(0, -1)$.

Equation of the circle passing through three given points

Find the lengths of the tangents to the following circles from the points stated:
18 (6, 8) to the circle $x^2 + y^2 = 9$.
19 (−3, −2) to the circle $x^2 + y^2 = 3$.
20 (5, 6) to the circle $x^2 + y^2 - 2x - 4y = 4$.
21 (6, 8) to the circle $2x^2 + 2y^2 + 8x - 12y = 5$.
22 (0, 0) to the circle $3x^2 + 3y^2 + 12x + 8y + 43 = 0$.
23 Prove that the tangents from the point (2, −3) to the circles
$$x^2 + y^2 + 8x - 4y + 13 = 0, \quad x^2 + y^2 - 2x - 10y + 15 = 0$$
are equal in length.

24 Find the angle between the tangents to the circle $x^2 + y^2 - y - 3 = 0$ at the points where $x = 1$ and also find their point of intersection.

25 Find the angles at which the circle $x^2 + y^2 - 6x - 4y + 3 = 0$ meets the y axis.

26 Find the coordinates of the point on the circle
$$x^2 + y^2 - 12x - 4y + 30 = 0$$
which is nearest the origin.

27 Find the equation of the tangent at the point (3, −4) to the circle $x^2 + y^2 = 25$. What are the equations of the two tangents parallel to the y axis?

Show that the first tangent intersects these two tangents in points which subtend a right angle at the origin.

28 Show that the circles
$$x^2 + y^2 + 6y + 8 = 0, \quad x^2 + y^2 - 12x - 10y - 60 = 0$$
touch each other and find the equation of the tangent at the point of contact.

29 Two circles have centres (0, 0), (4, 0) and radii 1, 2 respectively. Prove that their external common tangents meet at the point (−4, 0) and have gradients $\pm \dfrac{1}{\sqrt{15}}$.

30 A is the point (2, 9) and C the point (1, 2). A line AP is drawn through A parallel to the line $3x = 4y$, to touch at P a circle, centre C. Find the equation of AP and the equation of the circle.

To find the equation of the circle passing through three given points

Example 9 Find the equation of the circle through the points (1, 6), (3, 2), (2, 3).
Let the equation of the circle be $x^2 + y^2 + 2gx + 2fy + c = 0$.
Then, as (1, 6) lies on the circle,

$$1 + 36 + 2g + 12f + c = 0.$$
Similarly, $\qquad 9 + 4 + 6g + 4f + c = 0$
and $\qquad 4 + 9 + 4g + 6f + c = 0.$

Simplifying the equations,
$$2g + 12f + c = -37,$$
$$6g + 4f + c = -13,$$
$$4g + 6f + c = -13.$$

Solving these simultaneous equations, we get
$$f = g = -6, \quad c = 47.$$
Thus, the equation of the circle is $x^2 + y^2 - 12x - 12y + 47 = 0$.

156 | The circle

Alternative Method The same result can be obtained by finding the centre of the circle as the point of intersection of the perpendicular bisectors of any two of the lines joining the given points.

Orthogonal circles Circles which intersect at right angles are called orthogonal circles. It is readily seen that in the case of two circles cutting orthogonally, the square of the line joining the centres is equal to the sum of the squares of the two radii.

Example 10 Prove that the circles $x^2 + y^2 = 9$, $x^2 + y^2 + 8x + 9 = 0$ cut orthogonally.
For the first circle, the centre is $(0, 0)$; radius = 3.
For the second circle, the centre is $(-4, 0)$; radius = $\sqrt{16-9} = \sqrt{7}$.
(Line joining centres)^2 = 16.
Sum of squares of radii = $9 + 7 = 16$.

Hence, the circles are orthogonal.

System of circles passing through the points of intersection of two given circles Let the equations of the given circles be

$$x^2 + y^2 + 2gx + 2fy + c = 0, \dots\dots\dots\dots\dots\dots\dots \text{(i)}$$
$$x^2 + y^2 + 2g'x + 2f'y + c' = 0. \dots\dots\dots\dots\dots\dots\dots \text{(ii)}$$

Consider the equation,

$$x^2 + y^2 + 2gx + 2fy + c + \lambda(x^2 + y^2 + 2g'x + 2f'y + c') = 0. \dots \text{(iii)}$$

As the coefficients of x^2 and y^2 are equal and there is no term in xy, equation (iii) represents a circle for all values of λ.

Also it is clear that any pair of values of x and y which satisfy (i) and (ii) simultaneously will also satisfy (iii).

Thus, equation (iii) represents for all values of λ, a circle passing through the points of intersection of the circles (i) and (ii).

Example 11 Find the equation of the circle which passes through the points of intersection of the circles $x^2 + y^2 - 6x - y + 4 = 0$, $x^2 + y^2 + 7x + 4y - 1 = 0$ and through the point $(4, 1)$.

Any circle through the points of intersection of the given circles has the equation,

$$x^2 + y^2 - 6x - y + 4 + \lambda(x^2 + y^2 + 7x + 4y - 1) = 0. \dots\dots\dots\dots \text{(i)}$$

If this circle passes through the point $(4, 1)$,

$$16 + 1 - 24 - 1 + 4 + \lambda(16 + 1 + 28 + 4 - 1) = 0$$
$$-4 + 48\lambda = 0$$
$$\lambda = \tfrac{1}{12}.$$

So, replacing λ by $\tfrac{1}{12}$ in equation (i),

$$12(x^2 + y^2 - 6x - y + 4) + (x^2 + y^2 + 7x + 4y - 1) = 0$$

or $13(x^2 + y^2) - 65x - 8y + 47 = 0$ is the equation of the required circle.

[B] EXAMPLES 11c

Find the equations of the circles passing through the following points:
1. (1, 1), (1, 0), (3, 2).
2. (0, 2), (0, −3), (4, 1).
3. (2, 3), (4, −1), (2, −1).
4. (−3, 2), (−2, 5), (2, 1).

Show that the following pairs of circles cut orthogonally:
5. $x^2 + y^2 = 4$, $x^2 + y^2 - 2x + 4y + 4 = 0$.
6. $x^2 + y^2 + 6y - 5 = 0$, $x^2 + y^2 - 8x + 5 = 0$.
7. $x^2 + y^2 - x + 6y + 7 = 0$, $x^2 + y^2 + 2x + 2y - 2 = 0$.
8. $x^2 + y^2 - ax + c = 0$, $x^2 + y^2 + by - c = 0$.

9. Find the equation of the circle which passes through the origin and through the common points of the two circles

$$x^2 + y^2 - 25 = 0, \quad x^2 + y^2 - 8x + 5 = 0.$$

10. Find the coordinates of the centre and the radius of the circle through the common points of the circles

$$x^2 + y^2 - 7x - 12 = 0, \quad x^2 + y^2 + 8x - 12 = 0$$

and passing through the point (1, 2).

11. Prove that the quadrilateral with vertices (7, 1), (6, 4), (−2, 4), (5, 5) is cyclic.

12. The equation $x^2 + y^2 + kx - 5k - 16 = 0$, where k can take any constant value, represents a system of circles. Find (i) the equation of the circle of this system which passes through the point (6, 1), (ii) the equation of the circle of this system whose centre is at the point (−2, 0).

13. For what values of k is the radius of the circle

$$x^2 + y^2 + 6x - 6 + k(x^2 + y^2 - 8x + 7) = 0$$

equal to zero? What are the coordinates of the centres of these point circles?

14. Show that the equation of any circle passing through the points of intersection of the line $x + y - 2 = 0$ and the circle $x^2 + y^2 - 36 = 0$ can be written as $x^2 + y^2 - 36 + \lambda(x + y - 2) = 0$.

15. Find the equation of the circle passing through the origin which meets the circle $x^2 + y^2 - 4x - 6y - 10 = 0$ at the ends of the chord $x + y = 5$.

Miscellaneous examples

Example 12 Find the equations of the tangents drawn from the point (−4, 3) to the circle $x^2 + y^2 = 5$.

Any straight line passing through the point (−4, 3) has an equation of the form

$$y - 3 = m(x + 4) \quad \text{where } m \text{ is the gradient} \quad \ldots\ldots\ldots\ldots\ldots\ldots \text{(i)}$$

The line (i) meets the circle, $x^2 + y^2 = 5$, in two points whose x coordinates are the roots of the equation,

$$x^2 + \{m(x+4) + 3\}^2 = 5,$$
$$x^2 + m^2(x+4)^2 + 6m(x+4) + 9 = 5,$$

i.e. $\quad x^2(1 + m^2) + x(8m^2 + 6m) + 16m^2 + 24m + 4 = 0.$

The straight line (i) is a tangent to the circle, if this equation has equal roots,

i.e. if $\quad \{2m(4m+3)\}^2 = 16(1 + m^2)(4m^2 + 6m + 1)$
$$4m^2(16m^2 + 24m + 9) = 16(4m^4 + 6m^3 + 5m^2 + 6m + 1),$$
$$36m^2 = 80m^2 + 96m + 16,$$
$$44m^2 + 96m + 16 = 0,$$
$$11m^2 + 24m + 4 = 0. \quad \ldots\ldots\ldots\ldots\ldots\ldots\ldots\ldots\ldots\ldots\ldots\ldots\text{(ii)}$$

Factorising, $(11m+2)(m+2) = 0$.

Hence, $m = -2$ and $-\frac{2}{11}$.

Substituting these values for m in (i), it follows that the equations of the tangents from the point $(-4, 3)$ to the circle are

$$y - 3 = -2(x+4), \quad \text{i.e.} \quad y + 2x + 5 = 0;$$

and $\quad y - 3 = -\frac{2}{11}(x+4), \quad \text{i.e.} \quad 11y + 2x - 25 = 0.$

Example 13 Given two fixed points A (1, 1), B (2, 3), show that the locus of a point P which moves so that $AP:PB = 2:1$ is a circle. Verify that the centre of the circle is the midpoint of the points dividing AB internally and externally in the ratio $2:1$.

Let the coordinates of P be (α, β).

Then $\quad AP^2 = (\alpha - 1)^2 + (\beta - 1)^2$

and $\quad BP^2 = (\alpha - 2)^2 + (\beta - 3)^2.$

As $AP:PB = 2:1$, $AP^2 = 4PB^2$.

$$\therefore (\alpha - 1)^2 + (\beta - 1)^2 = 4\{(\alpha - 2)^2 + (\beta - 3)^2\}$$
$$3\alpha^2 + 3\beta^2 - 14\alpha - 22\beta + 50 = 0.$$

Replacing α by x and β by y, we get the locus of P is the curve,

$$3(x^2 + y^2) - 14x - 22y + 50 = 0.$$

This is the equation of a circle, centre $(\frac{7}{3}, \frac{11}{3})$.

If (x_1, y_1), (x_2, y_2) are the points dividing AB internally and externally in the ratio $2:1$,

$$x_1 = \frac{1+4}{1+2} = \frac{5}{3}; \quad x_2 = \frac{-1+4}{-1+2} = 3.$$

$$y_1 = \frac{1+6}{1+2} = \frac{7}{3}; \quad y_2 = \frac{-1+6}{-1+2} = 5.$$

Hence the midpoint of (x_1, y_1), (x_2, y_2) is the point $\left(\frac{\frac{5}{3}+3}{2}, \frac{\frac{7}{3}+5}{2}\right)$.

i.e. $(\frac{7}{3}, \frac{11}{3})$, the centre of the circle.

Example 14 Find the equation of the locus of the centre of a circle which touches the line $2x - y = 3$ and passes through the point (2, 3).

Let the coordinates of the centre be (α, β).

As the circle passes through (2, 3), its radius $= \sqrt{(\alpha - 2)^2 + (\beta - 3)^2}$. Also the radius is equal to the perpendicular distance of (α, β) from the line $2x - y = 3$.

i.e. $\quad$ Radius $= \dfrac{2\alpha - \beta - 3}{\sqrt{5}}.$

$$\therefore \sqrt{(\alpha - 2)^2 + (\beta - 3)^2} = \frac{2\alpha - \beta - 3}{\sqrt{5}},$$

$$5(\alpha^2 - 4\alpha + 4 + \beta^2 - 6\beta + 9) = 4\alpha^2 - 4\alpha\beta + \beta^2 - 12\alpha + 6\beta + 9,$$

or, $\quad \alpha^2 + 4\alpha\beta + 4\beta^2 - 8\alpha - 36\beta + 56 = 0.$

As this is the relation between the x and y coordinates of any point on the locus of the centre, the equation of the locus is

$$x^2 + 4xy + 4y^2 - 8x - 36y + 56 = 0.$$

Example 15 Find the general equation of a circle passing through the points of intersection of the circle $x^2 + y^2 + 2gx + 2fy + c = 0$ and the straight line $lx + my + n = 0$.

Consider the equation
$$x^2 + y^2 + 2gx + 2fy + c + \lambda(lx + my + n) = 0,$$
where λ is any constant.

As the coefficients of x^2 and y^2 are the same and there is no term in xy, this equation represents a circle.

Furthermore, any pair of values of x and y satisfying the two given equations (i.e. the coordinates of the points of intersection of the given line and circle), must also satisfy this equation.

Hence the equation of any circle through the common points of the given line and circle is
$$x^2 + y^2 + 2gx + 2fy + c + \lambda(lx + my + n) = 0.$$

[B] MISCELLANEOUS EXAMPLES

1 Find the equation of the circle which passes through the point $(2, -1)$ and meets the circle $x^2 + y^2 - 5x + 8y - 2 = 0$ at the ends of the chord $2y - x + 1 = 0$.

2 Show that the circle $x^2 + y^2 - 2ax - 2ay + a^2 = 0$ touches the coordinate axes. Find the equation of the circle passing through the points $(2, 3)$, $(4, 5)$, $(6, 1)$.

3 Find the equations of the tangents from the origin to the circle $x^2 + y^2 - 6x + 2 = 0$.

4 Verify that the point $(1, 1)$ is the centre of the inscribed circle of the triangle with vertices $(2, -1)$, $(2, 6)$, $(-\frac{2}{9}, \frac{2}{3})$.

Find the equation of the circle and show that it touches the coordinate axes.

5 Show that the circles
$$4(x^2 + y^2) - 12x - 16y - 11 = 0 \quad \text{and} \quad 4(x^2 + y^2) - 60x + 48y + 173 = 0$$
touch. Show also that the circle $36(x^2 + y^2) - 276x + 513 = 0$ cuts each of these circles orthogonally.

6 Find the equation of that diameter of the circle
$$x^2 + y^2 + 6x - 4y - 23 = 0$$
which passes through the point $(5, -2)$.

Find also the equation of the perpendicular diameter.

7 Find the acute angle between the two tangents which can be drawn from the point $(-3, 4)$ to the circle $x^2 + y^2 = 10$.

8 Find the equation of the circle with centre $(12, 5)$ which touches the circle $x^2 + y^2 = 16$ externally.

Show also that $119y - 120x = 676$ is a common tangent.

9 A point moves such that its distance from the point $(3, -1)$ is twice its distance from the point $(0, 5)$. Show that the locus of the point is a circle and find the centre and radius.

10 Find the equations of the tangents to the circle $x^2 + y^2 = 4$ which are parallel to the line $y = 2x$.

11 Find the equation of the circle which passes through the points $(2, 0)$, $(4, 0)$, and has the line $2y = 3x - 5$ as a diameter.

12 Find the centres of the two circles which can be drawn to pass through the points $(1, 1)$, $(2, 0)$ and touch the line $y = x + 4$.

13 Prove that, for all values of m, the lines $y = mx \pm a\sqrt{m^2 + 1}$ are tangents to the circle $x^2 + y^2 = a^2$.

Find the equations of the tangents to the circle $x^2 + y^2 = 25$ which are parallel to $y + x = 0$.

14 Find (i) the length of, (ii) the acute angle between, the tangents from the origin to the circle $3x^2 + 3y^2 + 12x + 8y + 8 = 0$.

15 A point moves so that the lengths of the tangents from it to the circles
$$x^2 + y^2 - 4x - 5y + 1 = 0, \quad x^2 + y^2 + 3x + 6y - 2 = 0$$
are equal. Show that the locus of the point is a line perpendicular to the line joining the centres of the circles.

16 Find the cosine of the acute angle in which the two circles
$$x^2 + y^2 - 3x = 0, \quad x^2 + y^2 - 2x - 2y - 2 = 0$$
intersect.

17 Prove that the equation of the common chord of the circles
$$S \equiv x^2 + y^2 + 2gx + 2fy + c = 0; \quad S' \equiv x^2 + y^2 + 2g'x + 2f'y + c' = 0,$$
is $S - S' = 0$.

Deduce the equation of the common chord of the circles
$$x^2 + y^2 = 16, \quad x^2 + y^2 - 10x - 6y + 9 = 0.$$

18 Without drawing a scale diagram, show that the circle
$$x^2 + (y - 1)^2 = 16$$
lies completely inside the circle $x^2 + y^2 - x = 26$.

19 Find the coordinates of the centres of the circles which touch the x axis and pass through the points $(2, 2)$, $(-5, 1)$.

20 If the straight line $y = mx + c$ touches the circle $x^2 + y^2 + 4x = 4$, prove that $4m^2 + 4mc - c^2 + 8 = 0$. Prove that the length of the tangents from the point $(2, 1)$ to this circle is equal to 3 and find the inclinations of these tangents to the x axis.

21 Show that for all values of c, $x(x - 1) + y(y - 1) = c(x + y - 1)$ represents a circle passing through the points $(1, 0)$, $(0, 1)$.

Find the equation of such a circle with centre on the line $x + 2y = 6$, and also find the value of c which gives a circle touching the x axis.

22 A, B, C are the points $(-2, -4)$, $(3, 1)$, $(-2, 0)$. Find the equation of the circle passing through A, B, C and show that the tangent at B is parallel to the diameter through C.

23 A point moves such that the length of the tangent from it to the circle $x^2 + y^2 = a^2$ is equal to its distance from the point $(2a, 2a)$. Prove that the locus of the point is the straight line $4x + 4y = 9a$.

24 A circle is drawn having as a diameter the chord of intersection of the circle $x^2 + y^2 - 2x + 8y + 2 = 0$ and the line $2y - x + 3 = 0$. Find the equation of the circle.

25 As θ varies, show that the locus of the point $(2 + 3\cos\theta, -3 + 3\sin\theta)$ is the circle $(x - 2)^2 + (y + 3)^2 = 3^2$.

Show that the line $3x + 2y = 0$ meets the circle at the points where $\tan\theta = -\frac{3}{2}$ and deduce the coordinates of these points.

26 Find the equations of the circles which touch both axes and pass through the point $(6, 3)$. Find the equations of the tangents to both these circles at the point $(6, 3)$ and prove that the lengths of the perpendiculars on these tangents from the origin are equal.

27 Prove that the circles
$$x^2 + y^2 + 2gx + 2fy + c = 0, \quad x^2 + y^2 + 2g'x + 2f'y + c' = 0$$
cut orthogonally if $2gg' + 2ff' = c + c'$.

28 Find the equation of the circle whose centre lies on the line
$$3y - 4x = 14$$
and which passes through the points $(3, -4)$, $(2, 1)$.

29 Prove that the point (x', y') lies inside or outside the circle.
$$x^2 + y^2 + 2gx + 2fy + c = 0,$$
according as $x'^2 + y'^2 + 2gx' + 2fy' + c$ is $<$ or > 0.

30 Prove that the equation of the circle having the line joining (x_1, y_1) (x_2, y_2) as a diameter is $(x - x_1)(x - x_2) + (y - y_1)(y - y_2) = 0$.

A, B, C are the points $(2, 3), (4, 6), (8, 2)$ respectively. Write down the equations of the circles on AB, AC as diameters and find the coordinates of the point of intersection of these circles other than A.

31 Find the equation and length of the common chord of the two circles
$$x^2 + y^2 - 4x + 8y - 30 = 0, \quad x^2 + y^2 + 8x - 16y + 30 = 0.$$

32 Three circles of radius a, have their centres at the points $(c, 0), (c, b), (0, b)$. Prove that the equation of the circle which cuts all three orthogonally is
$$x^2 + y^2 - cx - by + a^2 = 0.$$

33 Obtain the equation of the circle which has the points $(p, q), (0, 1)$ as the ends of a diameter.

Show that, if $p^2 > 4q$, the circle cuts the x axis in points whose abscissae are the roots of the equation $x^2 - px + q = 0$.

Hence, solve graphically the equation $x^2 - 3x - 1 = 0$.

34 Show that the locus of the point whose coordinates are given by $x = 3\cos\phi + 2$, $y = 3\sin\phi - 4$ is a circle. Show the circle on a diagram, and indicate the points on it corresponding to $\phi = \tfrac{1}{4}\pi$ and $\phi = \pi$.

35 Find the equation of the circle which has as diameter the common chord of the circles $x^2 + y^2 - 6x - 4y + 9 = 0$, $x^2 + y^2 - 10x - 10y + 45 = 0$.

36 Find the condition that the circle $x^2 + y^2 + 2g_1 x + 2f_1 y + c_1 = 0$ should cut the circle $x^2 + y^2 + 2gx + 2fy + c = 0$ at the ends of a diameter of the latter circle. Find the locus of the centre of a circle which cuts the circles $x^2 + y^2 = 25$, $x^2 + y^2 - 2x - 4y - 11 = 0$ at the ends of diameters of the latter circle.

12 | The parabola

Definition A parabola is defined as the locus of a point which moves so that its distance from a fixed point is always equal to its distance from a fixed line.

Simplest form of the equation of a parabola Let the fixed point be S and the fixed line DX. Take as origin the point O midway between S and DX and choose axes Ox, Oy perpendicular to and parallel to DX.

Let $DO = OS = a$

i.e. S is the point $(a, 0)$ and DX is the line $x = -a$.

If $P(x, y)$ is the moving point,

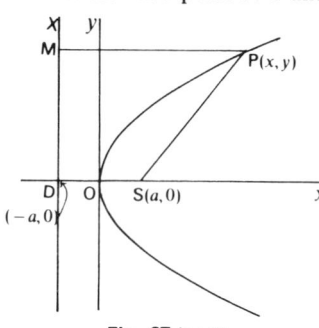

$$PS = PM.$$
$$\therefore (x-a)^2 + y^2 = (x+a)^2,$$

i.e. $\qquad y^2 = 4ax.$

Fig. 67 $(a + {}^{ve})$

So, with the axes chosen, the equation of the parabola is

$$y^2 = 4ax.$$

The fixed point S, $(a, 0)$ is called *the focus*; the fixed line DX, $x = -a$, is called *the directrix* of the parabola.

Shape of the parabola Using the equation $y^2 = 4ax$, it is seen that,

(i) the curve is symmetrical about Ox;
(ii) if a is positive, the curve is not defined for negative values of x; if a is negative, the curve is not defined for positive values of x;
(iii) the curve passes through the origin; for increasing values of x, y increases numerically;
(iv) as $2y\dfrac{dy}{dx} = 4a$, $\dfrac{dy}{dx} = \dfrac{2a}{y}$; at the origin $\dfrac{dy}{dx}$ is infinite and so the curve touches the y axis.

A sketch of the curve for a positive value of a is given in Fig. 67. The origin, O, is called *the vertex* of the parabola, Ox is called *the axis* and Oy *the tangent at the vertex*.

Example 1 Show that the equation $y^2 - 4y = 4x$ represents a parabola and make a sketch of the curve.

The equation
$$y^2 - 4y = 4x,$$
can be written
$$(y-2)^2 = 4(x+1).$$
Transferring the origin to the point $(-1, 2)$, the equation becomes
$$y^2 = 4x,$$
which is the equation of a parabola ($a = 1$).
Focus S is $(0, 2)$; directrix DX is $x = -2$.

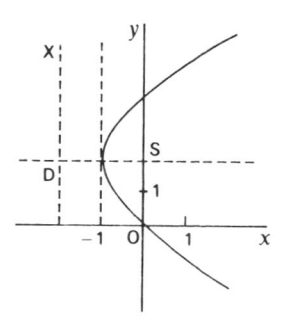

Fig. 68

Example 2 A chain hangs from two fixed points A, B on the same level, distance 100 m apart; the distance of the middle point of the chain below AB is 8 m. Assuming the chain hangs in the form of a parabola, find its equation.

Taking axes as shown in Fig. 69.
General equation of parabola is $x^2 = 4ay$.
The point B has coordinates (50, 8) and as it lies on the curve,
$$50^2 = 32a; \quad 4a = \tfrac{625}{2}.$$

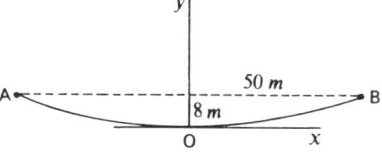

Fig. 69

∴ Equation of the parabola is $2x^2 = 625y$.

Example 3 Find the equation of the parabola with focus $(2, 1)$ and directrix $x + y = 2$.

Let (α, β) be any point on the parabola.
This point is equidistant from the focus and the directrix, therefore,
$$\sqrt{(\alpha - 2)^2 + (\beta - 1)^2} = \frac{\alpha + \beta - 2}{\sqrt{2}},$$
$$2(\alpha^2 - 4\alpha + 4 + \beta^2 - 2\beta + 1) = \alpha^2 + 2\alpha\beta + \beta^2 - 4\alpha - 4\beta + 4,$$
$$\alpha^2 - 2\alpha\beta + \beta^2 - 4\alpha + 6 = 0.$$

i.e. Equation of the parabola is $x^2 - 2xy + y^2 - 4x + 6 = 0$.

Equation of the tangent at (x_1, y_1) to the parabola $y^2 = 4ax$

$$y^2 = 4ax.$$

Differentiating w.r. to x,
$$2y \frac{dy}{dx} = 4a,$$
$$\frac{dy}{dx} = \frac{2a}{y}.$$

∴ Gradient of tangent at point $(x_1, y_1) = \dfrac{2a}{y_1}$.

Equation of tangent is

$$y - y_1 = \frac{2a}{y_1}(x - x_1),$$

$$yy_1 - y_1^2 = 2ax - 2ax_1.$$

As (x_1, y_1) lies on the curve, $y_1^2 = 4ax_1$,

hence
$$yy_1 - 4ax_1 = 2ax - 2ax_1,$$

$$\mathbf{yy_1 = 2a(x + x_1)}.$$

Example 4 Find the point of intersection of the tangent at the point $(2, -4)$ to the parabola $y^2 = 8x$, and the directrix.

Comparing $y^2 = 8x$ with the standard equation $y^2 = 4ax$, it follows that
$$a = 2.$$

Equation of tangent at $(2, -4)$ is
$$y(-4) = 4(x + 2),$$
i.e.
$$x + y + 2 = 0.$$

Directrix is the line $x = -2$.

∴ Tangent and directrix intersect at the point $(-2, 0)$.

[A] **EXAMPLES 12a**

Sketch the following parabolas, showing the foci and directrices:

1 $y^2 = 4x$. **2** $y^2 = 8x$. **3** $y^2 = 24x$.
4 $x^2 = 4y$. **5** $x^2 = 10y$. **6** $4x^2 = y$.
7 $y^2 + 4x = 0$. **8** $x^2 + 8y = 0$. **9** $(y-1)^2 = 4x$.
10 $y^2 = 6(x-2)$. **11** $x^2 + 4(y+1) = 0$. **12** $(y+2)^2 = x - 4$.

Show that each of the following equations represents a parabola and find (i) the vertex, (ii) the focus, (iii) the tangent at the vertex, and (iv) the directrix:

13 $y^2 - 6x = 12$. **14** $x^2 + 12y - 4 = 0$.
15 $y^2 - 2y = 6x$. **16** $y^2 + 4y + 2x = 0$.
17 $x^2 + x - 3y = 2$. **18** $2y^2 - 6y - 3x + 1 = 0$.

19 The parabola $y^2 = 4ax$ passes through the point $(2, 8)$. Find the coordinates of the focus.

20 A beam rests on two horizontal supports, 32 m apart and the maximum sag is 2 m. If the beam is in the shape of a parabola find the position of the focus relative to the line of the supports.

21 Find the equation of the tangent to the curve $y^2 = 12x$ at the point $(3, 6)$. What is the equation of the normal at this point?

22 Find the equation of the tangent to the parabola $y^2 = 8x$ which is parallel to the line $x + y = 0$.

23 The equation of a parabolic arch referred to horizontal and vertical axes through the vertex is $x^2 + 144y = 0$. If the height of the arch is 20 m, find the span.

24 Find the equation of the tangent to the parabola $y^2 = 4(x-1)$ at the point $(5, 4)$. Find the equations of the parabolas with the following foci and directrices:

25 Focus $(1, 1)$; directrix $x = -4$. **26** Focus $(2, 3)$; directrix $x = 6$.
27 Focus $(3, 0)$; directrix $y = 2$. **28** Focus $(-1, 4)$; directrix $y = 6$.

29 Focus $(0, 0)$; directrix $x + y = 4$.
30 Focus $(3, 1)$; directrix $y + 2x = 0$.

Parametric equations of a parabola The equation $y^2 = 4ax$ is always satisfied by the values,
$$x = at^2,$$
$$y = 2at,$$
where t is a parameter. These are the parametric equations of a parabola with respect to the axis and tangent at the vertex as the x and y axes.

The parametric coordinates of any point on the curve are $(at^2, 2at)$.

Example 5 Find the parametric equations of the parabolas,
(i) $y^2 = 12x$;   (ii) $x^2 = -12y$;   (iii) $y^2 = 4(x-2)$.

(i) In this case $a = 3$, and the parametric equations of the parabola are
$$x = 3t^2, \quad y = 6t.$$

(ii) For the parabola $x^2 = 4ay$, $x = 2at$ and $y = at^2$. As $a = -3$, the parametric equations are
$$x = -6t, \quad y = -3t^2.$$

(iii) Here $a = 1$; $x - 2 = t^2$, $y = 2t$. The parametric equations are
$$x = 2 + t^2, \quad y = 2t.$$

Focal chords A chord of a parabola is a straight line joining any two points on it. A chord passing through the focus S is a *focal chord*.

The focal chord perpendicular to the axis of the parabola (i.e. ZZ' in Fig. 70) is called *the latus rectum*; half the latus rectum, SZ, is the *semi-latus rectum*.

Taking the parabola with equation $y^2 = 4ax$ and remembering that the focus S is the point $(a, 0)$, it follows that the ordinate of Z is $2a$.

Hence SZ = ordinate of $Z = 2a$.

$\therefore$ The latus rectum $ZZ' = 4a$.

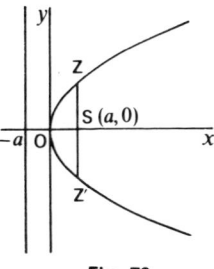

Fig. 70

Example 6 A focal chord is drawn through the point $(at^2, 2at)$ on the parabola $y^2 = 4ax$. Find the coordinates of the other end of the chord.

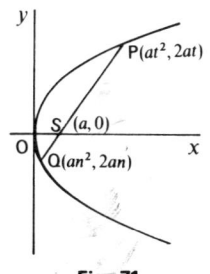

Let PQ be the focal chord with P the point $(at^2, 2at)$.
Let the coordinates of Q be $(an^2, 2an)$.
As PSQ is a straight line,

gradient SP = gradient QS.

$$\therefore \frac{2at}{at^2 - a} = \frac{2an}{an^2 - a},$$

or,
$$t(n^2 - 1) = n(t^2 - 1),$$
$$tn^2 - nt^2 - t + n = 0,$$
$$tn(n - t) + (n - t) = 0.$$

Fig. 71

As P and Q cannot be the same point, n does not equal t and $(n-t)$ is not zero.
Dividing by $(n-t)$,
$$tn+1=0,$$
$$n=-\frac{1}{t}.$$

$\therefore$ The coordinates of Q are $\left(\dfrac{a}{t^2}, -\dfrac{2a}{t}\right)$.

i.e. *The product of the parameters of the points at the extremities of a focal chord of a parabola is -1.*

[A] **EXAMPLES 12b**

Sketch the following parabolas:

1 $x=t^2$, $y=2t$.   **2** $x=5t^2$, $y=10t$.
3 $x=-3t^2$, $y=6t$.   **4** $x-1=4t^2$, $y=8t$.

Find the parametric coordinates of a point on each of the following parabolas:

5 $y^2=10x$.   **6** $y^2=24x$.   **7** $y^2+2x=0$.
8 $x^2=6y$.   **9** $y^2=8(x-1)$.   **10** $x^2=12(y-1)$.

11 Show that the loci of the following points for varying values of t are parabolas. Find the focus and directrix in each case:

(i) $(-8t, 4t^2)$.   (ii) $(6t, 3t^2)$.   (iii) $(t^2-1, 2t)$.
(iv) $(4t^2-1, 8t-2)$.   (v) $(3+t^2, 1+2t)$.   (vi) $(3-2t^2, 4t)$.

12 Find the length of the latus rectum of the locus $(6t^2, 12t)$.

13 Find the length of the semi-latus rectum of the parabola $y^2=8(2-x)$.

14 Find the equation of the focal chord of the parabola $y^2=12x$ which passes through the point $(2,3)$.

15 One extremity of a focal chord of the parabola $y^2=16x$ is the point $(1,4)$; find the coordinates of the other extremity.

16 Find the coordinates of the midpoint of the focal chord of the parabola $(8t^2, 16t)$, one extremity of which is the point $(2,8)$.

Tangent and normal at the point $(at^2, 2at)$ to the parabola $y^2 = 4ax$

$$y^2=4ax.$$

As before,
$$\frac{dy}{dx}=\frac{2a}{y}$$

$$=\frac{1}{t} \text{ at the point } (at^2, 2at).$$

Equation of tangent at $(at^2, 2at)$ is

$$y-2at=\frac{1}{t}(x-at^2),$$

i.e.
$$yt-x=at^2.$$

Gradient of normal at $(at^2, 2at) = -t$,

Equation of normal is
$$y - 2at = -t(x - at^2),$$
i.e. $$\mathbf{y + tx = 2at + at^3}.$$

Example 7 The normal at any point P on the parabola $y^2 = 8x$ meets the axis of the parabola at G. Find the locus of M, the midpoint of PG.

The parametric coordinates of P can be taken as $(2t^2, 4t)$. Equation of normal at P is
$$y + tx = 4t + 2t^3.$$
This meets the axis $y = 0$ where $x = 4 + 2t^2$. i.e. the coordinates of G are $(4 + 2t^2, 0)$.
$\therefore$ Coordinates of M are
$$\left(\frac{4 + 2t^2 + 2t^2}{2}, \frac{4t}{2} \right), \text{ i.e. } (2 + 2t^2, 2t).$$
The parametric equations of the locus of M are
$$x = 2 + 2t^2,$$
$$y = 2t.$$
Eliminating t, we obtain the Cartesian equation,
$$y^2 = 2x - 4.$$

Equation of a tangent in terms of its gradient The equation of the tangent at $(at^2, 2at)$ to the parabola $y^2 = 4ax$, is
$$yt - x = at^2.$$
Writing the gradient $\frac{1}{t}$, as m, this equation becomes
$$\frac{y}{m} - x = \frac{a}{m^2},$$
or, $$y = mx + \frac{a}{m},$$
i.e. *for all values of m, the straight line*
$$\mathbf{y = mx + \frac{a}{m}},$$
is a tangent to the parabola $y^2 = 4ax$.

Remembering that $m = \frac{1}{t}$, it follows that the point of contact of the tangent is
$$\left(\frac{a}{m^2}, \frac{2a}{m} \right).$$

Example 8 Find the equations of the tangents from the point (2, 3) to the parabola $y^2 = 4x$.

Any tangent to the parabola $y^2 = 4x$ is of the form,
$$y = mx + \frac{1}{m}.$$

This tangent passes through the point (2, 3), if

$$3 = 2m + \frac{1}{m},$$
$$2m^2 - 3m + 1 = 0,$$
$$(2m-1)(m-1) = 0,$$
$$m = \tfrac{1}{2}, 1.$$

∴ The tangents from the point (2, 3) are

$$y = \tfrac{1}{2}x + \frac{1}{\tfrac{1}{2}}; \quad \text{i.e.} \quad 2y = x + 4,$$

and $\quad y = 1x + \tfrac{1}{1}; \quad \text{i.e.} \quad y = x + 1.$

[A] EXAMPLES 12c

Find the equations of the tangents to the following parabolas at the points stated:

1 $x = 4t^2$, $y = 8t$; point $t = 3$.   **2** $x = 3t^2$, $y = 6t$; point $t = -2$.
3 $y^2 = 3x$; point (3, 3).   **4** $y^2 = 10x$; point (10, −10).
5 $y^2 = 16x$; point (9, 12).   **6** $y^2 = 4ax$; point $(a, 2a)$.

7 If the parabola $y^2 = 4ax$ passes through the point (3, 6), find the length of the latus rectum and the equation of the tangent at this point.

8 Find the equation of the chord joining the points with parameters 2 and −3 on the parabola $x = 4t^2$, $y = 8t$.

9 Find the equation of the tangent to the parabola $y^2 = 12x$ which is parallel to the line $y - 2x = 3$.

10 The tangent to $y^2 = 4x$ drawn parallel to the line $2y - x = 4$, meets the tangent at the vertex in P. Find the coordinates of P.

Find the equations of the normals to the following parabolas at the points stated:

11 $y^2 = 8x$; point (2, 4).   **12** $y^2 = 12x$; point (3, −6).
13 $2y^2 = x$; point (8, 2).   **14** $y^2 = 4x$; point $(t^2, 2t)$.

15 Find the intercepts on the coordinate axes of the tangent to the parabola $y^2 = 12x$ which makes an angle of 60° with the axis of the parabola.

16 What is the point of contact of the tangent $2y = x + 12$ to the parabola $y^2 = 12x$?

17 Write down the equation of the tangent to $y^2 = 14x$ which has a gradient m. What is the point of contact?

18 For the parabola $y^2 = 8(x - 2)$ show that $\dfrac{dy}{dx} = \dfrac{4}{y}$. Deduce the equations of the tangent and normal at the point (4, −4).

19 Two tangents to the parabola $y^2 = 16x$ have gradients of 2 and $\tfrac{1}{3}$. Find the coordinates of their point of intersection.

20 P is the point (4, 8) on the parabola $y^2 = 16x$ and S is the focus. The tangent and normal at P meet the axis of the parabola at T and G respectively. Find the coordinates of T and G and prove that (i) $PS = TS$, (ii) $PS = SG$.

21 Two perpendicular tangents are drawn to the parabola $y^2 = 20x$. If the gradient of one is $\tfrac{3}{2}$, determine their point of intersection and verify that it lies on the directrix.

22 The tangent at the point $P(2t^2, 4t)$ on the parabola $y^2 = 8x$ meets the tangent at the vertex at Q. If S is the focus, prove that SQ, PQ are perpendicular.

23 A focal chord of the parabola $y^2 = 16x$ is drawn with one extremity, the point $(4t^2, 8t)$. Prove that the tangents at the ends of the chord meet at right angles on the directrix.

24 The line $y = mx + \dfrac{3}{m}$ touches a certain parabola for all values of m. What is the equation of the parabola and what are the coordinates of the point of contact in terms of m?

25 Find the angle between the tangents at the points with parameters m and $-\dfrac{1}{m}$ on the parabola $y^2 = 4ax$.

26 Prove that the foot of the perpendicular drawn from the focus to the tangent at the point $(6t^2, 12t)$ on the parabola $y^2 = 24x$, lies on the tangent at the vertex.

27 Show that any point on the parabola $y^2 = 16(2-x)$ can be taken as $(2-4t^2, 8t)$. Obtain the equations of the tangent and normal at the point where $t = m$.

Geometrical properties of the parabola

Consider the parabola $y^2 = 4ax$. Any point P on the parabola can be taken as $(at^2, 2at)$. In the diagram, Fig. 72,

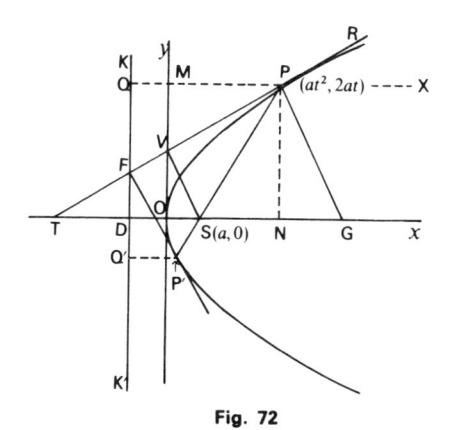

Fig. 72

O is the vertex of the parabola; Oy the tangent at the vertex;
S is the focus, i.e. the point $(a, 0)$;
$K'K$ is the directrix, i.e. the line $x = -a$;
PT, PG are the tangent and normal at P to the parabola.
The parabola has the following geometrical properties:
  (i) **SP = PQ.**
This result follows at once from the definition of a parabola.
  (ii) **OT = ON.**

Tangent at P is $\qquad yt - x = at^2$.

This line meets the x axis where $x = -at^2$.

∴ Length $OT = at^2$.

As ON is the x coordinate of P, $ON = at^2$.
$$\therefore OT = ON.$$

(iii) **TS = SP.**

$$TS = TO + OS = at^2 + a,$$
and
$$SP = PQ = PM + QM = at^2 + a.$$
$$\therefore TS = SP.$$

(iv) **PG bisects $\widehat{SPX}$.**

As $TS = SP$, $\quad\widehat{PTS} = \widehat{TPS}.$

But $\quad\widehat{PTS} = \widehat{RPX}\quad$ as PX is parallel to Ox.

$$\therefore \widehat{TPS} = \widehat{RPX},$$

hence $\quad\widehat{SPG} = \widehat{GPX},\quad$ as PG is perpendicular to PT,

i.e. PG, the normal, bisects $\widehat{SPX}$.

This is known as the *parabolic mirror property*; all rays of light emanating from a source at the focus S will be reflected in a direction parallel to the axis of the parabola.

(v) **SG = PS = ST.**

Equation of normal at P is
$$y + xt = 2at + at^3.$$

At G, $y = 0$,
$$\therefore x = 2a + at^2,$$
i.e.
$$OG = 2a + at^2.$$
$$\therefore SG = a + at^2.$$
i.e.
$$SG = SP = ST.$$

Geometrically, the result follows at once from the previous one, as $\widehat{SPG} = \widehat{GPX} = \widehat{PGS}$.

(vi) **$\widehat{SVP} = 90°$.**

Equation of tangent at P is
$$yt - x = at^2.$$

For V, $x = 0$ and so, $y = at$, i.e. V is the point $(0, at)$.

$$\text{Gradient of } SV = \frac{at - 0}{0 - a} = -t.$$

$$\text{Gradient of tangent } VP = \frac{1}{t}.$$

Hence, SV and VP are perpendicular, or $\widehat{SVP} = 90°$.

[B] **EXAMPLES 12d**

With reference to the diagram, Fig. 72, prove the results 1–11:

1 $\widehat{PFP'} = 90°.$ $\left(\text{Use the fact that the parameter of } P' \text{ is } -\frac{1}{t}\right).$

2 The subnormal NG is constant and equal to $2OS$.

3 FS is perpendicular to SP.

4 The circle on PP' as diameter touches the directrix.

5 F is the midpoint of QQ', where Q, Q' are the feet of the perpendiculars from P, P' on to the directrix.

6 PO produced meets the directrix at Q'.

7 QS is perpendicular to $Q'S$.

8 $SV^2 = SP \cdot OS$.

9 Quadrilateral $SFQP$ is cyclic and the normal at P is a tangent to the circle $SFQP$.

10 Quadrilateral $SVFD$ is cyclic. Deduce or prove otherwise, that $PS^2 = PV \cdot PF$.

11 QG, SP bisect each other.

12 The tangents at points L, L' on the parabola $y^2 = 4ax$ meet at N. If the parameters of L, L' are t, t' respectively, prove that the coordinates of N are $(att', a(t+t'))$. Deduce the result $SL \cdot SL' = SN^2$, S being the focus.

13 With the notation of the previous example, prove that $\widehat{NSL} = \widehat{NSL'}$.

14 Prove that $NL^2 : NL'^2 = SL : SL'$.

Miscellaneous examples on the parabola

Example 9 S is the focus of the parabola $y^2 = 12x$ and P is the point $(-3, 8)$. PS meets the parabola at Q and R. Prove that Q, R divide PS internally and externally in the ratio $5:3$.

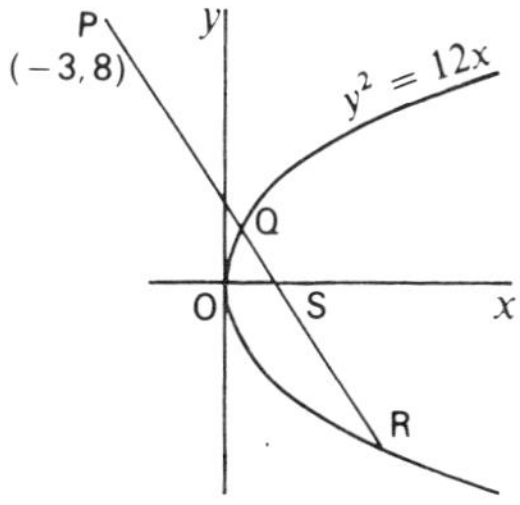

We will first solve the problem by a straight forward method and then illustrate a neater solution.

S, the focus, is the point $(3, 0)$.

Let Q be the point $(3t^2, 6t)$, then R is the point $\left(\dfrac{3}{t^2}, -\dfrac{6}{t}\right)$.

As $PQSR$ is a straight line,

gradient of QR = gradient of PS.

$$\therefore \frac{6t - \left(-\dfrac{6}{t}\right)}{3t^2 - \dfrac{3}{t^2}} = \frac{8 - 0}{-3 - 3},$$

i.e.
$$\frac{6\left(t + \dfrac{1}{t}\right)}{3\left(t^2 - \dfrac{1}{t^2}\right)} = -\frac{8}{6},$$

$$\frac{2}{t - \dfrac{1}{t}} = -\frac{4}{3}.$$

$$2t^2 + 3t - 2 = 0.$$

Factorising,
$$(2t - 1)(t + 2) = 0,$$
$$t = \tfrac{1}{2} \quad \text{and} \quad -2.$$

Fig. 73

Hence, as the parameter of Q is positive (upper branch of curve), Q is the point where $t = \frac{1}{2}$ and R the point where $t = -2$, i.e. Q is the point $(\frac{3}{4}, 3)$, R the point $(12, -12)$.

$$\frac{PQ}{QS} = \frac{x \text{ coord. of } Q - x \text{ coord. of } P}{x \text{ coord. of } S - x \text{ coord. of } Q} = \frac{\frac{3}{4}+3}{3-\frac{3}{4}} = \frac{15}{9} = \frac{5}{3}.$$

$$\frac{PR}{SR} = \frac{12+3}{12-3} = \frac{15}{9} = \frac{5}{3}.$$

i.e. Q, R divide PS internally and externally in the ratio $\frac{5}{3}$.

Alternative solution The point dividing the line PS in the ratio $\lambda:1$ has coordinates,

$$x = \frac{-3.1 + 3\lambda}{1 + \lambda} = \frac{3(\lambda - 1)}{\lambda + 1},$$

$$y = \frac{8.1 - 0\lambda}{1 + \lambda} = \frac{8}{\lambda + 1}.$$

This point lies on the locus $y^2 = 12x$ if

$$\left(\frac{8}{\lambda + 1}\right)^2 = 12 \cdot \frac{3(\lambda - 1)}{\lambda + 1},$$

$$64 = 36(\lambda - 1)(\lambda + 1),$$

$$16 = 9(\lambda^2 - 1),$$

$$\lambda^2 = \tfrac{25}{9}, \qquad \text{i.e.} \quad \lambda = \pm \tfrac{5}{3}.$$

Thus the points Q and R, common to the line PS and the parabola, are the points dividing the line in the ratios $5:3$ and $-5:3$.

Example 10 If the tangents at points P and Q on the parabola $y^2 = 4ax$ are perpendicular, find the locus of the midpoint of PQ.

Let P and Q be the points $(at_1^2, 2at_1)$, $(at_2^2, 2at_2)$.

$$\text{Gradient of tangent at } P = \frac{1}{t_1}.$$

$$\text{Gradient of tangent at } Q = \frac{1}{t_2}.$$

Hence, as the tangents are perpendicular,

$$\frac{1}{t_1}\frac{1}{t_2} = -1 \quad \text{or} \quad t_1 t_2 = -1.$$

The coordinates (α, β) of the midpoint of PQ are

$$\alpha = a\left(\frac{t_1^2 + t_2^2}{2}\right),$$

$$\beta = a(t_1 + t_2).$$

To eliminate t_1 and t_2, square the second equation, giving

$$\beta^2 = a^2(t_1 + t_2)^2 = a^2(t_1^2 + t_2^2 + 2t_1 t_2)$$

$$= 2a \cdot \frac{a}{2}(t_1^2 + t_2^2) + 2a^2(-1) \qquad \text{as } t_1 t_2 = -1,$$

$$= 2a\alpha - 2a^2.$$

So the locus of the midpoint of *PQ* has the equation
$$y^2 = 2a(x-a).$$

Example 11 Prove that the two tangents to the parabola $y^2 = 4ax$, which pass through the point $(-a, k)$, are at right angles.

Any tangent to the parabola is of the form,
$$y = mx + \frac{a}{m}.$$

This tangent passes through $(-a, k)$ if
$$k = -am + \frac{a}{m}.$$

Thus the gradients, m_1, m_2, of the two tangents through $(-a, k)$ are the roots of the equation,
$$am^2 + km - a = 0.$$

Product of roots, $\qquad m_1 m_2 = -\frac{a}{a} = -1.$

Hence the tangents from the point $(-a, k)$ to the parabola are at right angles.

[B] EXAMPLES 12e

1 Find the acute angle between the tangents from the point $(0, 4)$ to the parabola $y^2 = 16x$.

2 Find the equations of the tangents from the point $(-3, 2)$ to the parabola $y^2 = 4x$.

3 Tangents *PT*, *PT'* are drawn from the point P $(2, 5)$ to the parabola $y^2 = 8x$. Find the equation of the chord of contact *TT'*.

4 The tangents from the point P (α, β) to $y^2 = 12x$ are inclined at $45°$. Find the relationship between α and β and deduce the equation of the locus of *P*.

5 Write down the equation of a line, gradient *m*, which passes through the focus of the parabola $y^2 = 20x$. *P* is the point on this line with abscissa 20. The points of contact of the tangents from *P* to the parabola are *R* and *S*. Prove that the product of the ordinates of *R* and *S* is independent of *m*.

6 Any tangent to a parabola, focus *S*, meets the directrix in *D* and the latus rectum in *L*. Prove that $SD = SL$.

7 The straight line $2y - 3x + 1 = 0$ meets the parabola $y^2 = 16x$ in the points *P* and *Q*. Write down the equation whose roots are the ordinates of *P* and *Q* and hence find the coordinates of the midpoint of *PQ*.

8 Show that the equation of any straight line passing through the point $(2, 3)$ can be expressed in the form $y = 3 + m(x - 2)$.

Find the coordinates of the midpoint of any chord of the parabola $y^2 = 8x$ which passes through the point $(2, 3)$ and deduce the locus of the midpoints of chords of the parabola $y^2 = 8x$ which pass through the point $(2, 3)$.

9 Find the equation of the chord of the parabola $y^2 = 12x$ which is bisected at the point $(3, 2)$.

10 The coordinates of two points *P*, *Q* on the parabola $y^2 = 4x$ are $(m^2, 2m)$, $(4m^2, 4m)$. Prove that the line joining the midpoint of *PQ* and the point of intersection of the tangents at *P* and *Q*, is parallel to the axis of the parabola.

The parabola

11 Find the equation of the tangent to the parabola $y^2 = 4ax$ which is parallel to the line $2x + y = 0$ and find the coordinates of the point of contact.

12 A straight line through the focus of the parabola $y^2 = 16x$ meets the curve in the points P and Q. Prove that the product of the ordinates of P and Q is equal to the square on the semi-latus rectum.

13 If the normal at P, the point on $y^2 = 4ax$ with parameter $t = 2$, meets the parabola again at Q, prove that PQ subtends a right angle at the focus.

14 Find the equation whose roots are the ordinates of the points of intersection of the line $x + 2y = c$ and the parabola $y^2 = 10x$ and deduce the value of c when this line is a tangent to the parabola.

15 If the normal at any point P on the parabola $y^2 = 4ax$ meets the directrix at Q, prove that $aPQ^2 = SP^3$, S being the focus of the parabola.

16 Two points P, Q with coordinates $(2t_1^2, 4t_1)$, $(2t_2^2, 4t_2)$ are taken on the parabola $y^2 = 8x$. Show that the line joining the point of intersection of the tangents at P and Q and the midpoint of chord PQ is parallel to the axis of the parabola.

17 The tangent at $P(3, 6)$ on the parabola $y^2 = 12x$ meets the tangent at the vertex at K. Prove that the line through K parallel to the normal at P passes through the focus of the parabola.

18 If the tangents to the parabola $y^2 = 16x$ at the points $(16, 16)$, $(1, -4)$ intersect at T and the normals at these points intersect at R, prove that TR is parallel to the axis of the parabola.

19 Prove that the equation of the chord joining the points on the parabola $(at^2, 2at)$ with parameters t_1 and t_2 is $y(t_1 + t_2) - 2x = 2at_1t_2$.
Use this result to deduce the equation of the tangent at the point parameter t.

20 Any line through $(h, 0)$ meets the parabola $y^2 = 4ax$ in points P, Q, whose parameters are t_1, t_2 respectively. Show that the product of the ordinates of P and Q is constant.

21 Show that the ordinate of the other extremity of the normal chord of the parabola $y^2 = 4ax$, drawn through the point parameter t is

$$-\frac{2a(2 + t^2)}{t}.$$

22 P is a point on the parabola $y^2 = 16(x - 1)$, focus S; the normal at P meets the axis of the parabola at G. Find the coordinates of P if triangle SPG is equilateral.

23 P is any point on the parabola $y^2 = 8(x - 2)$; the normal at P meets the axis in G. Prove that the ordinate which bisects PG is equal to PG.

24 Find the equation of the parabola, focus $(2, -3)$ and directrix $4x + 3y = 7$. What are the coordinates of the vertex of this parabola?

25 PQ is a chord of a parabola perpendicular to the axis and R is any other point on the parabola. Prove that PR, QR meet the axis in points which are equidistant from the vertex.

26 The tangent at P any point on the parabola $y^2 = 12x$, meets the tangent at the vertex in T. Find the locus of the midpoint of TP.

27 The perpendicular from the vertex of the parabola $y^2 = 4(3 - x)$ on to any tangent, meets this tangent at P. Find the locus of P.

28 P is any point on the parabola $y^2 = 4ax$; the normal at P meets the axis of the parabola at G. Find the locus of the midpoint of PG.

29 Prove that the locus of the midpoint of a chord of the parabola $y^2 = 4ax$ which subtends a right angle at the vertex is the parabola

$$y^2 = 2a(x - 4a).$$

30 Draw the parabola with focus S (4, 0) and directrix $x - 2 = 0$. By accurate geometrical construction, obtain the tangent at the point (4, 2). [Use the mirror property of the parabola.]

31 P is any point on the parabola $y^2 = 4ax$, S is the focus and Z the point of intersection of the axis and the directrix. The line through S perpendicular to SP meets the normal at P in Q. If the perpendiculars to the axis from P, Q meet the axis in N, M, prove that $ZN = NM$.

32 A variable chord of the parabola $y^2 = 8x$ touches the parabola $y^2 = 2x$. Show that the locus of the point of intersection of the tangents at the end of the chord is a parabola.

33 TP, TP' are tangents to a parabola, prove that the line through T parallel to the axis bisects PP'.

13 | The ellipse

Definition The locus of a point P which moves such that the ratio of its distances from a fixed point S and from a fixed straight line ZQ is constant and less than one, is an ellipse.

The fixed point S is *the focus*, the fixed line ZQ *the directrix* and the constant ratio e, *the eccentricity* of the ellipse.

Simplest form of the equation of an ellipse Take ZS perpendicular to the directrix ZQ.

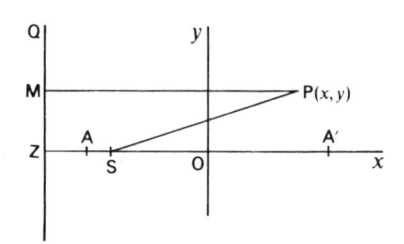

Fig. 74

Let A, A' divide SZ internally and externally in the ratio $e:1$ ($e < 1$). Then A, A' are points on the ellipse. Choose axes as shown in Fig. 74, where O is the midpoint of AA'.

Let
$$AA' = 2a.$$
$$SA = eAZ; \quad SA' = eA'Z.$$
$$\therefore SA' - SA = e(A'Z - AZ) = eAA',$$
$$(a + OS) - (a - OS) = 2ae,$$
$$OS = ae.$$

i.e. S is the point $(-ae, 0)$.

Also
$$SA' + SA = e(A'Z + AZ),$$
$$2a = e(\overline{a + OZ} + \overline{OZ - a}),$$
$$= 2eOZ,$$
$$OZ = \frac{a}{e}.$$

i.e. ZQ is the straight line $x = -\dfrac{a}{e}$.

Let $P(x, y)$ be any point on the ellipse.
Then
$$PS = ePM,$$
where PM is perpendicular to ZQ.

$$\therefore (x+ae)^2 + y^2 = e^2\left(x+\frac{a}{e}\right)^2,$$
$$x^2 + 2aex + a^2e^2 + y^2 = e^2x^2 + 2aex + a^2,$$
$$x^2(1-e^2) + y^2 = a^2(1-e^2),$$

i.e.
$$\frac{x^2}{a^2} + \frac{y^2}{a^2(1-e^2)} = 1.$$

Writing $b^2 = a^2(1-e^2)$, the equation becomes

$$\frac{x^2}{a^2} + \frac{y^2}{b^2} = 1.$$

Shape of the ellipse

1. The curve is symmetrical about both axes.

2. As
$$x^2 = a^2\left(1 - \frac{y^2}{b^2}\right), \quad |y| \leqslant b,$$

and as
$$y^2 = b^2\left(1 - \frac{x^2}{a^2}\right), \quad |x| \leqslant a.$$

3.
$$\frac{2x}{a^2} + \frac{2y}{b^2}\frac{dy}{dx} = 0,$$

$$\frac{dy}{dx} = -\frac{b^2 x}{a^2 y}.$$

$\therefore$ At the points $(\pm a, 0)$, the gradients are infinite; at the points $(0, \pm b)$, the gradients are zero.

A sketch of the curve is given in Fig. 75.

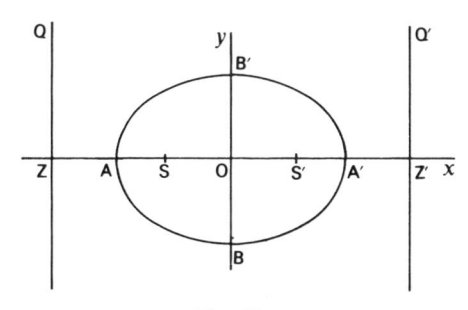

Fig. 75

178 | The ellipse

On account of the symmetry, it is clear that the curve could also be described by using the focus S' and the corresponding directrix $Z'Q'$.

The foci S, S' are the points $(-ae, 0)$, $(ae, 0)$.
The directrices ZQ, Z'Q' are the lines $x = -a/e$, $x = a/e$.
AA' is *the major axis*, BB' *the minor axis* and O *the centre* of the ellipse.

$$AA' = 2a; \qquad BB' = 2b.$$

The eccentricity e of the ellipse is given by the equation,

$$b^2 = a^2(1-e^2).$$

Example 1 Find (i) the eccentricity, (ii) the coordinates of the foci, and (iii) the equations of the directrices of the ellipse $\dfrac{x^2}{9} + \dfrac{y^2}{4} = 1$.

(i) Comparing the equation with $\dfrac{x^2}{a^2} + \dfrac{y^2}{b^2} = 1$, we have

$$a = 3, \quad b = 2.$$

As
$$b^2 = a^2(1 - e^2),$$
$$e^2 = \frac{a^2 - b^2}{a^2} = \frac{5}{9},$$

i.e.
$$e = \frac{\sqrt{5}}{3}.$$

(ii) Coordinates of the foci are $(\mp ae, 0)$, i.e. $(\mp\sqrt{5}, 0)$.

(iii) Equations of directrices are $x = \mp\dfrac{a}{e}$, i.e. $x = \mp\dfrac{9\sqrt{5}}{5}$.

Example 2 The centre of an ellipse is the point $(2, 1)$. The major and minor axes are of lengths 5 and 3 units and are parallel to the y and x axes respectively. Find the equation of the ellipse.

First find the equation of the ellipse with respect to the centre $(2, 1)$ as origin.
As the major axis is parallel to the y axis, the equation is

$$\frac{x^2}{b^2} + \frac{y^2}{a^2} = 1,$$

where $b = \tfrac{3}{2}$, $a = \tfrac{5}{2}$.

i.e.
$$\frac{4x^2}{9} + \frac{4y^2}{25} = 1.$$

Replacing x by $x - 2$ and y by $y - 1$, we get the equation with respect to the given origin;

$$\frac{4(x-2)^2}{9} + \frac{4(y-1)^2}{25} = 1.$$

Diameters A chord of an ellipse which passes through the centre is called a *diameter*.

By symmetry, if the coordinates of one end of a diameter are (x_1, y_1), those of the other end are $(-x_1, -y_1)$.

Equation of the tangent at the point (x_1, y_1) to the ellipse $x^2/a^2 + y^2/b^2 = 1$

$$\frac{x^2}{a^2} + \frac{y^2}{b^2} = 1.$$

Differentiating w.r. to x, $\quad \dfrac{2x}{a^2} + \dfrac{2y}{b^2}\dfrac{dy}{dx} = 0,$

$$\frac{dy}{dx} = -\frac{b^2 x}{a^2 y}.$$

$\therefore$ Gradient of tangent at (x_1, y_1) is $\dfrac{-b^2 x_1}{a^2 y_1}$.

Equation of tangent at (x_1, y_1) is

$$y - y_1 = -\frac{b^2 x_1}{a^2 y_1}(x - x_1),$$

$$\frac{y y_1}{b^2} - \frac{y_1^2}{b^2} = -\frac{x x_1}{a^2} + \frac{x_1^2}{a^2},$$

$$\frac{x x_1}{a^2} + \frac{y y_1}{b^2} = \frac{x_1^2}{a^2} + \frac{y_1^2}{b^2}$$

$$= 1 \quad \text{as } (x_1, y_1) \text{ lies on the ellipse,}$$

i.e. equation of tangent is

$$\frac{xx_1}{a^2} + \frac{yy_1}{b^2} = 1.$$

Example 3 Find the equation of the tangent at the point $(2, 3)$ to the ellipse $3x^2 + 4y^2 = 48$.

Write the equation of the ellipse as

$$\frac{x^2}{16} + \frac{y^2}{12} = 1.$$

i.e. $a^2 = 16$, $b^2 = 12$.

$\therefore$ Equation of tangent at $(2, 3)$ is

$$\frac{2x}{16} + \frac{3y}{12} = 1,$$

$$x + 2y = 8.$$

[A] EXAMPLES 13a

Make rough sketches of the following ellipses:

1 $\dfrac{x^2}{4} + y^2 = 1.$ 2 $\dfrac{x^2}{9} + \dfrac{y^2}{4} = 1.$ 3 $\dfrac{4x^2}{25} + \dfrac{4y^2}{9} = 1.$

4 $\dfrac{x^2}{9} + \dfrac{y^2}{25} = 1.$ 5 $4x^2 + y^2 = 36.$ 6 $\dfrac{(x-1)^2}{16} + \dfrac{(y-3)^2}{9} = 1.$

The ellipse

Find (i) the eccentricities, (ii) the coordinates of the foci, and (iii) the equations of the directrices of the following ellipses:

7 $\dfrac{x^2}{2} + y^2 = 1.$ **8** $\dfrac{x^2}{6} + \dfrac{y^2}{4} = 1.$ **9** $2x^2 + y^2 = 8.$

10 $4x^2 + 9y^2 = 16.$ **11** $x^2 + 16y^2 = 25.$ **12** $4x^2 + y^2 = 4.$

13 $\dfrac{x^2}{9} + \dfrac{(y+1)^2}{4} = 1.$ **14** $(x+2)^2 + 4(y-1)^2 = 4.$

Write down the equations of the tangents to the following ellipses at the points stated:

15 $\dfrac{x^2}{4} + y^2 = 1;$ point $(-2, 0).$ **16** $2x^2 + 3y^2 = 30;$ point $(3, 2).$

17 $x^2 + 4y^2 = 8;$ point $(2, -1).$ **18** $9x^2 + 4y^2 = 40;$ point $(-2, -1).$

19 Find the coordinates of the ends of the diameter $y = x$ of the ellipse $9x^2 + 16y^2 = 144.$

20 Find the equation of the normal to the ellipse $x^2 + 2y^2 = 22$ at the point $(2, 3).$

21 The foci of an ellipse are the points $(3, 0), (-3, 0).$ Find the lengths of (i) the major axis, (ii) the minor axis, if the eccentricity is $\tfrac{3}{4}.$

22 Find the coordinates of the points of intersection of the ellipse $x^2 + 4y^2 = 4$ and the circle $x^2 + y^2 = 2.$

23 Prove that the line $y = 2x + 3$ is a tangent to the ellipse $x^2 + 2y^2 = 2$ and find the coordinates of the point of contact.

24 The foci of an ellipse are the points $(2, 1), (6, 1).$ If the eccentricity is $\tfrac{2}{3},$ find the lengths of the major and minor axes and the equation of the ellipse.

Parametric equations of an ellipse The equation, $\dfrac{x^2}{a^2} + \dfrac{y^2}{b^2} = 1,$ is always satisfied by the values,

$$x = a \cos\theta,$$
$$y = b \sin\theta,$$

where θ is a parameter. These are the parametric equations of an ellipse referred to its major and minor axes as the axes of x and y respectively.

The parametric coordinates of any point on the curve are $(a \cos\theta, b \sin\theta).$

Example 4 Find the parametric coordinates of any point on each of the following ellipses: (i) $4x^2 + 9y^2 = 16,$ (ii) $(x-2)^2 + 4y^2 = 4.$

(i) The equation can be written

$$\frac{x^2}{4} + \frac{y^2}{\tfrac{16}{9}} = 1;$$

i.e. $a = 2, b = \tfrac{4}{3}.$

∴ Parametric coordinates of any point on the curve are $(2\cos\theta, \tfrac{4}{3}\sin\theta).$

(ii) The equation can be written

$$\frac{(x-2)^2}{4} + y^2 = 1.$$

So we can take $x - 2 = 2\cos\theta$, $y = \sin\theta$.
∴ Parametric coordinates of any point on the curve are $(2 + 2\cos\theta, \sin\theta)$.

Geometrical interpretation of the parameter θ In Fig. 76, let P be the point of the ellipse

$$\frac{x^2}{a^2} + \frac{y^2}{b^2} = 1,$$

with parameter θ.

Let the ordinate NP produced meet the circle $x^2 + y^2 = a^2$ at the point Q.

Then $ON = a\cos\theta$,
and $OQ = a$.
Hence $Q\hat{O}N = \theta$.

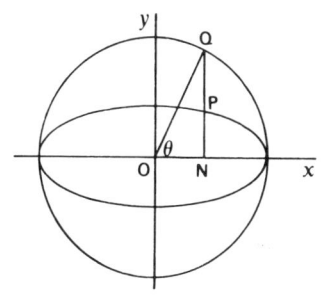

Fig. 76

The angle QON is called *the eccentric angle* of P and is equal to the parameter of the point.

The circle, $x^2 + y^2 = a^2$, is called *the auxiliary circle* of the ellipse.

Area of the ellipse $\dfrac{x^2}{a^2} + \dfrac{y^2}{b^2} = 1$ The area of the ellipse is four times the area in the positive quadrant.

$$\therefore \text{Area} = 4\int_0^a y\,dx.$$

The integral is simplified by making θ the independent variable, where
$$x = a\cos\theta, \qquad y = b\sin\theta.$$

The limits of θ are $\dfrac{\pi}{2}$ and 0.

$$\text{Area} = 4\int_{\frac{\pi}{2}}^0 b\sin\theta\,(-a\sin\theta\,d\theta)$$

$$= -2ab\int_{\frac{\pi}{2}}^0 (1 - \cos 2\theta)\,d\theta$$

$$= -2ab\left[\theta - \frac{\sin 2\theta}{2}\right]_{\frac{\pi}{2}}^0$$

$$= -2ab\left(-\frac{\pi}{2}\right)$$

$$= \pi ab.$$

i.e.   **Area of ellipse $\dfrac{x^2}{a^2} + \dfrac{y^2}{b^2} = 1$ is πab.**

The ellipse

Example 5 The semi-minor axis of an ellipse is of length k. If the area of the ellipse is $2\pi k^2$, find its eccentricity.

$$\text{Area} = \pi ab$$
$$= \pi ak \quad \text{as} \quad b = k.$$

i.e. $\pi ak = 2\pi k^2; \quad a = 2k.$

$$\therefore e^2 = \frac{a^2 - b^2}{a^2} = \frac{4k^2 - k^2}{4k^2}$$
$$= \frac{3}{4}.$$

$$\therefore \text{Eccentricity } e = \frac{\sqrt{3}}{2}.$$

[A] EXAMPLES 13b

Sketch the following loci for varying values of θ:

1 $(5\cos\theta, 3\sin\theta)$.   **2** $(2\cos\theta, \sin\theta)$.   **3** $(3\sin\theta, 4\cos\theta)$.
4 $(1 + 3\cos\theta, 2\sin\theta)$.   **5** $(2\cos\theta - 2, \sin\theta + 1)$.

Find (i) the eccentricity, (ii) the coordinates of the foci, and (iii) the equations of the directrices of the following ellipses:

6 $x = 6\cos\theta, y = 2\sin\theta$.   **7** $x = 4\cos\theta, y = 3\sin\theta$.
8 $x = \sqrt{2}\cos\theta, y = \sin\theta$.   **9** $x = 2\sin\theta, y = 3\cos\theta$.

10 Prove that the locus of the point $(2 + 3\cos\theta, 2\sin\theta - 3)$ is an ellipse with centre $(2, -3)$. What is the area of the ellipse?

11 Find the distance between the foci of the ellipse, $x = 1 + 4\cos\theta, y = 1 + \sin\theta$.

Obtain the parametric equations of the following ellipses:

12 $\dfrac{x^2}{9} + \dfrac{y^2}{4} = 1$.   **13** $x^2 + 4y^2 = 4$.   **14** $4x^2 + 9y^2 = 9$.

15 $\dfrac{(x-1)^2}{4} + y^2 = 1$.   **16** $\dfrac{(x+2)^2}{16} + \dfrac{(y-1)^2}{9} = 1$.

17 PQ is a diameter of the ellipse $x^2 + 4y^2 = 9$. If the eccentric angle of P is $\dfrac{\pi}{3}$, what is the eccentric angle of Q? Find the gradients of the tangents at P and Q.

18 Sketch the ellipse $4x^2 + 9y^2 = 36$. By using the auxiliary circle, find the points on the curve with eccentric angles: $\dfrac{\pi}{6}, \dfrac{2\pi}{3}, -\dfrac{\pi}{3}, -\dfrac{3\pi}{4}$.

19 If the ratio of the areas of an ellipse and its auxiliary circle is 5:9, find the eccentricity of the ellipse.

20 The distance between the foci of an ellipse of eccentricity $\frac{3}{4}$ is 8 cm. Find the area of the ellipse.

21 The coordinates of a point on the ellipse $b^2x^2 + a^2y^2 = a^2b^2$ are $(a\cos\phi, b\sin\phi)$. Find the coordinates of the corresponding point on the auxiliary circle.

Tangent and normal at the point (a cos φ, b sin φ) to the ellipse

$$\frac{x^2}{a^2} + \frac{y^2}{b^2} = 1$$

We have
$$\frac{dy}{dx} = -\frac{b^2 x}{a^2 y}.$$

∴ Gradient of the tangent at the point $(a \cos \phi, b \sin \phi)$

$$= -\frac{b^2}{a^2} \frac{a \cos \phi}{b \sin \phi}$$

$$= -\frac{b}{a} \cot \phi.$$

Equation of tangent at $(a \cos \phi, b \sin \phi)$ is

$$y - b \sin \phi = -\frac{b}{a} \cot \phi \,(x - a \cos \phi),$$

$$ay \sin \phi - ab \sin^2 \phi = -bx \cos \phi + ab \cos^2 \phi,$$

$$bx \cos \phi + ay \sin \phi = ab (\sin^2 \phi + \cos^2 \phi) = ab,$$

i.e.
$$\frac{x \cos \phi}{a} + \frac{y \sin \phi}{b} = 1.$$

Equation of normal at $(a \cos \phi, b \sin \phi)$ is

$$y - b \sin \phi = \frac{a}{b} \tan \phi \,(x - a \cos \phi),$$

$$yb \cos \phi - b^2 \sin \phi \cos \phi = ax \sin \phi - a^2 \sin \phi \cos \phi,$$

$$ax \sin \phi - by \cos \phi = (a^2 - b^2) \sin \phi \cos \phi,$$

i.e.
$$\frac{ax}{\cos \phi} - \frac{by}{\sin \phi} = a^2 - b^2.$$

Example 6 PP' is a double ordinate of the ellipse $9x^2 + 16y^2 = 144$. The normal at P meets the diameter through P' at Q. Find the locus of the midpoint of PQ.

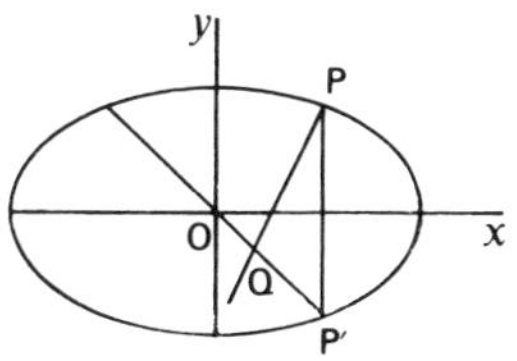

The equation of the ellipse is

$$\frac{x^2}{16} + \frac{y^2}{9} = 1.$$

Let P be the point $(4 \cos \phi, 3 \sin \phi)$; P' will be the point $(4 \cos \phi, -3 \sin \phi)$.

Equation of diameter OP' is

$$y = -\tfrac{3}{4} \tan \phi \, x.$$

Equation of normal at P is

$$\frac{4x}{\cos \phi} - \frac{3y}{\sin \phi} = 7.$$

Fig. 77

∴ At Q,
$$\frac{4x}{\cos\phi} - \frac{3}{\sin\phi}(-\tfrac{3}{4}\tan\phi\, x) = 7,$$
$$\frac{4x}{\cos\phi} + \frac{9x}{4\cos\phi} = 7,$$
$$25x = 28\cos\phi,$$
$$x = \tfrac{28}{25}\cos\phi.$$
$$y = -\tfrac{3}{4}\tan\phi\,\tfrac{28}{25}\cos\phi$$
$$= -\tfrac{21}{25}\sin\phi.$$

The coordinates of the midpoint of PQ are $(\tfrac{64}{25}\cos\phi, \tfrac{27}{25}\sin\phi)$.
The parametric equations of the locus of the midpoint of PQ are
$$x = \tfrac{64}{25}\cos\phi, \quad y = \tfrac{27}{25}\sin\phi.$$
Eliminating ϕ, $\quad (\tfrac{25}{64}x)^2 + (\tfrac{25}{27}y)^2 = 1,\quad$ the required locus.

Equation of a tangent in terms of its gradient The equation of the tangent at the point $(a\cos\phi, b\sin\phi)$ to the ellipse $\dfrac{x^2}{a^2} + \dfrac{y^2}{b^2} = 1$ is

$$\frac{x\cos\phi}{a} + \frac{y\sin\phi}{b} = 1,$$

i.e.
$$y = \left(-\frac{b}{a}\cot\phi\right)x + b\csc\phi.$$

Writing the gradient $\left(-\dfrac{b}{a}\cot\phi\right)$, as m, we have

$$\cot\phi = -\frac{am}{b},$$
$$\csc^2\phi = 1 + \cot^2\phi = \frac{a^2m^2 + b^2}{b^2}.$$
$$\therefore b\csc\phi = \pm\sqrt{a^2m^2 + b^2}.$$

So, in terms of its gradient m, the equation of a tangent to the ellipse is

$$\mathbf{y = mx \pm \sqrt{a^2m^2 + b^2}.}$$

Example 7 Find the equations of the tangents to the ellipse $\dfrac{x^2}{25} + \dfrac{y^2}{9} = 1$ which are parallel to the diameter $y = 2x$.
Any tangent has the equation
$$y = mx \pm \sqrt{25m^2 + 9}.$$
∴ The tangents with gradient $m = 2$ are
$$y = 2x \pm \sqrt{109}.$$

Example 8 Find the locus of the point of intersection of perpendicular tangents to the ellipse $\dfrac{x^2}{a^2} + \dfrac{y^2}{b^2} = 1$.

Let $P(\alpha, \beta)$ be the point of intersection of a pair of perpendicular tangents to the ellipse.
Any tangent has the equation $y = mx \pm \sqrt{a^2m^2 + b^2}$.
The gradients m_1, m_2 of the tangents through P are the roots of the equation,
$$\beta = m\alpha \pm \sqrt{a^2m^2 + b^2},$$
i.e. $\qquad m^2(a^2 - \alpha^2) + 2m\alpha\beta + b^2 - \beta^2 = 0.$

Product of roots, $\qquad m_1 m_2 = \dfrac{b^2 - \beta^2}{a^2 - \alpha^2}.$

But as the tangents are perpendicular, $m_1 m_2 = -1$,

$$\therefore \dfrac{b^2 - \beta^2}{a^2 - \alpha^2} = -1,$$

$$\alpha^2 + \beta^2 = a^2 + b^2.$$

Hence, the locus of the point of intersection of perpendicular tangents to the ellipse is the circle,
$$x^2 + y^2 = a^2 + b^2.$$
This circle is called *the director circle* of the ellipse.

[A] EXAMPLES 13c

Find the equations of the tangents to the following ellipses at the points stated:
1. $x = 2\cos\theta, y = \sin\theta;$ point $\theta = \pi/4$.
2. $x = 3\cos\theta, y = 2\sin\theta;$ point $\theta = \pi/3$.
3. $x = 4\cos\theta, y = 2\sin\theta;$ point $\theta = -\pi/3$.
4. $16x^2 + 25y^2 = 400;$ point $(5\cos\phi, 4\sin\phi)$.
5. $x^2 + 4y^2 = 4;$ point $(2\cos\phi, \sin\phi)$.

Write down the equations of the tangents to the following ellipses with the gradients stated:

6. $\dfrac{x^2}{4} + \dfrac{y^2}{2} = 1;$ gradient 3.
7. $\dfrac{x^2}{16} + \dfrac{y^2}{9} = 1;$ gradient -1.

8. $x^2 + 2y^2 = 16;$ gradient $\tfrac{1}{2}$.
9. $2x^2 + 3y^2 = 30;$ gradient $\dfrac{\sqrt{3}}{2}$.

10 For what values of c is the line $y = \tfrac{1}{2}x + c$ a tangent to the ellipse $x^2 + 16y^2 = 16$?

11 Find the equations of the tangents to the ellipse $4x^2 + 9y^2 = 36$ which are equally inclined to the axes.

12 Find the gradients of the tangents drawn from the point $(4, 6)$ to the ellipse $\dfrac{x^2}{48} + \dfrac{y^2}{4} = 1$.

13 Show that the line $y = x - 5$ is a tangent to the ellipse $9x^2 + 16y^2 = 144$ and find the equations of the perpendicular tangents.

14 Find the equations of the tangent and normal to the ellipse $9x^2 + 4y^2 = 36$ at the point $(\sqrt{3}, \tfrac{3}{2})$.

The ellipse

15 Prove that the line $2x\cos\alpha + 3y\sin\alpha = 1$ is a tangent to the ellipse $4x^2 + 9y^2 = 1$ for all values of α.

16 The normal to the ellipse $b^2x^2 + a^2y^2 = a^2b^2$ at the point with eccentric angle ϕ meets the x and y axes at P and Q. Find the coordinate of P and Q.

17 Find the equations of the tangents to the ellipse $3x^2 + 4y^2 = 12$ which are parallel to the chord $x + 3y = 1$.

18 Find the equation with roots equal to the gradients of the tangents to the ellipse $9x^2 + 16y^2 = 144$ from the point $(3, 2)$. By noting whether the roots of the equation are real or imaginary, determine whether the point $(3, 2)$ is inside or outside the ellipse.

19 Determine whether the point $(-3, 2)$ lies inside or outside the ellipse $x = 5\cos\theta, y = 3\sin\theta$.

20 Find the acute angle between the two tangents which can be drawn to the ellipse $4x^2 + 9y^2 = 1$ from the point $(1, 1)$.

Conjugate diameters Let the equation of the chord PQ of the ellipse $\dfrac{x^2}{a^2} + \dfrac{y^2}{b^2} = 1$ be $y = mx + c$.

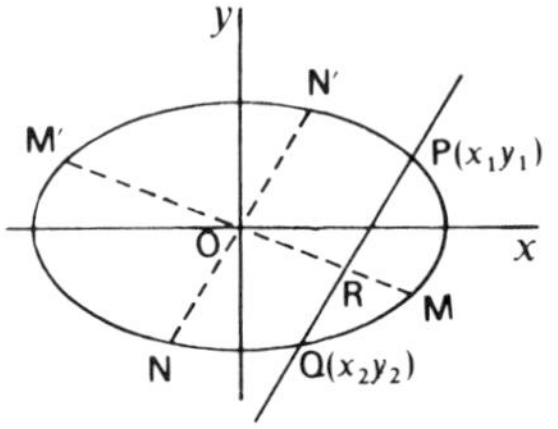

Then, if P, Q are the points (x_1, y_1), (x_2, y_2), x_1, x_2 are the roots of the equation,

$$\frac{x^2}{a^2} + \frac{(mx+c)^2}{b^2} = 1,$$

i.e. $x^2(b^2 + a^2m^2) + 2a^2mcx + a^2c^2 - a^2b^2 = 0$.

∴ Sum of roots,

$$x_1 + x_2 = -\frac{2a^2mc}{b^2 + a^2m^2}.$$

Fig. 78

Let $R, (\alpha, \beta)$ be the mid-point of PQ.

Then
$$\alpha = \frac{x_1 + x_2}{2} = -\frac{a^2mc}{b^2 + a^2m^2};$$

$$\beta = m\left(\frac{-a^2mc}{b^2 + a^2m^2}\right) + c = \frac{cb^2}{b^2 + a^2m^2}.$$

$$\therefore \frac{\beta}{\alpha} = -\frac{b^2}{a^2m},$$

$$\beta = -\frac{b^2}{a^2m}\alpha.$$

Thus the locus of R as c varies is the diameter

$$y = -\frac{b^2}{a^2m}x. \quad (MM' \text{ in Fig. 78}).$$

Consequently, the diameter MM' bisects all chords with gradient m. As one of these chords is the diameter NN', the diameter MM' bisects all chords parallel to the diameter NN'.

Such diameters are called *conjugate diameters*.

The equation of NN' is $\quad y = mx$;

the equation of MM' is $\quad y = m'x$,

where
$$m' = -\frac{b^2}{a^2 m},$$

or
$$\mathbf{mm'} = -\frac{\mathbf{b^2}}{\mathbf{a^2}}.$$

The gradients of all pairs of conjugate diameters are connected by this relationship.

Example 9 Find the equation of the chord of the ellipse $\frac{x^2}{4} + \frac{y^2}{2} = 1$, with midpoint $(1, \tfrac{1}{2})$.

Let the gradient of the diameter parallel to the chord be m.

As the conjugate diameter passes through the point $(1, \tfrac{1}{2})$, its gradient $m' = \tfrac{1}{2}$.

But
$$mm' = -\frac{b^2}{a^2} = -\tfrac{1}{2}.$$
$$\therefore m = -1.$$

The equation of the chord is
$$y - \tfrac{1}{2} = -1(x - 1),$$
$$2y + 2x = 3.$$

Parametric coordinates of the ends of conjugate diameters Let P be the point $(a\cos\phi, b\sin\phi)$ on the ellipse

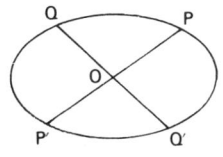

Fig. 79

$$\frac{x^2}{a^2} + \frac{y^2}{b^2} = 1.$$

Suppose $P'P$, $Q'Q$ are a pair of conjugate diameters. P' will be the point $(-a\cos\phi, -b\sin\phi)$.

Gradient of $P'P = \frac{b}{a}\tan\phi$,

$\therefore$ Gradient of $Q'Q = -\dfrac{b^2}{a^2\left(\dfrac{b}{a}\tan\phi\right)} = -\dfrac{b}{a}\cot\phi.$

Equation of $Q'Q$ is $y = -\dfrac{b}{a}\cot\phi\, x$ and this meets the ellipse where

$$\frac{x^2}{a^2} + \frac{b^2\cot^2\phi}{a^2}\frac{x^2}{b^2} = 1,$$
$$x^2(1 + \cot^2\phi) = a^2,$$
$$x^2\operatorname{cosec}^2\phi = a^2.$$
$$\therefore x = \pm a\sin\phi; \quad y = \pm b\cos\phi.$$

Consequently the coordinates of Q are $(-a\sin\phi, b\cos\phi)$, and the coordinates of Q' are $(a\sin\phi, -b\cos\phi)$.

Example 10 Prove that the sum of the squares of the lengths of two semi-conjugate diameters is constant.

Let OP, OQ be two semi-conjugate diameters.

Let P be the point $(a\cos\phi, b\sin\phi)$, then Q is the point $(-a\sin\phi, b\cos\phi)$.

$$\therefore OP^2 + OQ^2 = (a^2\cos^2\phi + b^2\sin^2\phi) + (a^2\sin^2\phi + b^2\cos^2\phi)$$
$$= a^2 + b^2 \quad \text{(a constant)}.$$

[A] EXAMPLES 13d

1 Find the equations of the diameters of the ellipse $4x^2 + 6y^2 = 24$, conjugate to the diameters:
    (i) $y = x$;   (ii) $y = 2x$;   (iii) $y = -x$;   (iv) $y = mx$.

2 Find the equations of the loci of the midpoints of systems of parallel chords of the ellipse $2x^2 + 8y^2 = 16$ with gradients, (i) 2, (ii) -1, (iii) m.

3 Find the coordinates of the midpoints of the following chords of the ellipse $x^2 + 4y^2 = 4$; (i) $y = x + 1$, (ii) $3y - x = 2$, (iii) $5y + 2x = 1$.

4 Obtain the equations of the chords of the ellipse $9x^2 + 16y^2 = 144$ with midpoints, (i) $(1, 1)$, (ii) $(2, -1)$, (iii) $(-2, 3)$, (iv) $(-1, 0)$.

5 What is the gradient of the tangents at the ends of the diameter $3y = 2x$ of the ellipse $2x^2 + 3y^2 = 6$?

6 If P is the point on the ellipse $3x^2 + 5y^2 = 15$ with eccentric angle $\pi/4$, find the equation of the diameter conjugate to that through P.

7 Find the gradients of the equal conjugate diameters of the ellipse $5x^2 + 8y^2 = 40$.

8 If the equal conjugate diameters of an ellipse are inclined at $120°$, find the eccentricity.

9 The length of a diameter of the ellipse $x^2 + 8y^2 = 8$ is 3 units, find the length of the conjugate diameter.

10 CP, CD are conjugate semi-diameters of the ellipse $4x^2 + 9y^2 = 36$. Prove that $CD^2 = SP \cdot S'P$, where S, S' are the foci.

11 CP, CQ are conjugate semi-diameters of the ellipse $x^2 + 4y^2 = 4$. Find the locus of the midpoint of PQ.

Simple geometrical properties of the ellipse

Consider the ellipse $\dfrac{x^2}{a^2} + \dfrac{y^2}{b^2} = 1$.

S, S' are the foci; ZQ, $Z'Q'$ the directrices.

PN, PT, PG are the ordinate, tangent and normal at any point P $(a\cos\phi, b\sin\phi)$. The following are some of the simple geometrical properties of the ellipse:

(i) $\left.\begin{array}{l}\mathbf{SP = ePM.} \\ \mathbf{S'P = ePM'.}\end{array}\right\}$ where e is the eccentricity of the ellipse.

Simple geometrical properties of the ellipse | 189

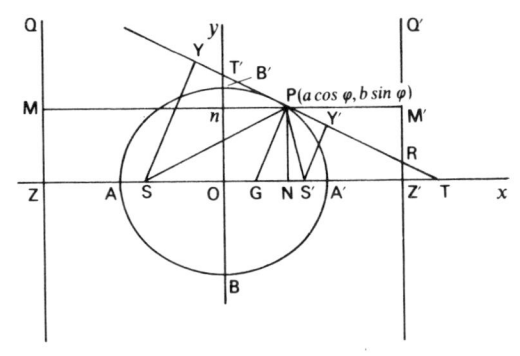

Fig. 80

These results follow at once from the geometrical definition of the ellipse.

(ii) **SP+S′P = 2a.**

Using (i), $\quad SP = ePM = e\left(a\cos\phi + \dfrac{a}{e}\right) \quad \left(OZ = \dfrac{a}{e}\right)$

$\qquad = a + ae\cos\phi.$

$\quad S'P = ePM' = e\left(\dfrac{a}{e} - a\cos\phi\right) \quad \left(OZ' = \dfrac{a}{e}\right)$

$\qquad = a - ae\cos\phi.$

Adding, $\quad SP + S'P = 2a = AA'$ — the major axis.

i.e. *The sum of the focal distances of any point on an ellipse is constant and equal to the length of the major axis.*

Conversely, the locus of a point P which moves such that the sum of its distances from two fixed points S, S' is constant and equal to k, is an ellipse with foci S, S' and major axis of length k.

(iii) **OG = e²ON.**

Equation of the normal PG is

$$\frac{ax}{\cos\phi} - \frac{by}{\sin\phi} = a^2 - b^2.$$

For G, $\qquad x = \dfrac{(a^2 - b^2)\cos\phi}{a}.$

i.e. $\qquad OG = \dfrac{a^2 - b^2}{a^2} a\cos\phi$

$\qquad\qquad = e^2 ON.$

(iv) SP, S'P are equally inclined to the tangent at P.

$$SG = SO + OG = ae + e^2 a \cos \phi$$
$$= e(a + ae \cos \phi)$$
$$= eSP. \quad \text{(from ii).}$$
$$S'G = S'O - OG = e(a - ae \cos \phi)$$
$$= eS'P.$$
$$\therefore \frac{SG}{S'G} = \frac{SP}{S'P},$$

hence PG bisects $\widehat{SPS'}$, and consequently SP, $S'P$ are equally inclined to the tangent at P.

[B] EXAMPLES 13e

In Fig. 80, prove the following results:
1. $ON.OT = a^2$.   2. $On.OT' = b^2$.   3. $SG = eSP$.
4. $S'G = eS'P$.   5. $SG = e^2 PM$.   6. $\widehat{GPS'} = \widehat{PM'S'}$.
7. $\widehat{PS'R} = 90°$.   8. $SY.S'Y' = b^2$.

9. The foci S, S' of an ellipse are 5 cm apart and the major axis is 7 cm long. Using ruler and compasses, construct the ellipse.

10. Construct the ellipse $(5 \cos \phi, 3 \sin \phi)$.

11. Any tangent to an ellipse meets the tangents at the ends of the major axis in M, M'. Prove that MM' subtends a right angle at either focus.

12. Construct the triangle $SS'P$ in which $SS' = 8$ cm, $SP = 6$ cm, $S'P = 4$ cm. If P is a point on an ellipse foci S, S', construct geometrically the tangent and normal at P. [Use result (iv).]

13. The perpendiculars from a focus of an ellipse on to the tangent and normal at any point meet the tangent at Y and the normal at Z. Prove that YZ passes through the centre of the ellipse.

14. Any tangent to the ellipse $\dfrac{x^2}{a^2} + \dfrac{y^2}{b^2} = 1$ meets the circle $x^2 + y^2 = a^2$ at L and M, prove that SL and $S'M$ are each perpendicular to the tangent, S, S' being the foci nearest to L, M respectively.

15. A latus rectum of an ellipse is defined as a focal chord which is perpendicular to the major axis. Show that the semi-latus rectum of the ellipse $(a \cos \phi, b \sin \phi)$ is of length $\dfrac{b^2}{a}$.

16. The tangent and normal at any point P on an ellipse meet the major axis at T and G respectively. If PN is the ordinate of P and C the centre, prove that $CT.NG = b^2$.

17. One focus and the corresponding directrix of an ellipse are given together with a straight line which is known to be a tangent. Show how to find geometrically, the other focus and directrix.

18. The tangents to an ellipse at P, P' meet at T. If PP' meets a directrix at K, prove that $\widehat{TSK}$ is a right angle where S is the focus corresponding to the directrix.

19. P is any point on an ellipse with major axis AA'. If PA, $A'P$ meet the directrix corresponding to focus S at K, K' respectively, prove

(i) $\widehat{KSK'} = 90°$;   (ii) $K'X.KX = XS^2$;   (iii) $PN:NA' = XK':XA'$ where

X is the point of intersection of the directrix and the x axis and PN the ordinate of P.

20 A circle, centre A, lies within a circle, centre B. Prove that the locus of the centre of a circle which touches both the given circles is an ellipse with foci A, B.

Miscellaneous examples

Example 11 A spring is in the shape of part of an ellipse. Referred to the principal axes of the ellipse, the coordinates of the ends of the spring are (16, 9) and (−16, 9). If the tangents at these points are perpendicular, find the eccentricity of the ellipse.
As the ends A, B of the spring are symmetrically placed with respect to the y axis (the minor axis), the tangents at the ends will be equally inclined to this axis,

i.e. gradient of tangent at $A = -1$.

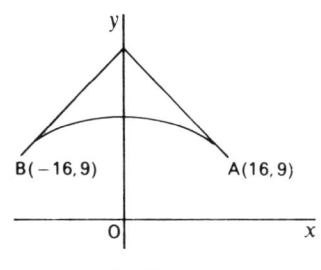

But the gradient at (x_1, y_1) on the ellipse $\dfrac{x^2}{a^2} + \dfrac{y^2}{b^2} = 1$ is $-\dfrac{b^2 x_1}{a^2 y_1}$.

$$\therefore -\frac{b^2}{a^2}\frac{16}{9} = -1,$$

$$16b^2 = 9a^2.$$

$$\therefore e^2 = \frac{a^2 - b^2}{a^2} = 1 - \frac{9}{16},$$

$$e = \frac{\sqrt{7}}{4}.$$

Fig. 81 [B(−16, 9), A(16, 9)]

Example 12 The tangent and normal to the ellipse $x^2 + 3y^2 = 2$ at the point $\left(1, \dfrac{1}{\sqrt{3}}\right)$ meet the x axis at P and the y axis at Q respectively. Prove that PQ touches the ellipse.

Equation of tangent at $\left(1, \dfrac{1}{\sqrt{3}}\right)$ is

$$\frac{x}{2} + \frac{3y}{2\sqrt{3}} = 1. \qquad \left(\frac{xx_1}{a^2} + \frac{yy_1}{b^2} = 1\right).$$

Equation of normal at $\left(1, \dfrac{1}{\sqrt{3}}\right)$ is

$$y - \frac{1}{\sqrt{3}} = \sqrt{3}(x - 1). \quad \left(\text{Gradient of tangent} = -\frac{1}{\sqrt{3}}\right).$$

For P, $\qquad \dfrac{x}{2} = 1$, i.e. $x = 2$.

P is the point $(2, 0)$.

For Q, $\qquad y - \dfrac{1}{\sqrt{3}} = -\sqrt{3}$, i.e. $y = -\dfrac{2}{\sqrt{3}}$.

Q is the point $\left(0, -\dfrac{2}{\sqrt{3}}\right)$.

Equation of PQ is
$$\frac{y-0}{-\frac{2}{\sqrt{3}}-0} = \frac{x-2}{0-2},$$

$$\sqrt{3}y = x - 2.$$

Substituting $x = 2 + \sqrt{3}y$ in the equation of the ellipse,
$$(2 + \sqrt{3}y)^2 + 3y^2 = 2,$$
$$6y^2 + 4\sqrt{3}y + 2 = 0,$$
$$3y^2 + 2\sqrt{3}y + 1 = 0,$$
$$(\sqrt{3}y + 1)^2 = 0.$$

i.e. the line PQ meets the ellipse in two coincident points, in other words, it is a tangent to the ellipse.

Example 13 The extremities of any diameter of an ellipse are L, L' and M is any other point on the curve. Prove that the product of the gradients of the chords LM, L'M is constant.

Let L, M be the points $(a\cos\phi, b\sin\phi)$, $(a\cos\phi', b\sin\phi')$ respectively.
L' will be the point $(-a\cos\phi, -b\sin\phi)$.

$$\text{Gradient of } LM = \frac{b(\sin\phi' - \sin\phi)}{a(\cos\phi' - \cos\phi)}$$

$$= \frac{b}{a} \cdot \frac{2\cos\frac{1}{2}(\phi'+\phi)\sin\frac{1}{2}(\phi'-\phi)}{-2\sin\frac{1}{2}(\phi'+\phi)\sin\frac{1}{2}(\phi'-\phi)}$$

$$= -\frac{b}{a}\cot\frac{1}{2}(\phi' + \phi).$$

Similarly, gradient of $L'M = \frac{b}{a}\tan\frac{1}{2}(\phi' + \phi).$

$\therefore$ Product of gradients $= -\frac{b^2}{a^2}.$

Example 14 Two conjugate diameters of the ellipse $\frac{x^2}{a^2} + \frac{y^2}{b^2} = 1$ meet the tangent at one end A of the major axis at L, M. Prove that $AL.AM = b^2$.

Let the conjugate diameters be $y = mx$ and $y = -\frac{b^2}{a^2 m}x$.

The tangent at A can be taken as $x = a$.
For L, $y = ma$; length $AL = ma$.
For M, $y = -\frac{b^2}{a^2 m}a = -\frac{b^2}{am}$; length $AM = \frac{b^2}{am}$.

$\therefore$ Product $AL.AM = b^2.$

[B] EXAMPLES 13f

1 Find the tangent of the acute angle between the two tangents to the ellipse $4x^2 + 9y^2 = 36$ which pass through the point $(-4, 2)$.

2 The equation of an ellipse is $9x^2 + 36y^2 = 324$. Find (i) the equation of the normal at the point parameter $\frac{\pi}{6}$; (ii) the perpendicular distance of the centre from the chord joining the points with parameters $\frac{\pi}{4}$, $-\frac{\pi}{3}$.

3 A diameter of the ellipse $9x^2 + 16y^2 = 144$ is inclined to the x axis at an angle whose tangent is 2. Find the parameters of (i) the ends of the diameter, (ii) the ends of the perpendicular diameter.

4 The normal at P, a point on the ellipse $b^2x^2 + a^2y^2 = a^2b^2$ passes through the lower end of the minor axis. Find the eccentric angle of P.

5 Find the eccentric angles of the ends of the double ordinates which pass through the foci of the ellipse $x^2 + 2y^2 = 2$.

6 Find the equations of the tangents to the ellipse $3x^2 + 4y^2 = 12$ which are parallel to the diameter $2y = 3x$. Obtain, in addition, the coordinates of the points of contact.

7 P is the point $(3\cos\phi, 2\sin\phi)$ on the ellipse $4x^2 + 9y^2 = 36$; the ordinate at P meets the major axis at N. If NP is produced to Q so that $PQ = PN$, what are the coordinates of Q. Deduce the locus of Q as P varies.

8 P and Q are the extremities of a pair of conjugate diameters of the ellipse $x^2 + 2y^2 = 4$, both points being above the axis. If P is the point $(2\cos\phi, \sqrt{2}\sin\phi)$, find the coordinates of Q.

9 M is the foot of the perpendicular drawn from the centre of the ellipse $4x^2 + 9y^2 = 36$ on to the tangent at the point parameter ϕ. Obtain the coordinates of M.

10 Prove that the tangents at the points on the ellipse $b^2x^2 + a^2y^2 = a^2b^2$ with eccentric angles ϕ, ϕ', intersect at the point with coordinates

$$a\frac{\cos\frac{1}{2}(\phi+\phi')}{\cos\frac{1}{2}(\phi-\phi')}, \quad b\frac{\sin\frac{1}{2}(\phi+\phi')}{\cos\frac{1}{2}(\phi-\phi')}.$$

11 If the straight lines joining a point P on an ellipse to the ends of the major axis meet the minor axis at L and M and the tangent at P meets the minor axis at N, prove that N is the midpoint of LM.

12 Find the locus of points from which the tangents to the ellipse $x^2 + 4y^2 = 4$ are inclined at $45°$.

13 The ordinate through any point P on an ellipse meets the auxiliary circle at Q. Prove that the tangents to the ellipse and circle at P and Q intersect on the major axis.

14 S is a focus of an ellipse and PS is a semi-latus rectum. If the tangent at P meets the minor axis at M, prove that the line joining M to the other focus is parallel to the normal at P.

15 If $PQ, P'Q'$ are conjugate diameters of the ellipse $b^2x^2 + a^2y^2 = a^2b^2$ and P has eccentric angle ϕ, prove that the eccentric angles of P', Q' are $\phi \pm \frac{\pi}{2}$.

16 CP, CP' are semi-conjugate diameters of the ellipse $9x^2 + 25y^2 = 225$. If the eccentric angle of P is ϕ, find the coordinates of the midpoint of PP'. What is the locus of this midpoint as ϕ varies?

17 The major axis of the ellipse $4x^2 + 9y^2 = 36$ meets a directrix at Z. Prove that the length of the tangent from Z to the director circle $x^2 + y^2 = 13$ is equal to the distance of Z from the corresponding focus.

18 If the normal at an end of a latus rectum of an ellipse passes through one end of the minor axis, prove that $e^4 + e^2 = 1$, e being the eccentricity.

19 The tangent at the point P $(a\cos\phi, b\sin\phi)$ on the ellipse
$$b^2x^2 + a^2y^2 = a^2b^2$$
meets the tangent at the point $(a, 0)$ at Q. Show that the line joining Q to the centre is parallel to that joining P to the point $(-a, 0)$.

20 R, R' are the points of contact of two fixed parallel tangents to an ellipse. The tangent at any point P meets the fixed tangents at T, T'. Prove that $RT:R'T' = PT:PT'$.

21 Two points P, P' on the ellipse $b^2x^2 + a^2y^2 = a^2b^2$ are such that the lines joining them to the centre C are perpendicular. Prove that the perpendicular from C to the chord PP' is of constant length.

22 OP, OP' are conjugate semi-diameters of the ellipse $x^2 + 4y^2 = 4$, OP having gradient m; perpendiculars are drawn from focus S to OP and from focus S' to OP'. Find the coordinates of the point of intersection of these perpendiculars.

23 PQ is a double ordinate of the ellipse $x^2 + 9y^2 = 9$; the normal at P meets the diameter through Q at R. Find the locus of the midpoint of PR.

24 OP, OP' are conjugate semi-diameters of an ellipse; the normals at P, P' meet the major axis at G, G' respectively. Prove that
$$PG^2 + P'G'^2 = b^2(2 - e^2).$$

25 A rod AB of length $(a+b)$ moves with A and B on perpendicular lines OA, OB respectively. Show that the locus of the point P which divides AB internally in the ratio $a:b$ is an ellipse and find the eccentricity $(a > b)$.

26 The ordinate through any point P on an ellipse meets the auxiliary circle at Q. If S is a focus, prove that SP is equal in length to the perpendicular from S on to the tangent at Q to the circle.

27 PP' is a normal chord of the ellipse $9x^2 + 16y^2 = 144$; the diameter through P meets the ellipse again at Q. If the tangents at P', Q meet at Y, prove that the diameter through Y is parallel to PP'.

28 Prove that a portion of a latus rectum of the ellipse
$$b^2x^2 + a^2y^2 = a^2b^2$$
intercepted between the ellipse and the auxiliary circle is of length $2b\left(1 - \dfrac{b}{a}\right)$.

29 QR, a diameter of an ellipse, subtends a right angle at a point P on the ellipse. Prove that PR is parallel to an axis of the ellipse.

14 | The hyperbola

Definition The locus of a point P which moves such that the ratio of its distances from a fixed point S and from a fixed straight line ZQ is constant and greater than one, is *a hyperbola*.

The fixed point S is *the focus*, the fixed line ZQ *the directrix* and the constant ratio e, *the eccentricity* of the hyperbola.

Simplest form of the equation of a hyperbola Take ZS perpendicular to the directrix ZQ. Choose axes Ox, Oy as shown in Fig. 82, where O is the mid-point of the line joining the points A, A' which divide SZ internally and externally in the ratio $e:1$.

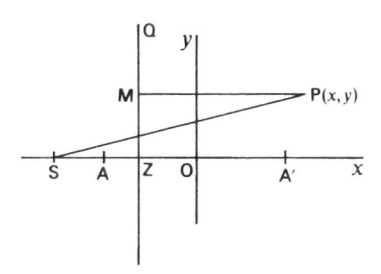

Following the method used for the ellipse,

$$OS = ae, \quad OZ = \frac{a}{e},$$

Fig. 82 where $AA' = 2a$.

i.e. S is the point $(-ae, 0)$ and ZQ the line $x = -\frac{a}{e}$.

Let $P(x, y)$ be any point on the hyperbola.
Then
$$PS = ePM,$$

$$(x + ae)^2 + y^2 = e^2\left(x + \frac{a}{e}\right)^2,$$

$$a^2(e^2 - 1) = x^2(e^2 - 1) - y^2,$$

i.e.
$$\frac{x^2}{a^2} - \frac{y^2}{a^2(e^2 - 1)} = 1.$$

Writing $b^2 = a^2(e^2 - 1)$, the equation becomes

$$\frac{x^2}{a^2} - \frac{y^2}{b^2} = 1.$$

Shape of the hyperbola
1. The curve is symmetrical about both axes.
2. As $x^2 = a^2\left(1 + \frac{y^2}{b^2}\right)$, the curve exists for all values of y;

as $y^2 = b^2\left(\dfrac{x^2}{a^2} - 1\right)$, the curve does not exist if $|x| < a$.

3.
$$\dfrac{2x}{a^2} - \dfrac{2y}{b^2}\dfrac{dy}{dx} = 0, \quad \dfrac{dy}{dx} = \dfrac{b^2 x}{a^2 y}.$$

∴ At the points $(\pm a, 0)$, the gradients are infinite.

4.
$$y^2 = b^2\left(\dfrac{x^2}{a^2} - 1\right).$$

As $|x|$ increases, $y^2 \to \dfrac{b^2 x^2}{a^2}$ or $y \to \pm\dfrac{bx}{a}$,

i.e. for increasing values of x the curve approaches closer and closer to the straight lines $y = \pm \dfrac{bx}{a}$.

Consequently, the lines $y = \pm \dfrac{bx}{a}$ are asymptotes of the hyperbola. A sketch of the curve is given in Fig. 83.

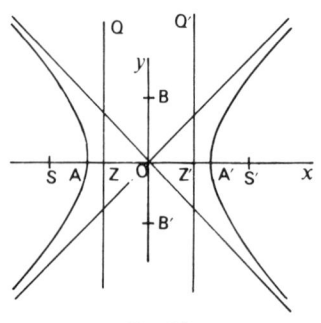

Fig. 83

On account of the symmetry, it is clear that the curve could also be described by using the focus S' and the directrix $Z'Q'$.

The foci S, S' are the points $(-ae, 0)$, $(ae, 0)$.
The directrices ZQ, Z'Q' are the lines $x = -a/e$, $x = a/e$.
AA' is called the *transverse axis*; $AA' = 2a$.
BB' is called the *conjugate axis*, where $OB = OB' = b$

and
$$b^2 = a^2(e^2 - 1).$$

Example 1 For the hyperbola $\dfrac{x^2}{9} - \dfrac{y^2}{4} = 1$, find (i) the eccentricity, (ii) the coordinates of the foci, (iii) the equations of the directrices and (iv) the equations of the asymptotes.

Comparing the equation with the standard form, it follows that,
$$a = 3, \quad b = 2.$$

(i) As
$$b^2 = a^2(e^2 - 1),$$
$$e^2 = \frac{a^2 + b^2}{a^2} = \frac{13}{9},$$

i.e.
$$e = \frac{\sqrt{13}}{3}.$$

(ii) Coordinates of the foci are ($\mp ae$, 0), i.e. ($\mp \sqrt{13}$, 0).

(iii) Equations of directrices are $x = \mp \dfrac{a}{e}$, i.e. $x = \mp \dfrac{9}{\sqrt{13}}$.

(iv) Equations of asymptotes are $y = \pm \dfrac{b}{a} x,$

i.e.
$$y = \pm \frac{2x}{3}.$$

[A] EXAMPLES 14a

Find (i) the eccentricities, (ii) the coordinates of the foci, (iii) the equations of the directrices, and (iv) the equations of the asymptotes of the following hyperbolas:

1 $\dfrac{x^2}{4} - y^2 = 1.$ **2** $x^2 - y^2 = 4.$ **3** $\dfrac{x^2}{12} - \dfrac{y^2}{4} = 1.$

4 $\dfrac{x^2}{144} - \dfrac{y^2}{25} = 1.$ **5** $x^2 - 4y^2 = 36.$ **6** $4x^2 - 4y^2 = 9.$

7 $\dfrac{(x-1)^2}{9} - \dfrac{y^2}{4} = 1.$ **8** $\dfrac{(x+1)^2}{64} - \dfrac{(y-2)^2}{36} = 1.$

9 Sketch the hyperbola $\dfrac{x^2}{4} - \dfrac{y^2}{2} = 1$ and find the acute angle included between its asymptotes.

10 Find the asymptotes of the hyperbola $(x-3)^2 - (y+1)^2 = 8.$

11 Show that $x^2 - y^2 + 2x + 4y = 0$ is the equation of a hyperbola. Find the centre of the hyperbola and prove that the asymptotes are perpendicular.

12 The foci of a hyperbola are the points (± 5, 0). Find the equation of the curve if $e = \frac{5}{4}$.

Find the equations of hyperbolas with the following foci, directrices and eccentricities:

13 Focus $(-2, 0)$; directrix $x = 0$; $e = \dfrac{3}{2}$.

14 Focus $(1, 0)$; directrix $x = 3$; $e = \dfrac{4}{3}$.

15 Focus $(1, 1)$; directrix $2x + 2y = 1$; $e = \sqrt{2}$.

Properties of the hyperbola Many results for the hyperbola are obtained from the corresponding results for the ellipse by merely writing $-b^2$ in place of b^2.

The student will be able to obtain the following results:

1. *The equation of the tangent to the hyperbola* $\dfrac{x^2}{a^2} - \dfrac{y^2}{b^2} = 1$ *at the point* (x_1, y_1) *is*

$$\frac{xx_1}{a^2} - \frac{yy_1}{b^2} = 1.$$

2. *The gradient form of the equation of the tangent to the hyperbola is*

$$y = mx \pm \sqrt{a^2 m^2 - b^2}.$$

3. *The locus of the midpoints of chords of the hyperbola with gradient m is the diameter*

$$y = \frac{b^2}{a^2 m} x.$$

4. *The gradients of conjugate diameters, one of which does not meet the curve in real points, are connected by the relationship,*

$$mm' = \frac{b^2}{a^2}.$$

Many of the geometrical properties of the hyperbola are similar to the corresponding properties of the ellipse.

E.g. *The difference of the focal distances of a point P is constant and equal to* $2a$,

i.e. $$SP \sim S'P = 2a.$$

The tangent at a point P bisects the angle $\widehat{S'PS}$.

Example 2 Show that there are two tangents to the hyperbola $x^2 - 4y^2 = 4$ parallel to the line $y = 2x - 3$ and find their distance apart.

$$\text{Gradient of tangents} = 2.$$

∴ Equations of tangents are

$$y = 2x \pm \sqrt{4 \cdot 4 - 1}$$
$$= 2x \pm \sqrt{15}.$$

The perpendicular distances of these tangents from the origin are $\pm \dfrac{\sqrt{15}}{\sqrt{5}}$, i.e. $\pm \sqrt{3}$.

Hence the distance between the tangents is $2\sqrt{3}$.

Example 3 Find the equation of the chord of the hyperbola $4x^2 - 9y^2 = 36$ with midpoint $(1, 1)$.

Let the gradient of the diameter parallel to the chord be m. As the conjugate diameter passes through $(1, 1)$, its gradient $m' = 1$.

But $$mm' = \frac{b^2}{a^2} = \frac{4}{9},$$

so, $m = \frac{4}{9}$.

∴ Equation of chord is $y - 1 = \frac{4}{9}(x - 1)$,

or $9y - 4x = 5$.

[B] EXAMPLES 14b

1 Find the equation of the tangent at the point (2, 1) on the hyperbola $x^2 - 2y^2 = 2$. Prove that the product of the perpendiculars from the foci to this tangent is 1.

2 Find the coordinates of the midpoint of the chord $2y = x + 1$ of the hyperbola $4x^2 - 9y^2 = 36$.

3 Find the equations of the tangents to the curve $x^2 - 2y^2 = 1$ parallel to the diameter $4y = 3x$.

4 Trace accurately, the hyperbola with foci $(-2, 0)$ $(3, 0)$ and semi-transverse axis of length 2 units.

5 Prove that the triangle formed by the asymptotes and any tangent to the curve $\frac{x^2}{4} - \frac{y^2}{2} = 1$ is of constant area.

6 Find the coordinates of points on the hyperbola $x^2 - 9y^2 = 9$, the normals at which are parallel to an asymptote.

7 Any straight line cuts the hyperbola $4x^2 - 8y^2 = 64$ at P, Q and the asymptotes at P', Q'. Show that $P'P = QQ'$.

8 Find the acute angle between the tangents which can be drawn from the point (2, 3) to the hyperbola $x^2 - y^2 = 2$.

9 An asymptote of the hyperbola $2x^2 - 8y^2 = 3$ meets a directrix in M and the tangent at a vertex in N. Prove that the lines joining the vertex to M and the corresponding focus to N are parallel.

10 The tangent at the point $P(2, 1)$ on the hyperbola $x^2 - 2y^2 = 2$ cuts an asymptote in T and the line joining P to a focus S meets the same asymptote in U. Prove $\widehat{STU} = \widehat{SUT}$.

11 Show that any point on the hyperbola $b^2x^2 - a^2y^2 = a^2b^2$ can be written as $(a \sec \phi, b \tan \phi)$. Find the equation of the tangent at this point.

12 Show that the locus of the point $\left(\frac{a}{2}t + \frac{1}{t}, \frac{b}{2}t - \frac{1}{t}\right)$ for varying values of t, is the hyperbola $\frac{x^2}{a^2} - \frac{y^2}{b^2} = 1$. Derive the equation of the tangent at the point parameter t.

The rectangular hyperbola A hyperbola with perpendicular asymptotes is a *rectangular hyperbola*.

For the asymptotes to be at right angles they must be inclined at 45° to the axis and consequently their gradients are ± 1.

$$\therefore \pm \frac{b}{a} = \pm 1,$$

i.e. $b = a.$

So, *the standard equation of a rectangular hyperbola referred to its principal axes is*

$$x^2 - y^2 = a^2.$$

Further, $$e^2 = \frac{a^2 + b^2}{a^2} = 2.$$

Eccentricity of a rectangular hyperbola $= \sqrt{2}$.

The rectangular hyperbola has many important applications. Its equation is simplified and its properties more easily obtained, by using the asymptotes as axes.

Equation of a rectangular hyperbola with respect to its asymptotes

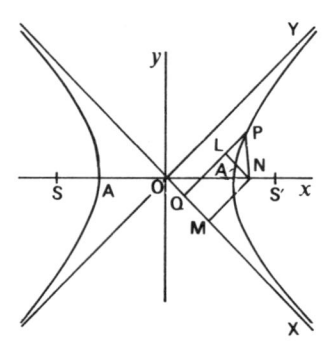

Let P be any point on a rectangular hyperbola. Let the coordinates of P be (x, y) referred to axes Ox, Oy and (X, Y) referred to the asymptotes OX, OY as axes (Fig. 84).

$$\begin{aligned} X = OQ &= OM - QM \\ &= OM - LN \\ &= ON\cos 45° - PN\cos 45°, \end{aligned}$$

i.e. $X = \dfrac{x}{\sqrt{2}} - \dfrac{y}{\sqrt{2}}.$

Similarly,

$$Y = PQ = \frac{x}{\sqrt{2}} + \frac{y}{\sqrt{2}}.$$

Fig. 84

$$\therefore \quad x - y = \sqrt{2}X; \qquad x + y = \sqrt{2}Y;$$
$$x^2 - y^2 = (x-y)(x+y) = 2XY.$$

As $\qquad x^2 - y^2 = a^2,$

$$XY = \frac{a^2}{2}.$$

Consequently, the equation of a rectangular hyperbola referred to its asymptotes as axes is

$$xy = c^2, \quad \text{where} \quad c^2 = \tfrac{1}{2}a^2.$$

Example 4 Find the coordinates of the vertices and the foci of the curve $xy = 2$.

Comparing the equation with

$$xy = \tfrac{1}{2}a^2,$$

it follows that $\quad a = 2,$
i.e. $\qquad OA = OA' = 2.$ (Fig. 85).

Parametric equations of a rectangular hyperbola

Also
$$OS = OS' = ae = 2\sqrt{2}.$$

As AA' is inclined at 45° to the axes, it follows that the coordinates of A, A' and S, S' are $(\mp\sqrt{2}, \mp\sqrt{2})$, $(\mp 2, \mp 2)$ respectively.

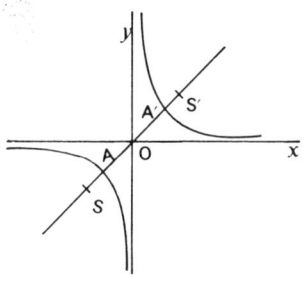

Fig. 85

Parametric equations of a rectangular hyperbola The equation $xy = c^2$, is always satisfied if

$$x = ct, \quad y = \frac{c}{t},$$

where t is a parameter. These are the parametric equations of the curve referred to the asymptotes, the parametric coordinates of any point being $(ct, c/t)$.

Tangent and normal at the point $(ct, c/t)$ to the curve $xy = c^2$

As
$$y = \frac{c^2}{x}, \quad \frac{dy}{dx} = -\frac{c^2}{x^2}.$$

Gradient of tangent at $(ct, c/t) = \dfrac{-c^2}{c^2 t^2} = -\dfrac{1}{t^2}$.

Equation of tangent at $(ct, c/t)$ is
$$y - \frac{c}{t} = -\frac{1}{t^2}(x - ct),$$

$$t^2 y + x = 2ct.$$

Equation of normal at $(ct, c/t)$ is
$$y - \frac{c}{t} = t^2(x - ct).$$

$$ty - t^3 x = c - ct^4.$$

Example 5 The tangent at any point P on the curve $xy = 4$ meets the asymptotes at Q and R. Show that P is the midpoint of QR.

Let P be the point $(2t, 2/t)$.

Equation of tangent of P is
$$t^2 y + x = 4t.$$

This line meets the asymptotes $y = 0$, $x = 0$ at the points $Q(4t, 0)$, $R\left(0, \dfrac{4}{t}\right)$.

The midpoint of QR has coordinates $\left(2t, \dfrac{2}{t}\right)$, i.e. P is the midpoint of QR.

[A] EXAMPLES 14c

Find the parametric coordinates of a point on each of the following curves:

1 $xy = 9$.   **2** $xy = 16$.   **3** $4xy = 25$.

4 $9xy = 1$.   **5** $xy = 2$.   **6** $xy = -4$.

7 $(x-1)y = 1$.   **8** $y = \dfrac{9}{x+2}$.   **9** $x+1 = \dfrac{25}{y-3}$.

For the following loci, find (i) the Cartesian equation, (ii) the coordinates of the vertices:

10 $\left(5t, \dfrac{5}{t}\right)$.   **11** $\left(3t, \dfrac{3}{t}\right)$.   **12** $\left(6t, \dfrac{6}{t}\right)$.

13 $\left(t, -\dfrac{1}{t}\right)$.   **14** $\left(1+2t, \dfrac{2}{t}\right)$.   **15** $\left(4t, \dfrac{4}{t}-1\right)$.

Sketch the following rectangular hyperbolas:

16 $xy = 8$.   **17** $y = \dfrac{2}{x}$.   **18** $2x = \dfrac{25}{y}$.

19 $x = 3t, y = \dfrac{3}{t}$.   **20** $x = 4t, y = \dfrac{4}{t}$.   **21** $xy = -2$.

22 $y = \dfrac{2}{x-2}$.   **23** $y = \dfrac{-2}{x+1}$.   **24** $(y-1)(2x+1) = 16$.

Find (i) the lengths of the transverse axes, (ii) the coordinates of the foci of the following curves:

25 $xy = 18$.   **26** $xy = 4$.   **27** $x = 8t, y = 8/t$.

28 $2xy = 25$.   **29** $y(x-1) = 2$.   **30** $x = 1+2t, y = 1+\dfrac{2}{t}$.

31 Find the equations of the tangents at the vertices of the rectangular hyperbola $xy = 8$.

32 Find the equation of the tangent and normal to the curve $xy = 16$ at the point $\left(4t, \dfrac{4}{t}\right)$.

33 Find the length of the diameter of the curve $xy = 4$ drawn through the point $(4, 1)$.

34 Sketch the locus $\left(\dfrac{t}{2}, \dfrac{1}{2t}\right)$ for varying values of t and show the points with parameters $t = 1$, -2 and $\tfrac{1}{2}$.

35 Find the equations of the tangents to the curve $xy = 3$ which are parallel to the line $y + 3x = 0$. What is the distance between the tangents?

36 Find the points of contact of the two tangents which can be drawn from the point $(-5, 1)$ to the curve $xy = 4$.

37 The normal at the point P $(8, 2)$ on the locus $(4t, 4/t)$ meets the curve again at Q. Find the length PQ.

38 Find the equations of the tangents to the curve $xy = 9$ which pass through the point $(-9, 3)$.

39 Sketch the locus $\left(1-3t, 2+\dfrac{3}{t}\right)$.

Miscellaneous examples

Example 6 A chord RS of the rectangular hyperbola $xy = c^2$ subtends a right angle at a point P on the curve. Prove that RS is parallel to the normal at P.

Let P be the point $\left(ct, \dfrac{c}{t}\right)$, R the point $\left(ct_1, \dfrac{c}{t_1}\right)$ and S the point $\left(ct_2, \dfrac{c}{t_2}\right)$.

Gradient of $PR = \dfrac{\dfrac{c}{t} - \dfrac{c}{t_1}}{ct - ct_1} = \dfrac{t_1 - t}{tt_1(t - t_1)} = -\dfrac{1}{tt_1}$.

Similarly, gradient of $PS = -\dfrac{1}{tt_2}$.

As $\widehat{RPS} = 90°$, $\left(-\dfrac{1}{tt_1}\right)\left(-\dfrac{1}{tt_2}\right) = -1$,

i.e. $-\dfrac{1}{t_1 t_2} = t^2$.

But the gradient of chord $RS = -\dfrac{1}{t_1 t_2}$, and the gradient of the normal at $P = t^2$.

∴ Chord RS is parallel to the normal at P.

Example 7 P and Q are points on the curve $xy = c^2$, centre O. Straight lines through Q parallel to the asymptotes meet the line through O perpendicular to OP in H and K. Prove that the circle on HK as diameter passes through P.

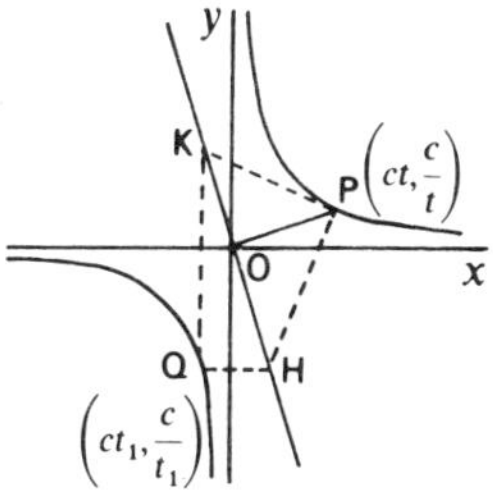

Referring to the diagram in Fig. 86, it is seen that the circle on HK as diameter passes through P if $\widehat{KPH}$ is a right angle.

Let P, Q be the points $\left(ct, \dfrac{c}{t}\right), \left(ct_1, \dfrac{c}{t_1}\right)$ respectively.

Gradient of $KH = \dfrac{-1}{\text{gradient of } OP} = -t^2$.

∴ Equation of KH is $y = -t^2 x$.

As QK is parallel to the y axis, the x coordinate of K is ct_1, so, K is the point $(ct_1, -ct^2 t_1)$.

Fig. 86

Similarly, H is the point $\left(-\dfrac{c}{t^2 t_1}, \dfrac{c}{t_1}\right)$.

Gradient of KP, $m_1 = \dfrac{\dfrac{c}{t} - (-ct^2 t_1)}{ct - ct_1} = \dfrac{1 + t^3 t_1}{t(t - t_1)}$.

Gradient of HP, $m_2 = \dfrac{\dfrac{c}{t} - \dfrac{c}{t_1}}{ct - \left(-\dfrac{c}{t^2 t_1}\right)} = \dfrac{t(t_1 - t)}{t^3 t_1 + 1}$.

Thus, as $m_1 m_2 = -1$, KP is perpendicular to HP and hence the semicircle on HK as diameter passes through P.

[B] EXAMPLES 14d

1 Find the equation of the normal at the point $\left(2a, \dfrac{a}{2}\right)$ on the curve $xy = a^2$.

2 The normal at the point $(5, 3)$ on the curve $xy = 15$ meets the asymptotes at A and B. Find the length AB.

3 The line $2y = x + 7$ intersects the locus $\left(2t, \dfrac{2}{t}\right)$ at A and B. Find the values of t corresponding to A and B and the coordinates of the point of intersection of the tangents at A and B.

4 Prove that the line $m^2 x + y = 8m$ is a tangent to the rectangular hyperbola $xy = 16$ for all values of m and find the coordinates of the point of contact.

5 Find the acute angle between the tangents which can be drawn from the point $(2, -3)$ to the curve $4xy = 25$.

6 If the lines joining the points $P\left(2t_1, \dfrac{2}{t_1}\right)$, $Q\left(2t_2, \dfrac{2}{t_2}\right)$ to the centre of the rectangular hyperbola $xy = 4$, are inclined at $45°$, find the relationship between t_1 and t_2.

7 Find the length of the chord of $xy = 9$ which is normal to the curve at the point $(-1, -9)$.

8 A straight line gradient m is drawn through the point $(4, 6)$ to meet the curve $xy = 4$ at P and Q. Find the coordinates of the midpoint of PQ and hence find the locus of the midpoints of chords which pass through $(4, 6)$.

9 Find the points of intersection of the rectangular hyperbolas $xy = 4$, $x^2 - y^2 = 15$. Show that the curves intersect at right angles.

10 Find the coordinates of the centre and the vertices of the rectangular hyperbola $x = 1 + 4t$, $y = 2 + \dfrac{4}{t}$. What are the equations of the asymptotes?

11 If the tangent at the point $P(2, 9)$ on the rectangular hyperbola $y(x-1) = 9$ meets the asymptotes at Q, R, show that $QP = PR$.

12 Find the asymptotes of the curve $6xy - 10x + 3y - 9 = 0$.

13 P, Q are any two points on a rectangular hyperbola. PK is drawn parallel to one asymptote meeting the other at K; QL is drawn parallel to the second asymptote meeting the first at L. Prove that PQ is parallel to LK.

14 Find the coordinates of the centre of the rectangular hyperbola whose vertices are the points $(0, 0)$ and $(3, 4)$. What are the equations of the asymptotes?

15 Any tangent to the curve $xy = 16$ meets the asymptotes at Q, R respectively. Lines through Q and R parallel to the asymptotes meet the curve at T and S. Show that TS is parallel to RQ.

16 The normal to the rectangular hyperbola $xy = 4$ at the point $P\left(2t, \dfrac{2}{t}\right)$ meets the lines $y = x$, $y = -x$ at Q, R respectively. If C is the centre of the curve, prove that $PQ = PR = PC$.

17 Two tangents to the rectangular hyperbola $xy = c^2$ at points on different branches of the curve, meet the asymptotes at P, P' and Q, Q' respectively. Show that PQ' is parallel to $P'Q$.

18 Show that the equation of the chord joining the points with parameters t_1, t_2 on the curve $x = ct$, $y = \dfrac{c}{t}$ is $t_1 t_2 y + x = c(t_1 + t_2)$.

Use this result to deduce the equation of the tangent to the curve at the point parameter t.

19 P is any point on the curve $y^2 - x^2 = 4$; the line $y = 4$ meets the curve at Q, R and PQ, RP meet one of the asymptotes at L, M respectively. Prove that LM is of constant length.

20 Prove that the straight line $lx + my = n$ touches the rectangular hyperbola $xy = c^2$, if $n^2 = 4lmc^2$. What are the coordinates of the point of contact?

21 The tangents at two points P, P' on a rectangular hyperbola meet an asymptote at Q, Q' and PP' meets it at K. Prove that $QK = Q'K$.

22 PP' is any chord of a rectangular hyperbola, vertices C, C'. Prove that the angles subtended by PP' at C and C' are either equal or supplementary.

23 Find the equations of the tangents to the curve $xy = -4$ which are perpendicular to the tangent at the point $\left(2t, \dfrac{2}{t}\right)$ to the curve $xy = 4$.

24 Prove that if the tangent at any point P on the rectangular hyperbola $xy = 9$ meets the asymptotes at L, L' and the normal at P meets the diameter through the vertices at G, then $\widehat{LGL'}$ is a right angle.

25 A straight line has its extremities on two fixed perpendicular straight lines and cuts off from them a triangle of constant areas. Prove that the locus of the midpoint of the line is a rectangular hyperbola having the fixed lines as asymptotes.

26 A, A' are the vertices of the rectangular hyperbola $xy = 16$ and P is any point on the curve. Show that the internal and external bisectors of angle APA' are parallel to the asymptotes.

27 Prove that the straight lines drawn from any point of the rectangular hyperbola $xy = c^2$ to the ends of any diameter are equally inclined to the asymptotes.

28 Find the locus of the middle points of all chords of the curve $xy = 4$ which are of length 2.

29 PN is the ordinate of the point P on a rectangular hyperbola; PG is the normal meeting the transverse axis at G. If NP produced meets an asymptote at Q, prove that QG is parallel to the other asymptote.

30 The chord joining the points $P\left(ct_1, \dfrac{c}{t_1}\right)$, $Q\left(ct_2, \dfrac{c}{t_2}\right)$ on the curve $xy = c^2$ meets the asymptotes when produced at P', Q'. Prove that $PP' = QQ'$.

[B] MISCELLANEOUS EXAMPLES

1 Prove that as t varies the point $(at + 2, 2t + b)$ moves along a straight line and find the values of a and b in order that the line has the equation $2x - 3y + 5 = 0$. Find the coordinates of the points on this line distant $\sqrt{2}$ from the origin.

2 Prove that the locus of the point $(3\cos\phi + 2, 3\sin\phi - 4)$ is a circle. Show the circle on a diagram and indicate the points on it corresponding to $\phi = \pi/4$ and $\phi = \pi$.

3 Find the equations of the tangent and normal to the parabola $y^2 = 4ax$ at the point $(at^2, 2at)$. Determine the length intercepted on the axis of the parabola by the ordinate and the normal at this point.

4 Find the locus of the foot of the perpendicular drawn from the focus of a parabola to a tangent to the curve.

If S is the focus of a parabola and SK the perpendicular to the tangent at P, prove that SK^2 is proportional to SP.

5 From a point P on a circle, a perpendicular PN is drawn to a fixed diameter HK

The hyperbola

of the circle. If Q divides PN in the ratio $2:3$, prove that the locus of Q is an ellipse and find its eccentricity.

Prove also that the tangent to the circle at P and the tangent to the ellipse at Q intersect on HK produced.

6 Find the eccentricity of the ellipse $7x^2 + 16y^2 = 56$ and prove that the distance of a directrix from the centre is about $3\cdot 77$.

Taking 2 cm as the unit along each axis, draw carefully the graph of the ellipse.

7 Prove that $x = \tfrac{1}{2}a\left(t + \dfrac{1}{t}\right)$, $y = \tfrac{1}{2}a\left(t - \dfrac{1}{t}\right)$ are the parametric equations of the rectangular hyperbola $x^2 - y^2 = a^2$.

Obtain the equation of the tangent at the point parameter t.

8 The line joining the point $P\,(at^2, 2at)$ on the parabola $y^2 = 4ax$ to the focus cuts the parabola again at Q, and the tangents at P and Q meet at R. Prove that

$$RP \cdot RQ = a^2\left(t + \frac{1}{t}\right)^3.$$

9 Obtain the equation of the normal at the point $P\,(ct, c/t)$ to the rectangular hyperbola $xy = c^2$ and prove that the normal meets the curve again at a point Q with parameter $-1/t^3$.

Prove that, unless P lies on the axis of the hyperbola, P is nearer the origin than Q.

10 Prove that the equation of the straight line joining the points on the ellipse $\dfrac{x^2}{a^2} + \dfrac{y^2}{b^2} = 1$ whose eccentric angles are $\alpha + \beta$ and $\alpha - \beta$ is

$$x \cos\alpha/a + y \sin\alpha/b = \cos\beta.$$

Show that if the tangent at a point P on the ellipse cuts a directrix in Z, then SZ, b and the perpendicular from the centre on to SP form a geometrical progression. S is the focus corresponding to the directrix.

11 The normal at P on the ellipse $(a\cos\phi, b\sin\phi)$ cuts the major axis at G. Prove that $OG = e^2 ON$ where O is the centre and NP the ordinate of P.

Prove also that, if S, S' are the foci,

$$SP/S'P = SG/S'G.$$

12 Find the equation of the chord joining the points $t = t_1$, $t = t_2$ on the parabola $x = at^2$, $y = 2at$.

Find the equation of the chord of this parabola through the point $(h, 0)$ which cuts the curve in two points, the algebraic sum of whose ordinates is $4ak$ and prove that it cuts the tangent at the vertex at a distance h/k from the vertex.

13 Prove that the gradient of the chord joining the points $(a\cos\phi_1, b\sin\phi_1)$, $(a\cos\phi_2, b\sin\phi_2)$ on the ellipse $\dfrac{x^2}{a^2} + \dfrac{y^2}{b^2} = 1$ is $-\dfrac{b}{a}\cot\dfrac{\phi_1 + \phi_2}{2}$.

The extremities of any diameter of the ellipse are L, L' and M is any other point on the curve. Prove that the product of the gradients of the chords LM, $L'M$ is constant.

14 Two parabolas have the origin as a common focus, their axes along the line $3x - 4y = 0$ and the point $(0, \tfrac{3}{2})$ in common. Find their equations and prove that the latus rectum of one is four times that of the other.

15 Find the perpendicular distance of the focus from the tangent to the parabola $y^2 = 4ax$ at the point $(at^2, 2at)$.

16 The parametric coordinates of a point are $(a\cos nt, b\cos 2nt)$. Prove that the locus of the point is a parabola. What are the coordinates of the vertex?

17 If the normal at P to the curve $xy = c^2$ meets the curve again at Q, prove that $PQ = OP^3/c^2$, O being the centre.

18 The normal at the point P ($a \cos \phi$, $b \sin \phi$) on an ellipse meets the radius of the auxiliary circle drawn to the corresponding point P', at R. Prove that the locus of R is the circle $x^2 + y^2 = (a+b)^2$.

19 Show that, if the tangent at the point P on a parabola meets the latus rectum at K, then SK is a mean proportional between the segments of the focal chord through P.

20 A tangent to the parabola $y^2 + 4x = 0$ meets the parabola $y^2 = 8x$ at P, Q. Prove that the locus of the middle point of PQ is $5y^2 = 16x$.

21 Prove that if a chord of the ellipse ($a \cos \phi$, $b \sin \phi$) subtends a right angle at the centre then it touches the circle $(x^2 + y^2)(a^2 + b^2) = a^2 b^2$.

22 Find the equations of the tangent and normal at the point (x_1, y_1) on the curve $y = 2\sqrt{ax}$.

Prove that the two tangents to this curve which pass through the point $(-a, k)$ are perpendicular.

23 The straight lines joining a point P on the ellipse $b^2 x^2 + a^2 y^2 = a^2 b^2$ to the foci meet the curve again at Q and R; the tangents at Q and R meet at T. Prove that T lies on the normal at P and that PT is bisected by the minor axis.

24 Prove that chords of the parabola $y^2 = 4ax$ which subtend a right angle at the origin all cut the x axis at the same point.

Find the locus of the midpoints of these chords.

25 Find the equation of the normal to the parabola $y^2 = 4ax$ at the point $P(ap^2, 2ap)$. If this normal cuts the curve again at Q prove that the lines joining the origin to P and Q are perpendicular if $p^2 = 2$.

26 Prove that any point whose x and y coordinates satisfy the equations

$$\frac{1-x/a}{t^2} = \frac{1+x/a}{1} = \frac{y/b}{t},$$

where t is a parameter, lies on the ellipse $x^2/a^2 + y^2/b^2 = 1$.

Show also that the coordinates of the point of intersection of the tangents to the ellipse at the points $t = t_1$, $t = t_2$ satisfy the equations

$$\frac{1-x/a}{t_1 t_2} = \frac{1+x/a}{1} = \frac{y/b}{\frac{1}{2}(t_1 + t_2)}.$$

27 Prove that the locus of the feet of the perpendiculars drawn from the origin to tangents to the parabola $(at^2, 2at)$ is the curve $x(x^2 + y^2) + ay^2 = 0$.

28 The straight line joining any point P on the curve $y^2 = 4ax$ to the origin and the perpendicular from the focus to the tangent at P, intersect at R. Find the equation of the locus of R.

29 The tangent at a point P on a rectangular hyperbola meets the asymptotes at V and W, and the normal at P meets the axes of the hyperbola at K and L. Prove that $VKWL$ is a square.

30 Prove that if the normal at P ($a \cos \phi$, $b \sin \phi$) on the ellipse $x^2/a^2 + y^2/b^2 = 1$ cuts the major axis in G, then $SG = eSP$ where S is a focus and e the eccentricity of the ellipse.

Hence prove that, if the normals at the ends P, Q of a focal chord meet at V, the line through V parallel to the major axis bisects PQ.

31 P, Q, R are three points on a parabola. The diameters (lines parallel to the axis) through P, Q meet QR, RP at D, E respectively. Prove that the tangents at P, Q intersect at the middle point of DE.

32 Find the equation of the normal at the point $(ct, c/t)$ on the rectangular hyperbola $xy = c^2$ and show that the length of the chord along it is $c(1+t^4)^{3/2}/t^3$.

33 Prove that the point $(a + r\cos\theta, r\sin\theta)$ lies on the circle
$$(x-a)^2 + y^2 = r^2$$
and find the equation of the tangent at this point.

Prove that the points of contact of the external common tangents of the two circles $(x-a)^2 + y^2 = r_1^2$, $(x+a)^2 + y^2 = r_2^2$ lie on the circle
$$x^2 + y^2 = a^2 + r_1 r_2.$$

34 The tangent at any point P on the parabola $y^2 = 4ax$ meets the axis of the curve at T. Through T a line TK is drawn perpendicular to TP to cut the line joining P to the vertex at K. Prove that the equation of the locus of K is $y^2(y^2 + 2x^2) + 8ax^3 = 0$.

35 A diameter POQ of an ellipse, centre O, focus S, meets the directrix corresponding to S in R; SR meets the auxiliary circle in P', Q'. Prove that OP', OQ' are parallel to SP, SQ respectively.

36 If a tangent to the rectangular hyperbola $\left(ct, \dfrac{c}{t}\right)$ forms with the axes of the hyperbola a triangle of area $\triangle_1$, and the corresponding normal forms with the asymptotes a triangle of area $\triangle_2$, prove that $\triangle_2 \triangle_1^2 = 8c^6$.

37 PN is the ordinate of a point P on the parabola $y^2 = 4ax$; NP is produced to Q so that $PQ = SP$ where S is the focus. Find the locus of Q.

Show that the tangent at Q to this locus intersects the tangent at P to the given parabola on the directrix of the latter.

38 PQ is a chord of a parabola drawn in a fixed direction. Prove that the locus of the point of intersection of the normals at P and Q is a straight line which is itself normal to the parabola.

39 A triangle is inscribed in the rectangular hyperbola $xy = c^2$. Prove that the perpendiculars to the sides at the points where they meet the asymptotes are concurrent.

If the point of concurrence is (x_1, y_1) for one asymptote and (x_2, y_2) for the other, prove that $x_2 y_1 = c^2$.

40 ABC is a triangle inscribed in a rectangular hyperbola and P, Q, R are the feet of the perpendiculars drawn to an asymptote from A, B, C respectively. Prove that the perpendiculars drawn from P, Q, R to BC, CA, AB respectively meet at a point on the other asymptote.

15 | Inverse circular and hyperbolic functions

Inverse circular functions
If $x = \sin y$ and $-\frac{1}{2}\pi \leqslant y \leqslant \frac{1}{2}\pi$, then y is said to be the *inverse sine* of x and is written

$$y = \sin^{-1} x \text{ or } \arcsin x;$$

i.e. $\sin^{-1} x$ or $\arcsin x$ is the angle between $-\frac{1}{2}\pi$ and $\frac{1}{2}\pi$ whose sine is x.
  Similarly, if $x = \cos y$ and $0 \leqslant y \leqslant \pi$, then $y = \cos^{-1} x$ or $\arccos x$;
  and if $x = \tan y$ and $-\frac{1}{2}\pi < y < \frac{1}{2}\pi$, then $y = \tan^{-1} x$ or $\arctan x$.
There are equivalent definitions for the inverse functions $\sec^{-1} x$, $\operatorname{cosec}^{-1} x$ and $\cot^{-1} x$.

N.B. The functions $\sin^{-1} x$, $\cos^{-1} x$ are only defined in the domain $-1 \leqslant x \leqslant 1$ whilst $\tan^{-1} x$ is defined for all real values of x.

Graphs of inverse circular functions
The inverse circular functions are closely related to the ordinary circular functions and their graphs are parts of those of the circular functions with the x and y axes interchanged.
Graphs of $y = \sin^{-1} x$, $y = \cos^{-1} x$ and $y = \tan^{-1} x$ are shown in Figs. 87 and 88.

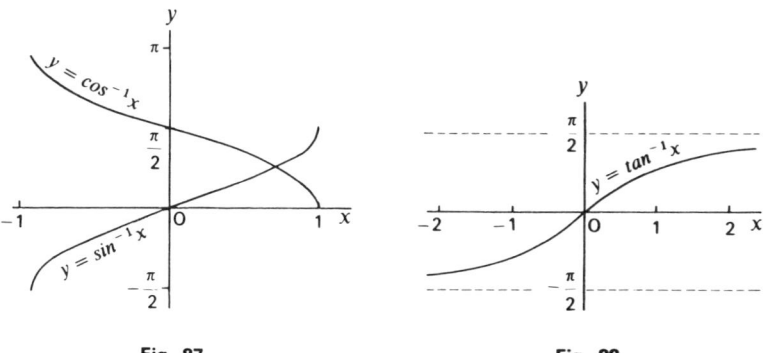

Fig. 87

Fig. 88

N.B. $\sin^{-1} x$ and $\tan^{-1} x$ are increasing functions and their derivatives with respect to x are positive; $\cos^{-1} x$ is a decreasing function and its derivative with respect to x is negative.

Example 1 State the values of (i) $\sin^{-1} 0.5$; (ii) $\cos^{-1} 0$; (iii) $\tan^{-1}(-1)$; (iv) $\cot^{-1}\sqrt{3}$.

(i) $\sin^{-1}(0.5)$, the angle in the interval $-\tfrac{1}{2}\pi$ to $\tfrac{1}{2}\pi$ whose sine is 0.5, $=\dfrac{\pi}{6}$;

(ii) $\cos^{-1} 0$, the angle in the interval 0 to π whose cosine is zero, $=\dfrac{\pi}{2}$;

(iii) $\tan^{-1}(-1)$, the angle in the interval $-\tfrac{1}{2}\pi$ to $\tfrac{1}{2}\pi$ whose tangent is -1, $=-\dfrac{\pi}{4}$;

(iv) $\cot^{-1}\sqrt{3} = \tan^{-1}\dfrac{1}{\sqrt{3}} = \dfrac{\pi}{6}$.

Identities connecting inverse circular functions

(i) $\cos^{-1} x + \sin^{-1} x = \dfrac{\pi}{2}$.

This result follows from the fact that $\sin A = \cos B$ if $A + B = \dfrac{\pi}{2}$.

Writing $\cos^{-1} x = B$ and $\sin^{-1} x = A$, then $x = \cos B = \sin A$.

$\therefore A + B = \dfrac{\pi}{2}$ or $\cos^{-1} x + \sin^{-1} x = \dfrac{\pi}{2}$.

(ii) If all three inverse tangents are positive acute angles,

$$\tan^{-1} a + \tan^{-1} b = \tan^{-1}\dfrac{a+b}{1-ab}.$$

We have $\tan(A + B) = \dfrac{\tan A + \tan B}{1 - \tan A \tan B}$.

i.e. $A + B = \tan^{-1}\left\{\dfrac{\tan A + \tan B}{1 - \tan A \tan B}\right\}$ ....(1)

Writing $A = \tan^{-1} a$, $B = \tan^{-1} b$ or $\tan A = a$, $\tan B = b$, it follows that

$$\tan^{-1} a + \tan^{-1} b = \tan^{-1}\dfrac{a+b}{1-ab}.$$

Note In a more general case the result (1) would have to be written as

$$A + B = n\pi + \tan^{-1}\left\{\dfrac{\tan A + \tan B}{1 - \tan A \tan B}\right\},$$

leading to $\tan^{-1} a + \tan^{-1} b = n\pi + \tan^{-1}\dfrac{a+b}{1-ab}$, $n = -1, 0$ or 1.

Example 2 Solve the equation $\tan^{-1} x + \tan^{-1} 2x = \tan^{-1}\sqrt{2}$, assuming that all the inverse tangents are positive acute angles.

We have, $\tan^{-1} x + \tan^{-1} 2x = \tan^{-1}\dfrac{x + 2x}{1 - x \cdot 2x}$

$= \tan^{-1}\dfrac{3x}{1 - 2x^2}.$

$$\therefore \tan^{-1}\frac{3x}{1-2x^2} = \tan^{-1}\sqrt{2},$$

i.e.
$$\frac{3x}{1-2x^2} = \sqrt{2},$$

$$2\sqrt{2}x^2 + 3x - \sqrt{2} = 0,$$
$$(x+\sqrt{2})(2\sqrt{2}x - 1) = 0.$$
$$\therefore x = -\sqrt{2}, \frac{1}{2\sqrt{2}}.$$

As $\tan^{-1}x$ and $\tan^{-1}2x$ have to be positive acute angles, the solution $x = -\sqrt{2}$ is clearly inadmissible.

Hence
$$x = \frac{1}{2\sqrt{2}}.$$

Differentiation of inverse circular functions Let $y = \sin^{-1}x$, then

$$\sin y = x.$$

Differentiating with respect to x,

$$\cos y \frac{dy}{dx} = 1,$$

$$\frac{dy}{dx} = \frac{1}{\cos y} = \frac{1}{\pm\sqrt{1-\sin^2 y}} = \pm\frac{1}{\sqrt{1-x^2}}.$$

From the graph, it is seen that $\sin^{-1}x$ is an increasing function and so its differential coefficient is positive.

$$\therefore \frac{d}{dx}(\sin^{-1}x) = \frac{1}{\sqrt{1-x^2}}; \quad \text{i.e.} \quad \int \frac{dx}{\sqrt{1-x^2}} = \sin^{-1}x + c.$$

Taking $\quad y = \cos^{-1}x \quad \text{or} \quad \cos y = x.$

Differentiating with respect to x,

$$-\sin y \frac{dy}{dx} = 1,$$

$$\frac{dy}{dx} = -\frac{1}{\sin y} = -\frac{1}{\pm\sqrt{1-\cos^2 y}} = \mp\frac{1}{\sqrt{1-x^2}}.$$

From the graph, it is seen that $\cos^{-1}x$ is a decreasing function so its differential coefficient is negative.

$$\therefore \frac{d}{dx}(\cos^{-1}x) = -\frac{1}{\sqrt{1-x^2}}.$$

Taking $\quad y = \tan^{-1}x \quad \text{or} \quad \tan y = x,$

212 | Inverse circular and hyperbolic functions

it follows that
$$\sec^2 y \frac{dy}{dx} = 1,$$

$$\frac{dy}{dx} = \frac{1}{\sec^2 y} = \frac{1}{1+\tan^2 y} = \frac{1}{1+x^2}.$$

$$\therefore \frac{d}{dx}(\tan^{-1} x) = \frac{1}{1+x^2}; \quad \text{i.e.} \quad \int \frac{dx}{1+x^2} = \tan^{-1} x + c.$$

More generally,
$$\frac{d}{dx}\left(\sin^{-1}\frac{x}{a}\right) = \frac{\frac{1}{a}}{\sqrt{1-\frac{x^2}{a^2}}}, \quad \text{if } a \text{ is a constant,}$$

$$= \frac{1}{\sqrt{a^2-x^2}},$$

and,
$$\frac{d}{dx}\left(\tan^{-1}\frac{x}{a}\right) = \frac{\frac{1}{a}}{1+\frac{x^2}{a^2}} = \frac{a}{a^2+x^2}.$$

Thus
$$\int \frac{dx}{\sqrt{a^2-x^2}} = \sin^{-1}\frac{x}{a} + c.$$

$$\int \frac{dx}{a^2+x^2} = \frac{1}{a}\tan^{-1}\frac{x}{a} + c.$$

These are two important standard types of integral.

Example 3 Differentiate (i) $\sec^{-1} 2x$, (ii) $\tan^{-1}\left(\frac{1-x}{1+x}\right)$.

(i) Let
$$y = \sec^{-1} 2x,$$
then
$$\sec y = 2x.$$

$$\therefore \sec y \tan y \frac{dy}{dx} = 2,$$

$$\frac{dy}{dx} = \frac{2}{\sec y \tan y} = \frac{2}{\pm \sec y \sqrt{\sec^2 y - 1}} = \pm \frac{2}{2x\sqrt{4x^2-1}}.$$

But $\sec^{-1} 2x$ is an increasing function as $\cos^{-1} 2x$ is decreasing.

$$\therefore \frac{d}{dx}\sec^{-1} 2x = \frac{1}{x\sqrt{4x^2-1}}.$$

(ii) Let
$$y = \tan^{-1}\left(\frac{1-x}{1+x}\right).$$

Differentiation of inverse circular functions | 213

$$\therefore \frac{dy}{dx} = \frac{1}{1+\left(\frac{1-x}{1+x}\right)^2} \times \frac{d}{dx}\left(\frac{1-x}{1+x}\right)$$

$$= \frac{(1+x)^2}{(1+x)^2+(1-x)^2} \times \frac{-1(1+x)-1(1-x)}{(1+x)^2}$$

$$= \frac{-2}{2+2x^2} = -\frac{1}{1+x^2}.$$

Note This result is easily obtained by writing $\tan^{-1}\left(\frac{1-x}{1+x}\right)$ as $\tan^{-1}1 - \tan^{-1}x$.

Then
$$\frac{d}{dx}\tan^{-1}\left(\frac{1-x}{1+x}\right) = \frac{d}{dx}\left(\frac{\pi}{4} - \tan^{-1}x\right)$$

$$= -\frac{1}{1+x^2}.$$

Example 4 Evaluate $\int_0^1 \frac{dx}{1+4x^2}$.

Write the denominator in the form $a^2 + x^2$,

$$\int_0^1 \frac{dx}{1+4x^2} = \frac{1}{4}\int_0^1 \frac{dx}{\frac{1}{4}+x^2} = \frac{1}{4}\int_0^1 \frac{dx}{(\frac{1}{2})^2+x^2}$$

$$= \frac{1}{4}\left[\frac{1}{\frac{1}{2}}\tan^{-1}\frac{x}{\frac{1}{2}}\right]_0^1 = \left[\frac{1}{2}\tan^{-1}2x\right]_0^1$$

$$= \frac{1}{2}\tan^{-1}2 - \frac{1}{2}\tan^{-1}0$$

$$= \frac{1}{2} \cdot 1 \cdot 1071 - 0$$

$$= 0 \cdot 5536.$$

[A] EXAMPLES 15a

Find the values of the following inverse functions:

1 $\sin^{-1}0.5$. 2 $\tan^{-1}1$. 3 $\cos^{-1}0.5$. 4 $\sin^{-1}\left(-\frac{\sqrt{2}}{2}\right)$.
5 $\cos^{-1}\frac{\sqrt{3}}{2}$. 6 $\tan^{-1}(-1)$. 7 $\cos^{-1}\left(-\frac{\sqrt{3}}{2}\right)$. 8 $\sin^{-1}\left(-\frac{1}{2}\right)$.
9 $\sec^{-1}2$. 10 $\csc^{-1}2$. 11 $\cot^{-1}(-1)$. 12 $\cot^{-1}\sqrt{3}$.
13 $\sec^{-1}\sqrt{2}$. 14 $\csc^{-1}\sqrt{2}$. 15 $\tan^{-1}2$.

16 Sketch the graphs of $\sec^{-1}x$, $\csc^{-1}x$, $\cot^{-1}x$.

17 If $\sin^{-1}x = \frac{\pi}{5}$, find the value of $\cos^{-1}x$.

18 If $2\sin^{-1}x = \cos^{-1}x$, find x.

19 Prove that $\frac{\pi}{4} + \tan^{-1}x = \tan^{-1}\left(\frac{1+x}{1-x}\right)$.

Differentiate with respect to x:

20 $\sin^{-1} 2x$. **21** $\tan^{-1}\dfrac{x}{2}$. **22** $\cos^{-1} 3x$. **23** $\cot^{-1} x$.

24 $\operatorname{cosec}^{-1} x$. **25** $\sin^{-1}\sqrt{x}$. **26** $\tan^{-1}\dfrac{1}{x}$. **27** $\cos^{-1} x^2$.

28 $\sin^{-1}\dfrac{1}{x+1}$. **29** $\tan^{-1}\dfrac{2+x}{1-2x}$. **30** $\sin^{-1}\dfrac{1-x}{1+x}$. **31** $\sec^{-1} 3x$.

32 $\tan^{-1}\sin x$. **33** $\sin^{-1}\cos x$. **34** $\sin^{-1}(2\sin x)$.

Evaluate the following integrals:

35 $\displaystyle\int \dfrac{dx}{\sqrt{4-x^2}}$. **36** $\displaystyle\int_0^1 \dfrac{dx}{\sqrt{2-x^2}}$. **37** $\displaystyle\int \dfrac{dx}{\sqrt{1-9x^2}}$.

38 $\displaystyle\int \dfrac{dx}{\sqrt{1-16x^2}}$. **39** $\displaystyle\int \dfrac{dx}{4+x^2}$. **40** $\displaystyle\int_0^3 \dfrac{dx}{9+x^2}$.

41 $\displaystyle\int \dfrac{dx}{4+9x^2}$. **42** $\displaystyle\int \dfrac{dx}{16+25x^2}$. **43** $\displaystyle\int_0^{\frac{1}{2}} \dfrac{dx}{\sqrt{1-2x^2}}$.

44 $\displaystyle\int_{-2}^0 \dfrac{dx}{1+(x+2)^2}$. **45** $\displaystyle\int \dfrac{dx}{\sqrt{16-(x-3)^2}}$. **46** $\displaystyle\int \dfrac{dx}{3+2(x-1)^2}$.

Hyperbolic functions The hyperbolic sine and cosine of a variable x are defined as follows:

Hyperbolic sine of x, $\quad \sinh x = \tfrac{1}{2}(e^x - e^{-x})$.

Hyperbolic cosine of x, $\quad \cosh x = \tfrac{1}{2}(e^x + e^{-x})$.

The hyperbolic tangent, secant, cosecant and cotangent are defined by the following relationships:

$$\tanh x = \dfrac{\sinh x}{\cosh x}; \qquad \operatorname{sech} x = \dfrac{1}{\cosh x};$$

$$\operatorname{cosech} x = \dfrac{1}{\sinh x}; \qquad \coth x = \dfrac{\cosh x}{\sinh x}.$$

Graphs of sinh x and cosh x The graphs of $\sinh x$ and $\cosh x$ are readily obtained from the graphs of e^x and e^{-x} (Fig. 89).

Formulae connecting hyperbolic functions Corresponding to formulae involving ordinary circular functions there are formulae involving the hyperbolic functions.

E.g. $\quad \cosh^2 x - \sinh^2 x = \tfrac{1}{4}(e^x + e^{-x})^2 - \tfrac{1}{4}(e^x - e^{-x})^2$

$\qquad\qquad\qquad\qquad = \tfrac{1}{4}(e^{2x} + 2 + e^{-2x}) - \tfrac{1}{4}(e^{2x} - 2 + e^{-2x}) = 1$.

i.e. $\quad \cosh^2 x - \sinh^2 x = 1$.

Hence, $\quad \operatorname{sech}^2 x = 1 - \tanh^2 x$,

and $\quad \operatorname{cosech}^2 x = \coth^2 x - 1$.

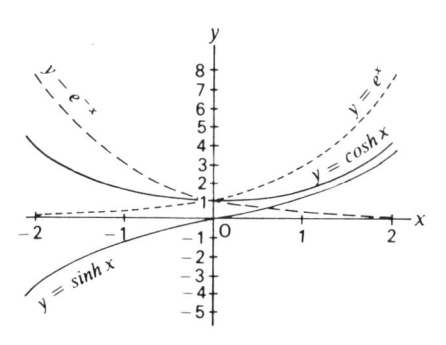

Fig. 89

Note The locus $x = a\cosh\theta$, $y = b\sinh\theta$, when expressed in Cartesian coordinates becomes

$$\left(\frac{x}{a}\right)^2 - \left(\frac{y}{b}\right)^2 = \cosh^2\theta - \sinh^2\theta = 1,$$

i.e.
$$\frac{x^2}{a^2} - \frac{y^2}{b^2} = 1, \quad \text{a hyperbola.}$$

In this course of work, a full knowledge of the properties of hyperbolic functions is unnecessary and further identities will not be proved. The student, however, should have little difficulty in obtaining the important results:

$$\sinh 2x = 2\sinh x \cosh x.$$
$$\cosh 2x = \cosh^2 x + \sinh^2 x,$$
$$= 2\cosh^2 x - 1,$$
$$= 2\sinh^2 x + 1.$$

Differentiation and integration of hyperbolic functions

$$\cosh x = \tfrac{1}{2}(e^x + e^{-x}).$$

$$\frac{d}{dx}\cosh x = \tfrac{1}{2}(e^x - e^{-x}) = \sinh x.$$

Similarly,
$$\frac{d}{dx}\sinh x = \cosh x.$$

$$\frac{d}{dx}\tanh x = \frac{d}{dx}\left(\frac{\sinh x}{\cosh x}\right)$$
$$= \frac{\cosh^2 x - \sinh^2 x}{\cosh^2 x} = \frac{1}{\cosh^2 x}$$
$$= \operatorname{sech}^2 x.$$

216 | Inverse circular and hyperbolic functions

i.e.
$$\frac{d}{dx}(\cosh x) = \sinh x; \quad \int \sinh x \, dx = \cosh x + c.$$

$$\frac{d}{dx}(\sinh x) = \cosh x; \quad \int \cosh x \, dx = \sinh x + c.$$

$$\frac{d}{dx}(\tanh x) = \operatorname{sech}^2 x; \quad \int \operatorname{sech}^2 x \, dx = \tanh x + c.$$

Example 5 Differentiate $\cosh 3x$ and evaluate $\int_0^{0.5} \sinh 3x \, dx$.

By the function of a function rule,

$$\frac{d}{dx} \cosh 3x = 3 \sinh 3x.$$

$$\therefore \int_0^{0.5} \sinh 3x \, dx = \left[\tfrac{1}{3} \cosh 3x \right]_0^{0.5}$$

$$= \tfrac{1}{3} \cosh 1{\cdot}5 - \tfrac{1}{3} \cosh 0$$
$$= \tfrac{1}{6}(e^{1{\cdot}5} + e^{-1{\cdot}5}) - \tfrac{1}{6}(e^0 + e^0)$$
$$= \tfrac{1}{6}(4{\cdot}482 + 0{\cdot}223) - \tfrac{1}{3}$$
$$= 0{\cdot}451.$$

[A] EXAMPLES 15b

Evaluate:
1. $\sinh 1$.
2. $\cosh 2$.
3. $\tanh 2$.
4. $\sinh(-2)$.
5. $\coth(-1)$.
6. $\operatorname{sech} 1{\cdot}5$.
7. Show that $\sinh(-x) = -\sinh x$ and $\cosh(-x) = \cosh x$.
8. Verify that $\cosh(A+B) = \cosh A \cosh B + \sinh A \sinh B$.

Differentiate with respect to x:

9. $\sinh 2x$.
10. $\cosh 3x$.
11. $\tanh 2x$.
12. $\sinh\left(\dfrac{x}{3}\right)$.
13. $\coth x$.
14. $\operatorname{sech} x$.
15. $\operatorname{cosech} x$.
16. $\sinh^2 x$.
17. $\cosh^2 x$.
18. $\ln \sinh x$.
19. $\ln \cosh x$.
20. $e^{\sinh x}$.

Integrate with respect to x:

21. $\cosh 2x$.
22. $\sinh 3x$.
23. $\sinh x/2$.
24. $\operatorname{sech}^2 3x$.
25. $\sinh x \cosh x$.
26. $\sinh^2 x$.

27. Prove that $\dfrac{1}{\cosh x + \sinh x} = \cosh x - \sinh x$ and evaluate

$$\int_0^1 \frac{dx}{\cosh x + \sinh x}.$$

28. If $y = \cosh 2x + \sinh 2x$, find $\dfrac{d^2 y}{dx^2}$.

29 If $y = a \cosh nx + b \sinh nx$, where a, b and n are constants, prove that
$$\frac{d^2y}{dx^2} - n^2 y = 0.$$

Inverse hyperbolic functions The inverse hyperbolic functions are defined in a similar manner to the inverse circular functions.
If $x = \sinh y$ for real values of y, $\quad y = \sinh^{-1} x$ or arcsinh x;
if $x = \cosh y$ for $y \geq 0$, $\quad y = \cosh^{-1} x$ or arccosh x.

The graphs of inverse hyperbolic functions are obtained from those of the hyperbolic functions by interchanging the x and y axes. Graphs of $\sinh^{-1} x$ and $\cosh^{-1} x$ are given in Fig. 90.

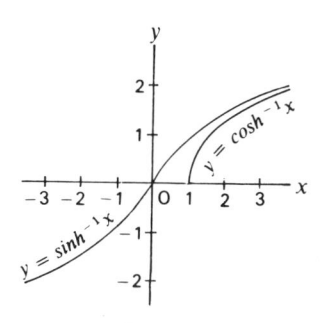

Fig. 90

N.B (i) $\sinh^{-1} x$ is defined for all real values of x and is a steadily increasing function; $\cosh^{-1} x$ is only defined for values of $x \geq 1$ and is also an increasing function;
(ii) $\sinh^{-1} 0 = 0$, $\cosh^{-1} 1 = 0$;
(iii) there are corresponding definitions for $\tanh^{-1} x$ and other inverse hyperbolic functions.

Inverse hyperbolic functions expressed in terms of logarithmic functions
We will now show that the inverse hyperbolic functions are not a new class of functions as they can be expressed in terms of logarithmic functions.

Suppose $y = \sinh^{-1} x$, then $\sinh y = x$.
i.e. $\qquad\qquad\qquad e^y - e^{-y} = 2x.$
Multiply by e^y, $\qquad e^{2y} - 2xe^y - 1 = 0.$
This is a quadratic equation in e^y.

$$\therefore e^y = \frac{2x \pm \sqrt{(2x)^2 - 4(-1)}}{2}$$
$$= x \pm \sqrt{x^2 + 1}.$$

But e^y is always positive and thus the negative sign is inadmissible.

Hence, $$e^y = x + \sqrt{x^2 + 1},$$
or, $$\mathbf{y = \sinh^{-1} x = \ln(x + \sqrt{x^2 + 1})}.$$

Similarly if $y = \cosh^{-1} x$, $\quad e^y = x \pm \sqrt{x^2 - 1}, \quad x \geq 1$
$$\mathbf{\cosh^{-1} x = \ln(x + \sqrt{x^2 - 1})}, \quad \text{as } \cosh^{-1} x \geq 0.$$

Example 6 Find the values of (i) $\sinh^{-1}(1\cdot5)$, (ii) $\cosh^{-1}(1\cdot5)$.

(i) $$\sinh^{-1}(1\cdot5) = \ln(1\cdot5 + \sqrt{(1\cdot5)^2 + 1})$$
$$= \ln 3\cdot303$$
$$= 1\cdot195.$$

(ii) $$\cosh^{-1}(1\cdot5) = \ln(1\cdot5 + \sqrt{(1\cdot5)^2 - 1})$$
$$= \ln 2\cdot618$$
$$= 0\cdot962.$$

Differentiation of inverse hyperbolic functions Let $y = \sinh^{-1} x$, or,
$$\sinh y = x.$$

Differentiating with respect to x,
$$\cosh y \frac{dy}{dx} = 1,$$
$$\frac{dy}{dx} = \frac{1}{\cosh y} = \frac{1}{\pm\sqrt{1 + \sinh^2 y}} = \pm\frac{1}{\sqrt{1 + x^2}}.$$

From the graph, it is seen that $\sinh^{-1} x$ is an increasing function for all values of x,
$$\therefore \mathbf{\frac{d}{dx}(\sinh^{-1} x) = \frac{1}{\sqrt{1 + x^2}}}; \quad \mathbf{\int \frac{dx}{\sqrt{1 + x^2}} = \sinh^{-1} x + c.}$$

Taking $\quad y = \cosh^{-1} x, \quad$ or $\quad \cosh y = x,$

then, $\quad \sinh y \dfrac{dy}{dx} = 1,$

$$\frac{dy}{dx} = \frac{1}{\sinh y} = \frac{1}{\pm\sqrt{\cosh^2 y - 1}} = \pm\frac{1}{\sqrt{x^2 - 1}},$$

i.e. $\quad \mathbf{\dfrac{d}{dx}(\cosh^{-1} x) = \pm\dfrac{1}{\sqrt{x^2 - 1}}}.$

The ambiguous sign arises from the fact that the relationship $\cosh y = x$, taken for all real values of y, gives a curve which is symmetrical about the x-axis. The

Differentiation of inverse hyperbolic functions | 219

ambiguity is eliminated as the inverse function $y = \cosh^{-1} x$, $y \geq 0$, is an increasing function and consequently its derivative is positive.

i.e. $\quad \dfrac{d}{dx}(\cosh^{-1} x) = \dfrac{1}{\sqrt{x^2-1}}; \quad \displaystyle\int \dfrac{dx}{\sqrt{x^2-1}} = \cosh^{-1} x + c.$

More generally, $\quad \dfrac{d}{dx}\left(\sinh^{-1}\dfrac{x}{a}\right) = \dfrac{1/a}{\sqrt{1+\dfrac{x^2}{a^2}}} = \dfrac{1}{\sqrt{a^2+x^2}},$

and $\quad \dfrac{d}{dx}\left(\cosh^{-1}\dfrac{x}{a}\right) = \dfrac{1/a}{\sqrt{\dfrac{x^2}{a^2}-1}} = \dfrac{1}{\sqrt{x^2-a^2}}.$

So the following important standard types of integral are obtained:

$$\int \dfrac{dx}{\sqrt{a^2+x^2}} = \sinh^{-1}\dfrac{x}{a} + c' = \ln\left(\dfrac{x+\sqrt{x^2+a^2}}{a}\right) + c'$$

$$= \ln(x+\sqrt{x^2+a^2}) + c, \qquad \text{as } a \text{ is constant.}$$

$$\int \dfrac{dx}{\sqrt{x^2-a^2}} = \cosh^{-1}\dfrac{x}{a} + c' = \ln(x+\sqrt{x^2-a^2}) + c.$$

Example 7 Differentiate $\sinh^{-1}(3x+2)$.

$$\dfrac{d}{dx}\sinh^{-1}(3x+2) = \dfrac{1}{\sqrt{1+(3x+2)^2}} \times \dfrac{d}{dx}(3x+2)$$

$$= \dfrac{3}{\sqrt{9x^2+12x+5}}.$$

Example 8 Evaluate $\displaystyle\int \dfrac{dx}{\sqrt{3x^2-1}}$.

First express the term under the root sign in the form $x^2 - a^2$.

$$\int \dfrac{dx}{\sqrt{3x^2-1}} = \dfrac{1}{\sqrt{3}}\int \dfrac{dx}{\sqrt{x^2-\tfrac{1}{3}}} = \dfrac{1}{\sqrt{3}}\cosh^{-1}\dfrac{x}{1/\sqrt{3}} + c$$

$$= \dfrac{1}{\sqrt{3}}\cosh^{-1}\sqrt{3}x + c.$$

[A] **EXAMPLES 15c**

Find the values of:
1. $\sinh^{-1} 2.$
2. $\cosh^{-1} 2.$
3. $\sinh^{-1}(-1).$
4. $\cosh^{-1} 3.$
5. $\sinh^{-1} 0.5.$
6. $\operatorname{cosech}^{-1} 0.25.$

7. If $\tanh y = \tfrac{1}{2}$, show that $\dfrac{e^{2y}-1}{e^{2y}+1} = \tfrac{1}{2}$ and deduce that $\tanh^{-1}\tfrac{1}{2} = \tfrac{1}{2}\ln 3.$

Inverse circular and hyperbolic functions

Express as logarithmic functions:

8 $\sinh^{-1} 2x$.

9 $\cosh^{-1}(x+1)$.

10 $\sinh^{-1}(2x-1)$.

11 $\cosh^{-1}\dfrac{x+1}{2}$.

12 $\sinh^{-1}\dfrac{x^2}{2}$.

13 $\operatorname{cosech}^{-1}\dfrac{1}{x}$.

14 $\operatorname{sech}^{-1}\dfrac{2}{x}$.

Differentiate with respect to x:

15 $\sinh^{-1} 2x$.

16 $\sinh^{-1}\dfrac{x}{3}$.

17 $\cosh^{-1}\dfrac{x}{2}$.

18 $\sinh^{-1}\dfrac{4x}{3}$.

19 $\cosh^{-1}\dfrac{x}{\sqrt{2}}$.

20 $\cosh^{-1}\sqrt{3}x$.

21 $\sinh^{-1}(1-3x)$.

22 $\cosh^{-1}\dfrac{x-1}{\sqrt{5}}$.

23 $\sinh^{-1}\dfrac{x-\sqrt{2}}{\sqrt{2}}$.

24 $\sinh^{-1}\sqrt{x}$.

25 $\cosh^{-1}\dfrac{1}{x}$.

26 $\cosh^{-1}\sec x$.

27 $\sinh^{-1} e^x$.

28 If $\tanh y = x$, prove that $\dfrac{dy}{dx} = \dfrac{1}{\operatorname{sech}^2 y}$. Deduce $\dfrac{d}{dx}(\tanh^{-1} x) = \dfrac{1}{1-x^2}$.

29 Find $\dfrac{d}{dx}\coth^{-1} x$.

30 Find $\dfrac{dy}{dx}$ if (i) $y = \tanh^{-1}\sin x$. (ii) $y = \coth^{-1}\dfrac{x+1}{x-1}$.

Evaluate the following integrals:

31 $\displaystyle\int \dfrac{dx}{\sqrt{x^2-4}}$.

32 $\displaystyle\int \dfrac{dx}{\sqrt{x^2+9}}$.

33 $\displaystyle\int_0^1 \dfrac{dx}{\sqrt{4+x^2}}$.

34 $\displaystyle\int_2^3 \dfrac{dx}{\sqrt{x^2-2}}$.

35 $\displaystyle\int \dfrac{dx}{\sqrt{4x^2-1}}$.

36 $\displaystyle\int_{-1}^1 \dfrac{dx}{\sqrt{9+4x^2}}$.

37 $\displaystyle\int \dfrac{dx}{\sqrt{1+3x^2}}$.

38 $\displaystyle\int_1^2 \dfrac{dx}{\sqrt{9x^2-4}}$.

39 $\displaystyle\int \dfrac{dx}{\sqrt{3x^2+2}}$.

40 $\displaystyle\int \dfrac{dx}{\sqrt{(x+2)^2+4}}$.

41 $\displaystyle\int_{\sqrt{2}}^{\sqrt{10}} \dfrac{dx}{\sqrt{5x^2-2}}$.

42 $\displaystyle\int \dfrac{dx}{\sqrt{2(x-3)^2+1}}$.

Integrals of the form $\displaystyle\int \dfrac{dx}{\sqrt{ax^2+bx+c}}$ This type of integral can always be reduced to one or other of the three standard forms,

$$\int \dfrac{dx}{\sqrt{a^2-x^2}}, \quad \int \dfrac{dx}{\sqrt{a^2+x^2}}, \quad \int \dfrac{dx}{\sqrt{x^2-a^2}}.$$

Standard forms of integral | 221

Example 9 Evaluate (i) $\int \dfrac{dx}{\sqrt{2x^2 - 4x + 5}}$; (ii) $\int_1^2 \dfrac{dx}{\sqrt{2 + 3x - 2x^2}}$.

(i) $\sqrt{2x^2 - 4x + 5} = \sqrt{2}\sqrt{\{x^2 - 2x + \tfrac{5}{2}\}} = \sqrt{2}\sqrt{\{(x-1)^2 + \tfrac{3}{2}\}}$.

$\therefore \int \dfrac{dx}{\sqrt{2x^2 - 4x + 5}} = \dfrac{1}{\sqrt{2}} \int \dfrac{dx}{\sqrt{(x-1)^2 + \tfrac{3}{2}}} = \dfrac{1}{\sqrt{2}} \sinh^{-1} \dfrac{\sqrt{2}(x-1)}{\sqrt{3}} + c$.

Note It is important to make the coefficient of x^2 unity, not only to simplify the process of completing the square but also to ensure that a constant multiplier is not omitted in the integration.

E.g. $\int \dfrac{dx}{\sqrt{(2x-1)^2 + 1}} = \tfrac{1}{2} \sinh^{-1}(2x - 1)$ *not* $\sinh^{-1}(2x - 1)$.

(ii) $\int_1^2 \dfrac{dx}{\sqrt{2 + 3x - 2x^2}} = \dfrac{1}{\sqrt{2}} \int_1^2 \dfrac{dx}{\sqrt{1 + \dfrac{3x}{2} - x^2}} = \dfrac{1}{\sqrt{2}} \int_1^2 \dfrac{dx}{\sqrt{\dfrac{25}{16} - \left(x - \dfrac{3}{4}\right)^2}}$

$= \left[\dfrac{1}{\sqrt{2}} \sin^{-1} \dfrac{(x - \tfrac{3}{4})}{5/4} \right]_1^2 = \left[\dfrac{1}{\sqrt{2}} \sin^{-1} \dfrac{4x - 3}{5} \right]_1^2$

$= \dfrac{1}{\sqrt{2}} (\sin^{-1} 1 - \sin^{-1} 0\cdot2) = 0\cdot968$.

Integrals of the form $\int \dfrac{dx}{ax^2 + bx + c}$ Although it is always possible to express this type of integral in terms of an inverse function, we will only take the case where the denominator does not split up into real factors, the condition for which is $b^2 < 4ac$. In the case where $b^2 > 4ac$ and the denominator can be expressed as a product of two linear factors, the integral is readily evaluated by the method of partial fractions discussed in the next chapter.

Example 10 Evaluate (i) $\int \dfrac{dx}{4x^2 + 2x + 1}$; (ii) $\int_{-\tfrac{1}{2}}^{1\tfrac{1}{2}} \dfrac{dx}{(2x-1)^2 + 4}$.

(i) $\int \dfrac{dx}{4x^2 + 2x + 1} = \dfrac{1}{4} \int \dfrac{dx}{x^2 + \dfrac{x}{2} + \dfrac{1}{4}} = \dfrac{1}{4} \int \dfrac{dx}{\left(x + \dfrac{1}{4}\right)^2 + \dfrac{3}{16}}$

$= \dfrac{1}{4} \dfrac{1}{\sqrt{3}/4} \tan^{-1} \dfrac{(x + \tfrac{1}{4})}{\sqrt{3}/4} + c$

$= \dfrac{1}{\sqrt{3}} \tan^{-1} \dfrac{4x + 1}{\sqrt{3}} + c$.

(ii) $\displaystyle\int_{-\frac{1}{2}}^{1\frac{1}{2}} \frac{dx}{(2x-1)^2+4} = \frac{1}{2^2}\int_{-\frac{1}{2}}^{1\frac{1}{2}} \frac{dx}{(x-\frac{1}{2})^2+1}$

$\displaystyle = \frac{1}{4}\left[\tan^{-1}(x-\tfrac{1}{2})\right]_{-\frac{1}{2}}^{1\frac{1}{2}}$

$\displaystyle = \frac{1}{4}\{\tan^{-1}1 - \tan^{-1}(-1)\}$

$\displaystyle = \frac{1}{4}\left\{\frac{\pi}{4} - \left(-\frac{\pi}{4}\right)\right\} = \frac{\pi}{8}.$

Example 11 Evaluate $\displaystyle\int \frac{(x+1)\,dx}{x^2+4x+6}$

This type of integral is evaluated by writing the numerator in the form, constant × differential coefficient of denominator + a second constant.

In this case, the D.C. of the denominator is $2x+4$ and we write the numerator as $\frac{1}{2}(2x+4) - 1$.

$$\int \frac{x+1}{x^2+4x+6}\,dx = \frac{1}{2}\int \frac{2x+4}{x^2+4x+6}\,dx - \int \frac{dx}{x^2+4x+6}.$$

The first integral on the R.H.S. is immediately integrable to $\frac{1}{2}\ln(x^2+4x+6)$ as the numerator is the D.C. of the denominator.

$$\therefore \int \frac{x+1}{x^2+4x+6}\,dx = \frac{1}{2}\ln(x^2+4x+6) - \int \frac{dx}{(x+2)^2+2}$$

$$= \frac{1}{2}\ln(x^2+4x+6) - \frac{1}{\sqrt{2}}\tan^{-1}\frac{x+2}{\sqrt{2}} + c.$$

[B] EXAMPLES 15d

Integrate with respect to x:

1. $\dfrac{1}{(x+2)^2+3}.$
2. $\dfrac{1}{(2x-1)^2+4}.$
3. $\dfrac{1}{\sqrt{4-(x-1)^2}}.$
4. $\dfrac{1}{\sqrt{x^2+2x+7}}.$
5. $\dfrac{1}{x^2+2x+5}.$
6. $\dfrac{1}{\sqrt{4-x-x^2}}.$
7. $\dfrac{1}{x^2-x+2}.$
8. $\dfrac{1}{\sqrt{x(x-1)}}.$
9. $\dfrac{1}{\sqrt{x(x+1)}}.$
10. $\dfrac{1}{2x^2+2x+5}.$
11. $\dfrac{1}{\sqrt{x(1-2x)}}.$
12. $\dfrac{1}{\sqrt{4+x-2x^2}}.$
13. $\dfrac{1}{\sqrt{3x^2-6x+8}}.$
14. $\dfrac{1}{(3x+2)^2+1}.$
15. $\dfrac{1}{x^2+x+1}.$
16. $\dfrac{1}{\sqrt{x^2-x+1}}.$
17. $\dfrac{1}{\sqrt{1+2x-3x^2}}.$

Miscellaneous examples | 223

18 Express $\dfrac{x+1}{\sqrt{x^2+1}}$ in the form $\dfrac{A\dfrac{d}{dx}(x^2+1)+B}{\sqrt{x^2+1}}$, where A and B are constants and hence prove that $\displaystyle\int \dfrac{x+1}{\sqrt{x^2+1}}\,dx = \sqrt{x^2+1} + \sinh^{-1} x + c$.

19 Express $\dfrac{3x-2}{\sqrt{3+2x-x^2}}$ in the form $\dfrac{A\dfrac{d}{dx}(3+2x-x^2)+B}{\sqrt{3+2x-x^2}}$, where A and B are constants. Hence evaluate $\displaystyle\int \dfrac{3x-2}{\sqrt{3+2x-x^2}}\,dx$.

20 Express $\dfrac{x^2+2x}{x^2+x+1}$ in the form $\dfrac{A(x^2+x+1)+B\dfrac{d}{dx}(x^2+x+1)+C}{x^2+x+1}$, where A, B and C are constants. Hence evaluate $\displaystyle\int_0^1 \dfrac{x^2+2x}{x^2+x+1}\,dx$.

Use the methods of Examples 18–20 to evaluate the following integrals:

21 $\displaystyle\int \dfrac{2x-1}{x^2+2x+3}\,dx.$ **22** $\displaystyle\int_0^{\frac{1}{2}} \dfrac{x+1}{\sqrt{1-x^2}}\,dx.$ **23** $\displaystyle\int \dfrac{2x-1}{\sqrt{x^2-1}}\,dx.$

24 $\displaystyle\int_0^1 \dfrac{3x-2}{\sqrt{x^2+4}}\,dx.$ **25** $\displaystyle\int \dfrac{x\,dx}{x^2+x+1}.$ **26** $\displaystyle\int \dfrac{x\,dx}{\sqrt{x^2-2x-3}}.$

27 $\displaystyle\int \dfrac{3x-4}{\sqrt{2x^2-4x+1}}\,dx.$ **28** $\displaystyle\int_{-1}^1 \dfrac{x^2}{1+x^2}\,dx.$ **29** $\displaystyle\int \dfrac{x^2+x+1}{x^2-x+1}\,dx.$

[B] MISCELLANEOUS EXAMPLES

1 If $\sin^{-1} a = x$, find the values of $\cos x$ and $\sin 2x$ in terms of a. Deduce that $2\sin^{-1} a = \sin^{-1}(2a\sqrt{1-a^2})$.

2 Differentiate with respect to x:

(i) $\cos^{-1} \dfrac{x}{\sqrt{1+x^2}}$; (ii) $\sinh^{-1} \sqrt{x}$; (iii) $\tan^{-1} \dfrac{4\sqrt{x}}{1-4x}$; (iv) $\dfrac{x\cos^{-1} x}{\sqrt{1-x^2}}$.

3 Find $\dfrac{dy}{dx}$ if (i) $\tan y = \dfrac{x^2+1}{x^2-1}$;

(ii) $\cos y = \dfrac{1-x^2}{1+x^2}$;

(iii) $\sin y = \dfrac{1}{\sqrt{1+x^2}}$

4 If $y = \tan^{-1} x$, $z = \tan^{-1} \dfrac{2x}{1-x^2}$, find $\dfrac{dy}{dz}$ as a function of x.

5 Assuming the inverse tangents refer to positive acute angles, solve the equation, $\tan^{-1} x + \tan^{-1}(x+1) = \tan^{-1} 2$.

Inverse circular and hyperbolic functions

6 Differentiate with respect to x, $\frac{1}{2}x\sqrt{x^2-a^2} - \frac{a^2}{2}\cosh^{-1}\frac{x}{a}$. Deduce the value of
$$\int \sqrt{x^2-4}\,dx.$$

7 By differentiating the function $\frac{1}{2}x\sqrt{a^2-x^2} + \frac{a^2}{2}\sin^{-1}\frac{x}{a}$, obtain the value of
$$\int_0^3 \sqrt{9-x^2}\,dx.$$

8 Show that $2\tan^{-1}x = \tan^{-1}\frac{2x}{1-x^2}$ and $3\tan^{-1}x = \frac{3x-x^3}{1-3x^2}$.

9 If $\tan y = x\tan^{-1}x$, find $\frac{dy}{dx}$.

10 Evaluate

(i) $\displaystyle\int_0^{\frac{1}{2}} \frac{x+1}{\sqrt{1-4x^2}}\,dx;$ (ii) $\displaystyle\int_1^2 \frac{x^2\,dx}{3x^2+6x+4};$ (iii) $\displaystyle\int_1^3 \frac{dx}{\sqrt{4x^2+x-3}};$

(iv) $\displaystyle\int_{-1}^1 \frac{x^2\,dx}{1+x^2};$ (v) $\displaystyle\int_{-1}^0 \frac{1-x}{\sqrt{2x^2+2x+3}}\,dx;$ (vi) $\displaystyle\int_0^1 \frac{1-x^2}{1+x^2}\,dx.$

11 Solve the following equations, assuming the inverse tangents refer to positive acute angles:

(i) $2\tan^{-1}\frac{x}{2} - \tan^{-1}x = \tan^{-1}\frac{1}{2};$

(ii) $\tan^{-1}3x + \tan^{-1}x = \tan^{-1}\frac{1}{2}.$

12 Writing the formula $\sin(A+B) = \sin A\cos B + \cos A\sin B$ in the form $A+B = \sin^{-1}(\sin A\cos B + \cos A\sin B)$, deduce the result
$$\sin^{-1}x + \sin^{-1}y = \sin^{-1}(x\sqrt{1-y^2} + y\sqrt{1-x^2}).$$

13 Prove that $\cos^{-1}x + \cos^{-1}y = \cos^{-1}(xy - \sqrt{1-x^2}\sqrt{1-y^2})$.

14 By rationalising the numerators of the integrands, evaluate

(i) $\displaystyle\int \sqrt{\frac{1+x}{1-x}}\,dx;$ (ii) $\displaystyle\int \sqrt{\frac{x}{x+1}}\,dx;$ (iii) $\displaystyle\int \sqrt{\frac{x-1}{x+1}}\,dx.$

15 Obtain the first three terms in the expansion of $\dfrac{1}{\sqrt{1-x^2}}$, where x is numerically less than 1.

Deduce that $\sin^{-1}x = x + \dfrac{1\,x^3}{2\,3} + \dfrac{1\cdot 3\,x^5}{2\cdot 4\,5} + \ldots$

16 Using the expansion of $\dfrac{1}{1+x^2}$, obtain the first four terms in the expansion of $\tan^{-1}x$ as a power series in x when $-1 < x < 1$.

17 If $y = (\sin^{-1}x)^2$, show that $(1-x^2)\dfrac{d^2y}{dx^2} = x\dfrac{dy}{dx} + 2$.

18 If $\ln y = \tan^{-1}x$, prove that $(2x-1)\left(\dfrac{dy}{dx}\right)^2 + y\dfrac{d^2y}{dx^2} = 0$.

19 Prove that $\dfrac{d}{dx}\tan^{-1}\left(\tanh\dfrac{x}{2}\right) = \dfrac{1}{2}\operatorname{sech} x$.

20 Prove that $\sinh^{-1} x = \ln(x + \sqrt{x^2+1})$.

If $\sinh u = \tan x$, prove that $\tanh u = \sin x$ and that $u = \ln\tan\left(\dfrac{\pi}{4} + \dfrac{x}{2}\right)$.

16 | Partial fractions and their applications

Partial fractions If the two fractions $\dfrac{2}{x+3}, \dfrac{3}{2x-1}$ are added together the result can be expressed as a single compound fraction.

We have $\dfrac{2}{x+3} + \dfrac{3}{2x-1} = \dfrac{2(2x-1) + 3(x+3)}{(x+3)(2x-1)} = \dfrac{7x+7}{(x+3)(2x-1)}.$

If the function $\dfrac{7x+7}{(x+3)(2x-1)}$ is to be expanded as a power series or integrated, it is clearly much simpler to express it as the sum of the two fractions $\dfrac{2}{x+3}, \dfrac{3}{2x-1}$. These fractions are called *the partial fractions* of the function and we will now consider the process of expressing a given algebraic fraction as a sum of simpler fractions, i.e. partial fractions.

A full discussion of the subject of partial fractions is beyond the scope of this book and it will be sufficient here to state several rules by means of which any rational integral fractional function can be expressed as a sum of partial fractions.

Expression of a fractional function in partial fractions

Rule 1 *Before a fractional function can be expressed directly in partial fractions the numerator must be of at least one degree less than the denominator.*

E.g. the function $\dfrac{2x^2 + 3}{x^3 - 1}$ can be expressed in partial fractions, whereas the function $\dfrac{2x^3 + 3}{x^3 - 1}$ cannot be expressed directly in partial fractions. However, by division,
$$\dfrac{2x^3 + 3}{x^3 - 1} = 2 + \dfrac{5}{x^3 - 1},$$
and the fraction $\dfrac{5}{x^3 - 1}$ can be expressed directly as a sum of partial fractions.

Example 1 Write the fraction $\dfrac{4x^3 - 3x + 2}{(2x-1)(x+2)}$ in a form suitable for expressing in partial fractions.

$$\dfrac{4x^3 - 3x + 2}{(2x-1)(x+2)} = \dfrac{4x^3 - 3x + 2}{2x^2 + 3x - 2}$$
$$= 2x - 3 + \dfrac{10x - 4}{(2x-1)(x+2)}.$$

$$\begin{array}{r}
2x - 3 \\
2x^2 + 3x - 2 \overline{)4x^3 - 3x + 2}\\
4x^3 + 6x^2 - 4x \\
\hline
-6x^2 + x + 2\\
-6x^2 - 9x + 6\\
\hline
10x - 4
\end{array}$$

Rule 2 *Corresponding to any linear factor $ax+b$ in the denominator of a rational fraction there is a partial fraction of the form $\dfrac{A}{ax+b}$, where A is a constant.*

Example 2 Express the function $\dfrac{2x}{(x-1)(2x+1)(x+2)}$ in partial fractions.

Noting that the degree of the numerator is less than that of the denominator, we can assume

$$\frac{2x}{(x-1)(2x+1)(x+2)} \equiv \frac{A}{x-1} + \frac{B}{2x+1} + \frac{C}{x+2},$$

i.e. $\quad 2x \equiv A(2x+1)(x+2) + B(x-1)(x+2) + C(x-1)(2x+1).$

Let $x = 1$. $\quad 2 = 9A; \quad A = \tfrac{2}{9}.$

Let $x = -\tfrac{1}{2}.\ -1 = -\dfrac{9B}{4}; \quad B = \tfrac{4}{9}.$

Let $x = -2.\ -4 = 9C; \quad C = -\tfrac{4}{9}.$

$$\therefore \frac{2x}{(x-1)(2x+1)(x+2)} \equiv \frac{1}{9}\left\{\frac{2}{x-1} + \frac{4}{2x+1} - \frac{4}{x+2}\right\}.$$

Rule 3 *Corresponding to a linear factor $ax+b$ repeated r times in the denominator, there will be r partial fractions of the form*

$$\frac{A_1}{ax+b}, \frac{A_2}{(ax+b)^2}, \dots \frac{A_r}{(ax+b)^r}.$$

Example 3 Express as a sum of four partial fractions, $\dfrac{2x^2-3}{(x-1)^3(x+1)}$.

Let $\quad \dfrac{2x^2-3}{(x-1)^3(x+1)} \equiv \dfrac{A}{x-1} + \dfrac{B}{(x-1)^2} + \dfrac{C}{(x-1)^3} + \dfrac{D}{x+1},$

i.e. $\quad 2x^2 - 3 \equiv A(x-1)^2(x+1) + B(x-1)(x+1) + C(x+1) + D(x-1)^3.$

Let $x = 1.\quad -1 = 2C; \quad C = -\tfrac{1}{2}.$

Let $x = -1.\quad -1 = -8D; \quad D = \tfrac{1}{8}.$

The remaining constants can be obtained by equating coefficients of powers of x. Equating coefficients of x^3,

$$0 = A + D.$$
$$\therefore A = -\tfrac{1}{8}.$$

Equating the constant terms,

$$-3 = A - B + C - D.$$
$$\therefore B = 3 - \tfrac{1}{8} - \tfrac{1}{2} - \tfrac{1}{8} = \tfrac{9}{4}.$$

$$\therefore \frac{2x^2-3}{(x-1)^3(x+1)} \equiv \frac{1}{8}\left\{\frac{-1}{x-1} + \frac{18}{(x-1)^2} - \frac{4}{(x-1)^3} + \frac{1}{x+1}\right\}.$$

Rule 4 *Corresponding to any quadratic factor $ax^2 + bx + c$ in the denominator there will be a partial fraction of the form $\dfrac{Ax + B}{ax^2 + bx + c}$.*

Example 4 Express in partial fractions, $\dfrac{x^3 - 2}{x^4 - 1}$.

As
$$x^4 - 1 = (x^2 + 1)(x^2 - 1) = (x^2 + 1)(x + 1)(x - 1),$$

let
$$\frac{x^3 - 2}{x^4 - 1} \equiv \frac{Ax + B}{x^2 + 1} + \frac{C}{x + 1} + \frac{D}{x - 1}.$$

$$\therefore x^3 - 2 \equiv (Ax + B)(x + 1)(x - 1) + C(x^2 + 1)(x - 1) + D(x^2 + 1)(x + 1).$$

Let $x = 1$. $-1 = 4D$; $D = -\tfrac{1}{4}$.
Let $x = -1$. $-3 = -4C$; $C = \tfrac{3}{4}$.
Equating coefficients of x^3,
$$1 = A + C + D; \qquad A = \tfrac{1}{2}.$$

Equating coefficients of x^2,
$$0 = B - C + D; \qquad B = 1.$$

$$\therefore \frac{x^3 - 2}{x^4 - 1} \equiv \frac{1}{4}\left\{\frac{2x + 4}{x^2 + 1} + \frac{3}{x + 1} - \frac{1}{x - 1}\right\}.$$

Repeated quadratic factors in the denominator are dealt with in a similar way to repeated linear factors.

E.g. corresponding to a factor $(ax^2 + bx + c)^2$ there will be partial fractions $\dfrac{Ax + B}{ax^2 + bx + c}$ and $\dfrac{Cx + D}{(ax^2 + bx + c)^2}$.

[A] EXAMPLES 16a

Express in partial fractions:

1. $\dfrac{x}{(1 - x)(2 + x)}$.
2. $\dfrac{2x - 1}{(2x + 1)(x - 3)}$.
3. $\dfrac{3x}{(x - 2)(x + 1)}$.
4. $\dfrac{2}{(x - 1)^2(x + 1)}$.
5. $\dfrac{3x^2 - 4}{x(x^2 + 1)}$.
6. $\dfrac{x^2 + 3x}{x^2 - 4}$.
7. $\dfrac{3}{x(x - 2)^2}$.
8. $\dfrac{1}{x(x^2 + 4)}$.
9. $\dfrac{x^2 - x + 1}{x^2 - x - 2}$.
10. $\dfrac{4x - 3}{x^3(x + 1)}$.
11. $\dfrac{x^2}{(x - 1)^2}$.
12. $\dfrac{5x - 3}{(x + 2)(x - 3)^2}$.
13. $\dfrac{3x^2 + 2x}{(x + 2)(x^2 + 3)}$.
14. $\dfrac{3}{1 - x^3}$.
15. $\dfrac{x^2 + 1}{x(x^2 - 1)}$.
16. $\dfrac{2x - 1}{(x - 2)(x + 1)(x + 3)}$.
17. $\dfrac{4x - 1}{x^2(x^2 - 4)}$.
18. $\dfrac{1}{x^2 - 2}$.

19 $\dfrac{x^2}{(x-3)^3}$.

20 $\dfrac{(x+3)^2}{(x-3)^2(x+5)}$.

21 $\dfrac{x^3+2}{x(x^2-3)}$.

22 $\dfrac{1-2x}{x^3+1}$.

23 $\dfrac{x}{x^4-16}$.

24 $\dfrac{2x^3}{(1+x^2)(1-x)^2}$.

25 $\dfrac{x^4+1}{x^3+2x}$.

26 $\dfrac{3x}{(x+1)(3-x^2)}$.

27 $\dfrac{2x^2-5x}{(x^2-1)(x^2-4)}$.

28 $\dfrac{1}{x^3(1-2x)}$.

29 $\dfrac{1}{x^2+2x-1}$.

30 $\dfrac{2x-7}{(x^2+4)(x-1)^2}$.

31 $\dfrac{1}{x(x^2-1)^2}$.

32 $\dfrac{1}{x(x^2+4)^2}$.

The expansion of rational algebraic fractions Resolution into partial fractions greatly simplifies the expansion of a rational algebraic fraction.

Example 5 Expand the function $\dfrac{5x+6}{(x+2)(1-x)}$ in ascending powers of x as far as the term in x^3 and state the range of values of x for which the expansion is valid

Let
$$\dfrac{5x+6}{(x+2)(1-x)} \equiv \dfrac{A}{x+2} + \dfrac{B}{1-x}.$$

i.e. $5x+6 \equiv A(1-x) + B(x+2)$.

Let $x = 1$. $\qquad\qquad 11 = 3B;\qquad B = \tfrac{11}{3}$.

Let $x = -2$. $\qquad\quad -4 = 3A;\qquad A = -\tfrac{4}{3}$.

$$\therefore \dfrac{5x+6}{(x+2)(1-x)} = \dfrac{1}{3}\left\{-\dfrac{4}{x+2} + \dfrac{11}{1-x}\right\},$$

$$= \dfrac{1}{3}\left\{-2\left(1+\dfrac{x}{2}\right)^{-1} + 11(1-x)^{-1}\right\},$$

$$= \dfrac{1}{3}\left\{-2\left(1-\dfrac{x}{2}+\dfrac{x^2}{4}-\dfrac{x^3}{8}\ldots\right) + 11(1+x+x^2+x^3+\ldots)\right\}$$

$$= 3 + 4x + \dfrac{7x^2}{2} + \dfrac{15x^3}{4} + \ldots .$$

$\left(1+\dfrac{x}{2}\right)^{-1}$ can be expanded in ascending powers of x if $\left|\dfrac{x}{2}\right| < 1$; $(1-x)^{-1}$ can be expanded if $|x| < 1$; hence both expansions are valid if $|x| < 1$.

Example 6 Resolve $\dfrac{2x}{(1-x)(1+x^2)}$ into partial fractions. If this expression can be expanded in a series of ascending powers of x, prove that the coefficients of x^{2n} and x^{2n+1} are $1+(-1)^{n+1}$ and $1+(-1)^n$ respectively.

Let
$$\dfrac{2x}{(1-x)(1+x^2)} \equiv \dfrac{A}{1-x} + \dfrac{Bx+C}{1+x^2}.$$

$$\therefore\ 2x \equiv A(1+x^2) + (Bx+C)(1-x).$$

Let $x = 1$. $\quad 2 = 2A; \quad A = 1$.
Equating coefficients of x^2, $\quad 0 = A - B; \quad B = 1$.
Equating constant terms, $\quad 0 = A + C; \quad C = -1$.

$$\therefore \frac{2x}{(1-x)(1+x^2)} = \frac{1}{1-x} + \frac{x-1}{1+x^2}$$

$$= (1-x)^{-1} + (x-1)(1+x^2)^{-1}.$$

$$(1-x)^{-1} = 1 + x + x^2 + x^3 + \ldots + x^r + \ldots,$$

i.e. Coefficients of x^{2n} and x^{2n+1} are each 1.

$$(x-1)(1+x^2)^{-1} = (x-1)(1 - x^2 + x^4 - x^6 + \ldots)$$
$$= (x - x^3 + x^5 - x^7 + \ldots (-1)^{r+1} x^{2r-1} \ldots)$$
$$+ (-1 + x^2 - x^4 + x^6 \ldots (-1)^{r+1} x^{2r} \ldots).$$

i.e. Coefficients of x^{2n} and x^{2n+1} are $(-1)^{n+1}$ and $(-1)^{n+2}$ or $(-1)^n$ respectively. Hence the coefficients of x^{2n} and x^{2n+1} in the expansion of the given expression are $1 + (-1)^{n+1}$ and $1 + (-1)^n$ respectively.

[B] EXAMPLES 16b

State the necessary conditions for expanding the following functions in ascending powers of x and find the first three terms of the expansions:

1 $\dfrac{1}{x^2 - 4}$. **2** $\dfrac{2x - 4}{(1-2x)(1+x)}$. **3** $\dfrac{x}{(1+2x)(1+3x)}$.

4 $\dfrac{x-1}{(2-x)(3-x)}$. **5** $\dfrac{1}{x^2 - x - 2}$. **6** $\dfrac{2x}{3 - 2x - x^2}$.

7 $\dfrac{x+1}{(1-x)^3}$. **8** $\dfrac{1+x^2}{(1-x^2)(3+x)}$. **9** $\dfrac{2}{(1+x^2)(1-x)}$.

10 $\dfrac{2x-1}{(1+2x^2)(2+x)}$. **11** $\dfrac{9x}{(x+2)(2x+1)^2}$. **12** $\dfrac{1}{(x+1)(2-x^2)}$.

If x is so small that powers above the second can be neglected, find approximate values of the following functions:

13 $\dfrac{1}{(1-x)(1-2x)}$. **14** $\dfrac{2x-1}{(x+1)^2(1-3x)}$. **15** $\dfrac{1-x}{x^2(x^2+4)}$.

16 What is the coefficient of x^n in the expansion of $(1 - ax)^{-1}$? Find the coefficient of x^n in the expansion of $\dfrac{1}{(1+2x)(1-x)}$.

17 Show that the coefficient of x^r in the expansion of $\dfrac{2-3x}{1-3x+2x^2}$ is $1 + 2^r$.

18 Find the coefficient of x^{2n+1} in the expansion of $\dfrac{1}{(1-2x)(1+x^2)}$. For what range of values of x is the expansion valid?

19 By writing $\dfrac{1}{x+2}$ in the form $\dfrac{1}{x}\left(1 + \dfrac{2}{x}\right)^{-1}$, expand $\dfrac{1}{x+2}$ as a series of descending powers of x. For what values of x is the expansion valid?

20 For what values of x can the function $\dfrac{2x}{(x+1)(x-2)}$ be expanded in powers of $\dfrac{1}{x}$?
Find the first three terms in the expansion.

The integration of rational algebraic fractions We have seen that expressions of the form $\dfrac{f(x)}{g(x)}$ where $f(x)$ and $g(x)$ are rational integral algebraic functions of x can be resolved into partial fractions, so long as the degree of $f(x)$ is less than that of $g(x)$ and $g(x)$ itself can be expressed in terms of linear and quadratic factors. Consequently, such an expression can be integrated if each of the separate partial fractions can be integrated.

The following types of partial fractions will arise:

$$\frac{A}{ax+b}, \quad \frac{A}{(ax+b)^r}, \quad \frac{Ax+B}{ax^2+bx+c} \quad \text{and} \quad \frac{Ax+B}{(ax^2+bx+c)^r}.$$

The first three of these functions can be readily integrated by methods already discussed. The integration of the fourth function is more difficult and beyond the scope of this book; we will restrict ourselves to the integration of functions which yield partial fractions of the first three types.

Example 7 Integrate $\dfrac{1}{x(x^2-1)}$ with respect to x.

Let
$$\frac{1}{x(x^2-1)} \equiv \frac{A}{x} + \frac{B}{x+1} + \frac{C}{x-1}.$$

i.e.
$$1 \equiv A(x+1)(x-1) + Bx(x-1) + Cx(x+1).$$

Let $x = 0$. $\quad 1 = -A; \quad A = -1.$
Let $x = 1$. $\quad 1 = 2C; \quad C = \tfrac{1}{2}.$
Let $x = -1$. $\quad 1 = 2B; \quad B = \tfrac{1}{2}.$

$$\therefore \int \frac{dx}{x(x^2-1)} = -\int \frac{dx}{x} + \frac{1}{2}\int \frac{dx}{x+1} + \frac{1}{2}\int \frac{dx}{x-1}$$

$$= -\ln x + \tfrac{1}{2}\ln(x+1) + \tfrac{1}{2}\log(x-1) + c.$$

$$= \tfrac{1}{2}\ln\left(\frac{x^2-1}{x^2}\right) + c.$$

Example 8 Find $\displaystyle\int \frac{dx}{(x^2+x+1)(x^2-x+1)}$.

Let
$$\frac{1}{(x^2+x+1)(x^2-x+1)} \equiv \frac{Ax+B}{x^2+x+1} + \frac{Cx+D}{x^2-x+1}.$$

i.e.
$$1 \equiv (Ax+B)(x^2-x+1) + (Cx+D)(x^2+x+1).$$

Equating coefficients of x^3, $\quad 0 = A + C$ ....(i)
Equating coefficients of x^2, $\quad 0 = -A + B + C + D$ ....(ii)
Equating coefficients of x, $\quad 0 = A - B + C + D$ ....(iii)
Equating constant terms, $\quad 1 = B + D$ ....(iv)

232 | **Partial fractions and their applications**

Subtracting (iii) from (ii), $0 = -2A + 2B$ or $A = B$.
Substituting $A = B$ in (ii), $C + D = 0$.
But from (i), $C = -A = -B$, hence $B = D$. Using (iv), $B = D = \frac{1}{2}$, and therefore $A = B = D = \frac{1}{2}, C = -\frac{1}{2}$.

$$\therefore \int \frac{dx}{(x^2+x+1)(x^2-x+1)} = \frac{1}{2}\int \frac{x+1}{x^2+x+1}\,dx - \frac{1}{2}\int \frac{x-1}{x^2-x+1}\,dx.$$

Now
$$\int \frac{x+1}{x^2+x+1}\,dx = \frac{1}{2}\int \frac{2x+1}{x^2+x+1}\,dx + \frac{1}{2}\int \frac{dx}{x^2+x+1}$$

$$= \frac{1}{2}\ln(x^2+x+1) + \frac{1}{2}\int \frac{dx}{(x+\frac{1}{2})^2+\frac{3}{4}}$$

$$= \frac{1}{2}\ln(x^2+x+1) + \frac{1}{\sqrt{3}}\tan^{-1}\frac{2x+1}{\sqrt{3}}.$$

Similarly, $\int \frac{x-1}{x^2-x+1}\,dx = \frac{1}{2}\ln(x^2-x+1) - \frac{1}{\sqrt{3}}\tan^{-1}\frac{2x-1}{\sqrt{3}}$.

$$\therefore \int \frac{dx}{(x^2+x+1)(x^2-x+1)} = \frac{1}{4}\ln\frac{x^2+x+1}{x^2-x+1}$$

$$+ \frac{1}{2\sqrt{3}}\left(\tan^{-1}\frac{2x+1}{\sqrt{3}} + \tan^{-1}\frac{2x-1}{\sqrt{3}}\right) + c$$

$$= \frac{1}{4}\ln\frac{x^2+x+1}{x^2-x+1} + \frac{1}{2\sqrt{3}}\tan^{-1}\frac{\sqrt{3}x}{1-x^2} + c.$$

Example 9   Evaluate $\displaystyle\int_1^2 \frac{3x+2}{(2x-1)^2(3-x)}\,dx$.

Let $\quad \dfrac{3x+2}{(2x-1)^2(3-x)} \equiv \dfrac{A}{(2x-1)^2} + \dfrac{B}{2x-1} + \dfrac{C}{3-x}.$

i.e. $3x+2 \equiv A(3-x) + B(2x-1)(3-x) + C(2x-1)^2.$

Let $x = 3$. $11 = 25C;$ $C = \frac{11}{25}.$
Let $x = \frac{1}{2}$. $\frac{7}{2} = \frac{5}{2}A;$ $A = \frac{7}{5}.$
Equating coefficients of x^2,
$$0 = -2B + 4C; \qquad B = \frac{22}{25}.$$

$$\therefore \int_1^2 \frac{3x+2}{(2x-1)^2(3-2x)}\,dx = \frac{7}{5}\int_1^2 \frac{dx}{(2x-1)^2} + \frac{22}{25}\int_1^2 \frac{dx}{2x-1} + \frac{11}{25}\int_1^2 \frac{dx}{3-x}$$

$$= \frac{7}{5}\left[-\frac{1}{2(2x-1)}\right]_1^2 + \frac{22}{25}\left[\frac{1}{2}\ln(2x-1)\right]_1^2$$

$$+ \frac{11}{25}\left[-\ln(3-x)\right]_1^2$$

$$= \frac{7}{5}\left(-\frac{1}{6}+\frac{1}{2}\right) + \frac{11}{25}\left(\ln 3 - \ln 1\right) + \frac{11}{25}\left[-\ln 1 + \ln 2\right]$$

$$= \frac{7}{15} + \frac{11}{25}\ln 6.$$

[B] EXAMPLES 16c

Evaluate the following integrals:

1. $\int \dfrac{dx}{x^2-1}$.

2. $\int_4^5 \dfrac{dx}{(x-2)(x-3)}$.

3. $\int \dfrac{x\,dx}{(2-x)(3+x)}$.

4. $\int \dfrac{dx}{x^2(x-2)}$.

5. $\int_0^1 \dfrac{x-1}{x^2-7x+12}\,dx$.

6. $\int \dfrac{dx}{x(x^2+1)}$.

7. $\int_{-1}^1 \dfrac{x^3\,dx}{4-x^2}$.

8. $\int \dfrac{(4x-1)\,dx}{(2x-1)^2(x+5)}$.

9. $\int \dfrac{x^2\,dx}{(2x+1)(x+2)}$.

10. $\int_2^4 \dfrac{dx}{x^3-1}$.

11. $\int \dfrac{dx}{2-x^2}$.

12. $\int \dfrac{2\,dx}{x^2+2x-1}$.

13. $\int_1^2 \dfrac{3\,dx}{x^2(x^2+2)}$.

14. $\int \dfrac{1+3x}{10-3x-x^2}\,dx$.

15. $\int_{2\sqrt{3}}^{4\sqrt{3}} \dfrac{dx}{x(x^2-3)}$.

16. $\int \dfrac{x-4}{(x-1)(x-2)(x-3)}\,dx$.

17. $\int \dfrac{x^2}{(x+1)^3}\,dx$.

18. $\int \dfrac{dx}{x^3+x^2+x}$.

19. $\int_2^3 \dfrac{dx}{(x-1)^2(x^2+1)}$.

20. $\int \dfrac{x^2\,dx}{(2x-1)^3(x+1)}$.

21. $\int_{-1}^0 \dfrac{x^2}{(x-2)^2}\,dx$.

22. $\int \dfrac{x\,dx}{(x-1)(2x^2+4x+5)}$.

23. $\int \dfrac{x^3\,dx}{x^2-x-3}$.

24. $\int \dfrac{3x^2+24}{(x+2)(x^2+5)}\,dx$

[B] MISCELLANEOUS EXAMPLES

1. If $\dfrac{(x+2)^2}{(x-3)^2(x+5)}$ is expressed in the form $\dfrac{A}{x-3}+\dfrac{B}{(x-3)^2}+\dfrac{C}{x+5}$, determine the values of A, B and C.

2. Express as the sum of three partial fractions, $\dfrac{x-3}{(x-1)^2(x^2+2)}$.

3. Evaluate (i) $\int_0^1 \dfrac{2x^2+1}{x^2+x+1}\,dx$; (ii) $\int_1^2 \dfrac{dx}{x^2+3x+2}$.

4. Find the first four terms in the expansion of $\dfrac{(1+2x-3x^2)^{\frac{1}{2}}}{(1-2x)^3}$ in ascending powers of x.

5. Split into partial fractions (i) $\dfrac{x^2}{x^4+x^2-2}$; (ii) $\dfrac{x^3}{(x+2)(x-3)}$.

6. Integrate with respect to x: (i) $\dfrac{x}{x^2+6x+8}$; (ii) $\dfrac{x}{x^2+6x+10}$.

7. Express $\dfrac{3x+3}{(x-1)(x+2)}$ in partial fractions and hence, or otherwise, find the third differential coefficient of this fraction.

8. For what values of x can the function $\dfrac{5x-3}{(x-1)^2(x+1)}$ be expanded as a power series in x? If n is even, prove that the coefficient of x^n is $n-3$.

234 | Partial fractions and their applications

9 Obtain the expansion of $\dfrac{1}{(1-2x)(1+x)}$ as far as the term in x^3 and deduce the first four terms in the expansion of $\dfrac{\sqrt{1-x}}{(1-2x)(1+x)}$.

10 Prove that $\displaystyle\int \dfrac{dx}{x^3+3x^2-4} = \dfrac{1}{3(x+2)} + \dfrac{1}{9}\ln\dfrac{x-1}{x+2} + c$.

11 Express as a sum of three partial fractions, $\dfrac{3(5x+1)}{(x+2)^2(2x^2+1)}$.

12 Express $\dfrac{x}{(2x-1)(x+2)}$ in partial fractions and hence find its fourth differential coefficient.

13 Under what conditions can the function $\dfrac{1}{x^2-4x+3}$ be expanded (i) as a power series in x; (ii) as a power series in $\dfrac{1}{x}$?
Find the coefficients of x^{n+1} and $\dfrac{1}{x^{n-1}}$ in the two expansions.

14 Find the first four terms in the expansion of $\dfrac{(1-x)^2}{1-x^3}$ as a power series in x.
What is the coefficient of x^{3p} in the expansion of $\dfrac{1-x}{1+x+x^2}$?

15 Evaluate (i) $\displaystyle\int_0^{\frac{1}{\sqrt{2}}} \dfrac{dx}{x^2-2}$; (ii) $\displaystyle\int_{-\frac{1}{2}}^{\frac{1}{2}} \dfrac{x^2}{(1+x)^2}\,dx$; (iii) $\displaystyle\int_{-1}^{2} \dfrac{x\,dx}{(x-3)(x+2)}$.

16 Express $\dfrac{x^2(x^2+1)}{(x+1)(x^3+1)}$ in partial fractions.

17 Prove that $\displaystyle\int \dfrac{dx}{a-x}$ may lead to either of the expressions $\ln\dfrac{1}{a-x}$ or $\ln\dfrac{1}{x-a}$ and explain how the two solutions may be reconciled. Find the value of $\displaystyle\int_2^4 \dfrac{dx}{1-x^2}$.

18 Express $\dfrac{2x^3}{(1+x^2)(1-x)^2}$ in partial fractions and obtain its expansion as a power series in x as far as the term involving x^5.

19 Prove that $\dfrac{d}{dx}\left(\dfrac{x}{x^2+1}\right) = \dfrac{2}{(x^2+1)^2} - \dfrac{1}{x^2+1}$.
Deduce the value of $\displaystyle\int_0^1 \dfrac{dx}{(x^2+1)^2}$.

20 Use the method of Example 19 to evaluate $\displaystyle\int \dfrac{dx}{(x^2+2)^2}$.

21 Show that the nth differential coefficient of $\dfrac{1}{ax+b}$ is $(-1)^n \dfrac{a^n n!}{(ax+b)^{n+1}}$.
Use partial fractions to find the nth differential coefficient of

$$\dfrac{3x}{(2x-1)(4x+3)}.$$

Miscellaneous examples | 235

22 Express in partial fractions $\dfrac{x^3}{(x-1)(x^2+2)^2}$.

23 Find the coefficient of x^n in the expansion of the function
$$\frac{x-2}{(x+1)^2(x-1)}.$$

24 Evaluate (i) $\displaystyle\int \frac{dx}{1-x^4}$; (ii) $\displaystyle\int \frac{dx}{x^3-x^2-x+1}$.

25 Find the nth derivative of the function $\dfrac{x^2}{(x-1)(x+2)^2}$.

26 Find the coefficient of x^3 in the expansion of $\dfrac{(1+3x-4x^2)^{\frac{1}{2}}}{(x+1)(1-2x)^2}$.

27 Prove that
$$\int_b^a \frac{x^2\,dx}{(x^2+a^2)(x^2+b^2)} = \frac{\pi}{4(a-b)} - \frac{1}{a^2-b^2}\left(a\tan^{-1}\frac{b}{a} + b\tan^{-1}\frac{a}{b}\right).$$

28 Express $\dfrac{1+3x^2}{(1+x)(1-x)^2}$ in partial fractions and prove that when the function is expanded in descending powers of x, the coefficient of x^{-n} is $2n+1$ or $2n-1$ according as n is odd or even. For what values of x is the expansion valid?

29 Express $\dfrac{1}{x^4+1}$ in the form $\dfrac{Ax+B}{x^2+x\sqrt{2}+1} + \dfrac{Cx+D}{x^2-x\sqrt{2}+1}$.

17 | Further methods of integration and standard types of integral

Standard forms Before considering further methods of integration and standard types of integral it will be useful to summarise the standard types of integral already dealt with. The following list of fundamental results given in terms of differentiation and integration should be memorised:

Differentiation   **Integration**

1   $\dfrac{d}{dx} x^n = nx^{n-1};$ $\displaystyle\int x^n\, dx = \dfrac{x^{n+1}}{n+1} + c.$ [except when $n = -1$].

$\dfrac{d}{dx}(ax+b)^n = na(ax+b)^{n-1};$ $\displaystyle\int (ax+b)^n\, dx = \dfrac{(ax+b)^{n+1}}{a(n+1)} + c.$

[except when $n = -1$].

2   $\dfrac{d}{dx} \ln x = \dfrac{1}{x};$ $\displaystyle\int \dfrac{dx}{x} = \ln x + c.$ $x > 0.$

$\dfrac{d}{dx} \ln(ax+b) = \dfrac{a}{ax+b};$ $\displaystyle\int \dfrac{dx}{ax+b} = \dfrac{1}{a}\ln(ax+b) + c.$ $ax+b > 0.$

3   $\dfrac{d}{dx} e^x = e^x;$ $\displaystyle\int e^x\, dx = e^x + c.$

$\dfrac{d}{dx} e^{mx} = me^{mx};$ $\displaystyle\int e^{mx}\, dx = \dfrac{e^{mx}}{m} + c.$

4   $\dfrac{d}{dx} \sin x = \cos x;$ $\displaystyle\int \cos x\, dx = \sin x + c.$

$\dfrac{d}{dx} \sin mx = m\cos mx;$ $\displaystyle\int \cos mx\, dx = \dfrac{\sin mx}{m} + c.$

5   $\dfrac{d}{dx} \cos x = -\sin x;$ $\displaystyle\int \sin x\, dx = -\cos x + c.$

$\dfrac{d}{dx} \cos mx = -m\sin mx;$ $\displaystyle\int \sin mx\, dx = -\dfrac{\cos mx}{m} + c.$

6   $\dfrac{d}{dx} \tan x = \sec^2 x;$ $\displaystyle\int \sec^2 x\, dx = \tan x + c.$

$\dfrac{d}{dx} \tan mx = m\sec^2 mx;$ $\displaystyle\int \sec^2 mx\, dx = \dfrac{\tan mx}{m} + c.$

	Differentiation	**Integration**
7	$\dfrac{d}{dx} \cot x = -\csc^2 x;$	$\int \csc^2 x \, dx = -\cot x + c.$
	$\dfrac{d}{dx} \cot mx = -m \csc^2 mx;$	$\int \csc^2 mx \, dx = -\dfrac{\cot mx}{m} + c.$
8	$\dfrac{d}{dx} \sinh x = \cosh x;$	$\int \cosh x \, dx = \sinh x + c.$
9	$\dfrac{d}{dx} \cosh x = \sinh x;$	$\int \sinh x \, dx = \cosh x + c.$
10	$\dfrac{d}{dx} \sin^{-1} x = \dfrac{1}{\sqrt{1-x^2}};$	$\int \dfrac{dx}{\sqrt{1-x^2}} = \sin^{-1} x + c.$
	$\dfrac{d}{dx} \sin^{-1} \dfrac{x}{a} = \dfrac{1}{\sqrt{a^2-x^2}};$	$\int \dfrac{dx}{\sqrt{a^2-x^2}} = \sin^{-1} \dfrac{x}{a} + c.$
11	$\dfrac{d}{dx} \tan^{-1} x = \dfrac{1}{1+x^2};$	$\int \dfrac{dx}{1+x^2} = \tan^{-1} x + c.$
	$\dfrac{d}{dx} \tan^{-1} \dfrac{x}{a} = \dfrac{a}{a^2+x^2};$	$\int \dfrac{dx}{a^2+x^2} = \dfrac{1}{a} \tan^{-1} \dfrac{x}{a} + c.$
12	$\dfrac{d}{dx} \sinh^{-1} \dfrac{x}{a} = \dfrac{1}{\sqrt{a^2+x^2}};$	$\int \dfrac{dx}{\sqrt{a^2+x^2}} = \sinh^{-1} \dfrac{x}{a} + c$
		$= \ln(x + \sqrt{x^2+a^2}) + c'.$
13	$\dfrac{d}{dx} \cosh^{-1} \dfrac{x}{a} = \dfrac{1}{\sqrt{x^2-a^2}};$	$\int \dfrac{dx}{\sqrt{x^2-a^2}} = \cosh^{-1} \dfrac{x}{a} + c$
		$= \ln(x + \sqrt{x^2-a^2}) + c'.$

Methods of integration The following are the most important methods of integration:

(i) Integration by means of standard forms.
(ii) Integration of rational algebraic functions by the use of partial fractions.
(iii) Integration by substitution.
(iv) Integration by parts.
(v) Integration by use of reduction formulae.

Methods (i) and (ii) have already been introduced; the other methods will now be considered.

Method of substitution The method of integration by substitution involves a change of variable and is analogous to the function of a function method of differentiation.

Take the indefinite integral $\int (3x-1)^4 \, dx$, which can of course be evaluated directly. However, consider a change of variable which simplifies the function to be integrated, $(3x-1)^4$.

An obvious change is obtained by letting $z = 3x - 1$.

Now, let
$$y = \int (3x-1)^4 \, dx,$$

i.e.
$$\frac{dy}{dx} = (3x-1)^4 = z^4.$$

In order to change the variable from x to z, we must obtain $\dfrac{dy}{dz}$ as a function of z.

$$\frac{dy}{dz} = \frac{dy}{dx} \times \frac{dx}{dz} = \frac{z^4}{3}.$$

$$\therefore y = \int \frac{z^4}{3} \, dz = \frac{z^5}{15} + c.$$

i.e.
$$y = \int (3x-1)^4 \, dx = \frac{(3x-1)^5}{15} + c.$$

Example 1 By using the substitution $z = \sqrt{x}$, evaluate $\int \dfrac{dx}{\sqrt{x}+1}$.

Let
$$y = \int \frac{dx}{\sqrt{x}+1},$$

i.e.
$$\frac{dy}{dx} = \frac{1}{\sqrt{x}+1}.$$

Take $z = \sqrt{x}$ or $x = z^2$.

$$\therefore \frac{dy}{dz} = \frac{dy}{dx} \times \frac{dx}{dz} = \frac{1}{z+1} \times 2z = \frac{2z}{z+1}.$$

$$y = \int \frac{2z}{z+1} \, dz = \int \frac{2(z+1) - 2}{z+1} \, dz.$$

$$= \int 2 \, dz - \int \frac{2}{z+1} \, dz = 2z - 2 \ln (z+1) + c.$$

$$\therefore \int \frac{dx}{\sqrt{x}+1} = y = 2\sqrt{x} - 2 \ln (\sqrt{x}+1) + c.$$

General case Suppose we wish to change the variable from x to z in the integral $\int f(x) \, dx$ by using the relationship $x = \phi(z)$.

Let
$$y = \int f(x) \, dx,$$

i.e. $$\frac{dy}{dx} = f(x).$$

On substituting $x = \phi(z)$, $f(x)$ will become some function of z, say $F(z)$.

Thus $$\frac{dy}{dx} = F(z) \quad \text{and} \quad \frac{dx}{dz} = \phi'(z).$$

$$\therefore \frac{dy}{dz} = \frac{dy}{dx} \times \frac{dx}{dz} = F(z)\phi'(z),$$

$$y = \int F(z)\phi'(z)\,dz.$$

i.e. $$\int f(x)\,dx = \int F(z)\phi'(z)\,dz.$$

So, in order to change the variable from x to z after choosing a suitable transformation $x = \phi(z)$, we replace $f(x)$ by the function of z obtained by substituting for x in terms of z and replace dx by $\phi'(z)\,dz$.

The replacement of the symbol dx by $\phi'(z)\,dz$ is facilitated by making use of *differentials*. By definition, differentials dz, dx can be considered as small quantities, either finite or infinitesimal, whose ratio is the differential coefficient $\dfrac{dz}{dx}$.

i.e. $$dx : dz = \frac{dx}{dz} \quad \text{or} \quad dx = \frac{dx}{dz}\,dz.$$

Thus, as $$\frac{dx}{dz} = \phi'(z), \quad dx = \phi'(z)\,dz.$$

E.g. if $x = \sin z$, $dx = \cos z\,dz$;

if $x^2 = \dfrac{1}{z}$, $2x\,dx = -\dfrac{1}{z^2}\,dz$.

Example 2 Evaluate $\displaystyle\int \frac{2x\,dx}{1+x^4}$.

The form of the numerator suggests a substitution $x^2 = z$, giving $2x\,dx = dz$.

Also $$\frac{1}{1+x^4} = \frac{1}{1+z^2}.$$

$$\therefore \int \frac{2x\,dx}{1+x^4} = \int \frac{dz}{1+z^2} = \tan^{-1} z + c.$$

$$= \tan^{-1} x^2 + c.$$

Change of variable in a definite integral Consider the integral

$$\int_0^1 \sqrt{1-x^2}\,dx.$$

First take the indefinite integral $\int \sqrt{1-x^2}\, dx$; this can be evaluated by the substitution $x = \sin\theta$, making $dx = \cos\theta\, d\theta$ and $\sqrt{1-x^2} = \cos\theta$.

i.e.
$$\int \sqrt{1-x^2}\, dx = \int \cos^2\theta\, d\theta = \int \frac{\cos 2\theta + 1}{2} d\theta$$
$$= \frac{\sin 2\theta}{4} + \frac{\theta}{2} + c.$$

In order to evaluate the definite integral we could replace θ by $\sin^{-1} x$ in the result of the indefinite integration and then substitute the limits for x in the usual way. This is laborious and unnecessary as all we need do is to alter the given limits for x into the corresponding limits for θ.

As x increases from 0 to 1, $\sin\theta$ increases from 0 to 1 and thus θ increases from 0 to $\frac{\pi}{2}$, i.e. the limits for θ are 0 and $\frac{\pi}{2}$.

$$\therefore \int_0^1 \sqrt{1-x^2}\, dx = \int_0^{\frac{\pi}{2}} \cos^2\theta\, d\theta = \left[\frac{\sin 2\theta}{4} + \frac{\theta}{2}\right]_0^{\frac{\pi}{2}}$$
$$= \frac{\pi}{4}.$$

Example 3 Evaluate $\int_0^{\frac{\pi}{2}} \sin^5 x\, dx$ using the substitution $\cos x = c$.

Let
$$\cos x = c.$$
$$\therefore -\sin x\, dx = dc.$$

$$\int \sin^5 x\, dx = \int \sin^4 x (\sin x\, dx) = -\int (1 - \cos^2 x)^2 (-\sin x\, dx)$$
$$= -\int (1 - c^2)^2\, dc.$$

As the limits for x are 0 to $\frac{\pi}{2}$, the limits for c are 1 to 0.

$$\therefore \int_0^{\frac{\pi}{2}} \sin^5 x\, dx = -\int_1^0 (1-c^2)^2\, dc = \int_0^1 (1-c^2)^2\, dc$$
$$= \int_0^1 (1 - 2c^2 + c^4)\, dc = \left[c - \frac{2c^3}{3} + \frac{c^5}{5}\right]_0^1 = \frac{8}{15}.$$

[A] **EXAMPLES 17a**

Use the given substitutions to evaluate the following integrals:

1 $\int \frac{dx}{(2x-1)^5}$; let $2x - 1 = z$.

2 $\int \frac{x\, dx}{\sqrt{x^2 - 1}}$; let $x^2 - 1 = z$.

3 $\int \sin^5 x \cos x\, dx$; let $\sin x = s$.

4 $\int \frac{\ln x}{x}\, dx$; let $\ln x = z$.

Some common types of substitutions | 241

5. $\int \dfrac{dx}{2+\sqrt{x}}$; let $x = z^2$.

6. $\int \dfrac{x+1}{x\sqrt{x-2}}\,dx$; let $x - 2 = z^2$.

7. $\int \dfrac{x^3}{(x+5)^2}\,dx$; let $x + 5 = z$.

8. $\int \dfrac{x\,dx}{\sqrt{1-x^2}}$; let $x = \sin\theta$.

9. $\int_0^{\pi/4} \tan^3\theta \sec^2\theta\,d\theta$; let $\tan\theta = z$.

10. $\int_0^1 \dfrac{x^2\,dx}{1+x^6}$; let $x^3 = z$.

11. $\int_{3/4}^1 \dfrac{x\,dx}{(2x-1)^4}$; let $2x - 1 = z$.

12. $\int \dfrac{dx}{e^x + 1}$; let $e^x + 1 = z$.

13. $\int_0^{1/2} \dfrac{1}{\sqrt{x}}\cos\sqrt{x}\,dx$; let $x = z^2$.

14. $\int_0^1 \dfrac{x\,dx}{\sqrt{1+x^2}}$; let $1 + x^2 = z^2$.

15. $\int_0^1 \dfrac{dx}{e^x + e^{-x}}$; let $e^x = z$.

16. $\int_0^{1/\sqrt{2}} \dfrac{x\,dx}{1-x^4}$; let $x^2 = z$.

17. $\int_0^2 \sqrt{4-x^2}\,dx$; let $x = 2\sin\theta$.

18. $\int_0^{\pi/3} \sin^3 x\,dx$; let $\cos x = c$.

19. $\int_1^2 \dfrac{dx}{x(x^4+1)}$; let $x^4 + 1 = z$.

20. $\int \dfrac{dx}{(1-x)\sqrt{x}}$; let $x = z^2$.

Some common types of substitutions

(i) Integrals containing a term of the form $(ax + b)^n$—let $ax + b = z$.

(ii) Integrals containing a term of the form $\sqrt{a^2 - x^2}$ or $(\sqrt{a^2 - x^2})^n$—let $x = a\sin\theta$.

(iii) As in (ii) but with $\sqrt{a^2 - x^2}$ replaced by $\sqrt{a^2 + x^2}$—let $x = a\sinh\theta$ or $x = a\tan\theta$.

(iv) As in (ii) but with $\sqrt{a^2 - x^2}$ replaced by $\sqrt{x^2 - a^2}$—let $x = a\cosh\theta$.

(v) Integrals of odd *powers of sine or cosine*—let $\cos x = c$ or $\sin x = s$ respectively.

(vi) Integrals of the form $\int \dfrac{dx}{a + b\cos x}$ or $\int \dfrac{dx}{a + b\sin x}$ —let $\tan\dfrac{x}{2} = t$.

(vii) Integrals of the form $\int \dfrac{dx}{x\sqrt{ax^2 + bx + c}}$ —let $x = \dfrac{1}{z}$;

or $\int \dfrac{dx}{(px + q)\sqrt{ax^2 + bx + c}}$ —let $px + q = \dfrac{1}{z}$.

Example 4 Evaluate $\int \dfrac{dx}{x^2\sqrt{4-x^2}}$, assuming $|x| < 2$.

Let $x = 2\sin\theta$, then $dx = 2\cos\theta\,d\theta$;

$$x^2\sqrt{4-x^2} = 4\sin^2\theta\sqrt{4(1-\sin^2\theta)} = 8\sin^2\theta\cos\theta.$$

$$\therefore \int \frac{dx}{x^2 \sqrt{4-x^2}} = \int \frac{d\theta}{4\sin^2\theta} = \frac{1}{4}\int \cosec^2\theta \, d\theta.$$
$$= -\tfrac{1}{4}\cot\theta + c.$$

But $\cot^2\theta = \cosec^2\theta - 1 = 4/x^2 - 1 = \dfrac{4-x^2}{x^2}$.

$$\therefore \cot\theta = \frac{\sqrt{4-x^2}}{x}.$$

$$\therefore \int \frac{dx}{x^2\sqrt{4-x^2}} = -\frac{\sqrt{4-x^2}}{4x} + c.$$

Example 5 Evaluate $\displaystyle\int_0^{\frac{\pi}{2}} \frac{dx}{2+\cos x}$.

Let $\tan\dfrac{x}{2} = t$, then $\cos x = \dfrac{1-t^2}{1+t^2}$, $\sin x = \dfrac{2t}{1+t^2}$.

$$\therefore \tfrac{1}{2}\sec^2\frac{x}{2}dx = dt \quad \text{or} \quad \tfrac{1}{2}(1+t^2)dx = dt.$$

i.e. $\quad dx = \dfrac{2dt}{1+t^2}\quad \text{and}\quad 2+\cos x = 2 + \dfrac{1-t^2}{1+t^2} = \dfrac{3+t^2}{1+t^2}.$

The limits for t are $\tan 0$ and $\tan\dfrac{\pi}{4}$, i.e. 0 and 1.

$$\therefore \int_0^{\frac{\pi}{2}} \frac{dx}{2+\cos x} = \int_0^1 \frac{2dt/(1+t^2)}{(3+t^2)/(1+t^2)} = 2\int_0^1 \frac{dt}{3+t^2}$$
$$= 2\left[\frac{1}{\sqrt{3}}\tan^{-1}\frac{t}{\sqrt{3}}\right]_0^1 = 2\frac{1}{\sqrt{3}}\tan^{-1}\frac{1}{\sqrt{3}} = \frac{2}{\sqrt{3}}\frac{\pi}{6}$$
$$= \frac{\pi}{3\sqrt{3}}.$$

Example 6 Evaluate $\displaystyle\int \frac{dx}{(x+1)\sqrt{x^2+1}}$.

Let $x+1 = \dfrac{1}{z}$, then $dx = -\dfrac{1}{z^2}dz$.

$$\therefore \int \frac{dx}{(x+1)\sqrt{x^2+1}} = \int \frac{-\frac{1}{z^2}dz}{\frac{1}{z}\sqrt{\left(\frac{1}{z}-1\right)^2+1}} = -\int \frac{dz}{z\sqrt{\frac{1}{z^2}-\frac{2}{z}+2}}$$
$$= -\int \frac{dz}{\sqrt{1-2z+2z^2}} = -\frac{1}{\sqrt{2}}\int \frac{dz}{\sqrt{(z-\tfrac{1}{2})^2 + \tfrac{1}{4}}}$$
$$= -\frac{1}{\sqrt{2}}\sinh^{-1}\left(\frac{z-\tfrac{1}{2}}{\tfrac{1}{2}}\right) + c = -\frac{1}{\sqrt{2}}\sinh^{-1}(2z-1) + c.$$

$$\therefore \int \frac{dx}{(x+1)\sqrt{x^2+1}} = -\frac{1}{\sqrt{2}} \sinh^{-1} \frac{1-x}{1+x} + c.$$

[B] **EXAMPLES 17b**

Integrate

1. $\dfrac{x+2}{x\sqrt{x-2}}.$
2. $\dfrac{x^2}{(x-1)^4}.$
3. $\dfrac{\sqrt{x}-1}{1+\sqrt{x}}.$
4. $\dfrac{x}{\sqrt[3]{3+2x}}.$
5. $\sqrt{1-4x^2}.$
6. $\dfrac{3}{x\sqrt{2x-1}}.$
7. $\dfrac{1}{3+4\cos x}.$
8. $\dfrac{x^2}{\sqrt{4-x^2}}.$
9. $\dfrac{1}{e^x-1}.$
10. $\dfrac{x^2}{1+2x^3}.$
11. $\cos^3 x.$
12. $\dfrac{1}{1+\sin x}.$
13. $\sec x.$
14. $\sin^2 x \cos^3 x.$
15. $\dfrac{x}{\sqrt{x+2}}.$
16. $x\sqrt{4x^2-1}.$
17. $\dfrac{x}{(2x-1)^5}.$
18. $(x+2)(x-3)^6.$

Evaluate the following integrals:

19. $\displaystyle\int_1^2 \dfrac{x-1}{\sqrt{x+1}}\,dx.$
20. $\displaystyle\int_0^1 x^2(x-1)^7\,dx.$
21. $\displaystyle\int_{\pi/2}^{\pi} \dfrac{dx}{1-\cos x}.$
22. $\displaystyle\int_0^{\pi/6} \cos^5 x\,dx.$
23. $\displaystyle\int_0^1 \dfrac{2x}{1+x^4}\,dx.$
24. $\displaystyle\int_1^4 \dfrac{dx}{2\sqrt{x}(1+x)}.$
25. $\displaystyle\int_{\pi/6}^{\pi/2} \operatorname{cosec} x\,dx.$
26. $\displaystyle\int_0^{\sqrt{3}/2} \dfrac{x^3}{\sqrt{1-x^2}}\,dx.$
27. $\displaystyle\int_2^4 \dfrac{x+2}{(x-1)^3}\,dx.$
28. $\displaystyle\int_0^1 \dfrac{1}{\sqrt{x}} \sin\sqrt{x}\,dx.$
29. $\displaystyle\int_0^1 (e^x+1)^4\,dx.$
30. $\displaystyle\int_0^2 x^2\sqrt{4-x^2}\,dx.$
31. $\displaystyle\int_0^{\pi/2} \dfrac{dx}{5+12\cos x}.$
32. $\displaystyle\int_{1/2}^{3/4} \dfrac{dx}{\sqrt{x}\sqrt{1-x}}.$
33. $\displaystyle\int_1^2 \dfrac{dx}{x\sqrt{x^2+1}}.$
34. $\displaystyle\int_0^1 (1-x^2)^{3/2}\,dx.$
35. $\displaystyle\int_0^{\pi/6} \sin^5 x \cos^3 x\,dx.$
36. $\displaystyle\int_0^{1/2} \dfrac{dx}{(4+x^2)^{3/2}}.$

37. Using the substitution $\tan\theta = t$, evaluate

   (i) $\displaystyle\int \dfrac{d\theta}{1+\cos 2\theta};$ (ii) $\displaystyle\int \dfrac{d\theta}{2+\sin 2\theta}.$

38. Evaluate (i) $\displaystyle\int_0^{\pi/2} \dfrac{dx}{4-2\cos^2 x};$ (ii) $\displaystyle\int_0^{\pi/4} \dfrac{dx}{3+2\sin^2 x}$ [let $\tan x = t$].

39. Use the substitution $1+x^n = z^2$ to evaluate the integral $\displaystyle\int \dfrac{dx}{x\sqrt{1+x^n}}.$

40. By using the substitution $x = \sin^2\theta$, evaluate $\displaystyle\int_0^1 \sqrt{x}\sqrt{1-x}\,dx.$

Integration by parts Since

$$\frac{d}{dx}(uw) = w\frac{du}{dx} + u\frac{dw}{dx}$$

where u and w are functions of x,

$$\int \frac{d}{dx}(uw)\,dx = \int w\frac{du}{dx}\,dx + \int u\frac{dw}{dx}\,dx,$$

or,

$$uw = \int w\frac{du}{dx}\,dx + \int u\frac{dw}{dx}\,dx.$$

Let $\dfrac{dw}{dx} = v$, so that $w = \int v\,dx$.

$$\therefore\ u\int v\,dx = \int \left\{\int v\,dx\right\}\frac{du}{dx}\,dx + \int uv\,dx,$$

i.e.

$$\int uv\,dx = u\int v\,dx - \int\left\{\int v\,dx\right\}\frac{du}{dx}\,dx.$$

This result gives the following rule for the integral of a product of two functions:

The integral of a product of two functions is equal to the integral of one function times the other function minus the integral of the product of the integral already found times the differential coefficient of the other function.

This method of integrating a product is known as *integration by parts*.

Example 7 Integrate $x \sin x$ with respect to x.

$$\int x \sin x\,dx = \left[\int \sin x\,dx\right]x - \int\left[\int \sin x\,dx\right]\frac{dx}{dx}\,dx \qquad \text{taking } x \text{ as the function to be differentiated.}$$

$$= [-\cos x]x - \int[-\cos x]1\,dx$$

$$= -x\cos x + \int \cos x\,dx = -x\cos x + \sin x + c.$$

Example 8 Evaluate $\int x^4 \ln x\,dx$.

Take x^4 as the function to be integrated and $\ln x$ as the function to be differentiated.

$$\int x^4 \ln x\,dx = \left[\frac{x^5}{5}\right]\ln x - \int\left[\frac{x^5}{5}\right]\frac{1}{x}\,dx$$

$$= \frac{x^5 \ln x}{5} - \int \frac{x^4}{5}\,dx = \frac{x^5 \ln x}{5} - \frac{x^5}{25} + c.$$

Example 9 Evaluate $\int_0^1 x^2 e^{-x} dx$.

$$\int_0^1 x^2 e^{-x} dx = \left[[-e^{-x}]x^2\right]_0^1 - \int_0^1 [-e^{-x}] 2x\, dx$$

$$= -e^{-1} + 2\int_0^1 xe^{-x} dx$$

$$= -\frac{1}{e} + 2\left\{\left[[-e^{-x}]x\right]_0^1 - \int_0^1 [-e^{-x}] 1\, dx\right\}$$

$$= -\frac{1}{e} - \frac{2}{e} + 2\int_0^1 e^{-x} dx$$

$$= -\frac{3}{e} + 2\left[-e^{-x}\right]_0^1 = -\frac{3}{e} - \frac{2}{e} + 2e^0$$

$$= 2 - \frac{5}{e}.$$

[A] EXAMPLES 17c

Find the following integrals:

1 $\int x \cos x\, dx.$ **2** $\int xe^x\, dx.$ **3** $\int x \sin 2x\, dx.$

4 $\int x \ln x\, dx.$ **5** $\int xe^{-x}\, dx.$ **6** $\int x^2 \ln x\, dx.$

7 $\int x \cos 3x\, dx.$ **8** $\int \frac{1}{x^2} \ln x\, dx.$ **9** $\int x^2 e^x\, dx.$

10 $\int x^2 \cos x\, dx.$ **11** $\int x \sin x \cos x\, dx.$ **12** $\int x^2 e^{2x}\, dx.$

Evaluate:

13 $\int_0^\pi x \sin x\, dx.$ **14** $\int_1^2 \frac{1}{x^3} \ln x\, dx.$ **15** $\int_0^1 xe^{-2x}\, dx.$

16 $\int_0^\pi x \cos^2 x\, dx.$ **17** $\int_0^\pi x^2 \sin \frac{x}{2} dx.$ **18** $\int_0^{\frac{\pi}{4}} x \sin\left(\frac{\pi}{4} - x\right) dx.$

19 If $I = \int e^x \sin x\, dx$, $I' = \int e^x \cos x\, dx$, show that

$$I = e^x \sin x - I'; \quad I' = e^x \cos x + I.$$

Deduce the values of I and I'.

20 Use the method of the previous example to evaluate:

(i) $\int e^{2x} \sin x\, dx;$ (ii) $\int e^{-x} \cos x\, dx;$ (iii) $\int e^x \sin 3x\, dx;$

(iv) $\int e^{-2x} \cos 2x\, dx.$

The method of integration by parts has important applications to the integration of various single functions, amongst which are ln x, the inverse circular and hyperbolic functions and $\sqrt{ax^2 + bx + c}$, unity being taken as the second function.

Example 10 Integrate ln x and $\tan^{-1} x$ with respect to x.
(i) Write ln x as 1 ln x.

$$\int \ln x \, dx = \int 1 \ln x \, dx = [x] \ln x - \int [x] \frac{1}{x} dx$$

$$= x \ln x - \int dx = x \ln x - x + c.$$

(ii)
$$\int \tan^{-1} x \, dx = \int 1 \tan^{-1} x \, dx = [x] \tan^{-1} x - \int [x] \frac{1}{1+x^2} dx$$

$$= x \tan^{-1} x - \int \frac{x \, dx}{1+x^2} = x \tan^{-1} x - \tfrac{1}{2} \int \frac{2x \, dx}{1+x^2}$$

$$= x \tan^{-1} x - \tfrac{1}{2} \ln(1+x^2) + c.$$

Example 11 Evaluate $\int_1^2 \sqrt{x^2 + 2x + 2} \, dx.$

Let $I = \int_1^2 1 \sqrt{x^2 + 2x + 2} \, dx = \left[[x] \sqrt{x^2 + 2x + 2} \right]_1^2 - \int_1^2 [x] \frac{x+1}{\sqrt{x^2 + 2x + 2}} dx$

$$= 2\sqrt{10} - \sqrt{5} - \int_1^2 \frac{x^2 + x}{\sqrt{x^2 + 2x + 2}} dx. \quad \text{(i)}$$

To evaluate $\int_1^2 \frac{(x^2 + x) \, dx}{\sqrt{x^2 + 2x + 2}}$, the numerator is written in the form

$A(x^2 + 2x + 2) + B$ (differential coefficient of $x^2 + 2x + 2) + C.$

We have $x^2 + x = 1(x^2 + 2x + 2) - \tfrac{1}{2}(2x + 2) - 1.$

$$\therefore \int_1^2 \frac{(x^2 + x) \, dx}{\sqrt{x^2 + 2x + 2}} = \int_1^2 \frac{x^2 + 2x + 2}{\sqrt{x^2 + 2x + 2}} dx$$

$$- \frac{1}{2} \int_1^2 \frac{2x + 2}{\sqrt{x^2 + 2x + 2}} dx - \int_1^2 \frac{dx}{\sqrt{x^2 + 2x + 2}}.$$

On the R.H.S., the 1st integral is the one to be evaluated, I, the 2nd integral is readily evaluated as the numerator is the D.C. of the term under the root sign and the 3rd integral is of a standard form.

Substituting back into (i),

$$I = 2\sqrt{10} - \sqrt{5} - I + \left[\tfrac{1}{2} 2\sqrt{x^2 + 2x + 2} \right]_1^2 + \int_1^2 \frac{dx}{\sqrt{(x+1)^2 + 1}}.$$

i.e. $2I = 2\sqrt{10} - \sqrt{5} + \sqrt{10} - \sqrt{5} + \left[\sinh^{-1}(x+1) \right]_1^2$

$$= 3\sqrt{10} - 2\sqrt{5} + \sinh^{-1} 3 - \sinh^{-1} 2$$

$$= 3\sqrt{10} - 2\sqrt{5} + \ln(3 + \sqrt{10}) - \ln(2 + \sqrt{5}),$$

i.e. $$\int_1^2 \sqrt{x^2+2x+2}\,dx = \frac{3}{2}\sqrt{10} - \sqrt{5} + \tfrac{1}{2}\ln\left(\frac{3+\sqrt{10}}{2+\sqrt{5}}\right).$$

[B] EXAMPLES 17d

Evaluate:

1. $\int \sin^{-1} x\, dx.$
2. $\int \cos^{-1} x\, dx.$
3. $\int \sinh^{-1} x\, dx.$

4. $\int \sqrt{4-x^2}\, dx.$
5. $\int \sqrt{9+x^2}\, dx.$
6. $\int x \tan^{-1} x\, dx.$

7. $\int \lg x\, dx.$
8. $\int x \sin^{-1}(x^2)\, dx.$
9. $\int \tan^{-1}\left(\dfrac{1}{x}\right) dx.$

10. $\int e^{3x} \sin 2x\, dx.$
11. $\int \cot^{-1} x\, dx.$
12. $\int \sqrt{3+2x-x^2}\, dx.$

13. $\int \dfrac{\sin^2 x}{e^x}\, dx.$
14. $\int x \sin^{-1} x\, dx.$
15. $\int \sqrt{3-4x^2}\, dx.$

16. $\int x^3 e^{2x}\, dx.$
17. $\int \dfrac{x}{\sqrt{1-x^2}} \sin^{-1} x\, dx.$
18. $\int \dfrac{x \tan^{-1} x}{\sqrt{1+x^2}}\, dx.$

19. $\int \ln x^2\, dx.$
20. $\int \sqrt{2x^2+3}\, dx.$
21. $\int x^6 \ln \dfrac{1}{x}\, dx.$

Find the values of the following integrals:

22. $\int_0^{\frac{1}{2}} \sin^{-1} 2x\, dx.$
23. $\int_0^1 x^3 e^{-2x}\, dx.$
24. $\int_0^{\frac{\pi}{4}} e^{3x} \cos x\, dx.$

25. $\int_1^{\sqrt{2}} \sqrt{2-x^2}\, dx.$
26. $\int_0^1 x^3 \tan^{-1} x\, dx.$
27. $\int_0^1 \sqrt{x(1+x)}\, dx.$

28. Show that $\int e^x \sin x\, dx = \dfrac{\sqrt{2}}{2} e^x \sin\left(x - \dfrac{\pi}{4}\right)$. Hence evaluate

$$\int x e^x \sin x\, dx.$$

29. Prove that $2\int \sinh x \sin x\, dx = \cosh x \sin x - \sinh x \cos x.$

30. Write the integral $\int e^x \sin 4x \cos 2x\, dx$ as the sum of two integrals and hence obtain its value.

Reduction formulae Consider the integral $\int \sin^n x\, dx,\ n > 1.$

Writing $\sin^n x$ as $\sin^{n-1} x \sin x$ and integrating by parts,

$$\int \sin^n x \, dx = \int \sin^{n-1} x \sin x \, dx$$

$$= [-\cos x] \sin^{n-1} x - \int [-\cos x](n-1) \sin^{n-2} x \cos x \, dx$$

$$= -\cos x \sin^{n-1} x + (n-1) \int \cos^2 x \sin^{n-2} x \, dx$$

$$= -\cos x \sin^{n-1} x + (n-1) \int (1 - \sin^2 x) \sin^{n-2} x \, dx$$

$$= -\cos x \sin^{n-1} x + (n-1) \int \sin^{n-2} x \, dx - (n-1) \int \sin^n x \, dx.$$

$$\therefore \; n \int \sin^n x \, dx = -\cos x \sin^{n-1} x + (n-1) \int \sin^{n-2} x \, dx.$$

Writing
$$I_n = \int \sin^n x \, dx, \quad I_{n-2} = \int \sin^{n-2} x \, dx,$$

$$nI_n = -\cos x \sin^{n-1} x + (n-1) I_{n-2}.$$

This result is called *a reduction formula*; by successive applications of it, the integral of any power of $\sin x$ can be obtained.

e.g.
$$\int \sin^4 x \, dx = -\frac{\cos x \sin^3 x}{4} + \frac{3}{4} \int \sin^2 x \, dx \quad [\text{putting } n = 4]$$

$$= -\frac{\cos x \sin^3 x}{4} + \frac{3}{4} \left\{ -\frac{\cos x \sin x}{2} + \frac{1}{2} \int \sin^0 x \, dx \right\}$$

$$= -\frac{\cos x \sin^3 x}{4} - \frac{3 \cos x \sin x}{8} + \frac{3}{8} x + c \quad \text{as } \sin^0 x = 1.$$

Example 12 Prove the reduction formula

$$(n+1) I_{n+2} = \tan x \sec^n x + n I_n,$$

where $I_n = \int \sec^n x \, dx$.

We have $\frac{d}{dx}(\tan x \sec^n x) = \sec^2 x \sec^n x + \tan x \, n \sec^{n-1} x \sec x \tan x$

$$= \sec^{n+2} x + n \sec^n x \tan^2 x$$
$$= \sec^{n+2} x + n \sec^n x (\sec^2 x - 1)$$
$$= (n+1) \sec^{n+2} x - n \sec^n x.$$

Integrating w.r. to x,

$$\tan x \sec^n x = (n+1) \int \sec^{n+2} x \, dx - n \int \sec^n x \, dx.$$

i.e. $\qquad (n+1) I_{n+2} = \tan x \sec^n x + n I_n.$

This reduction formula will enable the integration of any integral power of $\sec x$.

[B] EXAMPLES 17e

1 If $I_n = \int \sin^n x \, dx$, use the reduction formula
$$nI_n = -\cos x \sin^{n-1} x + (n-1)I_{n-2},$$
to evaluate (i) $\int \sin^3 x \, dx$; (ii) $\int_0^{\pi/2} \sin^8 x \, dx$; (iii) $\int_0^{\pi} \sin^6 x \, dx$.

2 If $I_n = \int \tan^n \theta \, d\theta$, use the reduction formula
$$I_n = \frac{\tan^{n-1} \theta}{n-1} - I_{n-2},$$
to evaluate (i) $\int \tan^4 \theta \, d\theta$; (ii) $\int \tan^3 \theta \, d\theta$; (iii) $\int_0^{\pi/4} \tan^6 \theta \, d\theta$.

3 If $I_n = \int \frac{dx}{(x^2+1)^n}$, use the reduction formula
$$2(n-1)I_n = (2n-3)I_{n-1} + \frac{x}{(x^2+1)^{n-1}},$$
to evaluate (i) $\int \frac{dx}{(x^2+1)^2}$; (ii) $\int_0^1 \frac{dx}{(x^2+1)^4}$; (iii) $\int \frac{dx}{(x^2+2x+2)^3}$.

4 If $I_{n,m} = \int \sin^n x \cos^m x \, dx$, use the reduction formula
$$(n+m)I_{n,m} = -\sin^{n-1} x \cos^{m+1} x + (n-1)I_{n-2,m},$$
to evaluate

(i) $\int \sin^4 x \cos^2 x \, dx$; (ii) $\int \sin^3 x \cos^3 x \, dx$; (iii) $\int_0^{\pi/2} \sin^6 x \cos^2 x \, dx$.

5 If $I_{m,n} = \int x^m (\ln x)^n \, dx$, use the reduction formula
$$(m+1)I_{m,n} = x^{m+1}(\ln x)^n - nI_{m,n-1},$$
to evaluate (i) $\int x^4 (\ln x)^2 \, dx$; (ii) $\int x^2 (\ln x)^3 \, dx$; (iii) $\int_1^e \frac{(\ln x)^2}{x^3} \, dx$.

6 If $I_n = \int \cos^n x \, dx$, prove that
$$nI_n = \cos^{n-1} x \sin x + (n-1)I_{n-2}.$$
Deduce the values of (i) $\int \cos^4 x \, dx$; (ii) $\int \cos^5 x \, dx$; (iii) $\int_0^{\pi/2} \cos^6 x \, dx$.

7 If $I_n = \int x^n e^{-x} \, dx$, obtain the relationship between I_n and I_{n-1}.

Deduce the values of (i) $\int x^3 e^{-x} \, dx$; (ii) $\int_0^1 x^4 e^{-x} \, dx$.

Further methods of integration

8 Prove the result $\int \cot^n x \, dx = -\dfrac{\cot^{n-1} x}{n-1} - \int \cot^{n-2} x \, dx$.

Hence evaluate (i) $\int \cot^3 x \, dx$; (ii) $\int \cot^4 x \, dx$; (iii) $\int_{\pi/4}^{\pi/2} \cot^5 x \, dx$.

9 If $u_n = \int x^n \cos x \, dx$, $v_n = \int x^n \sin x \, dx$, prove the results

$$u_n = x^n \sin x - n v_{n-1};$$
$$v_n = -x^n \cos x + n u_{n-1}.$$

Hence obtain the reduction formula connecting u_n and u_{n-2} and use it to evaluate

(i) $\int x^4 \cos x \, dx$; (ii) $\int x^3 \cos x \, dx$; (iii) $\int_0^{\pi/2} x^5 \cos x \, dx$.

10 Prove that $n \int_0^{\pi/2} \sin^n x \, dx = (n-1) \int_0^{\pi/2} \sin^{n-2} x \, dx$. Evaluate

(i) $\int_0^{\pi/2} \sin^7 x \, dx$; (ii) $\int_0^{\pi/2} \sin^{10} x \, dx$; (iii) $\int_0^{\pi} \sin^4 x \, dx$.

[B] MISCELLANEOUS EXAMPLES

Evaluate the following integrals:

1 $\int_0^2 (x-2)^3 \, dx$.

2 $\int \sqrt{4-x} \, dx$.

3 $\int \dfrac{x \, dx}{\sqrt{1-2x^2}}$.

4 $\int_0^{\pi/2} \sin^4 x \cos x \, dx$.

5 $\int e^{3x-1} \, dx$.

6 $\int x e^{-2x} \, dx$.

7 $\int_0^1 \dfrac{x \, dx}{(1+2x)^{1/3}}$.

8 $\int \dfrac{x+1}{x^2+1} \, dx$.

9 $\int_0^{5/4} \dfrac{dx}{\sqrt{25-4x^2}}$.

10 $\int_0^1 \dfrac{e^x}{(e^x+1)^2} \, dx$.

11 $\int \left(\dfrac{x-2}{x+2}\right)^2 dx$.

12 $\int \dfrac{\sin x}{a + b \cos x} \, dx$.

13 $\int \dfrac{x \, dx}{(1-3x)^3}$.

14 $\int_0^{\pi/2} \sin^2 x \cos^2 x \, dx$.

15 $\int \sin 4x \cos x \, dx$.

16 $\int (x^{1/2} - 2x^{1/3})^2 \, dx$.

17 $\int_3^4 \dfrac{dx}{1-x^2}$.

18 $\int_0^{\pi/4} \sin^4 x \, dx$.

19 $\int \dfrac{1+x}{\sqrt{1+x^2}} \, dx$.

20 $\int x^2 \cos 2x \, dx$.

21 $\int \dfrac{2x-3}{2x^2+5x-3} \, dx$.

22 $\int_{\pi/12}^{\pi/8} \cot 2x \, dx$.

23 $\int \operatorname{cosec}^2 3x \, dx$.

24 $\int_1^2 x^3 \ln x \, dx$.

25 $\int \dfrac{dx}{\sqrt{x(2-x)}}$.

26 $\int_2^3 \dfrac{2x^5}{x^3-1} \, dx$.

27 $\int \dfrac{\sqrt{1+2x}}{x} \, dx$.

28 $\int_{-1}^0 \dfrac{9 \, dx}{(x-1)(x+2)^2}$.

29 $\int_0^{\pi/3} \cos \dfrac{5x}{2} \cos \dfrac{x}{2} \, dx$.

30 $\int e^{-x} \sin x \, dx$.

Miscellaneous examples | 251

31 $\int \dfrac{dx}{\sqrt{2x^2+3}}$.

32 $\int_1^4 \dfrac{x^2-5}{2\sqrt{x}} dx.$

33 $\int \cos^2 3x\, dx.$

34 $\int_1^e \dfrac{dx}{x(1+\ln x)}.$

35 $\int \dfrac{dx}{(e^x+2)^4}.$

36 $\int_0^{\pi/6} \sin^3 2x \cos 2x\, dx.$

37 $\int_0^{1/2} \dfrac{2x+1}{(1-x^2)(1-2x)} dx.$

38 $\int \sqrt{\dfrac{1-2x}{1+2x}}\, dx.$

39 $\int 10^x\, dx.$

40 $\int_0^{\pi/2} \dfrac{dx}{5+4\cos x}.$

41 $\int \cos^{-1}\left(\dfrac{1}{x}\right) dx.$

42 $\int \dfrac{\sin(\ln x)}{x} dx.$

43 $\int_0^1 \dfrac{x}{(1+x)(1+4x^2)} dx.$

44 $\int \sqrt{e^x+4}\, dx.$

45 $\int_0^{\pi/4} \tan^2 x\, dx.$

46 $\int \dfrac{x\, dx}{\sqrt{1-x}}.$

47 $\int \log_2 x\, dx.$

48 $\int \dfrac{x^2}{e^{2x}} dx.$

49 $\int_{-1}^0 \dfrac{1-x+2x^2}{(1-x)^3} dx.$

50 $\int \sin 5x \sin 2x\, dx.$

51 $\int \sec^4 x\, dx.$

52 $\int \dfrac{dx}{1-\sin x}.$

53 $\int_0^\infty x^2 e^{-3x}\, dx.$

54 $\int_0^a x^2 \sqrt{a^2-x^2}\, dx.$

55 $\int_{-1}^0 \dfrac{x^3}{1-x} dx.$

56 $\int \dfrac{dx}{1-2x+2x^2}.$

57 $\int \dfrac{\sin x \cos x}{\cos^2 x + 2\sin^2 x} dx.$

58 $\int_0^{\pi/3} \dfrac{\sin x}{\cos^5 x} dx.$

59 $\int_1^2 \dfrac{dx}{\sqrt{4x-x^2}}.$

60 $\int \dfrac{x^2\, dx}{1-x^6}.$

61 If $I_n = \int \dfrac{\sin nx}{\sin x} dx$, prove that
$$I_n = \dfrac{2\sin(n-1)x}{n-1} + I_{n-2}. \quad (n \geq 2).$$
Deduce the value of $\int_{\pi/4}^{\pi/2} \dfrac{\sin 4x}{\sin x} dx.$

62 If $I_n = \int e^{ax} \cos^n x\, dx$, prove that
$$(a^2+n^2)I_n = e^{ax} \cos^{n-1} x\,(a\cos x + n\sin x) + n(n-1)I_{n-2}.$$
Evaluate $\int_0^{\pi/2} e^{2x} \cos^4 x\, dx.$

63 If $I_n = \int_0^{\pi/2} x^n \sin x\, dx$, prove that $I_n = n\left(\dfrac{\pi}{2}\right)^{n-1} - n(n-1)I_{n-2}.$

64 If $u_{m,n} = \int_0^{\pi/2} \cos^m x \sin nx\, dx$, prove that
$$(m+n)u_{m,n} = 1 + mu_{m-1, n-1}.$$
Evaluate $\int_0^{\pi/2} \cos^5 x \sin 2x\, dx.$

Further methods of integration

65 If $I_{m,n} = \int \dfrac{x^m\,dx}{(\ln x)^n}$, prove that

$$(n-1)I_{m,n} = -\dfrac{x^{m+1}}{(\ln x)^{n-1}} + (m+1)u_{m,n-1}$$

66 Prove that $\displaystyle\int_0^{\pi/2} \cos^n x\,dx = \dfrac{n-1}{n}\int_0^{\pi/2}\cos^{n-2}x\,dx$ and deduce the value of $\displaystyle\int_0^{\pi/2}\cos^8 x\,dx$.

67 Show geometrically that $\displaystyle\int_0^{\pi/2}\sin^n x\,dx > \int_0^{\pi/2}\sin^{n+1}x\,dx$ for positive integral values of n.

68 Without evaluating the integrals, prove that

$$\int_0^1 x^m(1-x)^n\,dx = \int_0^1 x^n(1-x)^m\,dx.$$

69 Prove that $\displaystyle\int_{-a}^{a}\phi(x^2)\,dx = 2\int_0^a \phi(x^2)\,dx$ and $\displaystyle\int_{-a}^{a}\phi(x^2)x\,dx = 0$.

What is the value of $\displaystyle\int_{-1}^{1}\dfrac{x^3}{1+x^2}\,dx$?

70 If $I_n = \displaystyle\int_0^{\pi}\dfrac{\cos nx}{5-4\cos x}\,dx$, express $I_n + I_{n-2}$ as a single integral. Deduce that $2(I_n + I_{n-2}) = 5I_{n-1}$ when $n > 1$. Prove that $I_3 = \dfrac{\pi}{24}$.

18 | Differential properties of plane curves
Curvature

Plane curves expressed in Cartesian coordinates Suppose $P(x, y)$ is any point on the curve $y = f(x)$. Let Q be a neighbouring point with coordinates $(x + \delta x, y + \delta y)$.

Taking A as any fixed point on the curve, let the arc length $AP = s$ and arc $PQ = \delta s$.

Considering $\triangle PQR$,

$$(\delta s)^2 \approx (\delta x)^2 + (\delta y)^2,$$

or

$$\left(\frac{\delta s}{\delta x}\right)^2 \approx 1 + \left(\frac{\delta y}{\delta x}\right)^2.$$

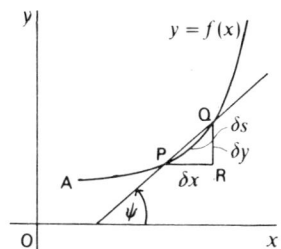

Fig. 91

In the limit as $Q \to P$,

$$\left(\frac{ds}{dx}\right)^2 = 1 + \left(\frac{dy}{dx}\right)^2 \quad \dots \dots \dots \dots \dots \dots \dots \dots \dots \text{(i)}$$

or, in terms of differentials, $ds^2 = dx^2 + dy^2$.

If $\widehat{QPR} = \psi$, $\tan\psi = \dfrac{\delta y}{\delta x}$, $\sin\psi \approx \dfrac{\delta y}{\delta s}$, $\cos\psi \approx \dfrac{\delta x}{\delta s}$. In the limit as $Q \to P$, ψ becomes the angle between the tangent at P and Ox and we have

$$\tan\psi = \frac{dy}{dx}, \quad \sin\psi = \frac{dy}{ds}, \quad \cos\psi = \frac{dx}{ds}.$$

If the curve is given in parametric coordinates, i.e. $x = f(t)$, $y = g(t)$, the result (i) becomes

$$\left(\frac{ds}{dt}\right)^2 = \left(\frac{dx}{dt}\right)^2 + \left(\frac{dy}{dt}\right)^2 \quad \dots \dots \dots \dots \dots \dots \text{(ii)}$$

Arc length and area of surface of revolution Using (i) and (ii) above, by integrating with respect to x,

Arc length AB

$$= \int_{x_1}^{x_2} \sqrt{1 + \left(\frac{dy}{dx}\right)^2}\, dx: \quad \text{(Fig. 92)}$$

$$= \int_{t_1}^{t_2} \sqrt{\left(\frac{dx}{dt}\right)^2 + \left(\frac{dy}{dt}\right)^2}\, dt,$$

where t_1, t_2 are the parameters of A and B.

Area of surface swept out by element of arc δs in one revolution about $Ox \approx 2\pi y\, \delta s$ (Fig. 92).

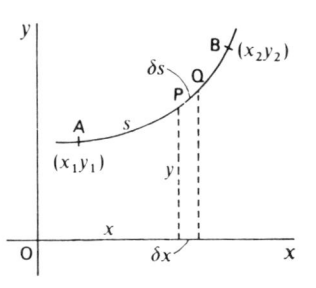

Fig. 92

254 | Differential properties of plane curves

$$\therefore \text{ Area of surface of revolution of arc AB} = \int_A^B 2\pi y \, ds,$$

$$= 2\pi \int_{x_1}^{x_2} y \sqrt{1 + \left(\frac{dy}{dx}\right)^2} \, dx.$$

If the curve is expressed in parametric coordinates,

$$\text{Area of surface of revolution} = 2\pi \int_{t_1}^{t_2} y \sqrt{\left(\frac{dx}{dt}\right)^2 + \left(\frac{dy}{dt}\right)^2} \, dt.$$

Example 1 Show that in the catenary $y = c \cosh \frac{x}{c}$, the length of arc from the vertex where $x = 0$, to any point is given by $s = c \sinh \frac{x}{c}$.

Arc length from vertex to point $(x, y) = \int_0^x \sqrt{1 + \left(\frac{dy}{dx}\right)^2} \, dx$.

As $y = c \cosh \frac{x}{c}$, $\frac{dy}{dx} = \sinh \frac{x}{c}$ and $\sqrt{1 + \left(\frac{dy}{dx}\right)^2} = \cosh \frac{x}{c}$.

$$\therefore \text{ Arc length } s = \int_0^x \cosh \frac{x}{c} \, dx = c \sinh \frac{x}{c}.$$

Example 2 Find the area of surface of a sphere, radius r, contained between parallel planes distance a and b from the centre where $a < b < r$.

Consider the sphere to be generated by the rotation of a circle, radius r, about a diameter, Ox (Fig. 93). If (x, y) is a point on the circle, $x^2 + y^2 = r^2$, i.e. equation of circle is $x^2 + y^2 = r^2$.

Differentiating, $2x + 2y \frac{dy}{dx} = 0$; $\frac{dy}{dx} = -\frac{x}{y}$.

Curved surface area contained between planes $x = a, x = b$

$$= 2\pi \int_a^b y \sqrt{1 + \left(\frac{dy}{dx}\right)^2} \, dx$$

$$= 2\pi \int_a^b \sqrt{x^2 + y^2} \, dx = 2\pi \int_a^b r \, dx$$

$$= 2\pi r (b - a).$$

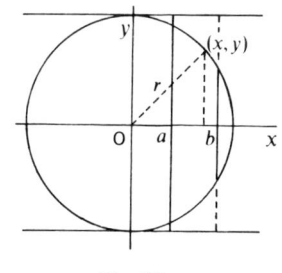

Fig. 93

Note $2\pi r(b - a)$ is equal to the area cut off by the two planes on the cylinder which circumscribes the sphere.

i.e. *the area of a zone of a sphere contained between two parallel planes is equal to the area cut off by the planes on the circumscribing cylinder.*

Example 3 Find the total length of the curve $x = a \cos^3 t, y = a \sin^3 t$.

The curve consists of four equal arcs of which AB is the arc in the first quadrant (Fig. 94).

At $A, t = 0$; at $B, t = \frac{\pi}{2}$.

$$\therefore \text{ Total length of curve} = 4 \int_0^{\frac{\pi}{2}} \sqrt{\left(\frac{dx}{dt}\right)^2 + \left(\frac{dy}{dt}\right)^2} \, dt.$$

But $\left(\dfrac{dx}{dt}\right)^2 + \left(\dfrac{dy}{dt}\right)^2$

$= (-3a \cos^2 t \sin t)^2 + (3a \sin^2 t \cos t)^2$
$= 9a^2 \cos^2 t \sin^2 t (\cos^2 t + \sin^2 t) = 9a^2 \cos^2 t \sin^2 t.$

$\therefore \text{ Total length of curve} = 4 \int_0^{\frac{\pi}{2}} 3a \cos t \sin t \, dt$

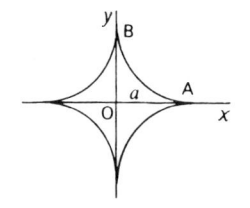

Fig. 94

$= 6a \int_0^{\frac{\pi}{2}} \sin 2t \, dt = 6a \left[-\frac{\cos 2t}{2} \right]_0^{\frac{\pi}{2}}$

$= 6a.$

[B] EXAMPLES 18a

1 Find the length of the arc of $x^3 = y^2$ from $x = 0$ to $x = 2$.

2 Writing the equation of a circle, radius r, as $x^2 + y^2 = r^2$, prove that the circumference is of length $2\pi r$.

3 Find the length of the curve $y = \dfrac{c}{2}(e^{\frac{x}{c}} + e^{\frac{-x}{c}})$ between the points where $x = -a$ and $x = a$.

4 What is the length of the arc of the curve $x^3 = 8y^2$ from $x = 1$ to $x = 3$?

5 For the curve $x = p + a \sin \theta$, $y = q + a \cos \theta$, show that $\dfrac{ds}{d\theta}$ is constant.

6 Show that the length of the arc of the cycloid, $x = a(\theta + \sin \theta)$, $y = a(1 - \cos \theta)$, between the points $\theta = 0$ and $\theta = \pi$, is $4a$.

7 A plane drawn perpendicular to the base of a hemisphere cuts off a portion whose curved surface area is half the area of the plane base. Show that the plane must be at a distance from the centre equal to half the radius.

8 Show that for the parabola $y^2 = 4ax$, $\dfrac{ds}{dx} = \sqrt{1 + \dfrac{a}{x}}$.

9 Find by integration, the curved surface area of a right circular cone, slant height l, radius r.

10 Prove that the surface area of the paraboloid obtained by the rotation of the arc of the parabola $y^2 = 4ax$ between $x = 0$ and $x = h$ about the axis of x, is $\frac{8}{3}\pi a^{\frac{1}{2}} \{(h+a)^{\frac{3}{2}} - a^{\frac{3}{2}}\}$.

11 What fraction of the surface of a sphere, radius r, is visible from a point distant kr from the centre ($k > 1$)?

12 Find the area of the surface generated when the curve $x = a\cos^3 t$, $y = a\sin^3 t$ is rotated about the x axis.

13 A hemispherical bowl, of radius 20 cm, contains water. Find the depth of the water when half of the surface is wetted.

14 In the curve $x = 3\cos\theta - 2\cos\dfrac{3\theta}{2}$, $y = 3\sin\theta - 2\sin\dfrac{3\theta}{2}$, prove that the length of the arc between the points with parameters 2π and θ is $24 \cos \dfrac{\theta}{4}$.

15 The portion of the catenary $y = c \cosh\dfrac{x}{c}$ between $x = 0$ and $x = c$, is rotated about the x axis, find the area of the surface generated.

16 A spherical surface, radius r, is divided into two parts by a plane at distance x from the centre; prove that the ratio of the total areas of the two parts is $(3r^2 + 2rx - x^2): (3r^2 - 2rx - x^2)$.

Curvature Let P, Q be neighbouring points on a curve (Fig. 95).

Denote the arc length AP measured from a fixed point A on the curve by s and the arc length PQ by δs.

Let the tangents at P, Q make angles ψ, $\psi + \delta\psi$ respectively with a fixed straight line Ox.

Then the angle $\delta\psi$ through which the tangent turns as the point of contact moves from P to Q is called *the total curvature* of the arc PQ. *The mean curvature* of the arc is defined as the total curvature divided by the arc length, i.e. $\dfrac{\delta\psi}{\delta s}$.

Fig. 95

In order to find *the curvature at the point* P we imagine the arc length PQ to become infinitesimally small and define the curvature at P as the limiting value of the mean curvature of the arc PQ.

i.e. $\qquad\qquad$ **Curvature at P** $= \underset{\delta s \to 0}{\text{Lt.}} \dfrac{\delta\psi}{\delta s} = \dfrac{d\psi}{ds}$.

Curvature of a circle

Using the diagram in Fig. 96, average curvature of arc $PQ = \dfrac{\delta\psi}{\delta s}$.

Clearly the angle $\delta\psi$ between the tangents is equal to the angle between the radii drawn to P and Q.

$\therefore \delta s = R\,\delta\psi$ where R is the radius. Hence the average curvature of $PQ = \dfrac{\delta\psi}{R\delta\psi} = \dfrac{1}{R}$, from which it follows that the curvature at any point P $= \dfrac{1}{R}$, i.e. the curvature of a circle is measured by the reciprocal of the radius.

Thus, if ρ is the radius of the circle which has the same curvature as a given curve at any point P, we have

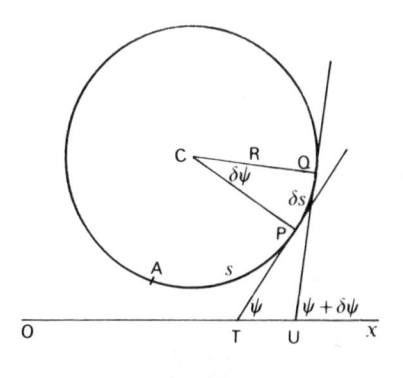

$$\rho = \dfrac{1}{\dfrac{d\psi}{ds}} = \dfrac{ds}{d\psi}.$$

Fig. 96

Radius of curvature. Cartesian coordinates | 257

Such a circle is called *the circle of curvature*, its radius ρ, *the radius of curvature* and its centre C, *the centre of curvature* of the given curve at the point P.

Hence, we have the general results,

$$\text{Curvature, } \kappa = \frac{d\psi}{ds}.$$

$$\text{Radius of curvature, } \rho = \frac{ds}{d\psi}.$$

Radius of curvature. Cartesian coordinates Let the equation of a curve be $y = f(x)$.

We have
$$\tan \psi = \frac{dy}{dx}.$$

Differentiating w.r. to x, $\sec^2 \psi \dfrac{d\psi}{dx} = \dfrac{d^2 y}{dx^2}.$

$$\frac{d\psi}{dx} = \frac{\dfrac{d^2 y}{dx^2}}{1 + \tan^2 \psi} = \frac{\dfrac{d^2 y}{dx^2}}{1 + \left(\dfrac{dy}{dx}\right)^2}.$$

$$\therefore \frac{d\psi}{ds} = \frac{d\psi}{dx} \times \frac{dx}{ds} = \frac{\dfrac{d^2 y}{dx^2}}{1 + \left(\dfrac{dy}{dx}\right)^2} \times \frac{1}{\sqrt{1 + \left(\dfrac{dy}{dx}\right)^2}}$$

$$= \frac{\dfrac{d^2 y}{dx^2}}{\left\{1 + \left(\dfrac{dy}{dx}\right)^2\right\}^{\frac{3}{2}}}.$$

i.e.
$$\rho = \frac{ds}{d\psi} = \frac{\left\{1 + \left(\dfrac{dy}{dx}\right)^2\right\}^{\frac{3}{2}}}{\dfrac{d^2 y}{dx^2}}, \quad \ldots\ldots\ldots\ldots\ldots\ldots (A)$$

If the curve is given by parametric equations $y = F(t)$, $x = f(t)$, then

$$\frac{dy}{dx} = \frac{dy}{dt} \div \frac{dx}{dt} = \frac{\dot{y}}{\dot{x}},$$

where a dot denotes differentiation w.r. to t.

$$\frac{d^2 y}{dx^2} = \frac{d}{dx}\left(\frac{\dot{y}}{\dot{x}}\right) = \frac{d}{dt}\left(\frac{\dot{y}}{\dot{x}}\right) \times \frac{dt}{dx}$$

$$= \frac{\dot{x}\ddot{y} - \dot{y}\ddot{x}}{(\dot{x})^2} \times \frac{1}{\dot{x}} = \frac{\dot{x}\ddot{y} - \dot{y}\ddot{x}}{(\dot{x})^3}.$$

$$\therefore \rho = \frac{\left\{1+\left(\frac{dy}{dx}\right)^2\right\}^{\frac{3}{2}}}{\frac{d^2y}{dx^2}} = \frac{\{\dot{x}^2+\dot{y}^2\}^{\frac{3}{2}}}{\dot{x}\ddot{y}-\dot{y}\ddot{x}} \quad \ldots \ldots \ldots \ldots \text{(B)}$$

Example 4 Find the radius of curvature of the curve $y = x + 3x^2 - 4x^3$ at the origin.

We have
$$\frac{dy}{dx} = 1 + 6x - 12x^2;$$

$$\frac{d^2y}{dx^2} = 6 - 24x.$$

$$\therefore \rho = \frac{\left\{1+\left(\frac{dy}{dx}\right)^2\right\}^{\frac{3}{2}}}{\frac{d^2y}{dx^2}} = \frac{\{1+(1+6x-12x^2)^2\}^{\frac{3}{2}}}{6-24x}.$$

So, when $x = 0$, radius of curvature $= \dfrac{2^{\frac{3}{2}}}{6} = \dfrac{\sqrt{8}}{6} = 0 \cdot 47$.

Example 5 For the curve $x = a\sin^3\theta$, $y = a\cos^3\theta$, prove that $\rho = 3a\sin\theta\cos\theta$

$$\frac{dy}{dx} = \frac{dy}{d\theta} \div \frac{dx}{d\theta} = -\frac{3a\cos^2\theta\sin\theta}{3a\sin^2\theta\cos\theta} = -\cot\theta.$$

$$\frac{d^2y}{dx^2} = \frac{d}{dx}(-\cot\theta) = \frac{d}{d\theta}(-\cot\theta) \times \frac{d\theta}{dx}$$

$$= \frac{\csc^2\theta}{3a\sin^2\theta\cos\theta} = \frac{1}{3a\sin^4\theta\cos\theta}$$

$$\therefore \rho = \frac{\left\{1+\left(\frac{dy}{dx}\right)^2\right\}^{\frac{3}{2}}}{\frac{d^2y}{dx^2}} = (1+\cot^2\theta)^{\frac{3}{2}} \, 3a\sin^4\theta\cos\theta$$

$$= (\csc^2\theta)^{\frac{3}{2}} \, 3a\sin^4\theta\cos\theta = 3a\sin\theta\cos\theta.$$

The result could also have been obtained by using formula (B).

[B] EXAMPLES 18b

Find the radii of curvature of the following curves at the stated points:

1. $y = x^2$; $x = 2$.
2. $y = \sin x$; $x = \dfrac{\pi}{4}$,
3. $y = e^x$; $x = 0$.
4. $y = \ln x$; $x = \frac{1}{3}$.
5. $y^2 = x^3$; point $(1, 1)$.
6. $y = 3x^2 + x^3$; $x = 0$.
7. $x = 2t^2$, $y = 4t$; $t = \sqrt{2}$.
8. $x = 4t$, $y = \dfrac{4}{t}$; $t = -2$.
9. $x = 4\cos\phi$, $y = 2\sin\phi$; $\phi = \dfrac{\pi}{2}$.

10 $x = 2\cosh\theta$, $y = \sinh\theta$; $\theta = 0$. **11** $x = 2\sin t$, $y = \sin 2t$; $t = \dfrac{\pi}{6}$.

12 $2x^2 - y^2 = 5$; point $(-2, \sqrt{3})$. **13** $3x^2 + 3y^2 = 10xy$: point $(3, 1)$.

14 $x^3 + 2y^3 = 3xy$; point $(1, 1)$.

15 Find the radius of curvature at the origin for the curve
$$y = x + 3x^2 - x^3.$$
Show that the radius of curvature at the point $(1, 3)$ is infinite.

16 Prove that the radius of curvature of the catenary $y = c\cosh\dfrac{x}{c}$ at any point (x, y) is $\dfrac{y^2}{c}$.

17 Prove that the radius of curvature at the point $\left(ct, \dfrac{c}{t}\right)$ on the rectangular hyperbola $xy = c^2$ is $\dfrac{c}{2}\left(t^2 + \dfrac{1}{t^2}\right)^{\frac{3}{2}}$.

18 For the cycloid $x = a(\theta + \sin\theta)$, $y = a(1 + \cos\theta)$, show that $\rho = 4a\cos\dfrac{\theta}{2}$.

19 Find the radius of curvature at the point θ on the curve $x = a\cos^3\theta$, $y = a\sin^3\theta$.

Deduce that the radius of curvature at the point (x, y) on the curve $\left(\dfrac{x}{a}\right)^{\frac{2}{3}} + \left(\dfrac{y}{a}\right)^{\frac{2}{3}} = 1$ is $3(axy)^{\frac{1}{3}}$.

20 Find the coordinates of the point on the curve $y = x^2(3 - x)$ at which the curvature is zero.

21 Find the points on the curve $y = 1 - 3x^3 + 2x^4$ where the curvature changes sign. What is the special significance of these points?

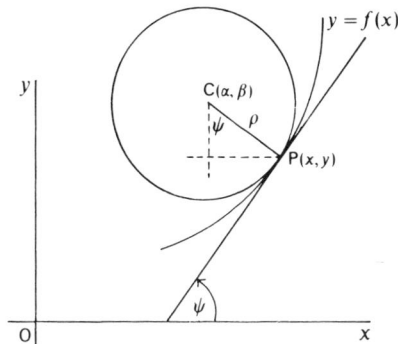

Fig. 97

Centre of curvature We have already defined the circle of curvature at a point P on a curve as the circle passing through P and having the same curvature as the curve at P (Fig. 97).

Let the centre of curvature C be the point (α, β).

Then $\alpha = x - \rho\sin\psi$,
$$\beta = y + \rho\cos\psi.$$

Substituting $\rho = \dfrac{\left\{1 + \left(\dfrac{dy}{dx}\right)^2\right\}^{\frac{3}{2}}}{\dfrac{d^2y}{dx^2}}$; $\sin\psi = \dfrac{dy}{ds} = \dfrac{\dfrac{dy}{dx}}{\sqrt{1 + \left(\dfrac{dy}{dx}\right)^2}}$;

$$\cos\psi = \frac{dx}{ds} = \frac{1}{\sqrt{1+\left(\frac{dy}{dx}\right)^2}},$$

$$\alpha = x - \frac{\frac{dy}{dx}\left\{1+\left(\frac{dy}{dx}\right)^2\right\}}{\frac{d^2y}{dx^2}};$$

$$\beta = y + \frac{1+\left(\frac{dy}{dx}\right)^2}{\frac{d^2y}{dx^2}}.$$

The locus of C as P moves along the curve is called *the evolute* of the curve.

Example 6 Find the centre of curvature at any point $(at^2, 2at)$ on the parabola $y^2 = 4ax$. Deduce the equation of the evolute.

$$\frac{dy}{dx} = \frac{2a}{y} = \frac{1}{t}; \quad \frac{d^2y}{dx^2} = \frac{d}{dx}\left(\frac{2a}{y}\right) = -\frac{2a}{y^2}\frac{dy}{dx} = -\frac{2a}{4a^2t^2}\frac{1}{t} = -\frac{1}{2at^3}.$$

$$\therefore \alpha = at^2 - \frac{\frac{1}{t}\left(1+\frac{1}{t^2}\right)}{-\frac{1}{2at^3}} = at^2 + 2a(t^2+1) = 3at^2 + 2a.$$

$$\beta = 2at + \frac{1+\frac{1}{t^2}}{-\frac{1}{2at^3}} = 2at - 2at(t^2+1) = -2at^3.$$

i.e. the centre of curvature at the point $(at^2, 2at)$ is $(3at^2 + 2a, -2at^3)$.

To find the equation of the evolute, t is eliminated from the equations,

$$\alpha = 3at^2 + 2a, \quad \beta = -2at^3.$$

We have $t^2 = \dfrac{\alpha - 2a}{3a}$,

$$\therefore \beta^2 = 4a^2t^6 = \frac{4a^2(\alpha - 2a)^3}{27a^3},$$

i.e. equation of evolute is $27ay^2 = 4(x-2a)^3$.

Curvature at the origin. Newton's formula Take the case of a curve touching the x axis at the origin. Let P, Q be points on the curve very close to O and symmetrically placed with respect to O. Consider the circle which can be drawn through the three points P, O, Q. Let C be the centre of this circle. In the limit as P and Q both tend to O, this circle becomes the circle of curvature at O because the circle and curve will intersect in three coincident points and consequently will have

Curvature at the origin. Newton's formula | 261

common values of $\dfrac{dy}{dx}$ and $\dfrac{d^2y}{dx^2}$, i.e. common values of ρ. Referring to Fig. 98, OC is clearly perpendicular to Ox and is therefore the axis of y; PQ is parallel to Ox.

Let (x, y) be the coordinates of Q, then as $NQ^2 = ON \cdot NM$,

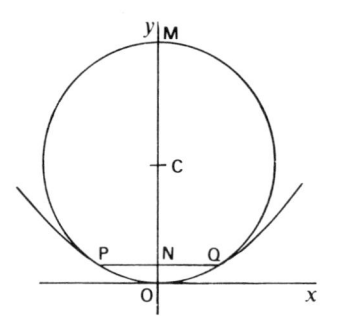

$$x^2 = y(2R - y), \quad \text{where } R = OC,$$

or
$$\dfrac{x^2}{y} = 2R - y.$$

Fig. 98

In the limit as both x and $y \to 0$, $R \to \rho$, the radius of curvature at O.

$$\therefore \textbf{Radius of curvature at } O = \underset{\substack{x \to 0 \\ y \to 0}}{\text{Lt}} \left(\dfrac{x^2}{2y}\right).$$

Similarly, if a curve touches the y axis at the origin, the radius of curvature at the origin is equal to $\underset{\substack{x \to 0 \\ y \to 0}}{\text{Lt}} \left(\dfrac{y^2}{2x}\right).$

Example 7 Find the radius of curvature of the parabola $y^2 = 4ax$ at the origin.

The curve $y^2 = 4ax$ touches the y axis at the origin and thus by Newton's formula,

$$\rho = \text{Lt} \dfrac{y^2}{2x}.$$

But
$$\dfrac{y^2}{2x} = 2a$$

$$\therefore \rho = 2a.$$

Example 8 Use Newton's method to determine the radius of curvature of the curve $a^2y^2 = x^3(x-a)$ at the point $(a, 0)$.

If we transfer the origin to the point $(a, 0)$, the equation of the curve becomes

$$a^2y^2 = (x+a)^3 x.$$

$$2a^2 y \dfrac{dy}{dx} = (x+a)^3 + 3x(x+a)^2.$$

$\therefore$ At the origin, $\dfrac{dy}{dx}$ is infinite and so the new y axis is a tangent to the curve.

$\therefore$ Radius of curvature at the new origin $= \text{Lt} \dfrac{y^2}{2x}$.

But
$$a^2y^2 = x^4 + 3x^3 a + 3x^2 a^2 + a^3 x$$

$$a^2 \left(\dfrac{y^2}{x}\right) = x^3 + 3x^2 a + 3xa^2 + a^3.$$

$$\therefore \underset{\substack{x \to 0 \\ y \to 0}}{\mathrm{Lt}} \left(\frac{y^2}{2x}\right) = \frac{a^3}{2a^2} = \frac{a}{2}.$$

i.e. Radius of curvature at the point $(a, 0) = \dfrac{a}{2}$.

[B] EXAMPLES 18c

Find the centres of curvature of the given curves at the points stated:

1 $y = x(1-x)$; $x = 1$.

2 $y = 2x^3$; $x = -2$.

3 $y = \sin x$; $x = \dfrac{\pi}{2}$.

4 $y^2 = 4x$; $y = 2$.

5 $y^2 = x^3$; $y = 1$.

6 $y = \dfrac{1}{x(x-1)}$; $x = \frac{1}{2}$.

7 $x = 4t$, $y = \dfrac{4}{t}$; $t = 2$.

8 $x = 5\cos\theta$, $y = 3\sin\theta$; $\theta = \dfrac{\pi}{2}$.

9 $x = t + \dfrac{1}{t}$, $y = t - \dfrac{1}{t}$; $t = \frac{1}{2}$.

10 $x = \cos 2t$, $y = 2\cos t$; $t = \dfrac{\pi}{3}$.

11 $x = -3 + 4\cos\theta$, $y = 1 + 4\sin\theta$; $\theta = \dfrac{\pi}{4}$.

12 $xy^2 + yx^2 = 2$; point $(1, 1)$.

13 $y = c\cosh\dfrac{x}{c}$; $x = 0$.

14 $y = 4\sin x - \sin 2x$; $x = \dfrac{\pi}{2}$.

15 Use Newton's method to find the radius of curvature of the curve $x^2 = 4y$ at the origin.

16 Find the radius of curvature of the curve $y^2 = x(2-x)$ at the origin.

17 What does the equation of the curve $y^2 = x(2-x)$ become if the origin is moved to the point $(2, 0)$? Use Newton's method to determine the radius of curvature at this point.

18 Find the radius of curvature of the curve $2y^2 = x^2(x+1)$ at the point $(-1, 0)$.

19 Find the radius of curvature of the curve $x^2 = 4y^3(y-3)$ at the point $(0, 3)$.

20 Prove that the radius of curvature of the curve $xy^2 = a^2(a-x)$ at the point $(a, 0)$ is $\dfrac{a}{2}$.

21 Prove that the centre of curvature at the point parameter ϕ on the ellipse $x = a\cos\theta$, $y = b\sin\theta$ has coordinates,

$$\alpha = \frac{a^2 - b^2}{a}\cos^3\phi, \quad \beta = -\frac{a^2 - b^2}{b}\sin^3\phi.$$

Deduce that the equation of the evolute is $(ax)^{\frac{2}{3}} + (by)^{\frac{2}{3}} = (a^2 - b^2)^{\frac{2}{3}}$. $a > b$.

22 Find the coordinates of the centre of curvature at the point parameter θ on the curve $x = a\cos^3\theta$, $y = a\sin^3\theta$.

[B] MISCELLANEOUS EXAMPLES

1 A sphere rests in a horizontal circular hole of radius 8 cm, with the lowest point of the sphere 4 cm below the plane of the hole. Calculate
(i) the area of the surface of that part of the sphere which is below the hole,
(ii) the volume of that part of the sphere.

2 A sphere passes through the eight vertices of a cube. If $ABCD$ is a face of the cube, find, correct to three significant figures, the ratio in which the area of the sphere is divided by the plane $ABCD$.

3 In the curve $x = e^\theta \sin\theta$, $y = e^\theta \cos\theta$, prove that $s = \sqrt{2}e^\theta + \text{constant}$.

4 Use Newton's method to show that the radius of curvature of the curve $ay^3 = x^4$ at the origin is zero.

5 Prove that the difference between the area of the curved surface of a segment of a sphere and the area of its plane base is equal to the area of a circle whose radius is the height of the segment.

6 Find the radius of curvature at the point parameter t, on the curve $x = at^2$, $y = 2at$. Prove that the centre of curvature at the point $t = \dfrac{\sqrt{2}}{3}$ is $\left(\dfrac{8a}{3}, -\dfrac{4a\sqrt{2}}{27}\right)$.

7 Use the results $\tan\psi = \dfrac{dy}{dx}$, $\dfrac{ds}{dx} = \sqrt{1 + \left(\dfrac{dy}{dx}\right)^2}$, to prove that for the catenary $y = c\cosh\dfrac{x}{c}$, $s = c\tan\psi$.

8 Two points, B, C on the same diameter of a sphere of radius a and outside it, are at distances r $(> a)$ and $2r$ from the centre on the same side. Prove that the difference between the areas of the sphere visible at the two points is $\dfrac{\pi a^3}{r}$.

9 Find the length of the arc of the cycloid
$$x = a(\theta + \sin\theta), \quad y = a(1 + \cos\theta)$$
between the points $\theta = \pm\pi$.

10 A plane cuts off from a sphere a volume equal to $\tfrac{7}{27}$ of the whole. Prove that the diameter perpendicular to the plane is divided in the ratio 1 to 2.

11 Two spheres, radii a, b, cut orthogonally; prove that the area of the surface of the first sphere external to the second is $2\pi a^2\left(1 + \dfrac{a}{\sqrt{a^2 + b^2}}\right)$.

12 For the curve $ay^2 = x^3$, if the arc length measured from the origin to the point (x, y) is s, prove that $\left(\dfrac{27s}{8a} + 1\right)^2 = \left(1 + \dfrac{9x}{4a}\right)^3$.

13 Find the radii of curvature of the curve $2y = 2x^4 - 9x^3 + 11x^2$ when (i) $x = 1$, (ii) $x = 0$.

14 Find the radius of curvature of the curve $y^2 = \dfrac{a^3 x}{a^2 - x^2}$ at the origin.

15 Show that the length of the arc of the parabola $y^2 = 4ax$ cut off by the line $3y = 8x$ is $a(\ln 2 + \tfrac{15}{16})$.

16 For the cycloid $x = a(\theta + \sin\theta)$, $y = a(1 - \cos\theta)$, prove that
$$s = 4a\sin\psi.$$

17 Express the curve $(x^2 + y^2)^2 - 4axy^2 = 0$ in polar coordinates and prove that the area of the loop in the positive quadrant is $\tfrac{1}{4}\pi a^2$.

18 The curve $x = a \ln (\sec \theta + \tan \theta) - a \sin \theta$,
$$y = a \cos \theta,$$
between $\theta = 0$ and $\dfrac{\pi}{2}$ is rotated about the x axis. Prove that the area of the curved surface is $2\pi a^2$.

19 Find the radius of curvature ρ at any point P on the curve
$$x = a \cos^3 t, \quad y = a \sin^3 t.$$
If the normal at P meets the x axis in G and the y axis in g, prove that
$$\rho^2 = 9PG \cdot Pg.$$

20 Prove that the coordinates of the centre of curvature C at any point (x, y) of the curve $ay^2 = x^3$ are $\left(-x - \dfrac{9x^2}{2a}, 4y + \dfrac{4ay}{3x}\right)$.

Prove that the radius of curvature of the locus of C at the origin is $\dfrac{8a}{9}$.

21 Find the radius ρ and the coordinates of the centre C of the smallest circle of curvature of the curve $y = e^x$.

If the tangent to the curve at its point of contact with the circle meets the x axis at P, show that $9CP^2 = 11\rho^2$.

19 | Differential equations

Formation of differential equations A differential equation is a relationship between an independent variable x, a dependent variable y and one or more of the derivatives of y with respect to x, $\dfrac{dy}{dx}$, $\dfrac{d^2y}{dx^2}$, etc.

E.g. $$x\frac{dy}{dx} - y^2 = 0; \qquad xy\frac{d^2y}{dx^2} - y^2 \sin x = 0$$

are differential equations.

Before considering methods of solution of the simplest types of differential equation, it is of importance to see how a differential equation is formed. We will find that a *differential equation of the first order* (i.e. an equation in which the highest derivative appearing is the first, $\dfrac{dy}{dx}$) always results from the elimination of an arbitrary constant A from an equation containing x, y and A. For example, the equation $y = x + A$, leads to the differential equation $\dfrac{dy}{dx} = 1$.

More generally, a *differential equation of the second order* (i.e. an equation in which the highest derivative is $\dfrac{d^2y}{dx^2}$) will always be obtained by the elimination of two arbitrary constants A, B from an equation containing x, y, A, B, and so on for higher order differential equations.

For example, if $$y = A \sin x + B \cos x,$$

$$\frac{dy}{dx} = A \cos x - B \sin x,$$

$$\frac{d^2y}{dx^2} = -A \sin x - B \cos x = -y.$$

Hence, if $y = A \sin x + B \cos x$, $\dfrac{d^2y}{dx^2} + y = 0$.

$y = A \sin x + B \cos x$, where A and B are any constants, is called *the complete primitive* or *general solution of the differential equation*.

Example 1 Form the differential equation of which $y = x + \dfrac{a}{x}$, where a is an arbitrary constant, is the complete primitive.

$$y = x + \frac{a}{x},$$

$$\therefore \frac{dy}{dx} = 1 - \frac{a}{x^2}.$$

Substituting for a from the first equation into the second

$$\frac{dy}{dx} = 1 - \frac{1}{x^2}(xy - x^2) = 1 - \frac{y}{x} + 1.$$

i.e.
$$x\frac{dy}{dx} + y = 2x.$$

Example 2 If $y = Ax^2 + Bx$, prove that $\dfrac{d^2y}{dx^2} - \dfrac{2}{x}\dfrac{dy}{dx} + \dfrac{2y}{x^2} = 0.$

$$\frac{dy}{dx} = 2Ax + B;$$

$$\frac{d^2y}{dx^2} = 2A.$$

Hence
$$\frac{dy}{dx} = x\frac{d^2y}{dx^2} + B,$$

and
$$y = \frac{1}{2}x^2\frac{d^2y}{dx^2} + Bx.$$

Eliminating B,
$$x\frac{dy}{dx} - y = x^2\frac{d^2y}{dx^2} - \frac{x^2}{2}\frac{d^2y}{dx^2}$$

$$= \frac{x^2}{2}\frac{d^2y}{dx^2}.$$

$$\therefore \frac{d^2y}{dx^2} - \frac{2}{x}\frac{dy}{dx} + \frac{2y}{x^2} = 0.$$

Example 3 Find the differential equation of which $v = e^{-kt}(A \sin \omega t + B \cos \omega t)$ is the general solution, A, B being arbitrary constants and k, ω given fixed constants.

$$v = e^{-kt}(A \sin \omega t + B \cos \omega t)$$

i.e.
$$ve^{kt} = A \sin \omega t + B \cos \omega t \quad \dotfill (i)$$

Differentiating with respect to t,

$$e^{kt}\frac{dv}{dt} + kve^{kt} = A\omega \cos \omega t - B\omega \sin \omega t.$$

Differentiating again

$$e^{kt}\frac{d^2v}{dt^2} + k\frac{dv}{dt}e^{kt} + k\frac{dv}{dt}e^{kt} + k^2ve^{kt}$$

$$= -A\omega^2 \sin \omega t - B\omega^2 \cos \omega t$$

$$= -\omega^2 ve^{kt} \qquad \text{(using (i))}.$$

Formation of differential equations | 267

i.e. $$\frac{d^2v}{dt^2} + 2k\frac{dv}{dt} + v(\omega^2 + k^2) = 0.$$

[B] EXAMPLES 19a

Eliminate the arbitrary constant A from the following equations:

1 $y = 2x + A$.
2 $y = Ax + 3$.
3 $y = Ax^2 - 1$.
4 $Ay = x + 1$.
5 $y = Ae^{3x}$.
6 $y = Ax + \ln x$.
7 $y = A(1 - e^{-x})$.
8 $x^2 + y^2 = A$.
9 $\frac{1}{2}y^2 = \frac{1}{x} + A$.
10 $x^2 + y^2 = Ay$.
11 $y = \sin x + \frac{A}{\sin x}$.
12 $x^2 + y^2 + 2Ay - 4 = 0$.

Eliminate the arbitrary constants A, B from the following equations:

13 $y = Ax + B$.
14 $y = Ax^3 + Bx$.
15 $y = \frac{A}{x} + Bx$.
16 $y = A \ln x + B$.
17 $y = Ae^x + Be^{-x}$.
18 $y = Ae^{3x} + Be^x$.
19 $y = A \sin x + B \cos x$.
20 $y = (Ax + B)e^{2x}$.
21 $y = e^{-x}(A \cos 2x + B \sin 2x)$.
22 $y = A \sin(2x + B)$.
23 $y = Ae^{2x} \cos(x - B)$.
24 $y = Ae^x + (x + 2)e^{2x}$.

25 Show that $y = Ax^2 + B$ is the complete primitive of the equation
$$x\frac{d^2y}{dx^2} - \frac{dy}{dx} = 0.$$

26 Find the differential equation of which $y = Ax + A^3$ is the general solution.

27 If $y^2 = C\frac{x-1}{x+1}$ where C is a constant, prove that $(x^2 - 1)\frac{dy}{dx} = y$.

28 If α is an arbitrary constant and a a fixed constant show that $x \cos \alpha + y \sin \alpha = a$ is the complete primitive of the equation
$$\left(y - x\frac{dy}{dx}\right)^2 = a^2\left(1 + \left(\frac{dy}{dx}\right)^2\right).$$

29 If C is an arbitrary constant and k, g are fixed constants, verify that the general solution of the equation $\frac{dv}{dt} + kv = g$ is $v - \frac{g}{k} = Ce^{-kt}$.

30 If $\phi = \frac{Ae^{kr} + Be^{-kr}}{r}$, prove that $\frac{d^2\phi}{dr^2} + \frac{2}{r}\frac{d\phi}{dr} - k^2\phi = 0$.

31 If $y = (\sin^{-1} x)^2 + A \sin^{-1} x + B$, prove that $(1 - x^2)\frac{d^2y}{dx^2} - x\frac{dy}{dx} = 2$.

32 Prove that $y = A \cos(\ln x) + B \sin(\ln x)$ is the general solution of the equation $x^2\frac{d^2y}{dx^2} + x\frac{dy}{dx} + y = 0$.

33 Show that the general solution of the equation $\frac{d^2x}{dt^2} + \mu x = f$, where μ, f are given constants, is $x = \frac{f}{\mu} + A \cos(\sqrt{\mu} t + \alpha)$, where A and α are arbitrary constants.

Differential equations of the first order and first degree It follows from the work we have done in the formation of differential equations, that *the complete primitive* or *general solution of an equation of the first order will contain one arbitrary constant.*

Consideration will be given to the following standard types of first order equations:

(i) Variables separable;
(ii) Homogeneous equations;
(iii) Linear equations.

Variables separable Equations in which the variables are separable are such that when $\dfrac{dy}{dx}$ is replaced by differentials dy, dx, all the terms in x can be collected on one side and all the terms in y on the other. The general solution is then obtained directly by integration.

The standard forms of such equations will be $f(x) + g(y)\dfrac{dy}{dx} = 0$, and

$$f(x)\frac{dy}{dx} + g(y) = 0.$$

Example 4 Solve the equation $\dfrac{dy}{dx} = \dfrac{y}{x^2 - 1}$.

The equation can be written $\dfrac{dy}{y} = \dfrac{dx}{x^2 - 1}$.

Integrating,
$$\int \frac{dy}{y} = \int \frac{dx}{x^2 - 1} + C$$

$$= \tfrac{1}{2} \int \left\{ \frac{1}{x-1} - \frac{1}{x+1} \right\} dx + C,$$

i.e.
$$\ln y = \tfrac{1}{2}\ln \frac{x-1}{x+1} + C,$$

$$\ln \frac{y^2(x+1)}{x-1} = 2C,$$

or
$$y^2 = A\frac{x-1}{x+1}, \text{ where } A \text{ is an arbitrary constant.}$$

Example 5 Find the complete primitive of $\dfrac{x^2 + 1}{y + 1} = xy\dfrac{dy}{dx}$.

Separating the variables, $\dfrac{x^2 + 1}{x} dx = y(y+1) dy$.

Integrating,
$$\int \left(x + \frac{1}{x} \right) dx = \int (y^2 + y) dy + C,$$

i.e.
$$\frac{x^2}{2} + \ln x = \frac{y^3}{3} + \frac{y^2}{2} + C,$$

where C is an arbitrary constant.

Example 6 Find the equations of the Cartesian curves which have a constant subtangent.

$$\text{Subtangent} = \frac{y}{\frac{dy}{dx}} = y\frac{dx}{dy}.$$

∴ The subtangent is constant in length if

$$y\frac{dx}{dy} = k,$$

where k is a given constant.

i.e.
$$\frac{dx}{k} = \frac{dy}{y}.$$

Integrating,
$$\frac{x}{k} = \ln y + C,$$

$$= \ln\frac{y}{A}, \quad C \text{ being replaced by } -\ln A$$

i.e.
$$y = Ae^{\frac{x}{k}}$$

where A is an arbitrary constant.

Example 7 For a particle describing S.H.M. in a straight line, the acceleration is towards the origin O and directly proportional to the displacement x from the origin. Find the velocity of the particle at displacement x.

$$\text{Acceleration} = \frac{dv}{dt} = \frac{dv}{dx}\frac{dx}{dt} = v\frac{dv}{dx}.$$

$$\therefore v\frac{dv}{dx} = -\omega^2 x,$$

where ω^2 is a given constant.

i.e.
$$v\, dv = -\omega^2 x\, dx.$$

Integrating,
$$\frac{v^2}{2} = -\omega^2 \frac{x^2}{2} + C.$$

If the amplitude of the S.H.M., i.e. the maximum displacement from O, is a, then $v = 0$ when $x = a$.

Hence
$$0 = -\omega^2 \frac{a^2}{2} + C.$$

$$C = \omega^2 \frac{a^2}{2}.$$

$$\therefore v^2 = \omega^2(a^2 - x^2).$$

[B] EXAMPLES 19b

Solve the following differential equations:

1 $x\,dx = y^2\,dy$. **2** $(x^2+1)\,dx = dy$. **3** $\dfrac{x}{dx} = \dfrac{y}{dy}$.

4 $\sec y\,dx = \sec x\,dy$. **5** $\dfrac{dx}{e^x} = y\,dy$. **6** $\dfrac{x\,dx}{x^2+1} = \dfrac{dy}{y}$.

7 $(x^2+1)\,dy = dx$. **8** $\dfrac{dy}{dx} = \dfrac{x+2}{y}$. **9** $x\dfrac{dy}{dx} = y^2 + 1$.

10 $\dfrac{dy}{dx} = \dfrac{x^2+x}{y^2+y}$. **11** $\dfrac{dy}{dx} = \dfrac{1+y}{1+x}$. **12** $\dfrac{dy}{dx} = \dfrac{y}{x(x+1)}$.

13 $\dfrac{x^2-1}{y-1} = xy\dfrac{dy}{dx}$. **14** $\dfrac{xy}{x+1} = \dfrac{dy}{dx}$. **15** $\cos^2 x \dfrac{dy}{dx} = \cos^2 y$.

16 $e^{x+y}\dfrac{dy}{dx} = 1$. **17** $\dfrac{dx}{dy} = x\tan y$.

18 $(\sin x + \cos x)\dfrac{dy}{dx} = \cos x - \sin x$. **19** $\tan x\,dy = \cot y\,dx$.

20 $2x\dfrac{dy}{dx} = 1 - 2y + y^2$. **21** $x\dfrac{dy}{dx} = \dfrac{y-1}{y+1} - y$.

22 Find the equation of the curve which satisfies the equation

$$\frac{dy}{dx} = \frac{1+y^2}{1+x^2},$$

and passes through the point (2, 1).

23 The gradient of a curve at any point (x, y) is given by the equation

$$\frac{dy}{dx} = \frac{y}{x^2 - 1}.$$

Find the equation of the curve if it passes through the point $(2, -1)$.

24 Obtain the solution of the equation $x^2\dfrac{dy}{dx} + y = 1$ for which $y = 1 + e$ when $x = 1$.

25 Find the general solution of the equation $(y^3+1)\dfrac{dy}{dx} - xy^2 = x$.

26 What is the complete primitive of the equation

$$\cos y \frac{dy}{dx} - x^3 = x^3 \cos^2 y?$$

27 Find the equation of the curves whose subnormals are constant and equal to a.

28 A curve is such that its gradient at any point is equal to twice the ordinate of the point. If the curve passes through the point (0, 2), find its equation.

29 Find the curves in which the portion of the tangent intercepted between the axes is bisected at the point of contact.

30 Show that the substitution $x + y = z$ reduces the equation $\dfrac{dy}{dx} = (x+y)^2$ to $\dfrac{dz}{dx} = z^2 + 1$. Hence solve the equation.

31 Use the substitution $x - y = z$ to solve the equation $\dfrac{dy}{dx} = 1 - \cos(x - y)$.

Homogeneous equations A homogeneous equation of the first order is one which can be written in the form

$$\frac{dy}{dx} = f\left(\frac{y}{x}\right).$$

To test whether a function of x and y can be written in the form $f\left(\dfrac{y}{x}\right)$, substitute $y = vx$ and the function should reduce to one of v only.

E.g. $\dfrac{dy}{dx} = \dfrac{x^2 y - y^3}{x^3 + xy^2}$ becomes $\dfrac{dy}{dx} = \dfrac{v - v^3}{1 + v^2}$

and so the equation is homogeneous.

To solve a homogeneous equation we use the substitution $y = vx$, giving $\dfrac{dy}{dx} = v + x\dfrac{dv}{dx}$; the resulting equation in v and x can be solved by separating the variables.

Example 8 Solve the equation $x\dfrac{dy}{dx} - y = \sqrt{x^2 + y^2}$, assuming $x > 0$.

The equation can be written $\dfrac{dy}{dx} = \dfrac{\sqrt{x^2 + y^2} + y}{x}$.

This equation is homogeneous, so let $y = vx$.

$$\therefore v + x\frac{dv}{dx} = \frac{\sqrt{x^2 + v^2 x^2} + vx}{x} = v + \sqrt{1 + v^2},$$

$$\therefore x\frac{dv}{dx} = \sqrt{1 + v^2}.$$

$$\frac{dv}{\sqrt{1 + v^2}} = \frac{dx}{x}.$$

Integrating, $\sinh^{-1} v = \ln x + C,$

or $\ln(v + \sqrt{1 + v^2}) = \ln x + C,$

$$\frac{v + \sqrt{1 + v^2}}{x} = e^C = A,$$

where A is an arbitrary constant.

But
$$v = \frac{y}{x},$$

$$\therefore \frac{y + \sqrt{y^2 + x^2}}{x^2} = A.$$

Example 9 Solve the equation $\dfrac{dy}{dx} = \dfrac{y(x+y)}{x(y-x)}$, x and $y > 0$.

As the equation is homogeneous, let $y = vx$.

$$\therefore v + x\frac{dv}{dx} = \frac{vx(x+vx)}{x(vx-x)} = \frac{v(1+v)}{v-1}.$$

$$x\frac{dv}{dx} = \frac{v(1+v)}{v-1} - v = \frac{2v}{v-1},$$

$$dv\left(\frac{v-1}{v}\right) = \frac{2dx}{x}.$$

Integrating, $\quad v - \ln v = 2\ln x + C,$

$$v = \ln Ax^2 v, \quad \text{where } \ln A = C.$$
$$\therefore Ax^2 v = e^v.$$

But $v = \dfrac{y}{x}$,

$$\therefore Axy = e^{\frac{y}{x}}.$$

[B] EXAMPLES 19c

Solve the following equations:

1 $\dfrac{dy}{dx} = \dfrac{x+y}{x}.$ **2** $\dfrac{dy}{dx} = \dfrac{x-2y}{x}.$ **3** $\dfrac{dy}{dx} = \dfrac{x^2+y^2}{x^2}.$

4 $2x^2\dfrac{dy}{dx} = x^2 - y^2.$ **5** $xy\dfrac{dy}{dx} = x^2 + y^2.$ **6** $xy^2\dfrac{dy}{dx} = x^3 - y^3.$

7 $x\dfrac{dy}{dx} = y + \sqrt{x^2 - y^2}.$ **8** $\dfrac{dy}{dx} = \dfrac{x+y}{x-y}.$ **9** $y\dfrac{dy}{dx} = -x + 2y.$

10 $(x^3 + y^2x)dy = (y^3 - yx^2)dx.$ **11** $2x^2\dfrac{dy}{dx} = x^2 + y^2.$ **12** $(x^2 + y^2)\dfrac{dy}{dx} = xy.$

13 If $z = x + y$, show that the equation $\dfrac{dy}{dx} = \dfrac{x+y+2}{x+y+1}$ reduces to $\dfrac{dz}{dx} = \dfrac{2z+3}{z+1}$.
Hence solve the equation.

14 Use the substitution $y - x = z$, to solve the equation $\dfrac{dy}{dx} = \dfrac{y-x+1}{y-x+5}$.

15 Solve the equation $(2x - y + 3)dy = (2x - y - 1)dx.$

Linear equations   An equation of the form

$$\frac{dy}{dx} + Py = Q,$$

where P and Q are functions of x, but not of y, is said to be *linear of the first order*.

Linear equations are solved by multiplying throughout by the function $e^{\int P dx}$, known as an integrating factor.

This gives
$$e^{\int P dx}\frac{dy}{dx} + Py e^{\int P dx} = Q e^{\int P dx}.$$

The L.H.S. is an exact differential coefficient being equal to $\dfrac{d}{dx}\left[y e^{\int P dx}\right]$.

$$\therefore \frac{d}{dx}\left[y e^{\int P dx}\right] = Q e^{\int P dx}.$$

Hence,
$$y e^{\int P dx} = \int Q e^{\int P dx} dx + C.$$

In dealing with linear equations, it is important to remember that $e^{\ln f(x)} = f(x)$.

Example 10 Solve the equation $\dfrac{dy}{dx} - y = x$.

This is of the form $\dfrac{dy}{dx} + Py = Q$, with $P = -1$, $Q = x$. Multiply throughout by $e^{\int P dx}$, i.e. e^{-x}.

$$\therefore e^{-x}\frac{dy}{dx} - y e^{-x} = x e^{-x}.$$

$$\frac{d}{dx}(y e^{-x}) = x e^{-x}.$$

Integrating,
$$y e^{-x} = \int x e^{-x} dx + C,$$
$$= -x e^{-x} - e^{-x} + C.$$

i.e.
$$y = C e^x - x - 1.$$

Example 11 If $\dfrac{dy}{dx} + 2y \tan x = \sin x$ and $y = 0$ when $x = \dfrac{\pi}{3}$, find y in terms of x.

First find the general solution of the equation which is of the form $\dfrac{dy}{dx} + Py = Q$ with $P = 2 \tan x$, $Q = \sin x$. Multiply throughout by
$$e^{\int P dx} = e^{\int 2 \tan x \, dx} = e^{2 \ln \sec x} = e^{\ln \sec^2 x} = \sec^2 x.$$

$$\therefore \sec^2 x \frac{dy}{dx} + 2y \sec^2 x \tan x = \sin x \sec^2 x,$$

i.e.
$$\frac{d}{dx}(y \sec^2 x) = \frac{\sin x}{\cos^2 x}.$$

Integrating,
$$y \sec^2 x = \int \frac{\sin x \, dx}{\cos^2 x} + A,$$
$$= \frac{1}{\cos x} + A,$$

i.e. $$y = \cos x + A\cos^2 x.$$

But $y = 0$ when $x = \dfrac{\pi}{3}$,

$$\therefore 0 = \frac{1}{2} + \frac{A}{4}; \quad A = -2.$$

Hence $$y = \cos x - 2\cos^2 x.$$

Example 12 A particle falls freely from rest under gravity in a resisting medium, the resistance varying as the velocity. Find the velocity after it has fallen for t secs.
At time t after the start, let the velocity be v. The forces acting on the particle are mg downwards and the resistance kmv upwards (the resistance per unit mass being taken as kv where k is a fixed constant).

$$\therefore \text{Acceleration of particle,} \quad \frac{dv}{dt} = g - kv.$$

i.e. $$\frac{dv}{dt} + kv = g, \quad \text{a linear equation.}$$

Multiply by $e^{\int k\,dt} = e^{kt}$.

$$\therefore ve^{kt} = \int ge^{kt}\,dt + C,$$

$$= \frac{g}{k}e^{kt} + C.$$

i.e. $$v = \frac{g}{k} + Ce^{-kt}$$

where C is an arbitrary constant.
But as the particle starts at rest, $v = 0$ when $t = 0$.

$$\therefore 0 = \frac{g}{k} + C; \quad C = -\frac{g}{k}.$$

$$\therefore v = \frac{g}{k}(1 - e^{-kt}).$$

As $e^{-kt} \to 0$ as $t \to \infty$, the velocity approaches a fixed or terminal value $\dfrac{g}{k}$, as t increases.

[B] EXAMPLES 19d

Simplify the following expressions:

1. $e^{\ln x^2}$.
2. $e^{\ln \sin x}$.
3. $e^{\ln \sec^2 x}$.
4. $e^{2\ln x}$.
5. $e^{\frac{1}{2}\ln \cos x}$.
6. $e^{-3\ln x}$.
7. $e^{x\ln 2}$.
8. $e^{\int \tan x\,dx}$.
9. $e^{\frac{1}{2}\int \cot 2x\,dx}$.
10. $e^{\int \frac{x\,dx}{x^2+1}}$.
11. $e^{\int \frac{x\,dx}{1-x^2}}$.
12. $e^{\int \frac{x^2\,dx}{1+x^3}}$.

Solve the following equations:

13. $\dfrac{dy}{dx} + y = x$.
14. $\dfrac{dy}{dx} - xy = 0$.
15. $\dfrac{dy}{dx} + \dfrac{y}{x} = \dfrac{1}{x}$.

Differential equations of the second order | 275

16 $x\dfrac{dy}{dx} - y = x^2.$ **17** $\dfrac{dy}{dx} + y = e^{-x}.$ **18** $\dfrac{dy}{dx} - 2y = x.$

19 $\dfrac{dy}{dx} + y\cot x = \cos x.$ **20** $x\dfrac{dy}{dx} - 2y = x^3.$ **21** $\dfrac{dy}{dx} - y\tan x = \cos^2 x.$

22 $(x^2 - 1)\dfrac{dy}{dx} + 2xy = x.$ **23** $(1 - x^2)\dfrac{dy}{dx} - xy = 1.$ **24** $\dfrac{dy}{dx} - \dfrac{5y}{x} = x^6.$

25 $\dfrac{dy}{dx} + y = \sin x.$ **26** $\dfrac{dy}{dx} + 3y = e^{2x}.$ **27** $(x+1)\dfrac{dy}{dx} - 3y = (x+1)^4.$

28 $(1 + \sin x)\dfrac{dy}{dx} + y\cos x = \tan x.$ **29** $\dfrac{dx}{dy} + x = y^2.$ **30** $y\dfrac{dx}{dy} - x = y^3.$

31 Show that if $z = \tan y$, the equation

$$\dfrac{dy}{dx} + x\sin 2y = x^3\cos^2 y,$$

reduces to

$$\dfrac{dz}{dx} + 2xz = x^3.$$

Hence solve the given equation.

32 By writing $z = \dfrac{1}{y}$, reduce the equation $\dfrac{dy}{dx} + \dfrac{y}{x} = y^2$ to a linear form and hence obtain its general solution.

33 Solve the equation $(2x - 10y^3)\dfrac{dy}{dx} + y = 0$ by treating y as the independent variable.

34 If $\sin y = z$, show that the equation

$$\dfrac{dy}{dx} + \dfrac{1}{x}\tan y = \dfrac{1}{x^2}\sec y$$

reduces to

$$\dfrac{dz}{dx} + \dfrac{z}{x} = \dfrac{1}{x^2}.$$

Hence solve the equation.

35 For a particle projected upwards in a resisting medium, the velocity v is given by the equation $\dfrac{dv}{dt} + kv + g = 0$, where k, g are constants.

If the initial velocity is v_0, show that the particle comes to rest after a time $\dfrac{1}{k}\ln\left(\dfrac{kv_0 + g}{g}\right).$

Differential equations of the second order We will investigate only the very simplest types of second order equation, especially those of most frequent occurrence in the applications of the calculus to geometrical, dynamical and physical problems.

The following types of equation will be considered:

(i) $\dfrac{d^2y}{dx^2} = f(x);$ (ii) $\dfrac{d^2y}{dx^2} = f(y);$

(iii) Linear equations of the form

$$a\frac{d^2y}{dx^2} + b\frac{dy}{dx} + cy = 0 \text{ or } f(x),$$

where a, b, c are constants and $f(x)$ is some simple function of x.

Equations of the form $\frac{d^2y}{dx^2} = f(x)$ Such equations are immediately solvable by integrating twice with respect to x.

Example 13 Solve the equation $B\frac{d^2y}{dx^2} - W(l-x) = 0$ subject to the conditions that $y = 0$ and $\frac{dy}{dx} = 0$ when $x = 0$; B, W and l being given constants.

$$B\frac{d^2y}{dx^2} = W(l-x).$$

Integrating, $\quad B\frac{dy}{dx} = Wlx - \frac{Wx^2}{2} + C.$

But $\frac{dy}{dx} = 0$ when $x = 0$, hence $C = 0$.

$$\therefore B\frac{dy}{dx} = Wlx - \frac{Wx^2}{2}.$$

Integrating, $\quad By = \frac{Wlx^2}{2} - \frac{Wx^3}{6} + D.$

But $y = 0$ when $x = 0$, hence $D = 0$.

$$\therefore y = \frac{Wx^2}{6B}(3l - x).$$

Equations of the form $\frac{d^2y}{dx^2} = f(y)$ This type of equation appears in dynamical problems where the acceleration of a particle is a function of its displacement from some fixed origin.

The first integration is obtained by multiplying both sides of the equation by $2\frac{dy}{dx}$, giving

$$2\frac{dy}{dx}\frac{d^2y}{dx^2} = 2f(y)\frac{dy}{dx}.$$

But $\quad 2\frac{dy}{dx}\frac{d^2y}{dx^2} = \frac{d}{dx}\left(\frac{dy}{dx}\right)^2,$

so integrating both sides with respect to x,

$$\left(\frac{dy}{dx}\right)^2 = \int 2f(y)\frac{dy}{dx}dx + C = 2\int f(y)dy + C$$

Differential equations of the second order | 277

When this integration has been performed, the second integration is accomplished by separating the variables.

Example 14 Find the general solution of the equation $\dfrac{d^2y}{dx^2} = \dfrac{1}{y^3}$.

Multiply both sides by $2\dfrac{dy}{dx}$ and integrate w.r. to x.

$$\left(\dfrac{dy}{dx}\right)^2 = \int \dfrac{2}{y^3}\,dy + A$$

$$= -\dfrac{1}{y^2} + A.$$

$$\therefore \dfrac{dy}{dx} = \pm\dfrac{\sqrt{Ay^2 - 1}}{y},$$

i.e. $$\dfrac{y\,dy}{\sqrt{Ay^2 - 1}} = \pm dx.$$

Integrating, $$\dfrac{1}{A}\sqrt{Ay^2 - 1} = \pm x + B,$$

or $$Ay^2 - 1 = A^2(B \pm x)^2$$

where A, B are arbitrary constants.

Example 15 The general equation of motion of a particle describing an orbit under a force of attraction P towards the origin is $\dfrac{d^2u}{d\theta^2} + u = \dfrac{P}{h^2u^2}$, where h is a given constant and u is the reciprocal of the distance of the particle from the origin. Solve the equation in the case where $P = \mu u^2$.

We have $$\dfrac{d^2u}{d\theta^2} + u = \dfrac{\mu}{h^2} = k, \text{ say}.$$

Multiply by $2\dfrac{du}{d\theta}$ and integrate w.r. to θ,

$$\left(\dfrac{du}{d\theta}\right)^2 = 2\int (k - u)\,du + \text{const.}$$

$$= 2ku - u^2 + \text{const.}$$

$$= \{A^2 - (u - k)^2\}$$

where A is an arbitrary constant.

$$\therefore \pm\dfrac{du}{\sqrt{A^2 - (u - k)^2}} = d\theta.$$

Integrating, $$\pm \sin^{-1}\dfrac{u - k}{A} = \theta + B$$

where B is a second arbitrary constant.

$$\therefore u - k = \pm A \sin(\theta + B),$$

$$u = \frac{\mu}{h^2} \pm A \sin(\theta + B).$$

[B] EXAMPLES 19e

Solve the following equations:

1 $\dfrac{d^2y}{dx^2} = 3x^2.$ **2** $x^2 \dfrac{d^2y}{dx^2} = 1.$ **3** $\dfrac{d^2y}{dx^2} \cos^2 x = 1.$

4 $\dfrac{d^2x}{dt^2} = -g$ (a constant).   **5** $\dfrac{d^2x}{dt^2} = f \sin nt.$ **6** $x \dfrac{d^2y}{dx^2} = 1.$

7 $\dfrac{d^2y}{dx^2} = 2\dfrac{dy}{dx}$ $\left(\text{let } \dfrac{dy}{dx} = p\right).$ **8** $x \dfrac{d^2y}{dx^2} = \dfrac{dy}{dx}.$ **9** $\dfrac{d^2v}{dr^2} + \dfrac{1}{r}\dfrac{dv}{dr} = 0.$

10 $(1+x^2)\dfrac{d^2y}{dx^2} + 2x\dfrac{dy}{dx} = 0.$ **11** $\dfrac{d^2y}{dx^2} + y = 0.$ **12** $y\dfrac{d^2y}{dx^2} = 2\left(\dfrac{dy}{dx}\right)^2$

$$\left[\dfrac{d^2y}{dx^2} = p\dfrac{dp}{dy} \text{ where } p = \dfrac{dy}{dx}\right].$$

13 Solve the equation $\dfrac{d^2y}{dx^2} = y^3 - y$, given that $\dfrac{dy}{dx} = 0$ when $y = 1$ and $y = 2$ when $x = \sqrt{2}$.

14 Solve the equation $\dfrac{d^2y}{dx^2} = \tfrac{1}{2}e^y$, given that $y = 0$ and $\dfrac{dy}{dx} = 1$ when $x = 0$.

15 Solve the equation $\dfrac{d^2x}{dt^2} - kx^2 = 0$, given that $\dfrac{dx}{dt} = \sqrt{\dfrac{2k}{3}}$ and $x = 1$ when $t = 0$.

16 Solve the equation $\dfrac{d^2x}{dt^2} + \dfrac{ga^2}{x^2} = 0$, given that $x = h$, $\dfrac{dx}{dt} = 0$ when $t = 0$.

17 Solve the equation $\dfrac{d^2u}{d\theta^2} + u = \dfrac{P}{h^2u^2}$ in the case where $P = \mu u^3$, given that $\theta = \dfrac{du}{d\theta} = 0$ when $u = \dfrac{1}{c}$; μ, h and c being constants with $\mu > h^2$.

18 Solve the equation $k\dfrac{d^4y}{dx^4} = w$, where k, w are constants. Determine the constants so that $y = 0$, $\dfrac{d^2y}{dx^2} = 0$ both for $x = 0$ and $x = l$.

19 Verify that $y = x$ is a solution of the equation

$$(x^2 - 1)\dfrac{d^2y}{dx^2} - 2x\dfrac{dy}{dx} + 2y = 0.$$

Show that the substitution $y = vx$ leads to the equation

$$x(x^2 - 1)\dfrac{dv_1}{dx} = 2v_1, \text{ where } v_1 = \dfrac{dv}{dx}.$$

Hence show that the general solution of the original equation is $y = A + Ax^2 + Bx$.

20 Verify that $y = x^3$ is a solution of the equation

$$x^2 \frac{d^2y}{dx^2} + x \frac{dy}{dx} - 9y = 0.$$

Deduce the general solution by putting $y = vx^3$.

Linear equations with constant coefficients

General form: $\quad a \dfrac{d^2y}{dx^2} + b \dfrac{dy}{dx} + cy = 0,$

where a, b, c are constants.

This equation has a solution $y = e^{mx}$ if

$$am^2 + bm + c = 0. \quad \dotfill (i)$$

There are three cases to consider according as the roots of (i) are real and different, real and equal or imaginary.

The equation (i) is usually called the *auxiliary equation*.

Case 1. Roots of auxiliary equation real and different Let the roots be $m = m_1, m_2$.

Then $y = e^{m_1 x}$, $y = e^{m_2 x}$ are solutions of the differential equation and the general solution will be

$$y = Ae^{m_1 x} + Be^{m_2 x},$$

where A, B are arbitrary constants.

Example 16 Find the general solution of the equation $2\dfrac{d^2y}{dx^2} + 5\dfrac{dy}{dx} + 2y = 0.$

Auxiliary equation is $\quad 2m^2 + 5m + 2 = 0,$

$(2m + 1)(m + 2) = 0,$

i.e. $\quad m = -\tfrac{1}{2}, -2.$

∴ General solution is $y = Ae^{-x/2} + Be^{-2x}$.

Case 2. Roots of auxiliary equation real and equal Let the roots of the auxiliary equation be $m = m_1, m_1$.

In this case it can be shown that the general solution is

$$y = e^{m_1 x}(Ax + B).$$

where A, B are arbitrary constants.

Example 17 Solve the equation $\dfrac{d^2s}{dt^2} - 4\dfrac{ds}{dt} + 4s = 0.$

Auxiliary equation is $m^2 - 4m + 4 = 0,$

$m = 2, 2.$

General solution is $\quad s = e^{2t}(At + B).$

Case 3. Roots of auxiliary equation imaginary

The roots of the auxiliary equation will be of the form, $m = p \pm iq$, where i is the complex number $\sqrt{-1}$ and p, q are real numbers.

E.g. if $m^2 + 2m + 3 = 0$,

$$m = \frac{-2 \pm \sqrt{-8}}{2},$$
$$= -1 \pm \sqrt{-2},$$
$$= -1 \pm i\sqrt{2},$$

i.e. $p = -1, q = \sqrt{2}$.

In this case, the general solution of the differential equation can be shown to be

$$y = e^{px}(A \cos qx + B \sin qx).$$

Example 18 Solve the equation $\dfrac{d^2y}{dx^2} - 2\dfrac{dy}{dx} + 4y = 0$.

Auxiliary equation is $m^2 - 2m + 4 = 0$,

$$m = \frac{2 \pm \sqrt{-12}}{2}$$
$$= 1 \pm i\sqrt{3}.$$

∴ General solution is $y = e^x(A \cos \sqrt{3}x + B \sin \sqrt{3}x)$.

Example 19 The equation of motion of a simple pendulum in a medium whose resistance varies as the velocity is given by the equation

$$\frac{d^2\theta}{dt^2} + k\frac{d\theta}{dt} + \mu\theta = 0,$$

where k and μ are constants such that $k^2 < 4\mu$. Solve the equation.

Auxiliary equation is $m^2 + km + \mu = 0$.

$$m = \frac{-k \pm \sqrt{k^2 - 4\mu}}{2}$$
$$= -\frac{k}{2} \pm i\omega \quad \text{where } 4\omega^2 = 4\mu - k^2.$$

General solution is

$$\theta = e^{-\frac{kt}{2}}(A \cos \omega t + B \sin \omega t)$$
$$= Ce^{-\frac{kt}{2}} \cos(\omega t + \varepsilon) \quad \text{where } C, \varepsilon \text{ are constants.}$$

This result shows that the motion is S.H.M. with period $\dfrac{2\pi}{\omega}$ and decreasing amplitude $Ce^{-\frac{kt}{2}}$. Such a motion is called a damped S.H.M.

[B] EXAMPLES 19f

Solve the following equations:

1. $\dfrac{d^2y}{dx^2} - \dfrac{dy}{dx} = 0.$
2. $2\dfrac{d^2y}{dx^2} + \dfrac{dy}{dx} = 0.$
3. $\dfrac{d^2y}{dx^2} - 4y = 0.$

4. $\dfrac{d^2y}{dx^2} - \dfrac{dy}{dx} - 2y = 0.$
5. $\dfrac{d^2y}{dx^2} - 2\dfrac{dy}{dx} + y = 0.$
6. $\dfrac{d^2y}{dx^2} - 5\dfrac{dy}{dx} + 6y = 0.$

7. $\dfrac{d^2y}{dx^2} + y = 0.$
8. $3\dfrac{d^2y}{dx^2} + 2\dfrac{dy}{dx} - y = 0.$
9. $4\dfrac{d^2y}{dx^2} + 4\dfrac{dy}{dx} + y = 0.$

10. $\dfrac{d^2y}{dx^2} - 3\dfrac{dy}{dx} - 10y = 0.$
11. $\dfrac{d^2y}{dx^2} - 9y = 0.$
12. $\dfrac{d^2y}{dx^2} + 9y = 0.$

13. $\dfrac{d^2y}{dx^2} + \dfrac{dy}{dx} + y = 0.$
14. $\dfrac{d^2y}{dx^2} - 6\dfrac{dy}{dx} + 9y = 0.$
15. $\dfrac{d^2y}{dx^2} + 2\dfrac{dy}{dx} + 4y = 0.$

16. $3\dfrac{d^2y}{dx^2} - 4\dfrac{dy}{dx} + y = 0.$
17. $a^2\dfrac{d^2y}{dx^2} - 2a\dfrac{dy}{dx} + y = 0.$

18. $\dfrac{d^2y}{dx^2} - 6\dfrac{dy}{dx} + 13y = 0.$
19. $\dfrac{d^2y}{dx^2} - (a+b)\dfrac{dy}{dx} + aby = 0.$

20. $\dfrac{d^2y}{dx^2} + m^2 y = 0.$

21. Solve the equation $\dfrac{d^2x}{dt^2} + k\dfrac{dx}{dt} + \dfrac{k^2}{2}x = 0$, given that $x = 0$, $\dfrac{dx}{dt} = u$, when $t = 0$.

22. Solve the equation $\dfrac{d^2x}{dt^2} + \mu x = 0$ with the initial conditions, $x = a$, $\dfrac{dx}{dt} = 0$ when $t = 0$.

23. Obtain the solution of the equation $l\dfrac{d^2\theta}{dt^2} + g\theta = 0$, given that $\theta = 0$ when $t = 0$ and $\dfrac{d\theta}{dt} = 0$ when $\theta = \alpha$.

24. Obtain the solution of the equation $\dfrac{d^2s}{dt^2} + 4\dfrac{ds}{dt} + 13s = 0$ for which $s = a$, $\dfrac{ds}{dt} = u$, when $t = 0$.

25. Solve the equation $\dfrac{d^2y}{dx^2} + 2a\dfrac{dy}{dx} + a^2 y = 0$ if $y = a$, $\dfrac{dy}{dx} = 0$ when $x = 0$.

26. Show that the result of eliminating y from the equations

$$\dfrac{dx}{dt} + y = 0; \quad \dfrac{dy}{dt} + x = 0,$$

is

$$\dfrac{d^2x}{dt^2} - x = 0.$$

Hence solve the original equations if $x = 1$, $y = 0$ when $t = 0$.

282 | Differential equations

27 If
$$\frac{dy}{dt} + x = 0; \quad \frac{dx}{dt} = x + y,$$

show that
$$\frac{d^2x}{dt^2} - \frac{dx}{dt} + x = 0.$$

Hence find x and y if $x = 0$, $y = \sqrt{3}$ when $t = 0$.

28 If $x = e^t$ and y is a function of x, prove the results
$$x\frac{dy}{dx} = \frac{dy}{dt}; \quad x^2\frac{d^2y}{dx^2} = \frac{d^2y}{dt^2} - \frac{dy}{dt}.$$

Use this substitution to reduce the homogeneous equation
$$x^2\frac{d^2y}{dx^2} - 2x\frac{dy}{dx} + 2y = 0,$$

to a linear form and hence find its general solution.

Linear equations with constant coefficients of the form

$$a\frac{d^2y}{dx^2} + b\frac{dy}{dx} + cy = f(x),$$

where a, b, and c are constants and $f(x)$ is a simple function of x.

Suppose that
$$y = u(x)$$

is the general solution of this equation when $f(x)$ is replaced by 0 and that
$$y = v(x)$$

is any particular solution of the given equation obtained, for example, by inspection.

Then it can readily be proved by substitution that
$$y = u(x) + v(x)$$

is also a solution of the given equation.

Moreover, as $y = u(x)$, is the general solution of a second order differential equation, the function $u(x)$ will contain two arbitrary constants.

Consequently the solution
$$y = u(x) + v(x)$$

of the given equation will contain two arbitrary constants and hence it will be the general solution.

It follows that the general solution of the equation
$$a\frac{d^2y}{dx^2} + b\frac{dy}{dx} + cy = f(x)$$

is made up of the sum of two parts, one being the general solution of the allied equation
$$a\frac{d^2y}{dx^2} + b\frac{dy}{dx} + cy = 0,$$

Determination of particular integrals | 283

usually called the *Complementary Function*, and the other being any particular solution of the given equation, usually called the *Particular Integral*.

The method used for the determination of the Complementary Function has already been discussed and the method of determining Particular Integrals in a few simple special cases will now be illustrated.

Determination of particular integrals.

Case (i). f(x) = a polynomial of degree n

A particular integral can be found by substituting

$$y = a_0 x^n + a_1 x^{n-1} + a_2 x^{n-2} + \ldots + a_n;$$

the constants $a_0, a_1, a_2, \ldots a_n$ being determined by equating coefficients.

Example 20 Find a particular integral of the equation $\dfrac{d^2 y}{dx^2} + 4y = x^2 + 2$ and hence write down the general solution.

Let
$$y = a_0 x^2 + a_1 x + a_2.$$

Then
$$\frac{dy}{dx} = 2a_0 x + a_1; \quad \frac{d^2 y}{dx^2} = 2a_0.$$

Substituting in the given equation,

$$2a_0 + 4(a_0 x^2 + a_1 x + a_2) = x^2 + 2.$$

Equating coefficients, $\quad 4a_0 = 1; \quad a_0 = \tfrac{1}{4}.$

$$4a_1 = 0; \quad a_1 = 0.$$

$$2a_0 + 4a_2 = 2; \quad a_2 = \tfrac{3}{8}.$$

∴ A particular integral is $y = \tfrac{1}{4} x^2 + \tfrac{3}{8}$.

The Complementary Function is the general solution of the equation

$$\frac{d^2 y}{dx^2} + 4y = 0,$$

and is
$$y = A \cos 2x + B \sin 2x.$$

Hence the general solution of the given equation is

$$y = A \cos 2x + B \sin 2x + \tfrac{1}{4} x^2 + \tfrac{3}{8},$$

where A, B are arbitrary constants.

Case (ii). f(x) = ke^{mx} where k and m are constants

In general, a particular integral can be found by substituting

$$y = p e^{mx};$$

the constant p being determined by equating coefficients. If the function e^{mx}

occurs in the Complementary Function, p will be indeterminate and it will be necessary to try

$$y = pxe^{mx} \text{ or possibly } y = px^2 e^{mx}.$$

Example 21 Find the general solutions of the equations

$$\text{(i) } \frac{d^2x}{dt^2} - \frac{dx}{dt} - 2x = e^{3t}; \quad \text{(ii) } \frac{d^2x}{dt^2} - \frac{dx}{dt} - 2x = e^{2t}.$$

In both cases, the Complementary Function is the general solution of the equation

$$\frac{d^2x}{dt^2} - \frac{dx}{dt} - 2x = 0,$$

i.e. $$x = Ae^{2t} + Be^{-t}.$$

To find a particular integral of equation (i),

let $$x = pe^{3t}.$$

This is a solution of the equation if

$$9pe^{3t} - 3pe^{3t} - 2pe^{3t} = e^{3t}$$

i.e. $$4p = 1; \quad p = \tfrac{1}{4}.$$

Hence, a particular integral is $x = \tfrac{1}{4}e^{3t}$, and the general solution is

$$x = Ae^{2t} + Be^{-t} + \tfrac{1}{4}e^{3t}.$$

To find a particular integral of equation (ii), where the function e^{2t} is already included in the Complementary Function,
let $$x = pte^{2t}.$$

$$\frac{dx}{dt} = pe^{2t} + 2pte^{2t}; \quad \frac{d^2x}{dt^2} = 4pe^{2t} + 4pte^{2t}.$$

Substituting in the equation,

$$4pe^{2t} + 4pte^{2t} - pe^{2t} - 2pte^{2t} - 2pte^{2t} = e^{2t}.$$

i.e. $3p = 1; \quad p = \tfrac{1}{3}.$
∴ A particular integral is $x = \tfrac{1}{3}te^{2t}$,
and the general solution is

$$x = Ae^{2t} + Be^{-t} + \tfrac{1}{3}te^{2t}.$$

Case (iii). $f(x) = k \cos rx$ or $k \sin rx$ where k and r are constants

In general, a particular integral can be found by substituting

$$y = p \cos rx + q \sin rx;$$

the constants p and q being determined by equating coefficients.

If the functions $\cos rx$ and $\sin rx$ occur in the Complementary Function it will be necessary to try

$$y = x(p \cos rx + q \sin rx).$$

Example 22 Solve the equation $\dfrac{d^2y}{dx^2} + y = 12 \sin 2x$.

The Complementary Function is $y = A \cos x + B \sin x$.
Assume a particular integral of the form
$$y = p \cos 2x + q \sin 2x.$$

$\dfrac{dy}{dx} = -2p \sin 2x + 2q \cos 2x; \quad \dfrac{d^2y}{dx^2} = -4p \cos 2x - 4q \sin 2x.$

$\therefore \; -4p \cos 2x - 4q \sin 2x + p \cos 2x + q \sin 2x = 12 \sin 2x.$
Equating coefficients of cos 2x and sin 2x,
$$-3p = 0; \quad p = 0.$$
$$-3q = 12; \quad q = -4.$$

$\therefore$ A particular integral is $y = -4 \sin 2x$,
and the general solution is
$$y = A \cos x + B \sin x - 4 \sin 2x.$$

Case (iv). f(x) = the sum or difference of functions of the types already discussed

A particular integral will be the sum or difference of the particular integrals of the separate functions.

Example 23 Obtain the solution of the equation $\dfrac{d^2y}{dx^2} - 2\dfrac{dy}{dx} + y = x^2 - 3x + 2 \sin x$,
for which $y = 0, \dfrac{dy}{dx} = 1$ when $x = 0$.

The Complementary Function is $y = e^x(Ax + B)$.
For a particular integral assume
$$y = a_0 x^2 + a_1 x + a_2 + p \cos x + q \sin x.$$

$\dfrac{dy}{dx} = 2a_0 x + a_1 - p \sin x + q \cos x; \quad \dfrac{d^2y}{dx^2} = 2a_0 - p \cos x - q \sin x.$

Substituting in the equation,
$2a_0 - p \cos x - q \sin x - 2(2a_0 x + a_1 - p \sin x + q \cos x)$
$\qquad\qquad\qquad + a_0 x^2 + a_1 x + a_2 + p \cos x + q \sin x$
$\qquad\qquad = x^2 - 3x + 2 \sin x$

$\therefore \qquad\qquad 2a_0 - 2a_1 + a_2 = 0,$
$\qquad\qquad\qquad -4a_0 + a_1 = -3,$
$\qquad\qquad\qquad\qquad a_0 = 1,$
$\qquad\qquad\qquad -p - 2q + p = 0,$
$\qquad\qquad\qquad -q + 2p + q = 2.$

i.e. $a_0 = 1, a_1 = 1, a_2 = 0, q = 0, p = 1.$
Hence a particular integral is $\qquad y = x^2 + x + \cos x,$
and the general solution is
$$y = e^x(Ax + B) + x^2 + x + \cos x.$$

Differential equations

But when $x = 0$, $y = 0$ and $\dfrac{dy}{dx} = 1$,

$\therefore \qquad 0 = B + 1,$

and $\qquad\qquad 1 = A + B + 1.$

i.e. $A = 1$, $B = -1$.

$\therefore$ The required solution is

$$y = e^x(x-1) + x^2 + x + \cos x.$$

[B] EXAMPLES 19g

Verify that the given functions are particular integrals of the following equations and find the general solutions:

1 e^x; $\dfrac{d^2y}{dx^2} - 3\dfrac{dy}{dx} + 3y = e^x$.

2 $2x+1$; $\dfrac{d^2y}{dx^2} + \dfrac{dy}{dx} - 2y = -4x$.

3 $2\sin 3x$; $\dfrac{d^2y}{dx^2} + 4y = -10\sin 3x$.

4 3; $\dfrac{d^2y}{dx^2} + 4\dfrac{dy}{dx} + 4y = 12$.

5 $2e^{2x}$; $\dfrac{d^2y}{dx^2} + 6\dfrac{dy}{dx} + 9y = 50e^{2x}$.

6 $-\tfrac{1}{4}x\cos 2x$; $\dfrac{d^2y}{dx^2} + 4y = \sin 2x$.

Find the values of the constants for which the given functions are particular integrals of the following equations:

7 $ax+b$; $\dfrac{d^2y}{dx^2} + y = 2x$.

8 ae^{bx}; $\dfrac{d^2y}{dx^2} + 2\dfrac{dy}{dx} - 8y = 2e^{-2x}$.

9 a; $\dfrac{d^2y}{dx^2} - 5\dfrac{dy}{dx} - 6y = 12$.

10 $a\sin px + b\cos px$; $\dfrac{d^2y}{dx^2} + 9y = 2\cos 2x$.

11 axe^{bx}; $\dfrac{d^2y}{dx^2} + 3\dfrac{dy}{dx} - 4y = e^{-4x}$.

12 $ax^2 + bx + c$; $\dfrac{d^2y}{dx^2} - 4y = x^2 - 3$.

13 $x(a\sin px + b\cos px)$; $\dfrac{d^2y}{dx^2} + 16y = 2\cos 4x$.

14 $ae^{bx} + c$; $\dfrac{d^2y}{dx^2} + \dfrac{dy}{dx} + y = 2e^x + 3$.

Obtain particular integrals of the following equations and write down the general solutions:

15 $\dfrac{d^2y}{dx^2} - 9y = 2e^{2x}$.

16 $4\dfrac{d^2y}{dx^2} - 3\dfrac{dy}{dx} - y = 2$.

17 $\dfrac{d^2y}{dx^2} + y = 3\cos 2x$.

18 $4\dfrac{d^2y}{dx^2} - y = x^2 - 3x$.

19 $\dfrac{d^2y}{dx^2} - 6\dfrac{dy}{dx} - 16y = e^{-x}$.

20 $\dfrac{d^2y}{dx^2} + 2\dfrac{dy}{dx} + y = \sin 3x$.

21 $2\dfrac{d^2y}{dx^2} + \dfrac{dy}{dx} - 3y = \tfrac{1}{2}x^3$.

22 $\dfrac{d^2y}{dx^2} - \dfrac{dy}{dx} + y = \cos x - \sin x$.

23 $\dfrac{d^2y}{dx^2} - 4y = e^{-2x}$.

24 $\dfrac{d^2y}{dx^2} + 2\dfrac{dy}{dx} + 5y = 15$.

Solve the following equations:

25 $\dfrac{d^2y}{dx^2} - 2\dfrac{dy}{dx} + 5y = 80e^{3x}$.

26 $\dfrac{d^2y}{dx^2} + 8\dfrac{dy}{dx} + 25y = 50$.

27 $\dfrac{d^2y}{dx^2} + 9y = 3\cos 2x - \sin 2x$.

28 $\dfrac{d^2y}{dx^2} - 2\dfrac{dy}{dx} + 2y = 1 - x^2$.

29 $4\dfrac{d^2y}{dx^2} + y = x + e^x$.

30 $2\dfrac{d^2y}{dx^2} - \dfrac{dy}{dx} = 4x + 1$.

31 $\dfrac{d^2y}{dx^2} + 16y = 3\sin 4x$.

32 $\dfrac{d^2y}{dx^2} + y = \cosh x$.

33 Solve the equation $\dfrac{d^2\theta}{dt^2} + 4\theta = \sin t$ if $\theta = \dfrac{\pi}{6}$, $\dfrac{d\theta}{dt} = 0$ when $t = 0$.

34 If $\dfrac{d^2y}{dx^2} - 3\dfrac{dy}{dx} + 2y = 4 + x$ and $y = 4$, $\dfrac{dy}{dx} = 2$ when $x = 0$, find y in terms of x.

35 Obtain the solution of the equation $\dfrac{d^2y}{dx^2} - 9\dfrac{dy}{dx} = 2\sin 3x$ which is such that $y = \dfrac{dy}{dx} = 0$ when $x = 0$.

[B] MISCELLANEOUS EXAMPLES

1 If $y = e^{ax}\sin bx$, prove that $\dfrac{d^2y}{dx^2} - 2a\dfrac{dy}{dx} + (a^2 + b^2)y = 0$. If $a = 1$, $b = 2$, find the values of x for which y has stationary values.

2 Eliminate A from the equation $y = \cos x + A\sec x$.

3 Find the differential equation whose general solution is

$$y = ax\cos\left(\dfrac{n}{x} + b\right),$$

where a, b are arbitrary constants and n a fixed constant.

Solve the following differential equations:

4 $x\dfrac{dy}{dx} - 2y = x^4$.

5 $\sin x \dfrac{dy}{dx} + \cos y = 0$.

6 $(2x - y)dy = (2y - x)dx$.

7 $\dfrac{dy}{dx} = e^{x-y}$.

8 $\dfrac{d^2y}{dx^2} + 4y = 0$.

9 $x^3\dfrac{dy}{dx} = y^3 + y^2\sqrt{y^2 - x^2}$.

10 $x\dfrac{dy}{dx} - y - y^2 = 0$.

11 $4\dfrac{d^2y}{dx^2} - 3\dfrac{dy}{dx} + y = 0$.

12 $\dfrac{dy}{dx} + 2y\tan x = \cos^3 x$.

13 $\dfrac{d^2y}{dx^2} + 4\dfrac{dy}{dx} + 4y = 0$.

14 $(1+\sin^2 x)\dfrac{dy}{dx} - \cos x = y\cos x.$ **15** $\sec x\dfrac{dy}{dx} + y - \sin x = 0.$

16 $(1+y^2)dx - xy(1+x^2)dy = 0.$ **17** $\dfrac{d^2y}{dx^2} - \dfrac{dy}{dx} = 12y.$

18 $(1+x^2)\dfrac{dy}{dx} + xy = 0.$ **19** $\dfrac{dy}{dx} = (x+y)^2.$

20 $\dfrac{dy}{dx} = y\tan x - 2\sin x.$ **21** $\dfrac{dy}{dx} = \cot x \cot y.$

22 $\dfrac{d^2y}{dx^2} + 3\dfrac{dy}{dx} + 5y = 0.$ **23** $\dfrac{dy}{dx} + 2xy = x^3.$

24 $\dfrac{d^2y}{dx^2} = 2y.$ **25** $x(x-1)\dfrac{dy}{dx} - (x-2)y = x^3(2x-1).$

26 Find the solution of the equation

$$x\dfrac{dy}{dx} + y = y^2 \ln x,$$

for which $y = 1$ when $x = 1$. $\left(\text{Let } z = \dfrac{1}{y}\right).$

27 Solve the equation $\dfrac{d^2y}{dx^2} - W(l-x) = 0$, where W and l are constants, given that y and $\dfrac{dy}{dx}$ are zero when $x = 0$.

28 Solve $\dfrac{dy}{dx} + y = x$, given that $y = 0$ when $x = 0$ and show that y is always positive.

29 Solve $2\dfrac{d^2y}{dx^2} + 5\dfrac{dy}{dx} + 2y = 0$, where $y = 1$, $\dfrac{dy}{dx} = 0$ when $x = 0$.

30 If $(1+x^2)\dfrac{dy}{dx} = 1+y^2$, prove that $1+xy = C(x-y).$

31 Find the differential equation whose complete primitive is

$$y = A\cos\dfrac{2}{x} + B\sin\dfrac{2}{x}.$$

32 Find the equation of the family of curves in which the subnormal bears a constant ratio k to the abscissa.

33 Solve $\dfrac{dx}{dt} + 2tx = t$ given $x = 2$ when $t = 0$.

34 Integrate $x\dfrac{dy}{dx} + y = x^3y^2$ by using the substitution $z = \dfrac{1}{y}$. Find the particular solution for which $y = 1$ when $x = 1$.

35 Solve $\dfrac{d^2y}{dx^2} + \left(\dfrac{dy}{dx}\right)^2 + \dfrac{dy}{dx} = 0$ by writing $p = \dfrac{dy}{dx}.$

36 Solve $L\dfrac{d^2Q}{dt^2} + R\dfrac{dQ}{dt} + \dfrac{Q}{C} = 0$, given $Q = Q_0$, $\dfrac{dQ}{dt} = 0$ when $t = 0$ and $CR^2 < 4L$.

37 Use the substitution $z = e^{\frac{1}{v}}$, to solve the equation
$$\frac{dz}{dx} + \frac{z}{x}\ln z = \frac{z}{x^2}(\ln z)^2.$$

38 Find the equation of the family of curves in which N, the foot of the ordinate from P, any point on one of them, bisects OG where O is the origin and G the point of intersection of the normal at P with the x axis.

39 Solve the equation $x(x-1)\dfrac{d^2z}{dx^2} - (x-2)\dfrac{dz}{dx} = x^3(2x-1)$, by writing $\dfrac{dz}{dx} = y$. If $z = 0$ and $\dfrac{dz}{dx} = 0$ when $x = 2$, show that when $x = 4$, $z = 44 - \ln 9$.

40 By differentiating with respect to t, show that the result of eliminating x from

$$3\frac{dx}{dt} + \frac{dy}{dt} + 2x = k; \quad \frac{dx}{dt} + 4\frac{dy}{dt} + 3y = 0,$$

is

$$11\frac{d^2y}{dt^2} + 17\frac{dy}{dt} + 6y = 0.$$

Hence solve the original equations given $x = 0$, $y = 0$ when $t = 0$.

41 If $p = \dfrac{dy}{dx}$, show that $\dfrac{d^2y}{dx^2} = p\dfrac{dp}{dy}$. Hence solve the equation

$$\frac{d^2y}{dx^2} + 2y\frac{dy}{dx} = 0,$$

assuming that y is not constant.

20 | Complex numbers

Definition *Real numbers* consist of positive and negative integers, rational numbers which are of the form p/q, where p, q are integers, and irrational numbers such as π, $\sqrt{2}$, $\sqrt[3]{4}$, ...

Complex numbers are defined as numbers of the form

$$a + b\sqrt{(-1)} \quad \text{or} \quad a + ib,$$

where $\sqrt{(-1)}$ is represented by the symbol i and a, b are real numbers. A complex number consists of two parts, a, referred to as *the real part* and ib, referred to as *the imaginary part*.

N.B. (i) If $b = 0$, the complex number $a + ib$ is wholly real, and it follows conversely that real numbers can be thought of as special cases of complex numbers.
(ii) If $a = 0$, the complex number $a + ib$ is wholly imaginary.
(iii) The complex number $a + ib$ is zero, if and only if, $a = b = 0$.

Conjugate complex numbers The complex numbers $a + ib$ and $a - ib$ are called *conjugate numbers*.

If $a + ib = z$, the conjugate number $a - ib$ is written as z^*

Example 1 Solve the quadratic equation $z^2 + z + 1 = 0$.

Using the formula, $z = \dfrac{-1 \pm \sqrt{(1-4)}}{2} = \dfrac{-1 \pm \sqrt{(-3)}}{2}$,

$= \tfrac{1}{2}(-1 \pm i\sqrt{3})$.

Example 2 Factorise $(x + y)^2 + z^2$.

As $i^2 = -1$, the expression can be written as a difference of two squares $(x + y)^2 - (iz)^2$.

Hence the factors are $(x + y + iz)(x + y - iz)$.

Geometrical representation of complex numbers The complex number $z(= a + ib)$ is represented geometrically by the point P with rectangular Cartesian coordinates (a, b), Fig. 99. The point P corresponds uniquely to the number z and is often referred to as the point z.

A diagram in which complex numbers are represented in this way is called *an Argand diagram*.

In the Argand diagram, Fig. 99, if P represents the complex number $z(= a + ib)$, P^* will represent the conjugate of z, and P', the complex number $-z$, defined as $-a - ib$.

Geometrical representation of complex numbers | 291

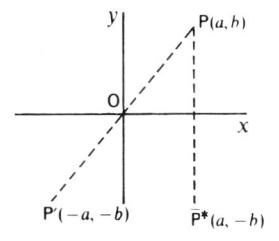

Fig. 99

Example 3 Represent the complex numbers $-2-i$, $-7+4i$, $-1+i$ on the Argand diagram by the points P, Q, R. Show that triangle PQR is right-angled.

Referring to the Argand diagram, Fig. 100, it is readily seen that the gradients of QR, PR are $-\frac{1}{2}$, 2 respectively. Hence the lines QR, PR are perpendicular and triangle PQR is right-angled at R.

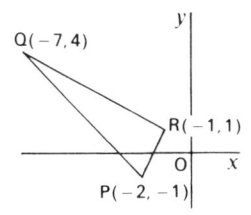

Fig. 100

[A] EXAMPLES 20a

1 Express each of the following complex numbers in the form $a+ib$;
(i) $2+\sqrt{(-1)}$; (ii) $3-\sqrt{(-2)}$; (iii) $1+\sqrt{(-9)}$; (iv) $c+\sqrt{(-k^2)}$;
(v) $a+\sqrt{(a^2-3a^2)}$.

2 Simplify (i) i^2; (ii) i^4; (iii) i^3; (iv) $1/i^2$; (v) $1/2i$.

3 Write down the conjugates of (i) $2+3i$; (ii) $3-4i$; (iii) $2i$; (iv) $1-i\sqrt{3}$; (v) $1+i$; (vi) $\cos\theta + i\sin\theta$.

4 Represent the following complex numbers on the Argand diagram:
(i) $2+i$; (ii) $-1+2i$; (iii) $3i$; (iv) $2-i\sqrt{3}$; (v) 4; (vi) $\frac{1}{2}(3-i)$; (vii) $-5i$.

5 Solve the following equations: (i) $z^2+1=0$; (ii) $z^2+9=0$; (iii) $z^2-z+1=0$;
(iv) $2z^2-3z+2=0$; (v) $z^2+8z+25=0$; (vi) $(z-1)^2+4=0$;
(vii) $(z^2+1)(z^2+2z+2)=0$.

6 Mark the point P on the Argand diagram which represents the complex number $z=3+4i$. Also mark the point P* which represents z^* and the point P' which represents $-z$.

7 Find the values of x and y in each of the following equations:
(i) $x+iy=0$; (ii) $(x-2)+i(y+1)=0$; (iii) $(x+y)+i(x-2)=0$;
(iv) $(x+2y-3)+i(4x-y-3)=0$.

8 If P, Q represent the complex numbers $2+i$, $4-3i$ in the Argand diagram, what complex number is represented by the mid-point of PQ?

9 Show that the four points representing the complex numbers $\pm 1 \pm i$ in the Argand diagram are the vertices of a square. Write down the complex number represented by the centre of the square.

10 Show that the points representing the complex numbers $1+2i$, $3+4i$, $-1+6i$ in the Argand diagram are the vertices of an isosceles triangle.

11 Show that the points representing the complex numbers
$$1, \quad \tfrac{1}{2}(-1\pm i\sqrt{3})$$
are the vertices of an equilateral triangle.

Fundamental processes

Equality   The complex numbers $x_1 + iy_1$, $x_2 + iy_2$ are said to be equal if, and only if, $x_1 = x_2$ and $y_1 = y_2$.

This definition leads to the rule:

When two complex numbers are equal, the real and imaginary parts can be equated.

E.g. if $\quad s + ic = e^x(\cos x + i\sin x) + e^{-x}(\sin x + i\cos x),$

then $\qquad s = e^x \cos x + e^{-x} \sin x \quad \text{and} \quad c = e^x \sin x + e^{-x} \cos x.$

Example 4   Find the values of x and y if $(x+2y) + i(x-y) = 1 + 4i$.
Equating real and imaginary parts, $x + 2y = 1$; $x - y = 4$.
Hence $\qquad\qquad\qquad\qquad x = 3, y = -1.$

Addition   *The sum $z_1 + z_2$ of the complex numbers $z_1 (= x_1 + iy_1)$, $z_2 (= x_2 + iy_2)$ is defined as the complex number*

$$(x_1 + x_2) + i(y_1 + y_2).$$

Geometrically, if P_1, P_2, P_3 represent the numbers $z_1, z_2, z_1 + z_2$, in the Argand diagram, then P_3 is the fourth vertex of the parallelogram determined by the points O, P_1, P_2. This result follows from the fact that the coordinates of P_1, P_2, P_3 are respectively $(x_1, y_1), (x_2, y_2), (x_1 + x_2, y_1 + y_2)$.

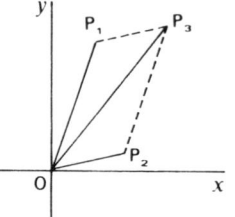

Fig. 101

Subtraction   *The difference $z_1 - z_2$ between the complex numbers $z_1 (= x_1 + iy_1)$, $z_2 (= x_2 + iy_2)$ is defined as the complex number*

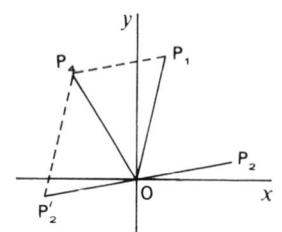

$$(x_1 - x_2) + i(y_1 - y_2).$$

Writing $z_1 - z_2$ as $z_1 + (-z_2)$ it follows that, in the Argand diagram the point P_4 representing this difference is the fourth vertex of the parallelogram determined by the points O, P_1, P'_2, where P'_2 represents the number $-z_2$.

Fig. 102

Example 5   If P represents a complex number z, find the points P_1, P_2, P_3 which represent (i) $2z$; (ii) $z + 2$; (iii) $z - 2i$.

(i) P_1 is the point on OP produced such that $OP = PP_1$.

(ii) P_2 is the fourth vertex of the parallelogram formed by the points O, P and the point representing the number 2.

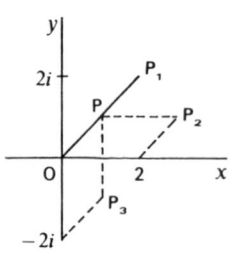

Fig. 103

(iii) P_3 is the fourth vertex of the parallelogram formed by the points O, P and the point representing the number $-2i$.

[A] EXAMPLES 20b

1 Write down the values of x and y in each of the following cases:
(i) $x + iy = 2 + 3i$; (ii) $x + iy = i\sqrt{3}$; (iii) $x + iy = 1 - i$;
(iv) $(x - 1) + i(y + 2) = 0$; (v) $(x + 1) + i(y - 1) = 3 + 2i$.

2 Write down the values of $z_1 + z_2$ and $z_1 - z_2$ in each of the following cases:
(i) $z_1 = 3 - 2i$, $z_2 = 1 + i$; (ii) $z_1 = 1 - i\sqrt{3}$, $z_2 = 2 + i\sqrt{3}$;
(iii) $z_1 = 3i$, $z_2 = 4 - i$; (iv) $z_1 = 2(1 - i)$; $z_2 = 3(-2 + i)$.

3 If P_1, P_2 represent the complex numbers $z_1 = 1 + 3i$, $z_2 = 3 + i$ in the Argand diagram, find the positions of the points which represent (i) $z_1 + z_2$; (ii) $z_1 - z_2$; (iii) $z_2 - z_1$; (iv) $2z_1$; (v) $2z_1 + z_2$; (vi) $z_2 - z_1{}^*$.

4 Find x and y in each of the following cases:
(i) $(x + y) + i(x - y) = 4 + 2i$; (ii) $(x - 2y) + i(x + y) = 3$;
(iii) $(x + iy) + 2(y - ix) = 3 + 4i$; (iv) $2(x + iy) = (y - ix) - 2(1 - i)$.

5 If z is any complex number, show that $z + z^*$ is wholly real and $z - z^*$ is wholly imaginary.

6 If u, v, x, y are real and $u + iv = \dfrac{x^2 + iy^2}{x^2 + y^2}$, write down the values of u and v in terms of x and y and deduce that $u + v = 1$.

7 If $z = x + iy$, solve the equations: (i) $z + 2z^* = 3 + i$; (ii) $3z - 2z^* = 2 + 5i$.

8 The points P_1, P_2 represent complex numbers z_1, z_2 in the Argand diagram, show how to construct the points which represent (i) $2z_1 + z_2$; (ii) $z_1 - 2z_2$; (iii) $z_1 - 2$; (iv) $z_1 + z_2 + 2i$.

9 The equation $z + a + i(z - 2) = 0$, where a is real, has a real solution for z. What is this solution and the value of a?

10 If $a^2 + 2iab + b^2 = 10 + 6i$, find the values of a and b.

11 The solutions of the equation $z^2 + z + 1 = 0$ are $z = z_1, z_2$. Find the values of (i) $z_1 + z_2$; (ii) $z_1 z_2$; (iii) $z^*{}_1 + z^*{}_2$.

Multiplication *The product of the complex numbers $x_1 + iy_1$, $x_2 + iy_2$ is defined by the relationship*

$$(x_1 + iy_1)(x_2 + iy_2) = (x_1 x_2 - y_1 y_2) + i(x_1 y_2 + x_2 y_1).$$

This definition makes it possible to use the ordinary rules of real algebra with the symbols x_1, y_1, x_2, y_2, i; for assuming these rules apply we have

$$(x_1 + iy_1)(x_2 + iy_2) = x_1 x_2 + ix_1 y_2 + iy_1 x_2 + i^2 y_1 y_2,$$
$$= (x_1 x_2 - y_1 y_2) + i(x_1 y_2 + x_2 y_1), \quad \text{as} \quad i^2 = -1.$$

Example 6 If $z = 2 + i$, find the values of (i) z^2; (ii) zz^*.

(i) $\qquad z^2 = z \cdot z = (2 + i)(2 + i) = 4 + 2i + 2i + i^2$
$\qquad\qquad\qquad = 3 + 4i, \quad \text{as} \quad i^2 = -1.$

(ii) $\qquad zz^* = (2 + i)(2 - i) = 2^2 - i^2 = 5.$

Complex numbers

Division *The quotient of the complex numbers $x_1 + iy_1$, $x_2 + iy_2$ is defined by the relationship,*

$$\frac{x_1 + iy_1}{x_2 + iy_2} = \frac{x_1 x_2 + y_1 y_2 + i(y_1 x_2 - x_1 y_2)}{x_2^2 + y_2^2}.$$

Again this definition makes it possible to apply the rules of real algebra, for

$$\frac{x_1 + iy_1}{x_2 + iy_2} = \frac{(x_1 + iy_1)(x_2 - iy_2)}{(x_2 + iy_2)(x_2 - iy_2)} = \frac{x_1 x_2 + y_1 y_2 + i(y_1 x_2 - x_1 y_2)}{x_2^2 + y_2^2}.$$

N.B. Note the use of the difference of two squares in making the denominator real.

Example 7 Express $(2+i)/(1+i)$ in the form $a + ib$.

We have $\quad \dfrac{2+i}{1+i} = \dfrac{(2+i)(1-i)}{(1+i)(1-i)} = \dfrac{2 - 2i + i - i^2}{1 - i^2}$

$$= \frac{3-i}{2} \quad \text{or} \quad \tfrac{1}{2}(3-i).$$

Example 8 If $(x + iy)(2 + i) = 3 - i$, find the values of x and y.

Expanding, $\quad (x + iy)(2 + i) = 2x - y + i(x + 2y)$.

So, equating real and imaginary parts,

$$2x - y = 3; \quad x + 2y = -1.$$

Hence $\quad x = 1, \quad y = -1$.

Example 9 If $z = 1 + 2i$ is a solution of the equation $z^3 + az + b = 0$ where a, b are real, find the values of a and b and verify that $z = 1 - 2i$ is also a solution of the equation.

Substituting $z = 1 + 2i$, we have

$$(1 + 2i)^3 + a(1 + 2i) + b = 0$$

But $\quad (1+2i)^3 = 1 + 3(2i) + 3(2i)^2 + (2i)^3 = 1 + 6i + 12i^2 + 8i^3,$

$$= -11 - 2i, \quad \text{as} \quad i^3 = -i.$$

$$\therefore \; -11 - 2i + a + 2ai + b = 0,$$

or $\quad (-11 + a + b) + i(2a - 2) = 0.$

Hence $\quad -11 + a + b = 0 \quad \text{and} \quad 2a - 2 = 0.$

$$\therefore \; a = 1, \quad b = 10.$$

Substituting $z = 1 - 2i$ and using the values of a and b,

$$z^3 + az + b = (1 - 2i)^3 + (1 - 2i) + 10,$$
$$= 1 + 3(-2i) + 3(-2i)^2 + (-2i)^3 + 11 - 2i,$$
$$= 1 - 6i - 12 + 8i + 11 - 2i = 0.$$

$\therefore z = 1 - 2i$ is also a solution of the equation.

The cube roots of unity If z is a cube root of 1,

$$z^3 = 1 \quad \text{or} \quad z^3 - 1 = 0.$$

So the cube roots of unity are the solutions of the equation $z^3 - 1 = 0$.
Factorising, $(z-1)(z^2 + z + 1) = 0$.

$$\therefore z = 1; \quad z^2 + z + 1 = 0.$$

The roots of the quadratic equation $z^2 + z + 1 = 0$ are

$$z = \frac{-1 \pm \sqrt{(1-4)}}{2} = \tfrac{1}{2}(-1 \pm i\sqrt{3}).$$

$\therefore$ The cube roots of unity are $1, \tfrac{1}{2}(-1 + i\sqrt{3}), \tfrac{1}{2}(-1 - i\sqrt{3})$.

It is easily verified that the complex roots have the property that one is the square of the other, for

$$\{\tfrac{1}{2}(-1 + i\sqrt{3})\}^2 = \tfrac{1}{4}(-2 - 2i\sqrt{3}) = \tfrac{1}{2}(-1 - i\sqrt{3});$$
and
$$\{\tfrac{1}{2}(-1 - i\sqrt{3})\}^2 = \tfrac{1}{4}(-2 + 2i\sqrt{3}) = \tfrac{1}{2}(-1 + i\sqrt{3}).$$

So the cube roots of unity can be expressed as

$$1, \;\omega, \;\omega^2$$

where $\omega = \tfrac{1}{2}(-1 \pm i\sqrt{3})$.

N.B. (i) As $z = \omega$ is a solution of $z^3 - 1 = 0$, $\omega^3 = 1$;
(ii) as $z = \omega$ is a solution of $z^2 + z + 1 = 0$, $\omega^2 + \omega + 1 = 0$.

Example 10 Solve the equation $(z-1)^3 = 1$.
Taking the cube root of both sides,

$$z - 1 = 1, \tfrac{1}{2}(-1 + i\sqrt{3}), \tfrac{1}{2}(-1 - i\sqrt{3}).$$
$$\therefore z = 2, \tfrac{1}{2}(1 + i\sqrt{3}), \tfrac{1}{2}(1 - i\sqrt{3}).$$

The equation can also be solved by expanding and factorising the resulting cubic expression.

Example 11 If ω is a cube root of unity, show that $\omega^4 + \omega^2 = -1$.
This result can be obtained by taking $\omega = \tfrac{1}{2}(-1 \pm i\sqrt{3})$ and working out the values of ω^2 and ω^4, but the following solution is much neater.

We have $\quad 1 + \omega + \omega^2 = 0, \quad$ so $\quad \omega^2(1 + \omega + \omega^2) = 0,$
i.e. $\quad \omega^2 + \omega^3 + \omega^4 = 0, \quad$ or $\quad \omega^2 + \omega^4 = -\omega^3.$
But $\omega^3 = 1,\quad$ so $\quad \omega^4 + \omega^2 = -1.$

[A] EXAMPLES 20c

1 Express in the form $a + ib$: (i) $(1 + 2i)(1 - i)$; (ii) $(2 - 3i)(2 + 3i)$;
(iii) $(4 - 3i)^2$; (iv) $\dfrac{1}{1-i}$; (v) $\dfrac{2i}{3+2i}$; (vi) $\dfrac{1-i}{2+i}$; (vii) $\dfrac{1+3i}{4-i}$; (viii) $\dfrac{1}{(1+i)(1-2i)}$;
(ix) $(1 - i)^3$; (x) $(2 + i)/(2 - i)^2$.

2 Find x and y in each of the following equations: (i) $x + iy = (3 - 2i)^2$;
(ii) $x - iy = 1/(2 - 5i)$; (iii) $x + iy = (1 + 2i)/(1 - i)$; (iv) $(x + iy)(2 + i) = (1 - i)^2$.

3 Verify that $z = \tfrac{1}{4}(3 + i\sqrt{7})$ is a solution of the equation $2z^2 = 3z - 2$.
Write down the other solution.

Complex numbers

4 If $\omega = \frac{1}{2}(-1 + i\sqrt{3})$, verify that $\omega^3 = 1$.

5 If $u + iv = 1/(x + iy)$, show that $u = x/(x^2 + y^2)$, $v = -y/(x^2 + y^2)$.

6 Simplify $(z - \overline{2 + 3i})(z - \overline{2 - 3i})$.

7 Given that $z = 2i$ is a solution of the equation $z^3 + 3z^2 + 4z = a$, find the value of the constant a. Verify that $z = -2i$ is also a solution.

8 If $z = \dfrac{2 - i}{1 + i}$, find the value of z^* in the form $a + ib$. Represent z and z^* by points in the Argand diagram.

9 Show that the three cube roots of unity are represented by the vertices of an equilateral triangle in the Argand diagram. What is the length of a side of the triangle?

10 Show that $z^3 + 1 = (z + 1)(z^2 - z + 1)$ and hence find the cube roots of -1.

11 Simplify $(1 + i\sqrt{3})^3 - (1 - i\sqrt{3})^3$.

12 If $\sqrt{(x + iy)} = a + ib$, where a, b are real, prove by squaring, that $x = a^2 - b^2$, $y = 2ab$. Find the values of a, b when $x = 3$, $y = 4$ and deduce the square roots of $3 + 4i$.

13 Solve the equation $(z - 2)^3 = 8$.

14 If $(z - 2)/(z + 2) = 1 + i$, where $z = x + iy$, find the values of x and y.

15 Given $(x + iy)^2 = 8 - 6i$, find the values of x and y.

16 Show that the factors of $x^2 + a^2$ are $x + ia$ and $x - ia$. Deduce the factors of $(x - 1)^2 + 4$ and $(x + a)^2 + y^2$.

17 Simplify $\dfrac{1}{x + ia} + \dfrac{1}{x - ia}$.

18 If ω, ω^2 are the complex cube roots of unity, find the values of

(i) $1 + \omega + \omega^2$; (ii) $\omega^3 + \omega^4 + \omega^5$; (iii) $1 + \omega + \omega^2 + \omega^3$.

19 Verify that $z = 1 - 2i$ is a solution of the equation $z^3 - 3z^2 + 7z - 5 = 0$.

20 If $z = 1 + i$ is a solution of the equation $z^3 + az + b = 0$, find the values of the real constants a and b.

21 Express $\dfrac{1}{\cos\theta - i\sin\theta}$ in the form $a + ib$ and hence show that

$$\frac{\cos\theta + i\sin\theta}{\cos\theta - i\sin\theta} = \cos 2\theta + i\sin 2\theta.$$

22 If ω is a complex cube root of unity, show that

$$x^3 - 1 = (x - 1)(x - \omega)(x - \omega^2).$$

The (r, θ) form of a complex number If P represents the complex number $z (= x + iy)$ in the Argand diagram, Fig. 104, then the length OP, denoted by r, is called the *modulus* of z and the angle XOP, measured in radians, is called the *argument* of z.

The modulus of z is written $|z|$ and is always positive. The argument of z, written $\arg z$, the angle turned through from the position OX to the position OP with the usual sign convention, is taken to be the value satisfying

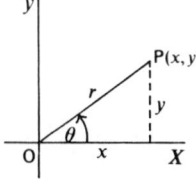

Fig. 104

The (r, θ) form of a complex number | 297

$-\pi < \theta \leqslant \pi$. Although $\tan\theta = y/x$, arg z cannot be expressed as $\tan^{-1}(y/x)$ as the inverse tangent only takes values between $-\frac{1}{2}\pi$ and $\frac{1}{2}\pi$.

As
$$x = r\cos\theta \quad \text{and} \quad y = r\sin\theta,$$
$$z = x + iy = r(\cos\theta + i\sin\theta),$$
where $r = |z| = \sqrt{(x^2 + y^2)}$ and $\theta = \arg z$ where $\tan\theta = y/x$.

The form $r(\cos\theta + i\sin\theta)$ is called the (r, θ) or modulus-argument form of the complex number.

This form is particularly useful in dealing with powers and roots of complex numbers.

Example 12 If $z = 3 + 4i$, find $|z|$ and arg z.
Referring to the diagram, Fig. 105, where P represents the number $3 + 4i$,
$$|z| = r = \sqrt{(3^2 + 4^2)} = 5;$$
$$\arg z = \theta = \tan^{-1}(4/3) = 0.927.$$

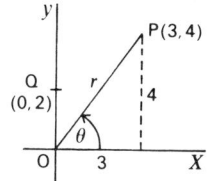

Fig. 105

Example 13 Express $2i$ in the (r, θ) form.
In Fig. 105, Q represents the number $2i$.
Clearly this number has a modulus (OQ) of 2 and an argument (angle XOQ) of $\frac{1}{2}\pi$.
$$\therefore 2i = 2(\cos\tfrac{1}{2}\pi + i\sin\tfrac{1}{2}\pi).$$

Example 14 Express $2/(-1+i)$ in the modulus-argument form.

We have
$$\frac{2}{-1+i} = \frac{2(-1-i)}{(-1+i)(-1-i)}$$
$$= \frac{2(-1-i)}{2} = -1 - i.$$

Representing $-1-i$ by the point $P(-1, -1)$ in the Argand diagram, Fig. 106, it follows that
$$r = OP = \sqrt{2} \quad \text{and} \quad \theta = -\tfrac{3}{4}\pi.$$

$$\therefore \frac{2}{-1+i} = \sqrt{2}\{\cos(-\tfrac{3}{4}\pi) + i\sin(-\tfrac{3}{4}\pi)\}.$$

Fig. 106

Example 15 Prove that $(\cos\theta + i\sin\theta)^3 = \cos 3\theta + i\sin 3\theta$ and deduce expressions for $\cos 3\theta$ and $\sin 3\theta$ in terms of $\cos\theta$ and $\sin\theta$.
$$(\cos\theta + i\sin\theta)^2 = \cos^2\theta + 2i\cos\theta\sin\theta + i^2\sin^2\theta,$$
$$= (\cos^2\theta - \sin^2\theta) + i(2\sin\theta\cos\theta),$$
$$= \cos 2\theta + i\sin 2\theta.$$
$$\therefore (\cos\theta + i\sin\theta)^3 = (\cos 2\theta + i\sin 2\theta)(\cos\theta + i\sin\theta),$$
$$= \cos 2\theta\cos\theta - \sin 2\theta\sin\theta + i(\sin 2\theta\cos\theta + \cos 2\theta\sin\theta),$$
$$= \cos(2\theta + \theta) + i\sin(2\theta + \theta) = \cos 3\theta + i\sin 3\theta.$$
But by expanding
$$(\cos\theta + i\sin\theta)^3 = \cos^3\theta + 3\cos^2\theta(i\sin\theta) + 3\cos\theta(i\sin\theta)^2 + (i\sin\theta)^3,$$
$$= \cos^3\theta - 3\cos\theta\sin^2\theta + i(3\cos^2\theta\sin\theta - \sin^3\theta).$$
$$\therefore \cos 3\theta + i\sin 3\theta = \cos^3\theta - 3\cos\theta\sin^2\theta + i(3\cos^2\theta\sin\theta - \sin^3\theta).$$

Complex numbers

Hence $\cos 3\theta = \cos^3 \theta - 3\cos \theta \sin^2 \theta$ and $\sin 3\theta = 3\cos^2 \theta \sin \theta - \sin^3 \theta$.

Example 16 If $z = \cos \theta + i \sin \theta$, show that
$$z + z^{-1} = 2 \cos \theta \text{ and } z^2 + z^{-2} = 2 \cos 2\theta.$$
Deduce the result $2\cos^2 \theta = \cos 2\theta + 1$.

$$z^{-1} = \frac{1}{z} = \frac{1}{\cos \theta + i \sin \theta} = \frac{\cos \theta - i \sin \theta}{(\cos \theta + i \sin \theta)(\cos \theta - i \sin \theta)},$$

$$= \frac{\cos \theta - i \sin \theta}{\cos^2 \theta + \sin^2 \theta} = \cos \theta - i \sin \theta.$$

$$\therefore z + z^{-1} = 2 \cos \theta.$$

Also
$$z^2 = (\cos \theta + i \sin \theta)^2 = \cos^2 \theta - \sin^2 \theta + 2i \sin \theta \cos \theta,$$
$$= \cos 2\theta + i \sin 2\theta,$$

and
$$z^{-2} = \frac{1}{z^2} = \frac{1}{\cos 2\theta + i \sin 2\theta} = \cos 2\theta - i \sin 2\theta.$$

$$\therefore z^2 + z^{-2} = 2 \cos 2\theta.$$

So
$$(2 \cos \theta)^2 = (z + z^{-1})^2 = z^2 + z^{-2} + 2,$$
$$= 2 \cos 2\theta + 2.$$
$$\therefore 2\cos^2 \theta = \cos 2\theta + 1.$$

[A] EXAMPLES 20d

1 Find the modulus of each of the following complex numbers:
(i) $3 + 4i$; (ii) $1 + 2i$; (iii) $4i$; (iv) $1 - i$; (v) $-1 + i$; (vi) $-5 - 12i$; (vii) $1 - i\sqrt{3}$; (viii) $-4 - 3i$; (ix) $2 + i\sqrt{5}$; (x) $a + ia$.

2 Find the argument of each of the complex numbers in the previous question.

3 Express each of the following complex numbers in the (r, θ) form:
(i) $4 - 3i$; (ii) $1 + i$; (iii) 1; (iv) -1; (v) i; (vi) $-i$; (vii) $-1 + i\sqrt{3}$; (viii) $\sqrt{3} - i$; (ix) $-2 - i\sqrt{5}$; (x) $a + ia$; $a > 0$.

4 Express the complex cube roots of unity in the modulus-argument form.

5 Find the roots of the equation $z^2 - z + 1 = 0$ and express them in the (r, θ) form.

6 Find the modulus and argument of (i) $(1 - i)(1 + 2i)$; (ii) $\dfrac{1-i}{1-2i}$.

7 If $z_1 = r_1(\cos \theta_1 + i \sin \theta_1)$, $z_2 = r_2(\cos \theta_2 + i \sin \theta_2)$, where θ_1, θ_2 are positive acute angles, show that $|z_1 z_2| = r_1 r_2$ and $\arg z_1 z_2 = \theta_1 + \theta_2$.

8 Prove the result $(\cos \theta + i \sin \theta)^n = \cos n\theta + i \sin n\theta$ for n equal to $1, 2, 3, -1, -2, -3$.

9 If $z = \cos \theta + i \sin \theta$, find the value of $z - z^{-1}$.

10 If P_1, P_2 represent the complex numbers z_1, z_2 in the Argand diagram and O is the origin, which lengths represent $|z_1|, |z_2|, |z_1 + z_2|$? Show geometrically that $|z_1 + z_2| \leq |z_1| + |z_2|$.

11 If P_1, P_2 represent the complex numbers z_1, z_2 in the Argand diagram, show that $|z_1 - z_2| = |z_2 - z_1| = P_1 P_2$,

12 If P represents the complex number z in the Argand diagram, what is the locus of P if $|z| = 1$?

13 If $|z_1| = |z_2|$ and $\arg z_1 = -\arg z_2$, show geometrically or otherwise, that z_1, z_2 are conjugate complex numbers.

14 By writing z_1 as $r_1(\cos\theta_1 + i\sin\theta_1)$ and z_2 as $r_2(\cos\theta_2 + i\sin\theta_2)$, where θ_1, θ_2 are positive acute angles, show that (i) $\arg z_1 z_2 = \arg z_1 + \arg z_2$; (ii) $\arg(z_1/z_2) = \arg z_1 - \arg z_2$.

15 Simplify $(\sin\tfrac{1}{6}\pi + i\cos\tfrac{1}{6}\pi)^3$.

16 Show that $\cos 4\theta + i\sin 4\theta = (\cos\theta + i\sin\theta)^4$ and deduce expressions for $\cos 4\theta$, $\sin 4\theta$ in terms of $\cos\theta$ and $\sin\theta$.

[B] MISCELLANEOUS EXAMPLES

1 Express $(2+i)^2/(1-i)$ in the form $a+ib$.

2 Simplify $(1+i\sqrt{3})^4 + (1-i\sqrt{3})^4$.

3 Solve the equation $(z-1)^3 = 8$.

4 If ω, ω^2 are the complex cube roots of unity, show that $\omega^3 + \omega^4 = -\omega^5$.

5 If one solution of the equation $z^2 + a + i(z+1) = 0$, where a is a real number, is real, find the value of a and complete the solution of the equation.

6 If P represents the complex number $3+2i$ in the Argand diagram, what complex numbers are represented by (i) the reflection of P in the y-axis; (ii) the reflection of P in the line $y = x$?

7 Given that $z = 2+i$ is one solution of the equation $z^3 + az^2 + z + b = 0$, find the values of the real constants a and b. Show that $z = 2-i$ is also a solution of the equation and find the third solution.

8 If A, B are the points $4, 2i$ in the Argand diagram, find the complex number represented by the circumcentre of triangle OAB, where O is the origin.

9 Find x and y if $\sqrt{(x+iy)} = 5 + 3i$.

10 Simplify $\dfrac{(\cos\theta - i\sin\theta)^3}{\cos 2\theta + i\sin 2\theta}$.

11 Plot the points $z_1 = 1+4i$, $z_2 = 2+3i$, $z_3 = -1+i$ and use your diagram to find the values of (i) $|z_2 - z_3|$; (ii) $|z_3 - z_1|$; (iii) $|z_1 - z_2|$.

12 If $w = z^2$ where $w = u+iv$ and $z = x+iy$, find u, v in terms of x, y.

13 Express the roots of the equation $z^3 + 1 = 0$ in the (r, θ) form and show that they will be represented by the vertices of an equilateral triangle in the Argand diagram.

14 Express $1/(1+z)$ where $z = \cos\theta + i\sin\theta$, in the form $a+ib$.

15 Find the modulus and argument of (i) $(1+i)(1+i\sqrt{3})$; (ii) $(1+i\sqrt{3})^3$.

16 Prove that the points representing the complex numbers $1, -1, iq, 1/iq$ are concyclic.

17 Express $\sqrt{(1+i\sqrt{3})}$ in the form $a+ib$ where $a > 0$.

18 Find the complex numbers z which satisfy the equation
$$(z+1)^3 = 8(z-1)^3.$$

19 If z is a complex number and z^* its conjugate, show that $zz^* + 2z + 2z^*$ is real.

20 The equation $z^4 - 2z^3 + az^2 - 8z + 4 = 0$ has a root which is wholly imaginary, find this root and the value of the real constant a.

21 If ω is a complex cube root of unity, show that $1 + \omega + \omega^2 = 0$ and deduce that $a^3 + b^3 + c^3 - 3abc = (a+b+c)(a+b\omega+c\omega^2)(a+b\omega^2+c\omega)$.

22 Show geometrically that $|z_2 - z_3| + |z_3 - z_1| \geq |z_1 - z_2|$. When does the equality arise?

23 If P_1, P_2 represent the complex numbers z_1, z_2, show that OP_1, OP_2 are perpendicular if z_1/z_2 is wholly imaginary.

24 Solve the equation $zz^* - 2z + 2z^* = 5 - 4i$, where z, z^* are conjugate complex numbers.

25 If ω is a complex cube root of unity and $x \neq 1$, show that

$$\frac{1}{1-x^3} = \frac{1}{3}\left(\frac{1}{1-x} + \frac{\omega}{\omega-x} + \frac{\omega^2}{\omega^2-x}\right).$$

Revision papers

PAPER A (1)

1 If α and β are the roots of the equation $3x^2 - 2x - 1 = 0$, find the value of $\alpha^3\beta + \alpha\beta^3$. Also obtain the equation which has roots $1/\alpha^2$ and $1/\beta^2$.

2 Expand $(x-y)^6$ by the binomial theorem and use the result to evaluate $(19\frac{3}{4})^6$ correct to the nearest thousand. (C.)

3 Using the same axes, draw the graphs of $xy = 2$ and $2y = x + 3$. Hence find the ranges of values of x for which $\dfrac{4}{x} > x + 3$.

4 In a triangle ABC, $a = 8$, $b = 7$ and $A = 43°$. Calculate (i) the side c, (ii) the area of the triangle. (C.)

5 Draw the graph of the function $1 - \cos 2x$ for values of x between 0 and 2π. Use your graph to solve the equation $6 \sin^2 x = x$, where x is measured in radians. (L.)

6 Find the length intercepted on the line $3x + 2y = 1$ by the lines $2x + 3y = 4$ and $x - 2y = 11$.

7 Show that, for all values of t, the point P given by $x = at(t+2)$, $y = 2a(t+1)$ lies on the curve $y^2 = 4ax + 4a^2$ and find the equation of the normal to the curve at P.

If this normal meets the x-axis at G and N is the foot of the perpendicular from P to the x-axis, prove that $NG = 2a$. (N.)

8 Differentiate with respect to x: (a) $(3x^2 + 7)^3$, (b) $\dfrac{1}{x^2 - 4}$.

Find $\displaystyle\int (2x^2 - 3)^2\, dx$ and $\displaystyle\int_0^{\pi/6} \cos 2\theta \cos 5\theta\, d\theta$. (L.)

9 Determine the maximum and minimum values of the function
$$\frac{x-3}{x^2(x-2)}.$$
(C.)

10 An arc APB of the curve $y = 7x - x^2$ is cut off by the straight line $y = 2x + 4$; AM and BN are the ordinates of A and B. Find (i) the area of the segment APB cut off by the chord AB, (ii) the volume generated by the revolution of the area $MAPBN$ about MN (answer as a multiple of π). (N.)

PAPER A (2)

1 (i) Solve the simultaneous equations:
$$x^2 - 2xy + 2y^2 - 20x + 5y - 4 = 0;\quad 3x + 2y + 1 = 0.$$
(ii) If α, β are the roots of $x^2 - 5x + 2 = 0$, find the values of $\alpha^3 + \beta^3$ and $\alpha^4 + \beta^4$. (C.)

2 (i) Find values of a so that $2x^2 + xy - y^2 + ax - 6y - 5$ can be factorised.

(ii) Find the least value of n such that the sum of n terms of the series $1 + 1\cdot03 + (1\cdot03)^2 + \ldots$ is greater than 10. (N.)

3 A man standing due south of a tower on level ground observes the angle of elevation of the top to be $67°$. After walking 120 m due east he finds that the angle of elevation is $51°$. Calculate the height of the tower. (L.)

4 (i) Prove that $\cos(A+B) = \cos A \cos B - \sin A \sin B$, where the angles A, B and $A+B$ are all acute.

(ii) Prove that $\cos\theta + \cos\left(\theta + \dfrac{2\pi}{3}\right) + \cos\left(\theta + \dfrac{4\pi}{3}\right) = 0$ and find the value of the expression $\sin\theta + \sin\left(\theta + \dfrac{2\pi}{3}\right) + \sin\left(\theta + \dfrac{4\pi}{3}\right)$. (C.)

5 ACB is a minor arc of a circle; the perpendicular bisector of AB meets the arc at C and D is the midpoint of AB. If $AB = 10$ cm and $CD = 3$ cm, calculate the area of the segment ACB.

6 The vertex A of a triangle ABC has coordinates $(3, 7)$, the equation of the side BC is $4x - 3y = 11$, the coordinates of the midpoint of BC are $(5, 3)$ and the area of the triangle is 40 sq. units. Calculate the length of the perpendicular from A to BC and deduce the length of BC. Hence calculate the coordinates of B and C. (N.)

7 Find the coordinates of the centre C and the radius of the circle which has the points $A(1, 4)$ and $B(7, 12)$ at the ends of a diameter. Show that the points $D(4, 13)$ and $E(1, 12)$ lie on this circle.
Find the equations of the tangents to the circle at D and at E and find the tangent of the acute angle between them.

8 (i) Find the maximum and minimum values of $\dfrac{x^2}{2x+1}$.

(ii) Find the area of the segment of the curve $y = 2x(3-x)$ which is cut off by the line $y = 2x$. (L.)

9 (a) Differentiate with respect to x: $\dfrac{x}{3-x}$ and $\sin x°$.

(b) Without using tables, find $\sqrt[3]{7\cdot98}$, correct to four decimal places. (C.)

10 A moving point P starts with a velocity of 8 metres per s from a point O and moves along a straight line OA so that its acceleration after t s is $(10 - 6t)$ metres per s per s. Find the distance of P from O after t s. Find also, (i) the value of t when the velocity begins to decrease, (ii) the distance of P from the starting point at the instant when it begins to return to O. (N.)

PAPER A (3)

1 Solve the following equations:

(a) $\sqrt{2x+3} - \sqrt{3x-5} = 1$; (b) $3^{x-2} = 1\cdot8$, correct to 3 significant figures.

2 Obtain the expansion of $(1+x)^{10}(1-2x)^2$ in ascending powers of x up to and including the term in x^4. Evaluate the expression to four decimal places when $x = 0\cdot02$. (C.)

3 The nth term of a series is $A(\tfrac{3}{2})^{n-1} + Bn$, where A and B are constants. If the first term is 3 and the third term is 6, find A and B and also the sum of the first ten terms as accurately as tables permit.

4 Prove the result $a^2 = b^2 + c^2 - 2bc \cos A$.

A quadrilateral $ABCD$ is such that $AB = 3$ m, $BC = CD = 5$ m, $DA = 6$ m, and the diagonal $AC = 7$ m. Calculate the angle BCD and the length BD. (C.)

5 Find the values of x between $0°$ and $180°$ inclusive, for which
(i) $\sin 3x = \sin x$; (ii) $2\cos^2 x - \sin^2 x = 1$; (iii) $\sin 2x + \cos x = 0$. (C.)

6 The points $P(2, 3), Q(-11, 8), R(-4, -5)$ are vertices of a parallelogram $PQRS$ which has PR as a diagonal. Find the coordinates of S.
Determine also the area of the parallelogram. (N.)

7 Find the coordinates of the centroid of the triangle formed by the lines $y = 4x - 3$, $3x + 2y = 16$ and $3y = x + 2$. Find also the coordinates of the circumcentre of the triangle. (L.)

8 (i) If $y = a\cos^2\theta + b\sin^2\theta$, prove that $\dfrac{d^2y}{d\theta^2} + 4y = 2(a+b)$.

(ii) A vessel is formed by the rotation of the curve $y = \frac{1}{4}x^2$ about the y-axis, the unit along both axes being 1 cm. If the vessel stands with the y-axis vertical and water is poured in at a steady rate of 5 cm^3 s^{-1}, find the rate at which the depth is increasing when the water is 1 cm deep. (L.)

9 (i) Find the maximum and minimum values of the function
$$x^2/(x+1)^3.$$

(ii) Find $\int \cos^2 x \, dx$ and $\int \cos^3 x \, dx$. (L.)

10 Show that the gradient of the curve $y = x(x-3)^2$ is zero at the point $P(1, 4)$, and sketch the curve.

The tangent at P cuts the curve again at Q. Calculate the area contained between the chord PQ and the curve. (C.)

PAPER A (4)

1 (i) Solve the simultaneous equations $x - 2y = 4$; $x^2 + 3xy + 4y^2 = 2$.
(ii) If the roots of the equation $x^2 + px + q = 0$ are α and β, find the equation with roots $\alpha^2 + 1$ and $\beta^2 + 1$. (N)

2 Prove the result $\log_a N = \log_b N \times \log_a b$.

If $\log_4 m = a$ and $\log_{12} m = b$, prove that $\log_3 48 = \dfrac{a+b}{a-b}$.

3 Prove that in the triangle ABC,
$$\tan \frac{B-C}{2} = \frac{b-c}{b+c} \cot \frac{A}{2}.$$

Find the values of B, C and a if $b = 2\cdot31$ m, $c = 3\cdot68$ m and $A = 46°$. (L.)

4 The plane of a rectangular target is vertical and lies east and west; the altitude of the sun is $60°$. At what angle is the sun from the south when the area of the shadow is one half that of the target?

5 If $\operatorname{cosec}\theta - \sin\theta = m$ and $\sec\theta - \cos\theta = n$, prove that
$$m^2n^2(m^2 + n^2 + 3) = 1.$$

6 The coordinates of the points A, B, C are $(-2, 1), (2, 7), (5, 5)$ respectively. Prove that these points form three vertices of a rectangle and that $AB = 2BC$. If D is the fourth vertex of the rectangle, calculate the distance of C from the diagonal BD. (C.)

7 P is a point on the curve $y = ax/(a+x)$; N is the foot of the perpendicular from P to the x-axis and T is the point where the tangent at P meets the x-axis. Prove that $a \cdot TO = ON^2$ and that the area of the triangle PNT is $\frac{1}{2}ON^2$, where O is the origin. (N.)

8 A curve is traced out by the point whose coordinates are
$$x = 2\cos\theta + \cos 2\theta, \quad y = 2\sin\theta - \sin 2\theta.$$
Show that the gradient at the point parameter θ is $-\tan\frac{1}{2}\theta$ and that the equation of the tangent to the curve at this point is
$$x \sin\tfrac{1}{2}\theta + y \cos\tfrac{1}{2}\theta = \sin\tfrac{3}{2}\theta. \tag{L.}$$

9 Evaluate $\displaystyle\int_0^{4/3} (3x-1)^3\,dx$ and $\displaystyle\int_1^2 \cos^2\left(\frac{\pi x}{4}\right) dx.$

Find $\displaystyle\int \frac{x\,dx}{\sqrt{2x^2 - 3}}.$ (C.)

10 Prove that the area enclosed between the line $y = 3$ and the portion of the curve $y = 6\sin x$, for which x lies between 0 and π, is $6\sqrt{3} - 2\pi$.

Find the volume generated by the complete revolution of this area about the line $y = 0$. (C.)

PAPER A (5)

1 Solve the equations

(i) $\sqrt{x+2} + \sqrt{x-5} = \sqrt{3x+7}$; (ii) $x + 2y = 2$; $x^3 + 8y^3 = 56$.

2 (i) State the theorem on the expansion, in powers of x, of $(1+x)^n$, where n is a positive integer.

(ii) Evaluate $(0.994)^8$, correct to the sixth place of decimals. (C.)

3 Express the function $12x - 8x^2 - 5$ in the form $-8(x-a)^2 + b$ and hence show that it is always negative. Sketch the graph of the function.

4 Prove that $\sin^2\alpha + \sin^2(60° - \alpha) + \sin^2(60° + \alpha) = 3/2$.

ABC is an equilateral triangle inscribed in a circle of radius R. If P is any point on the circumference prove that $PA^2 + PB^2 + PC^2 = 6R^2$.

5 Draw the graph of $y = \sin x° - \frac{1}{2}\sin 2x°$ for values of x from 0 to 180.

Use your graph to find approximately the positive solutions of the equation $x = 100 \sin x° - 50 \sin 2x°$. (C.)

6 Find the equation of the tangent to the curve $y = x^3$ at the point $P(t, t^3)$. Prove that this tangent cuts the curve again at the point $Q(-2t, -8t^3)$ and find the locus of the midpoint of PQ.

7 Find the equation of the reflection of the line $4x + 3y = 0$ in the line $3x + 4y = 0$. (C.)

8 Express $\dfrac{3x-5}{x^2-1}$ in partial fractions and show, by differentiation, that the function has one maximum value and one minimum, of which the latter is the greater. (L.)

9 Give a rough sketch of the curve $y^2 = x(x-1)^2$. Find the equation of the tangent at the point (4, 6) and the coordinates of the point in which it meets the curve again. (C.)

10 Evaluate to three places of decimals the definite integrals:

(i) $\int_0^{\frac{1}{4}} \cos^2 x \, dx$ and $\int_0^{\frac{1}{4}} \sin^2 x \, dx;$ (ii) $\int_0^{\frac{1}{4}} \frac{\cos \sqrt{x}}{\sqrt{x}} \, dx.$ (L.)

PAPER A (6)

1 (a) Given that $a+b = p$ and $ab = q$, find in terms of p and q the value of $\dfrac{a^2}{b} + \dfrac{b^2}{a}$.

(b) Find the equation whose roots are $\alpha - \dfrac{2}{\beta}, \beta - \dfrac{2}{\alpha}$, where α, β are the roots of the equation $x^2 - 4x + 8 = 0$.   (N.)

2 A railway carriage has eight seats, four of them with their backs to the engine. In how many ways can a party of seven people seat themselves in the carriage (a) without restriction, (b) if a chosen pair must sit together with their backs to the engine?
(C.)

3 For any triangle, prove the result $\dfrac{a}{\sin A} = \dfrac{b}{\sin B} = \dfrac{c}{\sin C}$.

Hence, or otherwise, show that
$$a \operatorname{cosec} \tfrac{1}{2} A = (b+c) \sec \tfrac{1}{2} (B-C).$$
In a triangle ABC, $a = 5\cdot 3$ m, $b+c = 11\cdot 8$ m, and $A = 46°$. Find angle C.   (L.)

4 The roof of a house is inclined at an angle of 25° to the horizontal and meets a horizontal plane in the line AB, A being due north of B and 30 m from it; the point X on the roof is due east of A, above it and 10 m from it; XY is a vertical chimney 4 m high. Calculate the bearing and elevation of Y from B.   (N.)

5 Find all the angles x between 0 and 2π such that $\tan x = \tan 4x$.   (C.)

6 The gradient of the side PQ of the rectangle $PQRS$ is $3/4$. The coordinates of the opposite corners Q, S are respectively $(6, 3)$ and $(-5, 1)$. Find the equation of PR.
(N.)

7 Prove that the perpendicular bisectors of the sides of the triangle formed by the lines $3y - x = 2$, $x+y = 10$, $5x - 3y = 2$ intersect at the point $(3\tfrac{3}{4}, 2\tfrac{3}{4})$.

Find, in its simplest form, the equation of the circumcircle of the triangle.   (L.)

8 Differentiate with respect to x, (i) $\sin \sqrt{x}$, (ii) $\sqrt{\sin x}$, (iii) 10^{2x}.   (C.)

9 Find the equation of the tangent to the curve $y = a^3/x^2$ at the point P with coordinates $(a/t, at^2)$.

The tangent to the curve at P meets the axes Ox, Oy in A, B respectively. Express AB^2 in terms of t and find the values of t for which the length AB is least.   (N.)

10 (i) Evaluate $\int_0^1 \dfrac{x \, dx}{(1+x^2)^2}, \quad \int_0^{\frac{1}{4}\pi} \cos^3 x \, dx.$

(ii) The surface of a peg top is obtained by rotating the curves
$$y^2 = 4(x+1) \quad \text{from } x = -1 \text{ to } x = 0,$$
and $\quad x/3 + y/2 = 1 \quad \text{from } x = 0 \text{ to } x = 3,$
about the x-axis. Find the volume of the top.   (N.)

PAPER A (7)

1 (a) Factorise $3x^2 - 10xy - 8y^2 + 5x + 22y - 12$.

(b) Solve the simultaneous equations:
$$3x - y + 2z = 1, \quad x + 3y - 3z = -4, \quad y - 2x + z = 7.$$

2 Determine n if in the expansion of $(2 + 3x)^n$ in ascending powers of x, the coefficient of x^{12} is four times that of x^{11}.

3 Expand $\ln(1 + 2x)$ as far as the term in x^4. For what values of x is the expansion valid?

Prove that $\ln(1 + x - 2x^2) = x - \dfrac{5x^2}{2} + \dfrac{7x^3}{3} - \dfrac{17x^4}{4}$ approximately, if x is small.

4 The vertex of a hollow right circular cone is A and BC is a diameter of the base; $BC = 6$ cm, $AB = AC = 8$ cm. The point P lies halfway along AB. By considering the cone to be unrolled into a sector of a circle, find the length of the shortest path on the cone from C to P. (N.)

5 Find, graphically, the value of θ between $0°$ and $90°$ which satisfies the equation $\tan\theta = 2 + \cos\theta$.

If $x = \tan\tfrac{1}{2}\theta$, express $\tan\theta$ and $\cos\theta$ in terms of x and hence find an approximate root of the equation $x^4 + 2x^3 + 2x^2 + 2x - 3 = 0$. (L.)

6 (a) Determine whether the point $(38, 24\tfrac{1}{2})$ lies inside or outside the triangle formed by the lines $x = 0, 8y = 5x, 8y = 160 + x$, explaining the method used.

(b) Find the coordinates of the point equidistant from the three vertices of the triangle. (C.)

7 Prove that the line $x\cos\alpha + y\sin\alpha - p = 0$ touches the circle $x^2 + y^2 - 2ax = 0$, if $p = a(1 + \cos\alpha)$.

8 (i) Differentiate $\sqrt{\dfrac{ax + b}{cx + d}}$ with respect to x, expressing your answer in its simplest form.

(ii) If $x = a(\theta - \sin\theta)$ and $y = a(1 - \cos\theta)$, where θ is a parameter, prove that
$\dfrac{dy}{dx} = \cot\tfrac{1}{2}\theta$ and $\dfrac{d^2y}{dx^2} = -\dfrac{1}{4a}\operatorname{cosec}^4 \tfrac{1}{2}\theta.$ (C.)

9 A light is h m above a point A on a horizontal table and a thin rod of length a cm held vertically with one end B on the table, is moving with velocity v m s^{-1} towards the vertical line through A. Find the rate at which the length of the shadow is decreasing when $AB = x$ m. (N.)

10 Evaluate the finite area cut off from the parabola $2y = x^2$ by the line $x + y = 4$, and show that the y-axis divides this area into two parts which are in the ratio $20:7$.

Determine the volume of the solid generated by a complete revolution of the smaller part about the x-axis. (N.)

PAPER A (8)

1 Express $\alpha^3 + \beta^3$ in terms of $(\alpha + \beta)$ and $\alpha\beta$. Find the quadratic equation whose roots are the cubes of the roots of $2x^2 - x - 7 = 0$. (C.)

2 Find how many three-letter code words can be made with the 26 letters of the alphabet (i) when no letter is repeated in the same word, (ii) when any letter can be repeated two or three times in the same word, (iii) when every word contains exactly one vowel and no letter more than once. (C.)

3 (i) Prove the identity
$$\cos\theta - 2\cos 3\theta + \cos 5\theta = 2\sin\theta(\sin 2\theta - \sin 4\theta).$$

(ii) Solve the equations (a) $\cos 2x = \sin x$, (b) $3\sec^2 x = \tan x + 5$, giving all solutions between $0°$ and $360°$. (C.)

4 A circle of radius r cm. and centre O is inscribed in a triangle ABC, touching the sides BC, CA, AB at D, E, F, respectively. The lengths of OA and OB are $2r$ cm and $r\sqrt{2}$ cm respectively. Show that the angle ABC is a right angle and that the area of the triangle ABC is $(3 + 2\sqrt{3})r^2$ cm^2.

Find the area bounded by the lines AE, AF and the minor arc EF. (N.)

5 Express $7\cos x + 24\sin x$ in the form $R\cos(x - \alpha)$.
Find
 (i) the maximum value of $7\cos x + 24\sin x$,
 (ii) the values of x between $-180°$ and $+180°$ for which
$$7\cos x + 24\sin x = 10.$$

6 The vertices of a triangle are $A(-2, 3)$, $B(3, 7)$ and $C(4, 0)$. Find the coordinates of the point D on the same side of AC as B such that the triangle ACD is right-angled at C and equal in area to triangle ABC. Calculate the area of triangle ACD. (C.)

7 Show that the perpendicular distance of the point (h, k) from the line $x\cos\alpha + y\sin\alpha = p$ is the numerical value of $(h\cos\alpha + k\sin\alpha - p)$.

Calculate the coordinates of the centres of the two circles of radius 5 which pass through the point $(-1, -1)$ and touch the line $3x - 4y + 24 = 0$. (C.)

8 (i) Differentiate $\sqrt{\dfrac{x}{1-x}}$ with respect to x.

(ii) If $y = ae^{-2x}\sin 3x$, prove that $\dfrac{d^2y}{dx^2} + 4\dfrac{dy}{dx} + 13y = 0$. (C.)

9 A curve whose equation has the form $y = x(x-2)(ax+b)$ touches the x-axis at the point where $x = 2$ and the line $y = 2x$ at the origin. Find the values of a and b, sketch the curve, and prove that the area enclosed by an arc of the curve and a segment of the line $y = 2x$ is $32/3$. (L.)

10 (a) Evaluate $\displaystyle\int_1^2 \frac{(x^4-1)^2}{x^2}\,dx$ and $\displaystyle\int_0^{\pi/6} \frac{\cos x\,dx}{1+\sin x}$.

(b) Find the ratio of the volumes formed by rotating the area enclosed by the curve $y = x^4$, the line $x = 1$ and the x-axis (i) about the x-axis, (ii) about the y-axis. (N.)

PAPER A (9)

1 (a) Sum to n terms the series $\ln a + \ln ab + \ln ab^2 + \ln ab^3 + \ldots$.

(b) The nth term of an A.P. is $\dfrac{3n-1}{6}$, prove that the sum of n terms is $\dfrac{n}{12}(3n+1)$.

2 Find the number of ways of arranging all the letters in the word *potato*. In how many of these ways will the arrangement begin with a *t*?

3 Prove that, if x is so small that its cube and higher powers can be neglected,
$$\sqrt{\frac{1+x}{1-x}} = 1 + x + \frac{x^2}{2}.$$
By taking $x = \frac{1}{9}$, prove that $\sqrt{5}$ is approximately equal to $\frac{181}{81}$. (C.)

4 (i) If $A + B + C = 180°$, prove that
$$\sin A \cos(B-C) + \sin B \cos(C-A) + \sin C \cos(A-B)$$
$$= 4 \sin A \sin B \sin C.$$
(ii) Prove that $\cos^2 \alpha + \cos^2(\alpha + 120°) + \cos^2(\alpha - 120°) = 1\frac{1}{2}$. (L.)

5 Two points P, Q are at a distance $2a$ apart. Two non-coplanar lines PR, QS are each perpendicular to PQ, each of length b and make an angle 2θ with one another. Prove that $RS^2 = 4(a^2 + b^2 \sin^2 \theta)$ and find the tangent of the angle between PQ and RS. (C.)

6 A point Q is distant r from a point $P(a, b)$, in a direction making an angle θ with the x-axis. Show that the coordinates of Q are
$$(a + r\cos\theta, b + r\sin\theta).$$
The centre of a rhombus is at the point $(5, 1)$; one vertex is at the point $(2, 5)$, and the area of the rhombus is 100 unit^2. Find the lengths of the diagonals and the coordinates of the other vertices. (N.)

7 Find the equation of the circle with centre $(2, 1)$ which touches the line $4x - 3y + 10 = 0$.
Find also the length and equation of the other tangent from the point where this line meets the line $y = 0$. (C.)

8 The tangent at the point $P(at, at^3)$ on the curve $a^2 y = x^3$ meets the x-axis at A and the y-axis at B. Prove that $PA = \frac{1}{2} AB$.
Show also that, if O is the origin, the area between the curve, the ordinate at P, and the x-axis is three-eighths of the area of triangle OAB. (L.)

9 (a) Differentiate with respect to x,
(i) $(x^3 - x^{-1})^2$, (ii) $\ln(\sin x - \cos x)$.
(b) Obtain the coordinates of the point of intersection of the tangents to the curve $y^2 = x^2(25 - x^2)$ at the points $(4, 12)$ and $(3, 12)$. (N.)

10 Find (a) $\int \sin^2 2x \, dx$, (b) $\int \frac{\sqrt{x}}{1-x} dx$. (C.)

PAPER A (10)

1 Draw the graph of $y = (x-1)^2(4-x)$ for values of x from 0 to 4.
Use the graph (i) to solve the equation $(x-1)^2(4-x) = x$, (ii) to estimate the range of values of k for which the equation $(x-1)^2(4-x) = kx$ has three real roots. (C.)

2 (a) Find the coefficient of x^5 in the expansion of $(1 + 4x)e^x$ in ascending powers of x.
(b) Write down the first four terms in the expansion of $\ln \frac{1+2x}{1-2x}$.
For what values of x is the expansion valid?

3 Prove that, in a triangle ABC, $\cos \frac{A}{2} = \sqrt{\frac{s(s-a)}{bc}}$.

Find the largest angle and the area of triangle ABC, in which $a = 10.6$ cm, $b = 11.8$ cm and $c = 15.6$ cm. (C.)

4 PQ is a vertical pole with its foot Q on ground level. Two points A and B on the ground are 50 m apart; the angle $QAB = 58°$, the angle $QBA = 49°$, and the angle $PAQ = 27°$. Calculate the height of the pole to the nearest metre. (L.)

5 Plot the points on the curve given by the equations $x = \cos t°$, $y = \cos 2t°$ for the values 0, 30, 60, ..., 180 of t, and sketch the curve.
Prove that the distance of any point of the curve from the point $(0, -7/8)$ is the same as its distance from the line $y = -9/8$. (L.)

6 Prove that the length of the perpendicular from the point (h, k) to the line $ax + by + c = 0$ is $(ah + bk + c)/\sqrt{a^2 + b^2}$ numerically.
Find whether the origin and the point $(52, -58)$ are on the same or opposite sides of the line $19x + 17y - 3 = 0$. (C.)

7 Find the coordinates of the centres of the circles which pass through the point $(1, 1)$ and touch the x-axis and the line $3x + 4y = 5$.

8 Obtain the equation of the tangent at the point $(3t^2, 2t^3)$ to the curve $27y^2 = 4x^3$.
Find the coordinates of the point where this tangent meets the curve again. (C.)

9 State and prove conditions for a function $f(x)$ to have a maximum value M when $x = a$. A geometrical argument is expected.
A circular cylinder is inscribed in a sphere of radius R. Find its maximum volume. (C.)

10 Sketch the curve $y = x/(1 + x)$ for positive values of x and find the area bounded by that portion of the curve corresponding to values of x from 0 to 2, the line $y = 0$, and the ordinate $x = 2$.
Find also the volume of revolution of this area about the line $y = 0$. (C.)

PAPER A (11)

1 (i) Find the number of ways of arranging four white, three black and two red marbles in line, assuming that marbles of the same colour are indistinguishable.
(ii) A bag contains six white and four black marbles. Find the chance that, if two marbles are drawn together, they are both black. (C.)

2 (a) Find the coefficient of x^7 in the expansion of $(2 - 3x)^5 (2 + 3x)^7$.
(b) Evaluate $\dfrac{1}{\sqrt[3]{7.98}}$, correct to four decimal places.

3 Prove that $1^2 + 2^2 + 3^2 + \ldots + n^2 = \tfrac{1}{6}n(n + 1)(2n + 1)$.
Find the sum of $2n$ terms of the series $1^2 + 2 + 3^2 + 4 + 5^2 + 6 + \ldots$. (C.)

4 Prove that $\tan^2(45° + \theta) = \dfrac{1 + \sin 2\theta}{1 - \sin 2\theta}$.
Prove also that $\tan 22\tfrac{1}{2}° = \sqrt{2} - 1$ and $\tan 67\tfrac{1}{2}° = \sqrt{2} + 1$. (C.)

5 A flagstaff 5 m. high is placed on the top of a hemispherical mound 60 m high. How far from the foot of the mound must a boy 1.2 m high stand so as just to see the top of the flagstaff?

6 Find formulae for the coordinates of the point which divides the straight line joining the points (x_1, y_1) and (x_2, y_2) externally in the ratio $p:q$.
The coordinates of three vertices of a parallelogram $ABCD$ are $A(-1, -2)$,

$B(2, -1)$, $C(3, 1)$. Find the coordinates of D, the other extremity of the diagonal BD, and find the tangent of the angle BDC. (N.)

7 Prove that the equation $(x - y + 2)^2 + 2(x + 3)(y - 4) = 0$ represents a circle, and find its centre and radius.
Prove also that the lines $x + 3 = 0$ and $y - 4 = 0$ are tangents to the circle and show in a sketch the relation of the line $x - y + 2 = 0$ to the tangents. (L.)

8 Sketch the curves represented by the equations

(i) $y = x(x - 2)$, (ii) $y^2 = x(x - 2)$.

9 (i) Differentiate with respect to x, $\sqrt{\dfrac{1-x}{x}}$ and $x^2 e^{-x}$.

(ii) Find the differential coefficient of $\lg x$ with respect to x. *Deduce* the value, correct to four decimal places, of $\lg 1\cdot002$, given that $\lg e = 0\cdot4343$. (C.)

10 Two points P, Q on the curve $y = 1/x$ have coordinates $(\theta t, 1/\theta t)$ and $(t, 1/t)$ respectively, where $0 < \theta < 1$. Parallels to the axes are drawn through P and Q to form a rectangle. Prove that the area of the smaller of the two parts into which the rectangle is divided by the curve is $\theta - 1 - \ln \theta$.
Prove also that the area enclosed by the chord PQ and the curve is independent of t, and find its numerical value when $\theta = 2/3$. (C.)

PAPER A (12)

1 (a) Solve the simultaneous equations $xy = 4$, $2x^2 + 3y^2 = 22$.
(b) Determine the range of values of k for which the roots of the equation $k(x^2 + 2x + 3) = 4x + 2$ are real and unequal.

2 In the expansion of $(1 + x^2)(1 + x)^n$ in ascending powers of x the ratio of the coefficients of x^3 and x is 6 to 1. Find n and the coefficient of x^4.

3 Draw the graph of $\sin 2x$ for values of x between 0 and π radians.
Use your graph to obtain approximate values of the solutions of the equation $2\pi(1 - \sin 2x) = 3x$, which lie between these limits, the angle being measured in radians. (N.)

4 A, B, C, D are points on an arc of a semi-circle, the chords AD and BC being parallel. If the arcs AB, BD subtend angles of α and β at the centre, show that the area of the portion of the semicircle between the parallel chords is $r^2(\alpha - \sin \alpha \cos \beta)$.

5 Find all solutions of the following equations lying between $-180°$ and $180°$;

(i) $1 - \cos x = 2 \sin^2 x$, (ii) $\sin(x - 10°)\cos(x - 24°) = 0\cdot5$.

6 The vertex of an isosceles triangle is the point $(3, 4)$, the equation of the base is $x - 2y + 2 = 0$, and the base angles are $\tan^{-1} 1/3$. Find the gradients and the lengths of the equal sides. (L.)

7 Obtain the equation of the locus of a point P which moves so that $PB = 2PA$, where A, B are respectively the points $(1, 0)$, $(-1, 0)$. Show that the locus is a circle and determine the radius and the coordinates of the centre.
Verify that $y = 4/3$ is a tangent to the circle and determine the equation of the other tangent that passes through the point $(-1, 4/3)$. (N.)

8 Find the derivatives with respect to x, of the functions

(i) $(15x^2 + 12x + 8)(1 - x)^{3/2}$, (ii) $(2x^2 + 10x + 11)e^{-2x}$,

and state in each case the values of x for which the derivative vanishes. (N.)

9 (i) Integrate with respect to x, $\left(2x - \dfrac{3}{\sqrt{x}}\right)^2$ and $\sin^3 x \cos^2 x$.

(ii) Evaluate
$$\int_0^3 \frac{2x^2}{\sqrt{4-x}}\,dx,$$
by means of the substitution $x = 4 - u^2$, or otherwise. (C.)

10 Sketch in the same figure the curves $y^3 = 4x$ and $2y^2 = x^3$.
Verify that the curves intersect only at the origin O and at the point P (2, 2).
Find the area enclosed between the arcs of these curves that join O and P and also the volume of the solid generated by a complete revolution of this area about the x-axis. (N.)

PAPER B (1)

1 Sum to n terms the following series

(a) $1.5 + 3.7 + 5.9 + \ldots$, (b) $\dfrac{1}{1.5} + \dfrac{1}{3.7} + \dfrac{1}{5.9} + \ldots$. (C.)

2 Express $\dfrac{3x^3 - 8}{x^4 - 16}$ as the sum of three partial fractions. (C.)

3 Find the positive value of a for which the sum of the first three terms in the expansion of $(1 + ax)e^{x + \frac{1}{2}x^2}$ in ascending powers of x is a perfect square. (N.)

4 A plane stratum of rock intersects a vertical cliff face in a line inclined at an angle α to the vertical and intersects a second vertical cliff face which is at right angles to the first, in a line inclined at the same angle α to the vertical. Prove that the angle between the two lines of intersection is $\cos^{-1} \cos^2 \alpha$, and find the angle of inclination of the stratum to the horizontal. (C.)

5 Write down the first four terms in the expansion of $\cos x$ in ascending powers of x.
Find the limit of $\dfrac{\cos x - \cos 2x}{x^2}$ as $x \to 0$.

6 (a) Write down the equation of the straight line through the point $(0, a)$ with gradient m.
A straight line through the point A $(0, 6)$ meets the line $y = 2x$ at P and the line $y = -2x$ at Q. If $QA : AP = 1 : 3$, obtain the equation of the straight line.

(b) Find the length of either tangent from $(1, 1)$ to the circle
$$4x^2 + 4y^2 - 4x - 12y + 17 = 0.$$ (N.)

7 Prove that the equation of the normal at the point $(at^2, 2at)$ on the parabola $y^2 = 4ax$ is $y + tx = 2at + at^3$.
A chord is drawn through the point P $(ap^2, 2ap)$ on the parabola $y^2 = 4ax$ and is a normal at its other end Q. Prove that the parameter of Q is a root of the equation $t^2 + pt + 2 = 0$. (C.)

8 Find the maximum and minimum values of $x/(1 + x^2)$, and the points of inflexion of the curve $y(1 + x^2) = x$. (C.)

9 (i) Use Maclaurin's series to expand $\tan x$ in ascending powers of x as far as the term in x^3.

(ii) Evaluate $\displaystyle\int_0^3 \frac{3x + 4}{\sqrt{x + 1}}\,dx$ and $\displaystyle\int_0^1 \cos^2\left(\frac{\pi x}{2}\right)\,dx$. (C.)

10 Sketch the curve $y^2 = x^2(4 - x)$ and find the area of the loop.

PAPER B (2)

1 If α, β are the roots of the equation $ax^2 + bx + c = 0$ and $\alpha : \beta = \lambda : \mu$, show that $\lambda \mu b^2 = (\lambda + \mu)^2 ac$.

Deduce (i) the condition for the equation $ax^2 + bx + c = 0$ to have equal roots; (ii) the condition for the roots of the equation $ax^2 + bx + c = 0$ to be in the same ratio as those of $a'x^2 + b'x + c' = 0$. (C.)

2 A pack of 52 playing cards contains 20 honours. Show that a selection of three cards from the pack can be made in 22,100 different ways and that 4960 of these contain no honour. Deduce the probability that a particular selection of three cards contains at least one honour. (C.)

3 (a) Find the values of the constants A, B, C, D in the identity

$$\frac{1}{2 + 3x + 4x^2} = A + B + \frac{Cx^2 + Dx^3}{2 + 3x + 4x^2}.$$

(b) Solve the simultaneous equations

$$xy = 6 + 3x, \quad yz = 5 + y, \quad zx = 2 + 2z. \quad (N.)$$

4 Prove that, if $t = \tan \frac{1}{2}\theta$, then $\sin \theta = \dfrac{2t}{1+t^2}$ and $\cos \theta = \dfrac{1-t^2}{1+t^2}$.

By expressing $\dfrac{3 + \cos \theta}{\sin \theta}$ in terms of t, show that this expression cannot have any value between $-2\sqrt{2}$ and $+2\sqrt{2}$. (C.)

5 The triangle ABC is right-angled at B; a circle is drawn with centre A and radius AB to cut AC at D. If the angle DBC is x radians, and the sum of the lengths of the sides AB, AD and the arc BD is twice the length of the side BC, prove that

$$\tan 2x = 1 + x.$$

Verify that $x = 0.5$ is an approximate root of this equation and obtain this root correct to two decimal places.

6 Prove that the circle which has as a diameter the common chord of the two circles

$$x^2 + y^2 + 2x - 5y = 0, \quad x^2 + y^2 + 6x - 8y = 1$$

touches the axes of coordinates. (L.)

7 The foci of an ellipse are S (4, 0) and S' (−4, 0) and any point P on the ellipse is such that $SP + S'P = 10$. Find the equation of the ellipse.

Find the equation of the tangent to the ellipse at the point in the first quadrant with an x-coordinate of 3. Also find the coordinates of the point in which this tangent meets the directrix associated with the focus S. (C.)

8 In the curve $x = at^2$, $y = at^3$, the tangent at P meets the curve again at Q. If m_1 and m_2 are the gradients of the tangents at P and Q, prove that

$$m_1 + 2m_2 = 0.$$

9 By examining the sign of the differential coefficient of $x - \ln (1 + x)$, prove that $x > \ln (1 + x)$ if (i) $x > 0$, or (ii) $-1 < x < 0$. Illustrate by a figure. (C.)

10 (i) Evaluate $\displaystyle\int_0^{\frac{1}{2}\pi} \sin 2x \cos 3x \, dx$.

(ii) Using the substitution $y = e^x + 1$, find $\displaystyle\int \frac{dx}{(e^x + 1)^2}$. (C.)

PAPER B (3)

1 The pth term of a progression is P, the qth term is Q, and the rth term is R. Show that, if the progression is arithmetical,
$$P(q-r) + Q(r-p) + R(p-q) = 0,$$
and that if it is geometrical,
$$(q-r)\log P + (r-p)\log Q + (p-q)\log R = 0. \tag{C.}$$

2 (i) Find the term independent of x in the expansion of $\left(\dfrac{1}{3x} - \dfrac{3}{2}x^2\right)^9$.

(ii) If x^3 and higher powers of x may be neglected, express
$$\left(1 + \frac{5x}{2} - \frac{3x^2}{2}\right)^8$$
in the form $1 + ax + bx^2$. (L.)

3 If $\tan\alpha = p$, $\tan\beta = q$, $\tan\gamma = r$, prove that
$$\tan(\alpha+\beta+\gamma) = \frac{p+q+r-pqr}{1-qr-rp-pq}.$$
Deduce a relation between p, q and r in each of the following cases:

(i) $\alpha+\beta+\gamma = 0$, (ii) $\alpha+\beta+\gamma = \dfrac{\pi}{2}$, (iii) $\alpha+\beta = \gamma$. (C.)

4 A pyramid vertex P, stands on a square base $ABCD$. The foot of the perpendicular from P to the base is a point on BD between B and D. The faces PAB, PCD make angles $\tan^{-1}\tfrac{3}{4}$, $\tan^{-1}\tfrac{3}{2}$ respectively with the base. Find (i) the ratio of the height of the pyramid to the side of the base, (ii) the angle APB.

5 Find all solutions between $0°$ and $360°$ of the equations

(i) $\cos x + \cos 3x = \cos 2x$; (ii) $3\cos x + 4\sin x = 4\cdot 2$. (N.)

6 Prove that the equation of any straight line passing through the point of intersection P of the lines $ax+by+c = 0$ and $a'x+b'y+c' = 0$ can be written in the form
$$ax+by+c+\lambda(a'x+b'y+c') = 0,$$
where λ is a constant.

Find the equations of the following lines:
   (i) the join of P to the origin,
   (ii) the line through P perpendicular to $ax+by+c = 0$. (C.)

7 Find the equation of the tangent to the parabola $y^2 = 4ax$ at the point $(at^2, 2at)$.

From any point P on a parabola with vertex A and focus S, the line PN is drawn perpendicular to the axis AS; the tangents at A and P intersect at Q. Prove that $PQ^2 = AN \cdot SP$. (N.)

8 (i) Differentiate with respect to x:

(a) $\dfrac{2x+1}{\sqrt{3-4x-4x^2}}$, (b) $e^{-x}\{(x+1)\cos x + x\sin x\}$.

(ii) If $y = \sin x \ln \tan\tfrac{1}{2}x$, show that $\dfrac{d^2y}{dx^2} + y = \cot x$. (N.)

9 Express $\dfrac{2(x+1)}{(x-1)(2x-1)}$ in partial fractions, and prove that

$$\int_2^5 \dfrac{2(x+1)}{(x-1)(2x-1)} dx = \ln \dfrac{256}{27}.$$ (C.)

10 If $f(x) = 2e^{-4x} - e^{-2x}$, show that $f(x) = 0$ for just one value of x. Denoting this value by a, show that $f'(2a) = 0$. Sketch the graph of the function.

Evaluate the area bounded by the x-axis, the ordinate $x = 0$, and the arc of the curve joining the points $x = 0$ and $x = a$. (C.)

PAPER B (4)

1 Sketch the curves with equations (i) $y = \dfrac{1}{x(x-2)}$, (ii) $y^2 = \dfrac{1}{x(x-2)}$. (C.)

2 Prove, by induction or otherwise, that

$$1^3 + 3^3 + 5^3 + \ldots + (2n-1)^3 = n^2(2n^2 - 1).$$ (C.)

3 (a) If α and β are the roots of the equation $ax^2 + bx + c = 0$, express in terms of α and β the sum and products of the roots of the equation $a^2 x^2 + (b^2 - 2ac)x + b^2 - 4ac = 0$.

(b) Find the values of n and k in order that the coefficients of x and of x^2 in the expansion of $(1-x)^n e^{kx + 6x^2}$ may both be zero. (N.)

4 Prove that, in a triangle ABC, $\tan \dfrac{B-C}{2} = \dfrac{b-c}{b+c} \cot \dfrac{A}{2}$.

Solve the triangle ABC, given that $b = 35 \cdot 31$ cm, $c = 16 \cdot 29$ cm, $A = 51° 34'$.

5 Assuming the formula for $\cos(A+B)$, prove that $\cos 2A = 2\cos^2 A - 1$ and that $\cos 3A = 4 \cos^3 A - 3 \cos A$.

Find the six angles between $0°$ and $360°$ which satisfy the equation $2\cos 3\theta = 1$, and use your result to obtain, correct to three significant figures, the three roots of the equation $8x^3 - 6x - 1 = 0$. (L.)

6 Find the equation of the circle S that passes through the point $(0, 6)$ and also through the points of intersection of the circle

$$x^2 + y^2 - 6x - 10y + 32 = 0$$

with the line $x - y + 2 = 0$.

If A, B are the points $(8, 4)$, $(0, 10)$ respectively, prove that AB touches the circle S and determine the coordinates of the point C which is such that the circle touches the sides of the triangle ABC. (N.)

7 Prove that the equation of the normal to the ellipse $x^2/a^2 + y^2/b^2 = 1$ at the point P $(a\cos\theta, b\sin\theta)$ is $ax\sin\theta - by\cos\theta = (a^2 - b^2)\sin\theta\cos\theta$.

If there is a value of θ between 0 and $\pi/2$ such that the normal at P passes through one end of the minor axis, show that the eccentricity of the ellipse must be greater than $1/\sqrt{2}$. (C.)

8 (i) Obtain the differential coefficients with respect to x of $\sqrt{5x^2 - 1}$ and $\ln(x^3 - 3x)$.

(ii) Prove that the curves $y = x + \cos x$, $y = e^x$ touch each other when $x = 0$ with $y = x + 1$ as the common tangent. Prove also that the curves have no other point in common. (N.)

9 Sketch the locus of the point $(\cos^3 t, \sin^3 t)$ where t is a parameter, and find the area enclosed by the locus. (C.)

10 (a) Find $\int \dfrac{dx}{x^2(x+1)}$.

(b) By means of the substitution $t = \tan\tfrac{1}{2}x$, find the value of
$$\int_0^{\frac{1}{2}\pi} \dfrac{dx}{3+5\cos x}.$$ (C.)

PAPER B (5)

1 (a) Solve the simultaneous equations $x(y+3) = 4$, $3y(x-4) = 5$.

(b) If p and q are real and not zero, find the condition that the roots of the equation $2p^2 x^2 + 2pqx + q^2 - 3p^2 = 0$ are real.
If the roots of this equation are α and β, prove that $\alpha^2 + \beta^2$ is independent of p and q. (N.)

2 (a) Find the first five terms of the expansion of $(1-4x)^{\frac{1}{2}}$ in ascending powers of x.

(b) Find the values of the constants a, b, c in the approximate formula
$$\ln\dfrac{1+x}{1-x} - \dfrac{2x(1-ax^2)}{1-bx^2} = cx^7,$$
in which x is so small that powers of x above the seventh can be neglected. (N.)

3 Draw the graph of $\cos\tfrac{1}{2}x$ for values of x between 0 and 2π radians. Use your graph to obtain an approximate root of the equation
$$4\pi\cos\tfrac{1}{2}x + x = 2\pi.$$ (N.)

4 Prove

(i) $\dfrac{\tan A - 3\tan 3A}{\cot A - 3\cot 3A} = \dfrac{\tan A - 3\cot A}{\cot A - 3\tan A}$.

(ii) $4\cos\theta\cos\left(\dfrac{2\pi}{3}+\theta\right)\cos\left(\dfrac{2\pi}{3}-\theta\right) = \cos 3\theta$.

5 Express $a\cos^2\theta + h\sin\theta\cos\theta + b\sin^2\theta$ in terms of $\cos 2\theta$ and $\sin 2\theta$ and hence in the form $\tfrac{1}{2}(a+b) + p\sin(2\theta+\gamma)$.

When $a = 4$, $b = 1$, $h = 4$, show that the maximum and minimum values of the expression are 5 and 0 respectively. (L.)

6 Two sides of a parallelogram lie along the lines
$$x - y + 1 = 0, \quad 2x + 3y - 6 = 0$$
and the diagonals meet at the point $(1, \tfrac{1}{2})$. Find the coordinates of the vertices of the parallelogram and the equations of the other two sides. (L.)

7 The tangents at points P and Q on a parabola intersect at R. The line through R parallel to the axis of the parabola intersects PQ at M. Prove that M is the midpoint of PQ. (C.)

8 If $x^2 + 2xy + 3y^2 = 1$, prove that $(x+3y)^3 \dfrac{d^2 y}{dx^2} + 2 = 0$.

9 Calculate the coordinates of the point of intersection of the curves $3y = 4e^x + 2e^{-x}$ and $y = e^{-x}$. Calculate also the coordinates of the point on the first curve for which y is a minimum.

With the same axes and scales, sketch the curves $y = e^x$, $y = e^{-x}$ and $3y = 4e^x + 2e^{-x}$. (C.)

10 Make a rough sketch of the parabola $4y = 3x^2 - 2x + 3$, and verify that the tangents at $P(1, 1)$, $Q(-1, 2)$ pass through the origin O. Determine the area bounded by these tangents and the arc of the parabola between the points of contact. Find also the volume generated by a complete revolution about the x-axis of that part of the above area that lies in the first quadrant. (N.)

PAPER B (6)

1 How many sets of 4 cards is it possible to select from an ordinary pack of 52 playing cards, each set of 4 being drawn from the complete pack?

How many of these sets will contain (a) one ace, (b) two aces, (c) at least one ace? (N.)

2 Express in partial fractions $\dfrac{2(x^2-1)}{x(2x-3)}$. Sketch the graph of the function $\dfrac{2(x^2-1)}{x(2x-3)}$.

3 Find the coefficients of x, x^2 and x^3 in the expansion of
$$e^{ax} + b\ln(1+x) - \sqrt{(1+2x)}$$
in ascending powers of x. Show that a and b can be determined so that all these coefficients are zero. (N.)

4 Prove that
 (i) $\sin(\alpha+\beta)\sin(\alpha-\beta) = \sin^2\alpha - \sin^2\beta$,
 (ii) if $A+B+C+D = 2\pi$, then
 $$\cos 2A + \cos 2B + \cos 2C + \cos 2D = 4\cos(A+B)\cos(A+C)\cos(A+D).$$ (L.)

5 State formulae for the area of a triangle in terms of (i) b, c, A, (ii) a, b, c.

Find the area of the quadrilateral $ABCD$ in which $AB = 5$ m, $BC = 3$ m, $CD = 10$ m, $DA = 8$ m and the internal angle $ABC = 120°$.

6 Show that the point P with coordinates $(\overline{1-t}x_1 + tx_2, \overline{1-t}y_1 + ty_2)$ lies on the line joining $P_1(x_1, y_1)$ and $P_2(x_2, y_2)$ and that $P_1P = tP_1P_2$.

The coordinates of points A, B, C are $(-a, 0), (-a, 4a), (3a, 4a)$ respectively. A point P is taken in BC so that $BP = tBC$ and a point Q in AP so that $AQ = tAP$. If t varies, show that the locus of Q is a parabola and obtain its focus. (N.)

7 Show that the equation of the tangent at the point $(ct, c/t)$ on the rectangular hyperbola $xy = c^2$ is $x + t^2 y = 2ct$.

NP is the ordinate of a point P on the hyperbola; the tangent at P meets the y-axis at M, and the line through M parallel to the x-axis meets the curve at Q. Show that NQ is the tangent at Q.

8 Find the maximum possible value of the volume of a cylinder which can be cut from a solid hemisphere of radius a and has its axis perpendicular to the base of the hemisphere. (N.)

9 (a) Use Maclaurin's series to obtain the first three terms in the expansion of $\ln(1+\sin x)$ in ascending powers of x.
 (b) In triangle ABC, angle A is calculated from known values of a, b, B. Show that the error in A due to a small error δb in b, is approximately
 $$\frac{a\sin B \cdot \delta b}{b^2 \cos A}.$$

10 Sketch the curve $y = \dfrac{x}{x+2}$ and find the area enclosed by the curve and the lines $x = 0$, $x = 1$, $y = 1$. Find also the volume obtained by revolving this area through four right angles about the line $y = 1$. (C.)

PAPER B (7)

1 If α, β are the roots of the quadratic equation $x^2 + 2px + q = 0$, prove that
$2 \log \{\sqrt{y-\alpha} + \sqrt{y-\beta}\} = \log 2 + \log \{y + p + \sqrt{y^2 + 2py + q}\}$. (N.)

2 (i) Write down the $(r+1)$th term in the expansion of $(1+x)^{12}$ in ascending powers of x, when $x = \cdot 03$. Show that if $r > 3$, the $(r+1)$th term is less than one-tenth of the rth term.
(ii) In the expansion of $(1 + x + px^2)^7$ the coefficient of x^2 is zero. Find the value of p. (L.)

3 The angular elevation of the summit of a mountain is measured from three points on a straight level road. From a point due south of the summit the elevation is α; from a point due east of it the elevation is β; and from the point of the road nearest to the summit the elevation is γ. If the direction of the road makes an angle θ east of north, prove that
(i) $\tan \theta = \tan \alpha \cot \beta$, (ii) $\tan^2 \gamma = \tan^2 \alpha + \tan^2 \beta$.
Find γ, if $\theta = 31°$ and $\alpha = 8°$. (C.)

4 (i) Find the greatest and least values of $\sin x + \cos x - 1$.
(ii) Prove that if x is positive, the expression $2x - \sin 2x$ is always positive.

5 Prove that the common chord of the circles
$$x^2 + y^2 = 50 \quad \text{and} \quad x^2 + y^2 - 2x - 4y = 40$$
is a diameter of the second circle, and that the circles cut at an angle of $\tan^{-1} \frac{1}{3}$. (C.)

6 Find the equation of the tangent to the ellipse $x^2/a^2 + y^2/b^2 = 1$ at the point $(a \cos \theta, b \sin \theta)$.
The tangent at any point on an ellipse cuts the minor axis at Q, and cuts the tangent at an end of the major axis at R. Prove that $QR = QS$, where S is a focus. (N.)

7 The curves $y^2 = 4ax$ and $xy = c^2$ intersect at right angles. Prove that (i) $c^4 = 32a^4$ and (ii) if the tangent and normal to either curve at the point of intersection meets the x-axis at T and G, then $TG = 6a$. (C.)

8 (i) On a certain curve for which $\dfrac{dy}{dx} = x + \dfrac{a}{x^2}$, the point $(2, 1)$ is a point of inflexion. Find the value of a and the equation of the curve.
(ii) A body moves in a straight line, so that its displacement s from a fixed point in the line after time t is given by $s = ae^{pt} + be^{-pt}$. If the initial displacement is zero and the initial velocity is u, find the values of a and b, and prove that the velocity v at displacement s is given by $v^2 = u^2 + p^2 s^2$. (C.)

9 Find $\displaystyle\int \frac{dx}{(x+1)\sqrt{x}}$ and $\displaystyle\int \sin^3 x \, dx$.

Show that $\displaystyle\int_1^3 \frac{x^2 + x - 3}{x(x^2 + 3)} dx = \frac{\pi \sqrt{3}}{18}$. (N.)

10 Draw a rough graph of the curve $y = \sec^2 x$ for values of x between $-\frac{1}{2}\pi$ and $\frac{1}{2}\pi$. The area enclosed by the curve, the line $y = 0$, and the lines $x = 0, x = \frac{1}{3}\pi$, is rotated about the x-axis. Find the volume of the solid of revolution so obtained. (C.)

PAPER B (8)

1 In how many ways can five a's and nine b's be arranged in a line? In how many ways can they be arranged if no two a's are to come together?

2 (a) Resolve $\dfrac{5x^2}{(x^2+1)(x-2)}$ into partial fractions.

(b) Simplify $\dfrac{(2^{2n} - 3 \times 2^{2n-2})(3^n - 2 \times 3^{n-2})}{3^{n-4}(4^{n+3} - 2^{2n})}$. (N.)

3 If x is real and $y = \dfrac{x^2 + 4x + 5}{(x+1)(x+3)}$, prove that y has no value numerically less than 1. Sketch the graph of the curve represented by the equation. (N.)

4 If G is the centroid of the triangle ABC and $\varDelta$ its area, prove that
$$9AG^2 = 2b^2 + 2c^2 - a^2 \quad \text{and} \quad \tan BGC = 12\varDelta/(b^2 + c^2 - 5a^2).$$ (L.)

5 (a) Find all the solutions between $0°$ and $180°$ of the equation
$$\sin 2x = \cos 3x.$$

(b) Simplify $\cos(\theta + \phi)\cos\phi + \sin(\theta + \phi)\sin\phi$.

(c) Prove that, if $\tan A = \dfrac{3 + 4x}{4 - 3x}$ and $\tan B = \dfrac{6 + 7x}{7 - 6x}$, the value of $\tan(A - B)$ is independent of x. (N.)

6 Show that the line $ax + by = c_1 + c_2$ lies midway between the lines $ax + by = 2c_1$, $ax + by = 2c_2$.

The gradient of the side AB of a rectangle $ABCD$ is 2. The point $(2, 3)$ lies on AB and the point $(-1, 2)$ on AD. The intersection of the diagonals is the point $(\tfrac{1}{2}, 5)$. Find the equations of the sides of the rectangle. (N.)

7 Find the equation of the chord joining the points
$$(at_1^2, 2at_1), \quad (at_2^2, 2at_2)$$
on the parabola $y^2 = 4ax$.

If P_1, P_2, P_3, P_4 are distinct points on the parabola, and if both the chords P_1P_2, P_3P_4 pass through the focus $(a, 0)$, prove that the chords P_1P_3, P_2P_4 meet on, or are parallel to, the directrix $x + a = 0$.

8 Find the values to three significant figures, of
$$\int_0^1 \frac{x}{1+x^2}\,dx, \quad \int_0^1 \frac{x}{\sqrt{1+x^2}}\,dx, \quad \int_0^1 \frac{\tan^{-1} x}{1+x^2}\,dx.$$ (C.)

9 (a) Differentiate with respect to x:

(i) $\sqrt{3 - 4x^2}$; (ii) $\sin^{-1} x + \ln\sqrt{1 - x^2}$; (iii) $\sin 3x - \cos^3 x$.

(b) Show that the equation of the tangent to the curve $y = e^x$ at the point where $y = 2$ is $y - 2x = 2 - \ln 4$. (N.)

10 Draw the graph of $\sqrt{\sin x}$ for values of x from 0 to $\pi/2$ and hence obtain an approximate value of $\int_0^{\pi/2} \sqrt{\sin x}\,dx$, giving your answer correct to two significant figures. (C.)

PAPER B (9)

1 Solve the equation $x^2 + 2yz = -11$, $y^2 + 2zx = -2$, $z^2 + 2xy = 13$.

2 (i) How many numbers greater than one million can be formed by using the figures 5, 5, 5, 5, 4, 4, 2? How many of these numbers are divisible by 4?

(ii) Write down the first four terms in the expansion of $(1-x)^{-\frac{1}{2}}$. By putting $x = \frac{1}{50}$, find the value of $\sqrt{2}$, correct to five decimal places. (N.)

3 OA, OB are perpendicular lines in a horizontal plane. P is a point vertically above A and Q is a point vertically above B. Prove that
$$PQ^2 = OP^2 + OQ^2 - 2AP \cdot BQ$$
and deduce that $\cos POQ = \sin AOP \sin BOQ$. (L.)

4 (a) Solve the equation $2\sin\theta + \sin 2\theta = \sin\frac{1}{2}\theta$, giving all solutions between $\pm 180°$.

(b) If $\sin 2A = \dfrac{2x}{1+x^2}$ and $\sin 2B = \dfrac{2y}{1+y^2}$, show that there are four possible values of $\tan(A-B)$, one of which is $\dfrac{y-x}{1+xy}$. (N.)

5 If a, b, c, d are the lengths of the sides AB, BC, CD, DA respectively of a cyclic quadrilateral, prove by using the Cosine Rule, that
$$AC^2 = \frac{(ac+bd)(ad+bc)}{ab+cd}; \qquad BD^2 = \frac{(ac+bd)(ab+cd)}{ad+bc}.$$

6 Find the equations of the tangent and normal to the ellipse
$$x^2 + 3y^2 = 2a^2$$
at the point $(a, a/\sqrt{3})$. If the tangent meets the x-axis at P and the normal meets the y-axis at Q, show that PQ touches the ellipse. (N.)

7 Points P, Q on a rectangular hyperbola are such that the tangent at Q passes through the foot of the ordinate at P. Show that the locus of the midpoint of the chord PQ is a rectangular hyperbola with the same asymptotes as the given hyperbola.

8 From a fixed point A, on the circumference of a circle of radius r, a straight line is drawn perpendicular to the tangent at a variable point P, cutting it at Y. Find the maximum area of the triangle APY.

If B is the other extremity of the diameter through A, find the maximum value of the sum of BP and PY. (N.)

9 Show, by use of the calculus, that for all positive values of x,

(i) $x - \dfrac{x^2}{2} < \ln(1+x) < x - \dfrac{x^2}{2} + \dfrac{x^3}{3}$,

(ii) $\dfrac{3\sin x}{2 + \cos x} < x$. (C.)

10 (i) Evaluate $\displaystyle\int_0^4 \frac{x\,dx}{\sqrt{2x+1}}$.

(ii) By means of the substitution $t = \tan x$, find $\displaystyle\int \frac{dx}{\cos^2 x + 4\sin^2 x}$. (C.)

PAPER B (10)

1 (a) If the roots of $ax^2 + 2bx + c = 0$ are real, show that the roots of
$$ax^2 + 2(ac+b)x + ac^2 + 2bc + c = 0$$
are also real.

(b) Show that the roots of the equation $ax^2 + 2bx + c = 0$ are

$$-\frac{b}{a}\left\{1 \pm \left(1 - \frac{ac}{b^2}\right)^{\frac{1}{2}}\right\}.$$

If ac is very small compared with b^2, deduce that the roots of the equation are approximately $-\dfrac{2b}{a} + \dfrac{c}{2b}$ and $-\dfrac{c}{2b}$. (N.)

2 Find, correct to two decimal places, the positive root of the equation $x^4 + 2x - 7 = 0$. (C.)

3 (i) Find a formula for all angles whose tangents are the same as that of a given angle α.

(ii) If $\tan(2A - 3B) = \cot(3A - 2B)$, and
$\tan(2A + 3B) = \cot(3A + 2B)$,

show that A and B are both multiples of $18°$ and that, if A is an odd multiple, then B is an even multiple.

(iii) Find the general solution of the equation $11 \sin x - 2 \cos x = 5$. (N.)

4 A straight pole AB makes an angle of $20°$ with the vertical; P is a point in the horizontal plane through A so that AP is perpendicular to AB and $AP = AB$. Find the angle which PB makes with the horizontal. (N.)

5 Prove that $\dfrac{1}{3} \geq \dfrac{\cot 3x}{\cot x} \geq 3$.

6 PQ is a variable focal chord of the parabola $y^2 = 4ax$; the normals at P, Q meet at N; show that the locus of N is the parabola $y^2 = a(x - 3a)$.

7 Obtain the equation of the normal to the ellipse $x^2/a^2 + y^2/b^2 = 1$ at the point $(a \cos \theta, b \sin \theta)$. The normal at P to the ellipse meets the x-axis at A and the y-axis at B. Show that the locus of Q, the midpoint of AB, is an ellipse with the same eccentricity as the original ellipse. Also, if the eccentric angle of P is $\pi/4$, show that PQ is a tangent to the second ellipse. (N.)

8 (a) Differentiate, with respect to x,

(i) $\dfrac{(x-2)(x-3)}{x-4}$, (ii) $\ln\{2x^2 + 1 + 2x\sqrt{x^2+1}\}$, (iii) $e^{-5x}\sin 3x$.

(b) If $x = a\cos^3\theta$, $y = b\sin^3\theta$, find in terms of θ, the values of $\dfrac{dy}{dx}$ and $\dfrac{d^2y}{dx^2}$. (N.)

9 The gradient of the curve whose equation is $y = ax^3 + bx^2 + c$ has the stationary value -6. When $x = -2$, y has the stationary value 2. Verify that the point $(-1, -2)$ lies on the curve, find the equation of the tangent at this point, and sketch the curve and tangent in the neighbourhood. (C.)

10 Evaluate the integrals:

(i) $\displaystyle\int_0^2 \frac{dx}{x^2+4}$; (ii) $\displaystyle\int_2^3 \frac{x}{(x+3)(x-1)}dx$; (iii) $\displaystyle\int_0^{\pi/3} \cos^3 x\, dx$; (iv) $\displaystyle\int_0^{\pi/4} \cos^2 2x\, dx$.

(N.)

PAPER B (11)

1 (a) Solve the equation $2^x \times 3^{2x} = 5$.

(b) Find the maximum and minimum values of $\dfrac{x+4}{2 - 3x - 2x^2}$.

Revision papers | 321

2 Find how many distinct selections of eight coins may be made from three pennies, five twopences, seven fivepences and nine tenpences, such that each selection includes at least one of each type of coin. (C.)

3 Write down the first four terms of the expansions of $\ln(1+x)$ and e^x in ascending powers of x, assuming the expansions are valid.

If x is so small that x^4 may be neglected, prove, by taking logarithms, the approximate formula $e^{-x}(1+x)^{1-\frac{1}{2}x} = 1 - x^2 + \frac{7}{12}x^3$. (N.)

4 (a) Prove the identities $\operatorname{cosec}\theta - \cot\theta = \tan\frac{1}{2}\theta$,
$$\sec\theta + \tan\theta = \tan(\tfrac{1}{2}\theta + \tfrac{1}{4}\pi).$$

(b) The median CN of a triangle ABC meets the side AB in the point N. If θ is the angle ANC, prove that
$$\frac{2c\cos\theta}{a^2 - b^2} = \frac{c\sin\theta}{2\Delta} = \frac{1}{CN},$$
where Δ is the area of $\triangle ABC$.

5 Eliminate θ between the equations
$$x = a\cos^2\theta + b\sin^2\theta, \quad y = (a-b)\sin\theta\cos\theta.$$

6 P is any point on the ellipse $x^2/a^2 + y^2/b^2 = 1$, $(a > b)$; A, A' are the ends of the major axis; AP meets the minor axis at Q. Prove that the line through Q parallel to $A'P$ is a tangent to one of the parabolas
$$ay^2 \pm 4b^2 x = 0. \tag{C.}$$

7 Obtain the equation of the tangent at the point (x_1, y_1) on the hyperbola $xy = c^2$ in the form $x/x_1 + y/y_1 = 2$.

The circle $2x^2 + 2y^2 = a^2 + b^2$ and the ellipse $b^2x^2 + a^2y^2 = a^2b^2$ intersect in the first quadrant at P. The line joining P to the origin meets the hyperbola $xy = c^2$ at Q and R. Show that the tangent at P to the ellipse and the tangents at Q and R to the hyperbola are all parallel. (N.)

8 (a) If a variable tangent to the curve $x^2y = c^3$ makes intercepts a, b on the axes of x and y respectively, prove that a^2b is constant.

(b) If the fraction $\frac{1}{2}$ is divided into two positive parts x, y, prove that the value of x^2y^3 is not greater than $0{\cdot}00108$.

9 Find the position of the centre of gravity of a plane lamina of uniform density in the form of a quadrant of an ellipse of semi-axes a and b. (C.)

10 Find: (i) $\displaystyle\int \frac{dx}{\sqrt{2x-x^2}}$; (ii) $\displaystyle\int \frac{1+3x}{(x-3)(x^2+1)}\,dx$; (iii) $\displaystyle\int_0^1 xe^x\,dx$.

PAPER B (12)

1 (i) Show that $x = 2$ is one root of the equation
$$x^3 - 4x^2 + (q+4)x - 2q = 0.$$
If the two remaining roots differ by 4, find the value of q and solve the equation completely.

(ii) Solve the equation $\sqrt{3(x-2)(x-3)} - \sqrt{(x-2)(x-5)} = x - 2$. (L.)

2 (a) If x is sufficiently small for powers of x above the third to be neglected, prove that
$$\ln\left(\frac{1+6x-6x^2}{1+x-6x^2}\right) = 5x\left(1 - \frac{7}{2}x + \frac{61}{3}x^2\right) \text{ approximately.}$$

(b) Prove that the coefficient of x^n in the expansion of $(1 - x + x^2)e^x$ in ascending powers of x is $(n-1)^2/n!$. (N.)

3 (a) If $\tan\theta = \tfrac{3}{4}$, calculate in fractional form, the values of $\sin 2\theta$ and $\sin 4\theta$.
Any point P is taken on the circumference of a circle of radius $5r$. With P as centre an arc of a circle radius $8r$ is drawn to cut the given circle. Show that the smaller of the two areas into which the circle is divided by the arc is $2r^2(12 - 7\theta)$ where $\tan\theta = \tfrac{3}{4}$.

(b) Find the least value of $\theta > 0$ which satisfies the equation
$$\tan 2\theta - 2\tan\theta = \sin 2\theta. \quad \text{(N.)}$$

4 $ABCD$ is a trapezium in which AB is parallel to CD; $AB = a$, $BC = b$, $CD = c$, $DA = d$; prove that the squares of the diagonals are
$$ac + \frac{ab^2 - cd^2}{a - c} \quad \text{and} \quad ac + \frac{ad^2 - cb^2}{a - c}.$$

5 Show that, if (x_1, y_1) is a point on the circle $x^2 + y^2 = a^2$, then for any value of k the equation
$$x^2 + y^2 - a^2 + k(xx_1 + yy_1 - a^2) = 0$$
represents a circle which touches the circle $x^2 + y^2 = a^2$ at the point (x_1, y_1).
Determine the equation of the circle C which passes through the point $(5, 1)$ and touches the circle $x^2 + y^2 = 2$ at the point $(1, 1)$, and find the equations of the tangents to C that are parallel to the line $y = x$. (N.)

6 A variable chord through the focus of the parabola $y^2 = 4ax$ cuts the curve at P and Q. The straight line joining P to $(0, 0)$ cuts the straight line joining Q to $(-a, 0)$ at R. Prove that the equation of the locus of R is
$$y^2 + 8x^2 + 4ax = 0. \quad \text{(N.)}$$

7 P_1, P_2 are points on the curve $xy = c^2$; the tangents at P_1, P_2 meet the line $y = 0$ at A_1, A_2 respectively and the line $x = 0$ at B_1, B_2 respectively. Prove that the chord $P_1 P_2$ passes through the midpoints of $A_1 A_2$ and $B_1 B_2$. (C.)

8 (i) Find the coordinates of the point P on the curve $y = 2x^2 + 8/x$, the tangent at which passes through the origin. Obtain also the coordinates of Q where the tangent at P meets the curve again.

(ii) Find the maximum value of the function $2\sin x + \sin 2x$. (N.)

9 Find (i) $\displaystyle\int \frac{dx}{x^2(x-1)}$ where $x > 1$; (ii) $\displaystyle\int \tan^4 x\, dx$;

and evaluate $\displaystyle\int_3^6 \frac{x(x-4)}{\sqrt{x-2}}\, dx.$ (N.)

10 (a) A line drawn parallel to the y-axis meets the curve $4y = e^{2x} + e^{-2x}$ at P and the x-axis at N. Show that the projection of PN on the normal to the curve at P is of constant length.

(b) Sketch the curve $y^2 = (x-3)(x-5)^2$, and find the area of the loop. (N.)

PAPER C (1)

1 If x is real and $y = \dfrac{5x^2 + 8x + 4}{x^2 + x}$, prove that y cannot lie between -4 and $+4$.

Find the coordinates of the turning points on the graph of y and the value of x at the point where the graph crosses its horizontal asymptote. Sketch the graph from $x = -3$ to $x = +3$. (N.)

2 Write down the first five terms of the series for $\ln(1+\tfrac{1}{2})$ and $\ln(1-\tfrac{1}{2})$, and deduce that $\ln\sqrt{3} = \dfrac{1}{2} + \dfrac{1}{3}\cdot\dfrac{1}{2^3} + \dfrac{1}{5}\cdot\dfrac{1}{2^5} + \ldots$

Show further that $\ln 2\sqrt{3} = 1 + \left(\dfrac{1}{2}+\dfrac{1}{3}\right)\dfrac{1}{2^2} + \left(\dfrac{1}{4}+\dfrac{1}{5}\right)\dfrac{1}{2^4} + \left(\dfrac{1}{6}+\dfrac{1}{7}\right)\dfrac{1}{2^6} + \ldots$ (N.)

3 Prove that the equation $4\cos^2\dfrac{\pi x}{2} = 3x+1$, where x is measured in radians, has a root lying between $x = 0.4$ and $x = 0.5$. Use Newton's method to obtain this root correct to two decimal places.

4 In any triangle ABC, prove that
  (i) $\sin\tfrac{1}{2}(A-B) = \dfrac{a-b}{c}\cos\dfrac{C}{2}$,
  (ii) $\sin^2 B + \sin^2 C = 1 + \cos(B-C)\cos A$. (N.)

5 Circles are drawn to touch the line $x + a = 0$ and to pass through the point $(a, 0)$. Find the equation of the locus of the centre.

Deduce the number of circles which pass through two given points and touch a given line. (C.)

6 The normal at a point P on the parabola $y^2 = 4ax$ meets the ellipse $2x^2 + y^2 = c^2$ in the points M, N. Prove that P is the midpoint of MN. Hence, or otherwise, prove that the conics cut at right angles. (C.)

7 (a) Differentiate with respect to x:
  (i) $x\tan^{-1}x - \ln\sqrt{1+x^2}$, (ii) $(2\sin 2x - \cos 2x + 5)e^{-x}$.

(b) The abscissa x of a point P moving along the x-axis is given in terms of the time t by the equation $x = 16(e^{-t} - e^{-2t})$.

Determine the acceleration when P reaches its maximum distance from O and the maximum speed in the subsequent motion. (N.)

8 Find the values of $\dfrac{dy}{dx}, \dfrac{d^2y}{dx^2}$ and the radius of curvature at the point $(1, 0)$ on the curve $3x^2 + 8xy - 3y^2 = 3$. Draw a figure showing the curve, the tangent, and the circle of curvature at this point. (C.)

9 (a) Find $\displaystyle\int \dfrac{1+x}{\sqrt{1-x}}\,dx$ and $\displaystyle\int \sin^3 x \cos^3 x \,dx$.

(b) Find the value of a for which $\displaystyle\int_0^1 \dfrac{x-a}{(x+1)(3x+1)}\,dx = 0$. (N.)

10 Make a rough sketch of the curve $xy^2 = a^2(a-x)$ and prove that the area between the curve and the axis of y is πa^2.

PAPER C (2)

1 Find λ in terms of a and b so that the values of x given by the equation
$$\dfrac{\lambda}{2x} = \dfrac{a}{x+1} + \dfrac{b}{x-1}$$
may be equal, a and b being positive and unequal.

If λ_1, λ_2 are the two values of λ for which this is true, and x_1, x_2 the corresponding values of x, prove that $\lambda_1\lambda_2 = (a-b)^2$ and $x_1 x_2 = 1$. (N.)

2 (a) Prove that, if n is any positive integer, the coefficient of x^n in the expansion of $\left(\dfrac{1+x}{1-x}\right)^3$ in ascending powers of x is $4n^2 + 2$.

(b) Prove the approximate formula
$$\ln\left(\dfrac{a}{b}\right) = \tfrac{1}{2}(a-b)\left(\dfrac{1}{a} + \dfrac{1}{b}\right),$$
where a, b are positive and $(a-b)/a$ is small, and show that the error is $\tfrac{1}{6}\{(a-b)/a\}^3$ approximately. (N.)

3 Find the values of θ between $0°$ and $360°$ which satisfy the equation
$$1 + \cos\theta + \cos 2\theta + 2(\sin\theta + \sin 2\theta) = 0. \quad \text{(C.)}$$

4 If $\cos\theta = \tan\theta$, find the value of $\sin\theta$.

In one diagram, sketch the graphs of $\cos\theta$ and $\tan\theta$ from $\theta = -2\pi$ to $\theta = 2\pi$. Show that, when the units of length on the axes are equal, the graphs intersect at right angles. (N.)

5 A right circular cone has base radius 3 cm and height 7 cm. A plane is drawn to cut at right angles a generator at a point 6 cm. from the vertex. Find the ratio of the areas into which this plane divides the base.

6 P is a variable point on the parabola $y^2 = 4ax$, latus rectum LL'. Through L, L' respectively, lines are drawn perpendicular to LP, $L'P$. Prove that these lines meet on a fixed straight line. Verify that the normal at L and the line through L' parallel to the axis of the parabola intersect on this line. (C.)

7 Find the equation of the tangent to the curve $3ay^2 = x^2(x+a)$ at the point $(2a, 2a)$.

Find the coordinates of the point P at which this tangent meets the curve again, and prove that it is normal to the curve at P. (C.)

8 (a) Obtain the equations of the tangents at the two points of inflexion on the curve $y = 3x^2 - 1/x^2$.

(b) A solid circular cylinder of volume V is placed with its axis coinciding with the axis of a hollow right circular cone of volume V_0 and with the circumference of one end in contact with the curved surface of the cone. Prove that, whatever the radius of the cylinder may be, part of the cylinder will be outside the cone if $V > \tfrac{4}{9} V_0$. (N.)

9 Evaluate
$$\int_1^2 \dfrac{dx}{\sqrt{12 + 8x - 4x^2}}, \quad \int_0^3 x \ln(1+x)\,dx \quad \text{and} \quad \int_2^3 \dfrac{dx}{x(1+x^2)}. \quad \text{(C.)}$$

10 If the velocity of a point moving in a straight line is v when its displacement from a fixed oigin on the line is x, prove that the acceleration is $v\dfrac{dv}{dx}$.

A particle P has an acceleration inversely proportional to the square of its distance from a fixed point O in a direction away from O. P is initially at rest at a distance 4 m from O. When P is 5 m from O its acceleration is $8\,\text{m s}^{-2}$. Prove that when OP is 8 m, the velocity of P is $\sqrt{50}\,\text{m s}^{-1}$.

PAPER C (3)

1 Find the greatest value, for a given value of n, of the number of combinations of n things, all different, taken r together (a) when n is even, (b) when n is odd.

Find the number of combinations of four letters which can be formed from the letters a, b, c, d, e any one of which may be used twice at most in any particular combination. (N.)

2 Sketch, in the same diagram, the graphs of e^x, e^{-x} and $\cosh x$.

If $y = c \cosh \dfrac{x}{c}$ and $y - c = k$, where k is small and positive, prove that
$$x^2 = 2ck - \tfrac{1}{3}k^2 \text{ approximately.} \quad \text{(C.)}$$

3 Obtain an expression for $\tan(A+B+C)$ in terms of the tangents of A, B, C. If A, B, C are the angles of a triangle, prove that
  (i) $\tan A + \tan B + \tan C = \tan A \tan B \tan C$,
  (ii) $\cot\tfrac{1}{2}A + \cot\tfrac{1}{2}B + \cot\tfrac{1}{2}C = \cot\tfrac{1}{2}A \cot\tfrac{1}{2}B \cot\tfrac{1}{2}C$. (N.)

4 In $\triangle ABC$, A, B, c are measured and a is calculated. If there is an error of x minutes in the measured value of A, show that the error in the calculated value of a is approximately $\dfrac{\pi x c \sin B}{10800 \sin^2(A+B)}$.

5 Show that, if the line $lx + my + n = 0$ touches the circle
$$x^2 + y^2 + 2gx + 2fy + c = 0,$$
then
$$(l^2 + m^2)(g^2 + f^2 - c) = (gl + fm - n)^2.$$

If $ab > 0$, find the equations of the common tangents of the circles
$$x^2 + y^2 = 2ax \quad \text{and} \quad x^2 + y^2 + 2bx = 0.$$

6 A point moves in a hyperbola so that the difference of its distances from the points $(4, 0)$ and $(-4, 0)$ is constant and equal to 6. Find the equation of the hyperbola. Prove that the line $y = mx + c$ will be a tangent to the hyperbola if $c^2 = 9m^2 - 7$, and hence obtain the equation of the locus of the meet of perpendicular tangents. (C.)

7 (a) Differentiate with respect to x, (i) $\tan^{-1}\left(\dfrac{x-6}{3x+2}\right)$, (ii) $\dfrac{x+2}{x-2}e^{-x}$.

  (b) If $xy = h - 9c^2 x + x^3 + k \ln x$, where h, k, c are constants, determine the values of x for which
$$x^2 \dfrac{d^2 y}{dx^2} + 3x \dfrac{dy}{dx} + y = 0. \quad \text{(N.)}$$

8 (a) Find the values of x for which $y = x \sin x + \cos x$ is stationary in the range $-\pi \leqslant x \leqslant \pi$, and determine in each case whether the stationary value corresponds to a maximum value of y, to a minimum value of y, or to a point of inflexion.

  (b) Find the *least* value of $y = a \sin x - b \cos x$ for values of x in the range $0 \leqslant x \leqslant \pi$, in the cases (i) $b > a > 0$, (ii) $b > 0 > a$. (C.)

9 Find $\displaystyle\int \dfrac{x+1}{x^2(x+2)} dx$ and $\displaystyle\int x^2 \ln x \, dx$.

Show that
$$\int_0^1 x^3 \sqrt{1-x^2}\, dx = 2/15. \quad \text{(N.)}$$

10 Solve the differential equations

  (i) $(1+x)^2 \dfrac{dy}{dx} + y^2 = 1$,

  (ii) $(1+x^2) \dfrac{dy}{dx} + xy = 1 + x^2$. (C.)

PAPER C (4)

1 Write down the series for $\sqrt{1+x}$ in ascending powers of x as far as the term in x^4.

Show also that the error in taking $\frac{1}{4}(6+x) - \dfrac{1}{2+x}$ as an approximation to $\sqrt{1+x}$ when x is small is approximately $x^4/128$. (C.)

2 If p and q are the roots of $\dfrac{1}{x+a} + \dfrac{1}{x+b} + \dfrac{1}{x} = 0$ and $a^2 + b^2 = 4ab$, prove that $p^2 + q^2 = 6pq$.

3 If $f(x) = (e^x - 1)/(e^x + 1)$, show that $f(-x) = -f(x)$.
If x is so small that x^5 and higher powers of x are negligible, find the values of the constants a, b, c, d in the approximation

$$\frac{e^x - 1}{e^x + 1} = ax + bx^2 + cx^3 + dx^4.$$ (N.)

4 Prove that $2\cos\dfrac{\theta}{2}(\cos\theta - \cos 2\theta) = \cos\dfrac{\theta}{2} - \cos\dfrac{5\theta}{2}$.

Solve the equation $\cos\theta - \cos 2\theta = \frac{1}{2}$, for values of θ in the range $0 \leqslant \theta \leqslant 2\pi$. (C.)

5 A straight line is drawn through the point $P(-2a, 0)$ to meet the parabola $y^2 = 4ax$ at the points Q, R. Prove that the normals to the parabola at Q, R meet on the parabola. (C.)

6 Prove that the equation of the chord of the ellipse $b^2x^2 + a^2y^2 = a^2b^2$ joining the points $(a\cos\alpha, b\sin\alpha)$ and $(a\cos\beta, b\sin\beta)$ is

$$\frac{x}{a}\cos\frac{\alpha+\beta}{2} + \frac{y}{b}\sin\frac{\alpha+\beta}{2} = \cos\frac{\alpha-\beta}{2}.$$

Through a point P on the major axis of an ellipse, a chord HK is drawn. Prove that the tangents at H and K meet the line through P at right angles to the major axis at points equidistant from P. (C.)

7 Show that, for real values of x, the function $\dfrac{4x - 13}{(x-1)(x-3)}$ cannot have any value between 1 and 4. For what values of x is the function positive?

8 Sketch the curve $(a^2 + x^2)y = a^2(a - x)$ where $a > 0$. Verify that the three points of inflexion are collinear. (N.)

9 Evaluate

(i) $\displaystyle\int_1^4 \frac{x^2 + 4x - 14}{(x+2)(x+5)(x+8)}\,dx,$ (ii) $\displaystyle\int_0^{\pi/4} \sin^2 x \sin 2x\,dx,$

(iii) $\displaystyle\int_1^e x \ln x\,dx.$ (N.)

10 Solve the differential equations

(i) $(x^2 - x)\dfrac{dy}{dx} = y$, where $y = 1$ when $x = 2$,

(ii) $\cos x \dfrac{dy}{dx} + ny \sin x = \cos^{n+3} x$, where $y = 0$ when $x = 0$. (C.)

PAPER C (5)

1 Show that, if $y = \dfrac{x^2+1}{x^2-a^2}$, y takes all real values twice except those for which $-\dfrac{1}{a^2} \leqslant y \leqslant 1$.

Sketch the curve $y = \dfrac{x^2+1}{x^2-4}$, indicating its asymptotes. (C.)

2 Sketch the graph of the function $x^3 - 3x$. Show that, if $|k| < 2$, the equation $x^3 - 3x = k$ has three roots which lie one in each of the intervals $-2 < x < -1$, $-1 < x < 1$, $1 < x < 2$. Show further that the root of smallest modulus lies between $-\frac{1}{3}k$ and $-\frac{1}{2}k$. (N.)

3 Prove that if $\tan x = k \tan(A - x)$, then $\sin(2x - A) = \dfrac{k-1}{k+1} \sin A$.

Find all the angles between $0°$ and $360°$ which satisfy the equation
$$2\tan x - \tan(30° - x) = 0.$$ (C.)

4 Two consecutive plane faces of a pyramid on a horizontal base make angles α and β with the horizontal; the corresponding base edges are inclined at an angle γ. If the line of intersection of the two faces is inclined at an angle θ to the horizontal, prove that
$$\cot^2\theta = \operatorname{cosec}^2\gamma\,(\cot^2\alpha + \cot^2\beta + 2\cot\alpha\cot\beta\cos\gamma).$$

5 Find the equation of the normal at the point $(at^2, 2at)$ of the parabola $y^2 = 4ax$. The point O is the origin, P is the point $(at^2, 2at)$ and the point Q on the parabola is such that OP and PQ are equally inclined in opposite senses to the axis of the parabola. Prove that the circumcentre of the triangle OPQ lies on the normal at P. (C.)

6 Normals are drawn from the point $\left(\dfrac{c}{4}, 2c\right)$ to the rectangular hyperbola $2xy = c^2$. Show that the feet of the only two real normals will lie on the line $4x - 2y + 3c = 0$.

7 Prove that the only maximum value of the function xe^{-x} occurs when $x = 1$. If a is a positive number less than $1/e$, show that the equation $xe^{-x} = a$ has two and only two real solutions. [It may be assumed that $xe^{-x} \to 0$ as $x \to \infty$.]

If a is chosen so that one of these solutions is $x = 2$, verify that the other root is about 0·4 and find its value correct to two decimal places. (C.)

8 Find the nth differential coefficient with respect to x of (i) $\cos ax$, (ii) $\ln(1+x)$. If $x = ay^2 + by + c$, where a, b and c are constants, prove that
$$\frac{d^2y}{dx^2} = -2a\left(\frac{dy}{dx}\right)^3.$$ (C.)

9 Prove that
(i) $\cosh^{-1} x = \ln(x + \sqrt{x^2 - 1})$;
(ii) the area enclosed by the curve $y^2(x^2 - 2x - 3) = 1$ and the lines $x = 3\frac{1}{2}, x = 5$ is $2\ln\dfrac{2+\sqrt{3}}{2}$. (C.)

10 The parametric equations of a curve are $x = a\cos^3\theta$, $y = a\sin^3\theta$. If P is the point with parameter θ, and the tangent at P meets the axes of coordinates at Q and R, prove (i) the length QR is constant and equal to a, (ii) the radius of curvature at P is $3a\sin\theta\cos\theta$. (C.)

PAPER C (6)

1 Write down the series in ascending powers of x for $(1-x)^{-n}$ and state the range of values of x for which the expansion is valid.

Sum to infinity the series $\dfrac{5}{12} + \dfrac{5.8}{12.18} + \dfrac{5.8.11}{12.18.24} + \ldots$ (C.)

2 Define the hyperbolic functions $\cosh x$ and $\sinh x$.
Prove that
  (i) $\cosh(u+v) = \cosh u \cosh v + \sinh u \sinh v$;
  (ii) $\sinh^2 u \cos^2 v + \cosh^2 u \sin^2 v = \tfrac{1}{2}(\cosh 2u - \cos 2v)$. (C.)

3 Find the general solutions of the following equations, where θ is measured in degrees:
  (i) $2\sin 3\theta - 7\cos 2\theta + \sin\theta + 1 = 0$,
  (ii) $\cos\theta - \sin 2\theta + \cos 3\theta - \sin 4\theta = 0$.

4 If
$$\frac{\sin(A+B)}{p} = \frac{\sin(A-B)}{q} = \frac{\sin A}{r},$$
prove that $\cos B = \dfrac{p+q}{2r}$ and $\sin B = \dfrac{p-q}{2r}\tan A$.

Hence, show that $\cos 2A = \dfrac{p^2 + q^2 - 2r^2}{2(r^2 - pq)}$.

5 P is a point with coordinates (X, Y). The feet of the perpendiculars from P to the lines $y = x\tan\alpha$ and $y = -x\tan\alpha$ are Q and R.
The midpoint of QR is M. Show that the angle POM, where O is the origin, is
$$\tan^{-1}\frac{XY\cos 2\alpha}{X^2\cos^2\alpha + Y^2\sin^2\alpha}.$$

6 Show that the coordinates of a point on the hyperbola $x^2/a^2 - y^2/b^2 = 1$ can be expressed in terms of a parameter t by the relations
$$\frac{x-a}{at^2} = \frac{x+a}{a} = \frac{y}{bt}.$$

Show also that if any point P on the hyperbola be joined to the ends A, A' of the axis along $y = 0$, the line through A at right angles to PA meets PA' on a fixed line. (C.)

7 Prove that, for $x > 0$, each of the functions $\ln\left(1+\dfrac{1}{x}\right) - \dfrac{2}{2x+1}$ and $1/\sqrt{x(x+1)} - \ln\left(1+\dfrac{1}{x}\right)$ decreases as x increases. To what limits do these functions tend as x tends to infinity?

Deduce that, if $x > 0$, $2/(2x+1) < \ln\left(1+\dfrac{1}{x}\right) < 1/\sqrt{x(x+1)}$. (N.)

8 (i) Find
$$\int\frac{1+x}{1+\cos x}\,dx \quad\text{and}\quad \int\frac{x+1}{x^2(x^2+1)}\,dx.$$

  (ii) Evaluate $\displaystyle\int_0^1 x^2(1-x)^{1/3}\,dx$. (C.)

9 An arc of a circle of radius a subtending an angle 2θ at the centre is rotated about its chord through an angle 2π. Prove that the area of the surface of revolution so formed is $4\pi a^2 (\sin\theta - \theta\cos\theta)$.

It is required to find the value of θ for which this area is $2\pi a^2$. Verify, using tables, that $\theta = 1\cdot 2$ radians is an approximate value, and obtain a closer approximation, correct to three decimal places. (C.)

10 Solve the differential equations

(i) $x\dfrac{dy}{dx} = 1 - y^2$; (ii) $\dfrac{dy}{dx} = 1 + \dfrac{y}{x}$; (iii) $\dfrac{d^2 y}{dx^2} - 3\dfrac{dy}{dx} + 2y = 0$.

PAPER C (7)

1 (a) Find the sum of all the odd numbers which are less than $6n + 1$ and are not multiples of 3.

(b) By expressing the general term in partial fractions, find the sum to n terms of the series

$$\frac{2}{1.3.4} + \frac{3}{2.4.5} + \frac{4}{3.5.6} + \ldots$$

(c) Sum the series $\displaystyle\sum_{n=1}^{\infty} \frac{n+1}{n} x^n$ if $|x| < 1$.

2 If $x + \dfrac{1}{x} = 1$, prove that $x^7 + \dfrac{1}{x^7} = 1$. (C.)

3 Write down the expansions of $\sin\theta$ and $\cos\theta$ in ascending powers of θ.

The area of a given circle is A and the areas of the inscribed and circumscribed regular polygons of n sides are denoted by A_1 and A_2.

Evaluate $\displaystyle\lim_{n\to\infty} \frac{A - A_1}{A_2 - A}$. (C.)

4 Show that if $a^2 + b^2 > c^2$ there are in general two solutions of

$$a\cos\theta + b\sin\theta = c$$

in the range $0 < \theta < 360°$.

Solve $\qquad 8\cos\theta - \sin\theta = 4$,

and if α and β are the values of θ in the above range, find the quadratic equation with integral coefficients, whose roots are $\sin\alpha$ and $\sin\beta$. (C.)

5 (a) Find the equation of the circle having as diameter the segment intercepted on the line $x\cos\alpha + y\sin\alpha = p$ by the circle $x^2 + y^2 = r^2$, given that $r > p > 0$.

(b) Two circles of radii r, R, where $r < R$, touch internally. Prove that the locus of the centres of circles which touch the larger circle internally and the smaller externally is an ellipse of eccentricity $(R - r)/(R + r)$. (N.)

6 Find the equation of the tangent at the point $(a\cos\theta, b\sin\theta)$ to the ellipse $x^2/a^2 + y^2/b^2 = 1$.

The chords AP and $A'P$, where AA' is the major axis of the ellipse and P is the point $(a\cos\phi, b\sin\phi)$, meet the tangents at A', A respectively at the points Q, R. Prove that QR is a tangent to the ellipse $x^2/a^2 + y^2/4b^2 = 1$. (C.)

7 Find the radius of curvature at the point (x_1, y_1) on the curve $y\sqrt{5} = x^3$.

Hence show that the radius of curvature of this curve has a minimum value at two points of the curve, and find the coordinates of these points. (C.)

8 Prove that the equation of the normal at any point on the curve given by $x = 2\cos t - \cos 2t$, $y = 2\sin t - \sin 2t$ is
$$x\cos\tfrac{3}{2}t + y\sin\tfrac{3}{2}t = \cos\tfrac{1}{2}t.$$
Prove that the normals at the points t and $t + \pi$ intersect at right angles, and find the locus of their point of intersection.

9 (a) What is the value of $\int \dfrac{dx}{x}$, (i) when x is positive, (ii) when x is negative?

Evaluate $\int_1^2 \dfrac{dx}{3x - 8}$.

(b) Find $\int \dfrac{dx}{\sin x}$ and $\int \dfrac{x^2}{1 - x^4} dx$.

10 Prove that the area bounded by the two curves $ay = 2x^2$, $y^2 = 4ax$ is $\tfrac{2}{3}a^2$, and find the coordinates of the centroid of this area. (C.)

PAPER C (8)

1 Determine the number $_nP_r$ of permutations and the number $_nC_r$ of combinations of n different things taken r at a time.

Show that, if two of the n things are alike, the number of combinations of r things is $_{n-1}C_r + {}_{n-2}C_{r-2}$ and the number of permutations of r things is
$$_{n-1}P_r + \frac{r(r-1)}{1 \cdot 2} {}_{n-2}P_{r-2}.$$
(C.)

2 Find the ranges of values of α in the interval $0 \leqslant \alpha \leqslant 2\pi$ for which the roots of the equation in x,
$$x^2 \cos^2\alpha + ax(\sqrt{3}\cos\alpha + \sin\alpha) + a^2 = 0,$$
are real. (C.)

3 Write down the series for $\ln(1 + x)$ in ascending powers of x and state the range of values of x for which it is valid.

Prove that, if $n > 1$,
$$\ln\frac{n}{n-1} > \frac{1}{n} > \ln\frac{n+1}{n};$$
and deduce that, if n is a positive integer,
$$1 + \ln n > 1 + \frac{1}{2} + \frac{1}{3} + \ldots + \frac{1}{n} > \ln(n + 1).$$
(C.)

4 Find the real general solutions of the equations in θ:
(i) $\tan\theta = a\cos 2\theta$, given that $\theta = 60°$ is one solution;
(ii) $\cos\theta = 2\tan 2\theta$. (C.)

5 Find the coordinates of the point of intersection of the tangents to the parabola $y^2 = 4ax$ at the points $(at_1^2, 2at_1)$ and $(at_2^2, 2at_2)$.

The line PQ is a tangent to the parabola $y^2 = 4ax$. The intercepts on PQ made (i) by the pair of tangents drawn from a given point A, and (ii) by the pair of tangents drawn from another given point B are equal in length. Prove that the intercepts on any other tangent by the pairs of tangents drawn from A and B are equal. (C.)

6 Two pairs of parallel tangents to the hyperbola $xy = c^2$ form a parallelogram $PQRS$. Prove that if one pair of opposite vertices lie on the hyperbola $xy = k^2$, the other pair lie on the hyperbola $xy(k^2 - c^2) = c^2 k^2$. (C.)

7 A straight line makes equal intercepts of $2a$ on the axes of x and y. Prove that its equation in polar coordinates is $r\cos(\theta - \pi/4) = a\sqrt{2}$, θ being measured in a counter-clockwise sense from the axis of x.

Prove that the equation $r = 4a\cos\theta$ represents a circle.

Obtain the values of θ for the two points of intersection of the straight line and the circle. (C.)

8 A particle is projected from a point on a straight line with velocity u and moves in that line in such a way that when it has traversed a distance s its velocity is $u/(1 + ksu)$, where k is a constant. Prove that its retardation varies as the cube of its velocity and find the time taken to reduce its velocity to $u/2$.

9 Find (i) $\int x(\ln x)^2 \, dx$, (ii) $\int \dfrac{dx}{\cos x - \cos^2 x}$.

Evaluate $\displaystyle\int_0^{a/2} x^2 \sqrt{a^2 - x^2}\, dx$. (C.)

10 Solve the equations

(i) $2y(x+1)\dfrac{dy}{dx} = 4 + y^2$, given that $y = 2$ when $x = 3$,

(ii) $y\cos^2 x \dfrac{dy}{dx} = \tan x + 2$, given that $y = 2$ when $x = \pi/4$. (C.)

PAPER C (9)

1 If $S \equiv 5x^2 + 4x + 2$ and $S' \equiv x^2 + 1$, find the values of λ for which $S + \lambda S'$ is a multiple of the square of a linear expression in x.

Express S and S' in the forms
$$S \equiv p(x-\alpha)^2 + q(x-\beta)^2; \quad S' \equiv p'(x-\alpha)^2 + q'(x-\beta)^2$$
determining the values of p, q, p', q', α and β. (C.)

2 (a) Express in partial fractions $6x/(x^6 - 1)$.

(b) Find the sum of the first n terms of the series
$$1 + 2^2 x + 3^2 x^2 + 4^2 x^3 + \ldots.$$
If $|x| < 1$, find also the sum to infinity. (N.)

3 Prove that

(i) $\cot\dfrac{\theta}{2} \geq 1 + \cot\theta$ if $0 < \theta < \pi$.

(ii) $(1 - \sin\theta)(5\sin\theta + 12\cos\theta + 13) \geq 0$ for all real values of θ. (C.)

4 A lighthouse A is at a distance a due south of a tower B on the coast. A tower C is at a distance a due east of B. From a ship S at sea eastward of the line AB and southward of the line BC the difference in the bearings of A and B is α and the difference in the bearings of B and C is β. If the angle ABS is γ, show that
$$\cot\gamma = \frac{\cot\alpha - 1}{\cot\beta - 1},$$
provided that neither α nor β is equal to $\pi/4$.

Find the distance of the ship from B in terms of a, α and β. (N.)

5 Show that, if $a > 1$, the roots of the equation $a \sin \theta + \cos \theta = a$, lying between 0 and 2π are $\frac{1}{2}\pi$ and an acute angle α.

Show also that, if t is very small, the corresponding roots of the equation $a \sin \theta + \cos \theta = a + t$ are $\frac{1}{2}\pi - t$ and $\alpha + t$.

6 Find the equation of the normal to the parabola $y^2 = 4ax$ at the point $(at^2, 2at)$.

Prove that, if $p^2 > 8$, two chords can be drawn through the point $(ap^2, 2ap)$ which are normal to the parabola at their second points of intersection, and that the line joining these points of intersection meets the axis of the parabola in a fixed point, independent of p. (C.)

7 Sketch the curve $x^3 = 27ay^2$ and find the equation of the tangent at the point $(3at^2, at^3)$. Prove that the locus of the point of intersection of perpendicular tangents to the curve is $y^2 = 4a(x - 4a)$. (C.)

8 If a curve touches the y-axis at the origin, show that its radius of curvature at the origin is $\lim_{x \to 0} y^2/2x$.

Find the value of a, assumed positive, if the radius of curvature of $y^2 = ax + 5x^3$ at the origin is equal to the radius of curvature of $y = \sin^2 x$ at the point $(\pi/6, 1/4)$. (C.)

9 Evaluate (i) $\int_0^\infty e^{-2x}(1 + x^2)dx$, (ii) $\int_0^1 \frac{dx}{(1+x^2)^{3/2}}$, (iii) $\int_0^3 \frac{1-x}{\sqrt{1+x}}dx$.

10 (i) Obtain the solution of the equation $\sin x \frac{dy}{dx} + 2y \cos x = x$, which is such that $y = 1$ when $x = \pi/2$.

(ii) The number N of bacteria in a culture increased at a rate proportional to N. The value of N was 100 initially and increased to 332 in one hour. What was the value of N after 80 minutes?

PAPER C (10)

1 (a) Prove that the sum to infinity of the series

$$1 + \frac{2x}{1!} + \frac{3x^2}{2!} + \frac{4x^3}{3!} + \ldots$$

is $(x + 1)e^x$.

(b) If $0 < |x| < 1$, find the sum of the infinite series

$$\frac{1}{1 \cdot 2} + \frac{x}{2 \cdot 3} + \frac{x^2}{3 \cdot 4} + \ldots \qquad (C.)$$

2 (i) If $y = x + \frac{1}{x}$, express $x^2 + \frac{1}{x^2}$ in terms of y, and solve the equation

$$x^4 - 2x^3 - 6x^2 - 2x + 1 = 0.$$

(ii) If $x^3 + x + G = 0$ and $y = x + \frac{1}{x}$, prove that

$$Gy^3 + (y - G)^2 = 0.$$

3 A lamp casts a shadow of a vertical rod on a horizontal floor; the length of the shadow is a, and the vertical and horizontal distances of the lower end of the rod from the near end of the shadow are b and c respectively. If the rod subtends equal angles at the two ends of the shadow, prove that the height of the lamp above the floor is $abc/(b^2 - c^2)$. (N.)

Revision papers | 333

4 Find the limiting value of the function $\dfrac{e^x - 1 + \ln(1-x)}{\sin^3 x}$ as x tends to zero.

5 A point P moves in a plane so that the ratio of its distances from two fixed points A, B in the plane is constant. Prove that its locus is a circle.

If $PA/PB = \lambda > 1$, and $AB = 2a$, prove that the radius of the circle is $2a\lambda/(\lambda^2 - 1)$, and find the length of the tangent to the circle from the mid-point of AB.   (C.)

6 Obtain the equations and points of contact of the tangents to
$$x^2/a^2 - y^2/b^2 = 1$$
which are parallel to $y = mx$. Interpret the case in which $m^2 a^2 = b^2$.

Show that the intercept made on any tangent by the asymptotes of the curve is bisected by the point of contact of the tangent.   (C.)

7 Find $\dfrac{dy}{dx}, \dfrac{d^2y}{dx^2}, \dfrac{dx}{dy}, \dfrac{d^2x}{dy^2}$ in terms of x and y when $x^3 + 3xy + y^3 = 0$.   (N.)

8 Sketch the curve $x = 3t - 5\cos t$, $y = 4\sin t$ for values of t from 0 to π.

Find the area of the surface of revolution formed when this arc of the curve is rotated about the x-axis through an angle of 2π.   (C.)

9 Find the indefinite integrals of
$$\frac{x}{(x^2+1)(x-1)}, \quad \frac{1}{e^x+1}, \quad e^x \cos x$$
and evaluate
$$\int_0^\pi x^2 \sin x \cos x \, dx.$$

10 The equation of motion of a particle of unit mass moving vertically downwards against a resistance of kv^2, where k is a constant and v is the velocity at time t, is
$$g - kv^2 = \frac{dv}{dt}.$$

If the particle starts from rest, show that
$$v = \frac{\alpha(e^{2k\alpha t} - 1)}{e^{2k\alpha t} + 1},$$
where $\alpha = \sqrt{g/k}$. Find the limiting velocity of the particle.   (C.)

PAPER C (11)

1 If x, y are positive, show that $\left(\dfrac{x^3+y^3}{2}\right)^2 \geq \left(\dfrac{x^2+y^2}{2}\right)^3$ and that
$$\left(\frac{2x^3+y^3}{3}\right)^2 \geq \left(\frac{2x^2+y^2}{3}\right)^3.$$

By the substitution $x^2 + y^2 = 2w^2$ or otherwise, show that if x, y, z are positive,
$$\left(\frac{x^3+y^3+z^3}{3}\right)^2 \geq \left(\frac{x^2+y^2+z^2}{3}\right)^3.$$ (N.)

2 (i) Express $\dfrac{3x^2 - 12x + 11}{x^3 - 6x^2 + 11x - 6}$ in partial fractions.

(ii) If $27x^3 \equiv (ax^2 + bx + c)(x^2 - 1) + (ex + f)(x + 2)^3$, find the constants a, b, c, e, f and hence or otherwise express $\dfrac{x^3}{(x+2)^3(x^2-1)}$ in partial fractions.

3 (i) Prove that $\tan^{-1}\dfrac{q}{p+q} + \tan^{-1}\dfrac{p}{p+2q} = \pi/4$.

(ii) Solve the equation $\tan^{-1} x + \tan^{-1}(1-x) = \tan^{-1}(9/7)$, assuming all the inverse tangents are acute angles.

(iii) Find a value of x between 0 and $\pi/2$ such that
$$\sin\{(\pi^2 - 4x^2)^{1/2}\} = \cos x.$$

4 The hypotenuse of a variable right-angled triangle is of constant length $2c$. Prove that the radius of the inscribed circle never exceeds $c(\sqrt{2}-1)$. (N.)

5 The straight line $lx + my = 1$ meets the circle $x^2 + y^2 + 2gx + c = 0$ in the points M and N, and O is the origin. Prove that, if OM and ON are perpendicular,
$$c(l^2 + m^2) + 2gl + 2 = 0,$$
and that in this case, the foot of the perpendicular from O on the line MN lies on the circle $2(x^2 + y^2) + 2gx + c = 0$. (C.)

6 A variable tangent to the parabola $y^2 = 4ax$ meets the circle $x^2 + y^2 = r^2$ at P and Q. Prove that the locus of the midpoint of PQ is
$$x(x^2 + y^2) + ay^2 = 0. \quad \text{(N.)}$$

7 (a) Differentiate $\sin^{-1}\left(\dfrac{2x}{1+x^2}\right)$, $\left(\dfrac{x^n - 1}{x^n + 1}\right)^n$.

(b) Find the nth derivative of $\dfrac{2x}{(x+2)(x-2)}$.

(c) If $y = \ln(1 + \cosh x)$, show that $\dfrac{dy}{dx} = \sinh x \dfrac{d^2 y}{dx^2}$.

8 Evaluate

(i) $\displaystyle\int_0^1 \tan^{-1} x\, dx$, (ii) $\displaystyle\int_0^1 \sqrt{\dfrac{2-x}{2+x}}\, dx$, (iii) $\displaystyle\int_0^{2\pi} e^{-x}|\sin \dot{x}|\, dx$. (N.)

9 The normal at P to the parabola $y^2 = 4ax$ meets the curve again at Q. Find the minimum length of PQ. (N.)

10 (i) If $y = A\tan\tfrac{1}{2}x + B(2 + x\tan\tfrac{1}{2}x)$, where A and B are constants, prove that
$$(1 + \cos x)\dfrac{d^2 y}{dx^2} = y.$$

(ii) Solve $\dfrac{d^2 y}{dx^2} + 2a\cos\alpha \dfrac{dy}{dx} + a^2 y = 0,$

where a and α are constants. (C.)

PAPER C (12)

1 (i) If a is positive, prove that $2 \leqslant a + \dfrac{1}{a} \leqslant a^3 + \dfrac{1}{a^3}$.

(ii) If x, y, z are positive variables with a constant sum c, show that the least value of $(x^2 + y^2 + z^2)$ is $c^2/3$.

2 Give definitions of the functions sinh x and cosh x.
Verify that sinh $(x+y) = $ sinh x cosh $y + $ cosh x sinh y and state the corresponding expression for cosh $(x+y)$.
Solve the equation $\tanh^{-1} 3x + \tanh^{-1} x = \tanh^{-1} \frac{8}{13}$. (C.)

3 (i) Sketch the curve $y = e^{-x^2} \cos 2\pi x$.
(ii) Show that, for all positive values of x, $xe^{1/x} \geq e$.

4 By multiplying throughout by $2 \sin \frac{1}{2}x$, find the sum of the series
$$\cos x + \cos 2x + \ldots + \cos (n-1)x.$$

$P_1 P_2 \ldots P_n$ is a regular polygon, each side being of length a. Prove that

$$(P_1 P_2)^2 + (P_1 P_3)^2 + \ldots (P_1 P_n)^2 = \tfrac{1}{2}a^2 n \operatorname{cosec}^2 \frac{\pi}{n}.$$

5 The normals PP', QQ' at the ends of a variable focal chord of a parabola, meet the parabola again in P', Q'. Prove that $P'Q'$ is parallel to PQ and that the ratio of their lengths is constant. Find this ratio. (N.)

6 A variable tangent to the hyperbola $xy = a^2$ meets the hyperbola $xy = b^2$ in points P, Q. Prove that the area between the chord PQ and the arc of the second hyperbola is constant. (N.)

7 Show that, if $x > 0$, $\ln (1+x) - 2x/(x+2)$ is an increasing function of x, and prove that, when n is a positive integer,
$$\left(1 + \frac{1}{n}\right)^n < e < \left(1 + \frac{1}{n}\right)^{n + \frac{1}{2}}.$$

8 (a) Prove that $2 \sin x + \tan x > 3x$ for $0 < x < \frac{1}{2}\pi$.

(b) Evaluate $\displaystyle\int_0^1 \frac{6x}{x^3 + 8} dx$ and $\displaystyle\int_{-\pi/6}^{\pi/6} \frac{dt}{3 \cos t + \cos^3 t}$. (N.)

9 The portion of the curve $y^2 = 4ax$ from $(a, 2a)$ to $(4a, 4a)$ revolves round the tangent at the origin. Prove that the volume bounded by the curved surface so formed and plane ends perpendicular to the axis of revolution is $\dfrac{62\pi a^3}{5}$ and find the square of the radius of gyration of this volume about its axis of revolution.

10 Prove that $\displaystyle\int \sec^n \theta \, d\theta = \frac{\sec^{n-2} \theta \tan \theta}{n-1} + \frac{n-2}{n-1} \int \sec^{n-2} \theta \, d\theta.$
By using the parameter θ given by $y = \tan^3 \theta$ or otherwise, find the area enclosed between the curve $x^{2/3} = y^{2/3} + 1$, the ordinate $x = 2\sqrt{2}$ and the line $y = 0$. (N.)

Answers

EXAMPLES 1a (Page 3)

1	15	2	30	3	120	4	9	5	240
6	24	7	120	8	24	9	729	10	11 880; 20 736
11	17 576	12	60	13	90 000	14	374	15	74

EXAMPLES 1b (Page 5)

1. 360 2. 24 3. 60 4. 720
5. 60 6. 40 320 7. 210 8. 336
9. 120; 120 10. 5040; 10; 11 880; 126; 9240; 840; 3600
11. $4!; \dfrac{6!}{3!}; \dfrac{9!}{5!}; \dfrac{n!}{(n-3)!}; \dfrac{8 \cdot n!}{(n-3)!}$
12. $20; 6; 840; 24; 720; \dfrac{r!}{(n-r)!}$.
13. $n(n-1)(n-2); (n-1)(n-2); n!; \tfrac{1}{2} \cdot 2n!$
14. 40 320 15. 6720 16. 720 17. 40 320
18. 3 628 800 19. 40 320

EXAMPLES 1c (Page 8)

1. 84 2. 1365 3. 120 4. 35 5. 1365
6. 1287 7. 45 8. 36; 120; 455; $5\tfrac{1}{4}$ 10. $_{11}C_6$; 462
11. 20 12. 9 13. 30 14. 11 662 000 15. 41 800
16. 1770 17. 44 352 18. 1008 19. 56 20. 6435; 360 360
21. 13 860

EXAMPLES 1d (Page 9)

1. $12 \cdot 6!$ 2. 144 3. 105 4. 2508 5. 1512; 336
6. 270 725; 242 164 7. 30; 60 8. $7 \cdot 8!$ 9. 508
10. 4320 11. 5760 12. 9! 13. $\dfrac{20!}{(10!)^2}$ 14. 120
15. 4752; 3168 16. $8 \cdot 9!$ 17. 120 18. 385 19. 130; 93

EXAMPLES 1e (Page 11)

1	12	**2**	20	**3**	30	**4**	10	**5**	280	**6**	20 160
7	90 090	**8**	210	**9**	3360	**10**	420	**11**	166 320	**12**	360
13	2520; 72			**14**	9	**15**	1260	**16**	30; 250		

EXAMPLES 1f (Page 13)

1	125	**2**	12^8	**3**	256	**4**	125	**5**	255	**6**	27 000
7	15	**8**	64	**9**	14	**10**	3^8	**11**	20 735	**12**	23

MISCELLANEOUS EXAMPLES (Page 14)

1 $\dfrac{13!}{(3!)^2 (2!)^2}$; $\dfrac{11 \cdot 12!}{(3!)^2 (2!)^2}$ **2** 76 145   **3** 27 720   **4** 1 209 600

5 $\dfrac{25!}{(5!)^6}$ **6** 120 960   **7** 5040   **8** 78; 390   **9** 240

11 945   **12** 181 440   **13** 17 280   **14** $25^5 \cdot 24^3 \cdot 23$ **15** 8
16 191; 96   **17** 120   **18** 14.8!   **19** 9   **20** 60
21 1022   **22** 360   **24** 420.8!.6!   **25** 2.10!
26 3!.4!.5!.6!   **27** $\dfrac{4n!}{(4!)^n n!}$; $\dfrac{3^n 4n!}{(4!)^n n!}$ **28** 175; 2
29 30,030   **30** 180

EXAMPLES 1g (Page 17)

1	$\frac{2}{5}$	**2**	$\frac{1}{13}$	**3**	$\frac{1}{2}$	**4**	$\frac{2}{35}$	**5**	$\frac{1}{4}$		
6	$\frac{1}{2}$	**7**	$37\frac{1}{2}$	**8**	3 to 1	**9**	126	**10**	$\frac{1}{4}$; $\frac{1}{2}$		
11	$\frac{3}{5}$	**12**	$\frac{1}{4}$	**13**	$\frac{1}{6}$	**14**	$\frac{2}{27}$	**15**	9 to 4		

EXAMPLES 1h (Page 18)

1 $\frac{5}{33}$	**2** 5 to 1	**3** $\dfrac{1}{270\,725}$	**4** 13; 716 to 13
5 $\frac{2}{7}$	**6** 44 to 1	**7** 4 to 3	**8** 9 to 11
9 7 to 2	**10** $\frac{5}{36}$	**11** $\frac{3}{8}$	**12** 2 to 1, 5 to 4
13 $\frac{44}{91}$	**14** $\frac{19}{23}$	**15** $\frac{1}{12}$	**16** 33 to 2
17 $\frac{1}{12}$	**18** 7 to 1	**19** $\frac{4}{9}$; $\frac{5}{8}$; $\frac{5}{18}$	**20** $\frac{1}{9}$; $\frac{5}{18}$

EXAMPLES 2a (Page 22)

1 3; $\frac{1}{2}$; -1; $-\frac{3}{2}$ **2** 8; $\frac{1}{8}$
3 4; 3; $\frac{1}{2}$; -1; 2; $\frac{3}{2}$; $\frac{5}{2}$; $\frac{7}{3}$; $\frac{2}{3}$; $-\frac{5}{2}$ **4** 2; 7; $\sin x$; x^2
5 (i) $3 \log a + \log b - 2 \log c$; (ii) $\frac{1}{2}(\log a - \log b)$; (iii) $-\log a - 3 \log b$;
  (iv) $\frac{1}{2}(\log a + 2 \log b + \log c)$; (v) $\frac{1}{3}(-2 \log a + 3 \log b - \log c)$;
  (vi) $\frac{2}{3}\log a + \frac{3}{2}\log b$; (vii) $-\frac{2}{3}\log a + \frac{3}{2}\log b + \frac{2}{3}\log c$

338 | Answers

6 (i) $\log \frac{175}{4}$; (ii) $\frac{1}{2}\log 2$; (iii) $\log \frac{x^2(x+1)}{(1-x)^3}$; (iv) $\frac{1}{9}\log \frac{(2x-1)^3(x+3)}{(x+1)^2}$;
(v) $\log(x+1)$

7   1·58(5); 0·813; 8·63; 0·642; 0, 1·85
10   1·94; 0·763; 3·42; 0·141; 10·5; 17·1; 1·98; 1·19
11   1·079181; 1·924279; 1·107210; 0·158412(5)   12   5; 6
13   1·26; 3·32; 2·20; 1·29   14   1·25; 2·46; 3·59

EXAMPLES 2b (Page 23)

1   0·573   2   3·50   3 $\frac{1}{2}$ 4   0·814
5   −0·00316   6   2·028   7   −1·50; −0·195
8   $a = 26·5, n = -1·323$ 9   2·63   10   1, 1·585
11   3·627; 2·352; 0·521; 1·020   12   0·881   13 $\pm 1·317$
17   −0·6932; −0·1439   18   0·144   19   155

EXAMPLES 2c (Page 27)

1   $\frac{1}{x}; \frac{1}{x+6}; \frac{3}{3x-2}; \frac{-1}{1-x}; \frac{2x+1}{x^2+x+1}; \frac{-3x^2}{2-x^3}; -\frac{1}{x}; \frac{1}{2(1+x)}; \frac{2}{1-x^2};$
$\frac{6}{2x+1}; -\tan x; \frac{1}{\sin x \cos x}; 2\cot x; \frac{-3}{2(1-2x)(x+1)}; \frac{x^2}{x^3+1}$

2   $\frac{1}{x}\lg e$; $\cot x \log_a e$; $\frac{2x}{x^2+1}$ 3 $\ln x + 1$; $x \ln x - x$

4   $2\ln x + 3$ 6   1   10 $2e^{2x}$

EXAMPLES 2d (Page 29)

1   $3\ln x$; $\frac{1}{2}\ln x$; $x - \ln x$; $x + 2\ln x - \frac{1}{x}$; $\ln(x+2)$; $\frac{1}{2}\ln(2x-5)$;
$-\ln(2-x)$; $-\frac{1}{3}\ln(1-3x)$; $\frac{2}{3}\ln(x+1)$; $-\frac{4}{9}\ln(1-x)$;
$-\frac{1}{p}\ln(q-px)$

2   $\ln(x^2+1)$; $\ln(x^3-x^2+4)$; $\frac{1}{4}\ln(1+x^4)$; $\frac{1}{2}\ln(x^2+2x+2)$;
$\frac{1}{6}\ln(x^6+1)$; $\ln(2+\sin x)$; $-\frac{1}{2}\ln(1+2\cos x)$; $\ln \sec x$; $\ln \sin x$;
$\ln(x \sin x)$; $\ln(\ln x)$; $\ln(2+\tan x)$.

3   $\frac{1}{2}\ln 3$; $\frac{1}{2}\ln \frac{4}{3}$; $\frac{1}{2}\ln 2$; $\ln 3$; $\frac{1}{3}\ln \frac{19}{4}$ 4 $2\ln 2$.

5   $\ln \tan x$ 6 $\ln(\sec x + \tan x)$ 7 $\frac{2x}{3} - \frac{2}{9}\ln(3x+1)$

8   $x - \ln(x+1)$; $x - \ln(x+2)$; $-3x - 3\ln(1-x)$; $-\frac{2x}{3} + \frac{17}{9}\ln(3x+7)$

9 $\dfrac{x^3}{6} + \dfrac{x^2}{8} + \dfrac{x}{8} + \dfrac{1}{16}\ln(2x-1)$

10 $2 + \ln 2;\ \tfrac{10}{3} - 8\ln\tfrac{3}{2};\ \tfrac{313}{12} + \ln 2$

EXAMPLES 2e (Page 32)

1 $2e^x;\ -\dfrac{1}{e^x};\ e^x - \dfrac{1}{e^x};\ 2e^{2x};\ -4e^{-4x};\ \tfrac{1}{2}e^{x/2};\ -\dfrac{1}{2\sqrt{e^x}};\ 2e^{2x} - \dfrac{2}{e^{2x}}$

2 $2e^x;\ -\dfrac{1}{e^x};\ e^x - \dfrac{1}{e^x};\ \dfrac{e^{2x}}{2};\ -\dfrac{e^{-4x}}{4};\ 2e^{x/2};\ -\dfrac{2}{\sqrt{e^x}};\ \dfrac{e^{2x}}{2} - 2x - \dfrac{1}{2e^{2x}}$

3 $3^x \ln 3;\ -10^{-x} \ln 10;\ 2xe^{x^2};\ 2a^{2x}\ln a;\ \sec^2 x\, e^{\tan x};\ -3e^{-3x}$

4 0·632; 1·06; 0·697 **7** $e^x(x^2 + 2x)$

8 $4e^{2t}$ **10** $e^{x^2};\ e^{\tan x};\ 2e^{\sqrt{x}}$ **11** 1

MISCELLANEOUS EXAMPLES (Page 34)

1 $n = 1\cdot 849;\ c = 1170$ **2** $-12\cdot 4$ **3** $x = 0\cdot 122$

4 $x = 0\cdot 5,\ 1\cdot 77$ **5** $x = -0\cdot 42,\ 1\cdot 51$

6 (i) $(6x-1)\ln(2x+1) + \dfrac{2x(3x-1)}{2x+1}$; (ii) $\dfrac{2x}{\ln x} - \dfrac{x^2+2}{x(\ln x)^2}$; (iii) $e^x\left(\ln 2x + \dfrac{1}{x}\right)$;

(iv) $-2e^{-2x}(\cos 4x + 2\sin 4x)$; (v) $-3\tan 3x$; (vi) $\dfrac{e^x}{1+e^x}$;

(vii) $\dfrac{4}{(e^x + e^{-x})^2}$; (viii) $\dfrac{-1 + \cos x - \sin x}{(1+\sin x)(1-\cos x)}$; (ix) $-\dfrac{1+x}{(1-x)(1+x^2)}$;

(x) $\dfrac{1}{\sqrt{x}(1-x)}$; (xi) $e^{x\sin x}(\sin x + x\cos x)$; (xii) $2^{\tan x}\sec^2 x \ln 2$;

(xiii) $e^{3\ln x}\dfrac{3}{x} = 3x^2$; (xiv) $\dfrac{1}{2}\left(\dfrac{e^x - 2}{e^x - 1}\right)$; (xv) $\dfrac{1}{x} + \dfrac{1}{\sqrt{x^2+1}}$; (xvi) $\dfrac{2}{\sqrt{x^2+1}}$

7 (i) $-\dfrac{x^2}{2} - x - 2\ln(1-x)$; (ii) $\dfrac{x^5}{5} + \dfrac{x^4}{4} + \dfrac{x^3}{3} + \dfrac{x^2}{2} + x + \ln(x-1)$;

(iii) $\tfrac{1}{2}\ln(x^2 + 2x - 1)$; (iv) $2\ln\sin\dfrac{x}{2}$; (v) $2^x \log_2 e$;

(vi) $\ln(x\cos x)$; (vii) $2(e^{x/2} + e^{-x/2})$; (viii) $-\tfrac{1}{2}e^{-x^2}$

8 20 **10** $a = 3\cdot 6,\ n = 1\cdot 15$ **11** $a = 4\cdot 1,\ n = 2\cdot 45$

12 (i) $\dfrac{(3x+4)^6}{\sqrt{1-x^3}}\left(\dfrac{18}{3x+4} + \dfrac{3x^2}{2(1-x^3)}\right)$; (ii) $\dfrac{x^3 \sin^5 x}{\cos^3 x}\left(\dfrac{3}{x} + 5\cot x + 3\tan x\right)$;

(iii) $x^{\sin x}\left(\cos x \ln x + \dfrac{\sin x}{x}\right)$; (iv) $x^{\ln x}\dfrac{2\ln x}{x}$.

340 | Answers

13 $\ln 3$ **14** $\ln \dfrac{3+\sqrt{8}}{2+\sqrt{3}}$

17 $a = -\sqrt{29},\ b = -\tan^{-1}\tfrac{5}{2};\ -29^{\frac{5}{2}} e^{-2x} \cos(5x - 5\tan^{-1}\tfrac{5}{2})$

18 (i) $(-1)^{n-1}\dfrac{(n-1)!}{(1+x)^n}$;   (ii) $-\dfrac{(n-1)!}{(1-x)^n}$;

(iii) $(-1)^{n-1}(n-1)!\left\{\dfrac{2^n}{(1+2x)^n} - \dfrac{1}{(1+x)^n}\right\}$;   (iv) $xe^x + ne^x$;

(v) $x^2 e^x + 2nxe^x + n(n-1)e^x$

19 $\text{Max}\left(\dfrac{1}{\sqrt{2}}, \dfrac{e^{-1/2}}{\sqrt{2}}\right)$; $\text{Min}\left(-\dfrac{1}{\sqrt{2}}, \dfrac{-e^{-1/2}}{\sqrt{2}}\right)$

20 $13^{n/2} e^{3x} \sin(2x + n\tan^{-1}\tfrac{2}{3})$.   **21** $\dfrac{\sqrt{e}}{2}$ cm   **25** $\pi(\tfrac{9}{2} + 2\ln 2)$

26 (i) $x = 0,\ 0{\cdot}631$;   (ii) $x = 1{\cdot}76,\ y = -0{\cdot}52$ **27** (i) $\dfrac{(\ln x)^{\ln x}}{x}(\ln(\ln x) + 1)$

28 (ii) $x = 1,\ -3$

EXAMPLES 3a (Page 41)

1 (i) $1 - 1 + \dfrac{1}{2!} - \dfrac{1}{3!} + \ldots$;   (ii) $1 + \dfrac{1}{2} + \dfrac{1}{2!}\left(\dfrac{1}{2}\right)^2 + \dfrac{1}{3!}\left(\dfrac{1}{2}\right)^3 + \ldots$;

(iii) $1 + 3 + \dfrac{3^2}{2!} + \dfrac{3^3}{3!} + \ldots$;   (iv) $1 + \dfrac{1}{3} + \dfrac{1}{2!}\left(\dfrac{1}{3}\right)^2 + \dfrac{1}{3!}\left(\dfrac{1}{3}\right)^3 + \ldots$;

(v) $2\left(1 + \dfrac{1}{2!} + \dfrac{1}{4!} + \ldots\right)$;   (vi) $1 + \dfrac{1}{3!}\left(\dfrac{1}{2}\right)^2 + \dfrac{1}{5!}\left(\dfrac{1}{2}\right)^4 + \ldots$;

2 (i) $1 - x + \dfrac{x^2}{2!} - \dfrac{x^3}{3!} + \ldots$;   (ii) $1 + 3x + \dfrac{(3x)^2}{2!} + \dfrac{(3x)^3}{3!} + \ldots$;

(iii) $1 + \dfrac{1}{x} + \dfrac{1}{2!}\left(\dfrac{1}{x}\right)^2 + \dfrac{1}{3!}\left(\dfrac{1}{x}\right)^3 + \ldots$;   (iv) $1 - x^2 + \dfrac{x^4}{2!} - \dfrac{x^6}{3!} + \ldots$

3 (i) $0{\cdot}3679$;   (ii) $1{\cdot}6487$;   (iii) $0{\cdot}7165$

4 (i) $1 + 2x + \dfrac{3x^2}{2} + \dfrac{2x^3}{3} \ldots$;   (ii) $-1 + x + \dfrac{x^2}{2} - \dfrac{5x^3}{6} \ldots$;

(iii) $1 + x - \dfrac{x^2}{2} - \dfrac{x^3}{6} \ldots$;   (iv) $2\left(1 + \dfrac{x^2}{2!} + \dfrac{x^4}{4!} + \dfrac{x^6}{6!} \ldots\right)$

5 (i) $\dfrac{e}{4!}$;   (ii) $\dfrac{e^2}{4!}$;   (iii) $\dfrac{e^2 3^4}{4!}$ **6** $\left(x + \dfrac{x^3}{3!} + \dfrac{x^5}{5!} + \ldots\right)$; $\left(1 + \dfrac{x^2}{2!} + \dfrac{x^4}{4!} + \ldots\right)$

8 $-3 - x$.   **10** $\dfrac{8}{3 \cdot 5!}$ **11** $1 + x - \dfrac{x^2}{2}$

12 $1 + x\ln 10 + \dfrac{(x\ln 10)^2}{2!} + \dfrac{(x\ln 10)^3}{3!}$ **13** $\dfrac{(\ln 2)^3}{3!}$

Answers | 341

EXAMPLES 3b (Page 45)

1. $0 \cdot 00995$ 2. $0 \cdot 4055$ 3. $0 \cdot 08004$; $2 \cdot 5649$
4. $-\frac{1}{3} < x \leqslant \frac{1}{3}$ 5. $2x - 2x^2 + \frac{8x^3}{3}$; $-\frac{1}{2} < x \leqslant \frac{1}{2}$
6. $-\frac{x}{2} - \frac{x^2}{8} - \frac{x^3}{24}$; $-2 \leqslant x < 2$ 7. $\frac{x}{3} - \frac{x^2}{18} + \frac{x^3}{81}$; $-3 < x \leqslant 3$
8. $2x - x^2 + \frac{2x^3}{3}$; $-1 < x \leqslant 1$ 9. $-x^2 - \frac{x^4}{2} - \frac{x^6}{3}$; $-1 < x < 1$
10. $-3x + \frac{3x^2}{2} - 3x^3$; $-\frac{1}{2} < x \leqslant \frac{1}{2}$ 11. $\frac{5x}{2} - \frac{37x^2}{8} + \frac{215x^3}{24}$; $-\frac{1}{3} < x \leqslant \frac{1}{3}$
12. $x^2 - \frac{x^4}{2} + \frac{x^6}{3}$; $-1 \leqslant x \leqslant 1$ 13. $5x - \frac{17x^2}{2} + \frac{65x^3}{3}$; $-\frac{1}{4} < x \leqslant \frac{1}{4}$
14. $x + \frac{x^2}{2} + \frac{4x^3}{3}$; $-1 < x < 1$ 15. $-2x + x^2 - \frac{8x^3}{3}$; $-\frac{1}{2} < x \leqslant \frac{1}{2}$
16. $x + \frac{x^2}{2} - \frac{2x^3}{3} + \frac{x^4}{4} + \frac{x^5}{5} - \frac{x^6}{3}$
17. $-x + \frac{x^2}{2} + \frac{2x^3}{3} + \frac{x^4}{4} - \frac{x^5}{5} - \frac{x^6}{3}$
18. $-\frac{2}{3}$ 19. $-\frac{17}{4}$

EXAMPLES 3c (Page 48)

1. $0 \cdot 000\,073$ 2. $1 \cdot 000\,000$ 3. $0 \cdot 000\,218$ 4. $0 \cdot 017\,453$
5. $0 \cdot 999\,962$ 6. $0 \cdot 004\,363$ 7. $0 \cdot 0003$ 8. $0 \cdot 0007$
9. $\theta = 0 \cdot 0245$ rad 11. $\theta = 0 \cdot 196$ rad
12. $\frac{1}{2}\left(2 - \frac{(2x)^2}{2!} + \frac{(2x)^4}{4!} - \frac{(2x)^6}{6!} \ldots \right)$ 13. $(-1)^{n+1} \frac{x^{2n-2}}{(2n-2)!}$ (nth term)
14. $x^3 - \frac{x^5}{2} + \frac{13x^7}{120}$ 15. $\frac{1}{2\sqrt{2}}\left(2 + 2x - \frac{(2x)^2}{2!} - \frac{(2x)^3}{3!} + \frac{(2x)^4}{4!} + \frac{(2x)^5}{5!} - \ldots \right)$
16. $\theta = 27° \, 42' = 0 \cdot 4835$ rad 17. $\sin 1 = 0 \cdot 8415$ 18. $\cos \frac{1}{2} = 0 \cdot 8776$
19. $2 - \sin 2$ 20. $\left(\frac{4}{\pi}\right)^2 \left(1 - \frac{\sqrt{2}}{2}\right)$ 21. $\frac{1}{6}$ 22. $\frac{1}{2}$
23. $\frac{1}{3}$ 24. 3

MISCELLANEOUS EXAMPLES (Page 51)

1. $\frac{7e^2}{6!}$ 2. $4x^3$; $-\frac{1}{2} < x < \frac{1}{2}$ 3. $\frac{1641}{8!}$ 4. $2 \cdot 397\,895$
6. $-2\left(2x + \frac{(2x)^3}{3!} + \frac{(2x)^5}{5!} \ldots \right)$ 7. $x + x^2 + \frac{x^3}{3} - \frac{x^5}{30}$; all values
8. $1 + x - \frac{x^3}{3} - \frac{x^4}{6}$; all values 9. $x + \frac{x^2}{2} + \frac{x^3}{3} + \frac{3x^5}{40}$; $-1 < x \leqslant 1$

10 $-x+\dfrac{x^2}{2}-\dfrac{x^3}{3}-\dfrac{3x^5}{40}$; $-1 \leqslant x < 1$ **11** $x+\dfrac{x^2}{2}+\dfrac{5x^3}{6}+\dfrac{7x^4}{12}$; $-1 < x < 1$

12 $x-\dfrac{x^2}{2}+\dfrac{5x^3}{24}-\dfrac{11x^4}{48}$; $-1 < x < 1$ **13** $1-2x+3x^2-\dfrac{10x^3}{3}$; $-1 < x < 1$

14 0·023 717; 2·302 585   **16** $-\dfrac{2}{3n}$ **17** $-\dfrac{3}{4n}$

18 $\dfrac{\theta}{1}+\dfrac{\theta^3}{3}+\dfrac{2\theta^3}{15}$ **21** $\ln 2 = 0.693\,15$; $\ln 3 = 1.098\,61$; $\ln 5 = 1.609\,44$

22 $y = \ln\dfrac{1+2x}{1-x}$; $3x - \dfrac{3x^2}{2} + 3x^3 - \dfrac{15x^4}{4}$; $-\tfrac{1}{2} < x \leqslant \tfrac{1}{2}$

23 $(a+b-1)$, $\tfrac{1}{2}(a^2-b+1)$, $\tfrac{1}{6}(a^3+2b-3)$; $a = -1$, $b = 2$

26 $\dfrac{1}{n}\left(\dfrac{1}{2^n}-\dfrac{1}{3^n}\right)$ **27** 2·302 58   **28** 0   **29** 2

30 0   **31** $\dfrac{n+m}{2}$ **32** $\tfrac{1}{8}$ **33** $-\tfrac{1}{2}$

34 $2\left\{\dfrac{n-1}{n+1}+\dfrac{1}{3}\left(\dfrac{n-1}{n+1}\right)^3+\ldots\right\}$

36 (i) 1;   (ii) $\tfrac{1}{2}\ln 2$;   (iii) $2\ln 2 - 1$;   (iv) $5e$

EXAMPLES 4a (Page 56)

1 $\tfrac{7}{18}$ **2** 1024   **3** $-3618a^5$; $7290a^6$ **4** 30; 51

5 $4x^3$; $4x^{10}$ **6** $1-2x-24x^3+48x^4+192x^6-384x^7-512x^9+1024x^{10}$

7 40   **8** 11   **9** $\tfrac{9}{14}$

10 (i) 9th;   (ii) 3rd;   (iii) 4th;   (iv) 3rd   **11** (i) $\tfrac{55}{27}.2^{14}$;   (ii) 10.3^{15}

12 (i) 6th;   (ii) 4th and 5th;   (iii) $(n+1)$th and $(n+2)$th

13 (i) $3^5.5^{10}\,_{15}C_{10}$;   (ii) $\tfrac{7}{144}$ **14** $2^{n-1}n^2$; $2^{n-1}.\dfrac{n(n^2-1)}{3}$

15 $1-nx+\dfrac{n(n+1)}{2}x^2+\dfrac{n(n-1)(n+4)}{6}x^3\ldots$ **16** 16

EXAMPLES 4b (Page 58)

1 $1+2+3$ **2** $2(1^2+2^2+3^2+4^2)$ **3** $1.2+2.3+\ldots 8.9$

4 $1^3+2^3+\ldots+20^3$ **5** $\dfrac{1}{1}+\dfrac{1}{2}+\dfrac{1}{3}+\ldots+\dfrac{1}{10}$ **6** $\dfrac{1}{1.3}+\dfrac{1}{2.4}+\ldots\dfrac{1}{15.17}$

7 $1^4+2^4+\ldots+n^4$ **8** $2.3+3.4+\ldots+(n+1)(n+2)$

9 $\dfrac{1}{1!}+\dfrac{1}{2!}+\ldots+\dfrac{1}{n!}$ **10** $\dfrac{1}{130}$ **11** $\dfrac{9}{7!}$

12 $2n$; n^2+3n+3 **13** $3r-2$; $\sum\limits_{1}^{30}(3r-2)$ **14** $(2r-1)^2$; $\sum\limits_{1}^{20}(2r-1)^2$

Answers | 343

15 $\dfrac{1}{r}$; $\sum_1^n \dfrac{1}{r}$ **16** $\dfrac{1}{(2r-1)^2}$; $\sum_1^n \dfrac{1}{(2r-1)^2}$

17 $(r+2)(r+3)$; $\sum_1^n (r+2)(r+3)$ **18** $\dfrac{1}{(r+1)(r+2)}$; $\sum_1^{2n} \dfrac{1}{(r+1)(r+2)}$

19 $\dfrac{r+1}{(2r-1)2r(2r+1)}$; $\sum_1^n \dfrac{r+1}{(2r-1)(2r)(2r+1)}$ **20** $\dfrac{-2}{6.7}$

21 $\sum_1^n \dfrac{(-1)^{r+1}}{2r-1}$ **22** $\tfrac{11}{6}$ **23** $\ln 24$ **24** $\tfrac{769}{3600}$

25 2870 **26** 1296 **27** 234 **28** 2450
29 1480 **30** 11 130 **31** 11 **32** 3563
33 1950 **34** 1420 **35** 41 075 **36** 10 660

37 (i) $\dfrac{2n(n+1)(2n+1)}{3}$; (ii) $\dfrac{n}{3}(4n^2-1)$

38 (i) $\dfrac{n(n+1)(3n^2+11n+4)}{12}$; $\tfrac{1}{3}n(2n+1)(6n^2+11n+2)$ **39** $\dfrac{n(n^2+2)}{3}$

EXAMPLES 4c (Page 61)

1 $n(2n-1)$ **2** $n^2(n+1)$ **3** $\tfrac{1}{4}n(n+1)(n^2+n+6)$

4 $\tfrac{1}{6}n(4n^2+15n+17)$ **5** $\tfrac{1}{2}n(n+1)(n^2+3n+1)$

6 $\dfrac{n}{2}(1-n)$ **7** $\dfrac{3^{n+1}}{2}-2^{n+1}+\tfrac{1}{2}$ **8** $\dfrac{n}{n+1}$ **9** $\dfrac{n(3n+5)}{2(n+1)(n+2)}$

10 5; 11; $6n-1$. **11** $\dfrac{1-a^n}{(1-a)^2}-\dfrac{na^n}{1-a}$ **12** $\dfrac{n}{4(n+4)}$

13 $\dfrac{2031}{256}$ **14** $\dfrac{25}{8}-\dfrac{1}{8.5^{n-1}}(4n+5)$; $\dfrac{25}{8}$

15 $\dfrac{n}{3(2n+3)}$ **16** $3n^2-2n-1$; $6n+1$; $9n^2+4n$

MISCELLANEOUS EXAMPLES (Page 62)

1 -8; -392 **3** 495 **4** $\dfrac{35}{16}-\dfrac{1}{16.5^{n-1}}(12n+7)$; $\dfrac{35}{16}$

5 $\dfrac{n}{3}(4n^2+12n-1)$ **6** $7x^3$

7 $1+10x+45x^2$; $1+14x+84x^2$; 6th term $\dfrac{7.2^{15}}{3^4}$

9 $\dfrac{n}{3n+1}$ **10** $40.3^4.7^7$

344 | Answers

11 (i) $\frac{1}{12}n(n+1)(n+2)(3n+1)$; (ii) $\dfrac{1+x-(2n+1)x^n+(2n-1)x^{n+1}}{(1-x)^2}$

12 $\dfrac{2(1-x^n)}{1-x}+\dfrac{1-(2x)^n}{1-2x}$ 14 $a=-3, b=\frac{1}{2}$ 15 $\frac{1}{2}n^2$

16 4th term $\dfrac{11.3^{12}.2^3}{5^2}$ 18 Common difference $\frac{1}{3}$ 19 2268

20 $\frac{1}{24}(n-1)n(n+1)(3n+2)$ 21 (i) 3^n; (ii) $\frac{1}{2}(3^n+1)$

22 $2^{n+1}-n-2$ 23 $\dfrac{3-(n^2+4n+1)a^n+n(n+2)a^{n+1}}{(1-a)^2}+\dfrac{2a(1-a^{n-1})}{(1-a)^3}$; 9

24 5th term $\frac{495}{16}$

EXAMPLES 5a (Page 68)

1 $\frac{1}{11}$ 2 0 3 2 4 $22\frac{2}{3}$
5 $\ln\frac{5}{4}$ 6 $\frac{1}{2}$ 7 $6\frac{3}{4}$ 8 $e-\dfrac{1}{e}$
9 $\frac{1}{2}\ln 5$ 10 $7\frac{11}{24}$ 11 $\dfrac{\pi^2}{32}-\dfrac{\sqrt{2}}{2}+1$ 12 $\dfrac{1}{2}\left(e^2-\dfrac{1}{e^2}\right)$
13 $\frac{1}{2}\ln 5$ 14 0 15 $\sqrt{2}-1$ 16 $\frac{7}{8}+\ln 2$
17 e^2-e 18 $2\ln 3$ 19 $\frac{121}{5}$ 20 $2\left(\dfrac{1}{e}-\dfrac{1}{e^2}\right)$
21 $2\log_2 e$

EXAMPLES 5b (Page 71)

1 0·4056 2 0·7850 3 6·060 4 2·889
5 0·5235 6 2·958 7 1·319 8 2·358
9 0·3425 10 0·7469 11 1745 joules 12 425 sec.

EXAMPLES 5c (Page 76)

1 π 2 $5\frac{25}{27}$ 3 4 4 $\frac{32}{3}$
5 $4\ln\frac{5}{3}$ 6 e^3-e 7 $30\frac{3}{20}$ 8 $\dfrac{5\pi}{6}$
9 $\dfrac{\pi}{4}\left(e^4-\dfrac{1}{e^4}\right)$ 10 $\dfrac{8\pi}{3}$ 11 $\dfrac{64\pi}{3}$ 12 $\dfrac{\pi}{4}(\pi-2)$
13 $\dfrac{2\pi}{15}$ 14 $\dfrac{\pi}{2}(\pi-2)$ 15 $\dfrac{1}{6};\dfrac{\pi}{5}$ 16 54
17 $\dfrac{4\sqrt{2}}{\pi}$ 18 $\dfrac{1}{\sqrt{2}}$ 19 $\dfrac{8}{3};\dfrac{1}{1}$ 20 $\dfrac{2(\pi+2)}{\sqrt{6\pi^2+16\pi}}$
21 $321\frac{2}{3}$ 22 $\dfrac{2a\omega}{\pi};\dfrac{a\omega}{\sqrt{2}}$ 23 $4\sqrt{2}$ 24 $\dfrac{e^4-1}{\sqrt{e^8+8e^4-1}}$

MISCELLANEOUS EXAMPLES (Page 77)

1 2
2 $\sqrt{3}+1$
3 $\frac{21}{128}$
4 $\frac{1}{2}$
5 $\frac{5}{24}$
6 $\frac{\sqrt{3}}{2}$
7 $1-3\ln\frac{5}{4}$
8 $\frac{1}{2}(1+\ln 2)$
9 $\frac{8}{3}$
10 $\ln 2 - \frac{5}{6}$
11 $e-1$
12 0
13 8 cm
14 $\frac{256}{15}$
18 $4\frac{1}{2}$
20 $(a, \pm 2\sqrt{2}a)$
21 1120; 1490
22 $\frac{\pi}{2}-1; \pi(\ln 2 - \frac{1}{2})$
24 $\sqrt{\frac{3}{8}} I_0^2 R$
25 $2\rho_0$
26 $2\sigma_0$

EXAMPLES 6a (Page 84)

1 (i) $\frac{a}{2}$; (ii) $\frac{5a}{9}$
2 (i) $\frac{h}{3}$; (ii) $\frac{3h}{8}$
3 $\frac{10}{7}$
4 $\bar{x}=0, \bar{y}=2\frac{2}{5}$
5 $\bar{y}=\frac{12}{5}; \bar{x}=\frac{3}{4}$

EXAMPLES 6b (Page 85)

1 $\bar{x}=\frac{8}{5}, \bar{y}=\frac{16}{7}$
2 $\bar{x}=\frac{13}{6}, \bar{y}=\frac{109}{30}$
3 $\bar{x}=\frac{6}{5}; \bar{y}=0$
4 $\bar{x}=\frac{3}{2}, \bar{y}=\frac{9}{10}$
5 $\left(\frac{3}{16\ln 2}, \frac{3}{2\ln 2}\right)$
6 $(\frac{3}{7}, 0)$
7 $(\frac{4}{3}, 0)$
8 $(\frac{5}{2}, 0)$
9 $(\frac{16}{5}, 0)$
10 $\left(\frac{3\ln 3}{2}, 0\right)$
11 $\frac{9}{2}; (\frac{3}{2}, \frac{13}{5})$
12 $(\frac{9}{20}, \frac{9}{20})$
13 $(\frac{5}{8}, 0)$
15 $\frac{h}{2}(\sqrt{13}-1)$
16 2 cm above centre of base
18 On axis, $2\frac{5}{14}$ cm from larger end
19 $\frac{11}{16}$ cm
20 $(\frac{63}{34}, 0)$
21 $\frac{4}{3}\rho_0 a(3-\sqrt{2}); \frac{3a}{35}(11-\sqrt{2})$
22 Distance $\frac{2a}{\pi+2}$ from centre
23 Distance $\frac{r}{2}$ from centre
24 $5a^3 + 5a^2 b + 5ab^2 - 3b^3 = 0$
25 Distance $\frac{2h}{5}$ from the base
26 Distance $\frac{r}{66}(53-16\sqrt{5})$ from the base of the hemisphere
27 3·62 cm from centre of hemisphere.

EXAMPLES 6c (Page 91)

1 $\frac{1}{3}Ma^2; \frac{4}{3}Mb^2, \frac{1}{3}Mb^2$
2 $\frac{7}{12}Ma^2$
3 $\frac{2}{3}Ma^2$
4 $\frac{cx}{h}$

5 $M\left(b^2 + \dfrac{a^2}{3}\right)$ **6** $10\sqrt{5}$ cm; $10\sqrt{6}$ cm **7** (i) 1500; (ii) 3000 kg cm
8 Ma^2; $\frac{1}{2}Ma^2$ **9** 9 kg m^2; $\sqrt{\frac{3}{2}}$ m **10** (i) 400 kg m^2; (ii) $\sqrt{\frac{8}{3}}$ m
11 36 kg m^2; 4 m **12** $\frac{32}{3}$; $\frac{212}{3}$; $\frac{244}{3}$ kg m^2 **13** $\frac{32}{3}$ kg m^2

EXAMPLES 6d (Page 93)

1 $832\frac{16}{21}$; $96\frac{4}{5}$ **2** $\dfrac{16}{15}\sqrt{2}$; $\dfrac{32\sqrt{2}}{7}$ **3** $\dfrac{4}{33}$; $\dfrac{4}{9}$ **4** $\dfrac{2\sqrt{6}}{5}$ m

5 $316\frac{1}{4}$ kg m^2 **6** 99·7 cm **8** 5·81 cm **10** $\frac{2}{7}Ma^2$

11 $2a\sqrt{\dfrac{4\sqrt{2}-3}{7}}$ **13** $\dfrac{M}{12}(3a^2+4h^2)$ **14** $\dfrac{M}{20}(2h^2+3a^2)$ **15** $h\sqrt{\dfrac{2}{3}}$

16 $\frac{2}{5}M\dfrac{(b^4+b^3a+b^2a^2+ba^3+a^4)}{b^2+ab+a^2}$; $\sqrt{\frac{2}{3}}a$, $\sqrt{\frac{5}{3}}a$ **17** $\frac{2}{5}Ma^2$; $\frac{83}{320}Ma^2$

18 $\dfrac{3a}{2\sqrt{5}}$; $\dfrac{3a}{2\sqrt{10}}$ **19** $\frac{2}{3}\cdot 4\pi x^4 \rho_0\left(1+\dfrac{cx}{a}\right)\delta x$ **20** $\frac{4}{3}Ma^2\sin^2\alpha$

EXAMPLES 7a (Page 98)

1 $(2,7)$, $(-\frac{8}{19}, -5\frac{2}{19})$ **2** $(1,2)$, $(-\frac{7}{5}, -\frac{14}{5})$ **3** $(1,1)$, $(-\frac{53}{88}, -\frac{25}{22})$
4 $x=-2, y=3, z=1$. **5** $p=0, q=2, r=1$ **6** $x=-2, y=-1, z=5$

7 $x=\pm 4, y=\pm 5$ **8** $x=\pm\dfrac{1}{\sqrt{3}}, y=\pm\dfrac{4}{\sqrt{3}}$; $x=\pm 3, y=\mp 1$

9 $x=0, \pm 3, y=\pm 1, \pm 2$ **10** $x=\frac{1}{2}, y=\frac{1}{5}, z=\frac{1}{3}$ **11** $x=\pm\dfrac{1}{\sqrt{2}}, \pm 1$

12 $m=\pm\dfrac{1}{\sqrt{5}}, \pm\sqrt{3}$ **13** $x=\pm 1, y=\pm 2$

14 $x=3, y=4$; $x=20, y=-30$

EXAMPLES 7b (Page 98)

1 $x=2, y=-\frac{1}{3}$ **2** $x=a+b, y=b-a$; $x=a-b, y=a+b$
3 $x=\pm 2, y=+1$; $x=\pm\sqrt{3}, y=\mp\sqrt{3}$ **4** $x=4, y=9$; $x=9, y=4$
5 $x=-2, y=\frac{5}{2}$; $x=5, y=-1$ **6** $x=1, y=2$; $x=2, y=1$
7 $x=0, y=0$; $x=1, y=\frac{1}{3}$; $x=-1, y=3$; $x=\frac{1}{2}, y=0$
8 $x=0, y=0$; $x=2, y=4$; $x=-2, y=-12$ **9** $x=2, -8$

10 (i) $x=-\dfrac{1}{a}, (a+b)$; (ii) $x=a, -\dfrac{1}{2}\left(a+\dfrac{1}{a}\right)$

11 $x = -1, y = 2, z = 2$; $x = \frac{100}{41}, y = -\frac{12}{41}, z = -\frac{189}{164}$
12 $x = \pm 1, -0.22, 1, 4.39$

EXAMPLES 7c (Page 101)

1 $n\pi + (-1)^n \frac{\pi}{3}$ **2** $2n\pi \pm \frac{\pi}{3}$ **3** $n\pi + \frac{\pi}{4}$ **4** $n\pi + (-1)^{n+1} \frac{\pi}{4}$

5 $2n\pi \pm \frac{3\pi}{4}$ **6** $n\pi - \frac{\pi}{3}$ **7** $n\pi + (-1)^{n+1} \frac{\pi}{6}$ **8** $2n\pi \pm \frac{\pi}{4}$

9 $n\pi - \frac{\pi}{6}$ **10** $2n\pi + (-1)^n \frac{\pi}{4}$ **11** $n\pi \pm \frac{\pi}{12}$ **12** $3n\pi \pm \frac{\pi}{8}$

13 $\frac{n\pi}{3} - \frac{\pi}{12}$ **14** $2n\pi + \frac{3\pi}{4}$ **15** $n\pi \pm (-1)^n \frac{\pi}{2}$

16 $(2n-1)\frac{\pi}{2}$ **17** $n\pi \pm \frac{\pi}{3}$ **18** $n\pi \pm (-1)^n \frac{\pi}{6}$

19 $2n\pi, 2n\pi - \frac{\pi}{2}$ **20** $n\pi + (-1)^n 0.7298, n\pi + (-1)^{n+1} \frac{\pi}{6}$

21 $2n\pi + 1.287, 2n\pi - 0.6434$ **22** $n\pi, 2n\pi \pm \frac{\pi}{3}; 0, \frac{\pi}{3}, \pi, \frac{5\pi}{3}, 2\pi$

23 $\frac{2}{5}n\pi, 2n\pi; 0, \frac{2\pi}{5}, \frac{4\pi}{5}, \frac{6\pi}{5}, \frac{8\pi}{5}, 2\pi$ **24** $\frac{n\pi}{2}; 0, \frac{\pi}{2}, \pi, \frac{3\pi}{2}, 2\pi$

25 $\frac{n\pi}{5 + (-1)^{n+1}3}; 0, \frac{\pi}{8}, \frac{3\pi}{8}, \frac{5\pi}{8}, \frac{7\pi}{8}, \pi, \frac{9\pi}{8}, \frac{11\pi}{8}, \frac{13\pi}{8}, \frac{15\pi}{8}, 2\pi$

26 $\frac{2n\pi}{5}, \frac{2n\pi}{3}; 0, \frac{2\pi}{5}, \frac{2\pi}{3}, \frac{4\pi}{5}, \frac{4\pi}{3}, \frac{6\pi}{5}, \frac{8\pi}{5}, 2\pi$

27 $2n\pi; 0, 2\pi$ **28** $\frac{n\pi}{2}; 0, \frac{\pi}{2}, \pi, \frac{3\pi}{2}, 2\pi$

29 $\frac{n\pi + (-1)^n \frac{\pi}{2}}{4 + (-1)^n}; \frac{\pi}{10}, \frac{\pi}{6}, \frac{\pi}{2}, \frac{5\pi}{6}, \frac{9\pi}{10}, \frac{13\pi}{10}, \frac{3\pi}{2}, \frac{17\pi}{10}$

30 $\frac{2n\pi}{5} + \frac{\pi}{10}, 2n\pi - \frac{\pi}{2}; \frac{\pi}{10}, \frac{\pi}{2}, \frac{9\pi}{10}, \frac{13\pi}{10}, \frac{3\pi}{2}, \frac{17\pi}{10}$

31 $\frac{n\pi}{4} + \frac{\pi}{8}; \frac{\pi}{8}, \frac{3\pi}{8}, \frac{5\pi}{8}, \frac{7\pi}{8}, \frac{9\pi}{8}, \frac{11\pi}{8}, \frac{13\pi}{8}, \frac{15\pi}{8}$

32 $\frac{n\pi}{2} + \frac{\pi}{4}; \frac{\pi}{4}, \frac{3\pi}{4}, \frac{5\pi}{4}, \frac{7\pi}{4}$ **33** $-\frac{19\pi}{30}, -\frac{11\pi}{30}, \frac{\pi}{30}, \frac{3\pi}{10}, \frac{7\pi}{10}, \frac{29\pi}{30}$

34 $\pm \frac{3\pi}{5}$ **35** $-\frac{2\pi}{3}, -\frac{\pi}{6}, \frac{\pi}{3}, \frac{5\pi}{6}$ **36** $\pm \frac{\pi}{12}, \pm \frac{11\pi}{12}$.

348 | **Answers**

37 $0, \pm\dfrac{\pi}{10}, \pm\dfrac{3\pi}{10}, \pm\dfrac{\pi}{2}, \pm\dfrac{7\pi}{10}, \pm\dfrac{9\pi}{10}, \pm\pi.$

38 $\pm\dfrac{\pi}{14}, \pm\dfrac{3\pi}{14}, \pm\dfrac{5\pi}{14}, \pm\dfrac{\pi}{2}, \pm\dfrac{9\pi}{14}, \pm\dfrac{11\pi}{14}, \pm\dfrac{13\pi}{14}$

EXAMPLES 7d (Page 105)

1 $\pm 70°32', \pm 101°32'$
2 $-116°34', -18°26', 63°26', 161°34'$
3 $-135°, -116°34', 45°, 63°26'$
4 $-16°15', -90°$
5 $\pm 90°, 19°28', 160°32'$
6 $0, \pm 60°, \pm 90°, \pm 180°$
7 $0, \pm 30°, \pm 150°, \pm 180°$
8 $9°58', 124°48'$
9 $64°40', -138°24'$
10 $23°25', -156°35'$
11 $0, -108°26', 71°34', \pm 180°$
12 $-135°, -24°18', 45°, 114°18'$
13 $2n\pi \pm \dfrac{\pi}{2}, 2n\pi \pm \dfrac{\pi}{3}$
14 $n\pi + (-1)^n \dfrac{\pi}{4}, n\pi + (-1)^{n+1}\dfrac{\pi}{4}$
15 $\dfrac{n\pi}{3}, \dfrac{n\pi}{2} \pm \dfrac{\pi}{12}$
16 $n\pi + 1·1071 + (-1)^n 0·6388$
17 $\pi(n+\tfrac{1}{2}), (2n+1)\dfrac{\pi}{4}$
18 $n\pi \pm \dfrac{\pi}{10}, n\pi \pm \dfrac{3\pi}{10}$
19 $\dfrac{n\pi}{4} + \dfrac{\pi}{8}$
20 $n\pi \pm \dfrac{\pi}{4}, 2n\pi \pm \dfrac{2\pi}{3}$
21 $n\pi + (-1)^n \dfrac{\pi}{6} - \dfrac{\pi}{3}$
22 $\dfrac{2n\pi}{3}$
23 $(2n+1)\dfrac{\pi}{2}, n\pi + 0·6435$
24 $n\pi + (-1)^n 0·7115, n\pi + (-1)^n 0·1536$
25 $\left(n-\dfrac{1}{4}\right)\pi, \left(n-\dfrac{1}{4}\right)\dfrac{\pi}{4}$
26 $\dfrac{2n\pi}{5} \pm \dfrac{\pi}{10}, n\pi \pm \dfrac{\pi}{2}$
27 $\dfrac{n\pi + 0·5884 + (-1)^n 0·281}{2}$
28 $0·1736, 0·7660, -0·9397$
29 $0·9263, -0·1368, -0·7895$
30 $0·3090, 1, -0·8090$

EXAMPLES 7e (Page 107)

1 (i) $-1·26$; (ii) $2·1$; (iii) $-3, 0·63, 2·4$ **2** $0·86, -1·2$
3 $0·45, 2·9$ **4** (i) $0·516$; (ii) $0·59$ **5** $1·45$
6 (i) $1·76$; (ii) $0·41$ **7** $1·23$ **8** 5
9 $0·91$ **10** $50°42'$ **11** $2·6$ **12** $1·89$
13 $0·86$ rad **14** $2·1$ **15** $0·481$

Answers | 349

EXAMPLES 7f (Page 110)

7 $m > 1.89$ **9** 1.41 **10** $0.703(5)$ **11** 2.85 **12** 0.51
13 $2.064(5)$ **14** $2.09(5)$ **15** -3.59 **16** $-\frac{1}{2}$

EXAMPLES 8a (Page 114)

1 $x > 2$ **2** $x > \frac{4}{3}$ **3** $x > \frac{2}{5}$ **4** $x > -3$
5 $x > \frac{13}{10}$ **6** $0 < x < 1$ **7** $-\frac{1}{2} < x < 2$ **8** $-1 < x < \frac{1}{3}$
9 $-1 < x < 3$ **10** $-\frac{1}{2} > x > 2$ **11** $x > 0$ **12** $-2 > x > 3$
13 $-6 < x < 4$ **14** $-\frac{1}{2} > x > 3$ **15** all values of x **16** all values of x
17 all values of x **18** $-2 < x < 0, x > 2$ **19** $5, 3, 3, 1$
21 $|x| < 3$ **22** $|x| < 4$ **23** $|2x-1| < 1$ **24** $|x-4| < 2$
25 $|x+2| < 2$ **26** $|2x-1| < 4$ **27** $|x+1| > 2$ **28** $|x-2| < \sqrt{3}$
29 $|x-2| > \sqrt{3}$ **30** $\left|x + \frac{5}{2}\right| < \frac{\sqrt{7}}{2}$
31 (i) $-5 < x < -1$; (ii) $-\frac{1}{2} > x > 1$; (iii) $-2 > x > \frac{8}{3}$; (iv) $1 < x < 4$
32 $-2\pi < \theta < -\frac{5\pi}{3}, -\frac{4\pi}{3} < \theta < -\frac{2\pi}{3}, -\frac{\pi}{3} < \theta < \frac{\pi}{3},$
$\frac{2\pi}{3} < \theta < \frac{4\pi}{3}, \frac{5\pi}{3} < \theta < 2\pi.$

EXAMPLES 8b (Page 116)

1 $0 < x < 1, x > 2$ **2** $x > 4, -\frac{3}{2} < x < 1$ **3** $-\frac{1}{2} < x < 1, x < -1$
4 $x > 2, -1 < x < \frac{3}{2}$ **5** $x < -\frac{1}{3}, 3 < x < 4$ **6** $|x| > 2, |x| < 1$
7 $x < -1, x > 2, 0 < x < \frac{1}{2}$ **8** $x < -\frac{2}{3}, 0 < x < \frac{1}{2}$ **9** $-4 < x < 1$
10 $-\frac{5}{3} < x < 0, x > \frac{3}{2}$ **11** $-\frac{5}{2} < x < 1, x > 2$ **12** $|x| > 1$
13 $-3 < x < 1, x > 2$ **14** $x < -3, 0 < x < 2$ **15** $-6 < x < 0, x > 6$
16 $-7 < x < -1, x > 4$ **17** $x < -3, 0 < x < \frac{1}{3}$ **18** $x < 0$
19 $-\frac{11\pi}{12} < x < -\frac{7\pi}{12}, -\frac{5\pi}{12} < x < -\frac{\pi}{12}, \frac{\pi}{12} < x < \frac{5\pi}{12}, \frac{7\pi}{12} < x < \frac{11\pi}{12}$
20 $-\pi < x < -2.0345, 1.1071 < x < \pi$ **21** $-\frac{2\pi}{3} < x < \frac{\pi}{3}$
22 $-\pi < x < -\frac{\pi}{4}, 0 < x < \frac{\pi}{4}$ **23** $|x| < 1.824$ **24** All values
25 $1 < k < 4$ **36** $x > \frac{1}{2}$

MISCELLANEOUS EXAMPLES (Page 117)

1 (i) $x = \pm \frac{1}{\sqrt{13}}, y = \pm \frac{6}{\sqrt{13}}; x = \pm \frac{1}{2}, y = \mp \frac{3}{2}$ (ii) $x = 1, y = 2, z = 3.$
2 $-2, -1; 1, 2; 3, 4$ **3** $\pm 138° 36', \pm 221° 24', 0°, \pm 360°$ **4** $n\pi - \frac{\pi}{2}$

Answers

5 Positive **6** 22° 30′, 202° 30′ **7** 0·704 **8** 2·095
10 $-\frac{1}{2} < x < \frac{1}{4}$ **11** $x = 2n\pi - \frac{\pi}{2}, \frac{2n\pi}{7} - \frac{\pi}{14}$ **12** 5 roots
13 (i) $x = 2, y = -1, z = 1$; (ii) $x = -1, y = 1, x = 1·39, y = 0·16$, $x = -1·59, y = -0·69$ **14** 2·047
15 (i) $\pm\frac{2\pi}{9}, \pm\frac{4\pi}{9}, \pm\frac{\pi}{2}, \pm\frac{8\pi}{9}$; (ii) $-131°34′, -11°34′, 48°26′, 168°26′$
16 $x > -1$ **17** $x = y = \sqrt[3]{\frac{2}{9}}; x = \frac{2}{\sqrt[3]{15}}, y = \frac{1}{\sqrt[3]{15}}; x = \frac{1}{\sqrt[3]{15}}, y = \frac{2}{\sqrt[3]{15}}$
18 6·6 rad **19** (i) $x > 4, -1 < x < 3$; (ii) $x > 2, \frac{2}{3} < x < 1$
20 (i) $x = a, \frac{4-3a}{3-2a}$; (ii) $x = -1, y = 1, z = 2; x = -\frac{67}{75}, \frac{131}{75}, \frac{110}{75}$
22 $-1·80, -0·45, 1·25$
23 (i) $x = 270°, y = 120°; x = 300°, y = 90°; x = 330°, y = 160°; x = 360°, y = 150°$ (ii) $x = 120°, y = 0°$ or $360°; x = 0°$ or $360°, y = 120°$.
24 0·642 rad **27** $0 < x < 1$ **28** 0·7660, 0·1736, $-0·9397$
30 $\frac{\pi}{8}\left(1 + \frac{\varepsilon}{4}\right)$ **32** (i) $1 < x < 2$; (ii) $1 < x < 2, 3 < x < 4$ **33** 28°, 77°
34 $-1 < x < 3$ **36** 62°; 0·601
37 (i) $x = n\pi, x = (4n-1)\frac{\pi}{8}$; (ii) $x = n\pi, x = n\pi - \frac{\pi}{4}$
38 $m = 0·16; x = 2·72$.
39 $x = 0, y = 0, x = 2, y = 1; x = 2, y = 2; x = 12, y = -3$
40 $x = 2n\pi + \frac{\pi}{4} \pm 1·2095, x = 2n\pi + \frac{\pi}{4} \pm 1·9321$

EXAMPLES 9a (Page 121)

1 $(0, 2), (3, 0), \left(\frac{3}{\sqrt{2}}, \sqrt{2}\right), \left(\frac{3\sqrt{3}}{2}, -1\right), \left(-\frac{3\sqrt{3}}{2}, 1\right)$ **2** $\sqrt{2}$ **3** 2
4 $\frac{2}{5}$ **5** $\left(\frac{1}{4}, \frac{9\sqrt{3}}{8}\right), \left(\frac{1}{\sqrt{2}}, \frac{3}{2\sqrt{2}}\right), \left(\frac{3\sqrt{3}}{4}, \frac{3}{8}\right), (-2, 0)$ **7** $(\frac{1}{2}, \frac{3}{2})$
8 $(0, 16), (0, 25)$ **11** $5y = x + 13$ **12** $y - 4 = (2 - x)^3$
13 $xy = 16$ **14** $\frac{x^2}{4} + \frac{y^2}{9} = 1$ **15** $\frac{(x-2)^2}{25} + \frac{(y-1)^2}{9} = 1$
16 $y^2 = 12x$ **17** $x = (y-1)^2(3-y)$ **18** $y^2 = x^2 + 4$
19 $(x-3)^2 + y^2 = 4$ **20** $y = 1 - 2x^2$ **21** $x^{\frac{2}{3}} + y^{\frac{2}{3}} = 1$
22 $y + tx = 2t + t^3$ **23** $t^2y + x = 2ct$ **25** $t = 2, -1$

EXAMPLES 9b (Page 124)

2 (i) 2·172; (ii) 7; (iii) 15·86 **3** $\dfrac{\pi}{2}$

4 (i) $\dfrac{5\sqrt{2}}{2}$; (ii) 1; (iii) 1; (iv) $\tfrac{1}{2} r_1 r_2 \sin(\theta_1 \sim \theta_2)$

5 1·294, 1·941, 1·5; 1·735

7 $(2, 2\sqrt{3})$; $(0, 2)$; $\left(-\dfrac{1}{\sqrt{2}}, \dfrac{1}{\sqrt{2}}\right)$; $\left(\dfrac{5}{\sqrt{2}}, -\dfrac{5}{\sqrt{2}}\right)$; $(-3\sqrt{3}, 3)$

8 $\left(5, \tan^{-1} \dfrac{4}{3}\right)$; $\left(5, \pi - \tan^{-1}\dfrac{4}{3}\right)$; $\left(13, -\tan^{-1}\dfrac{12}{5}\right)$; $\left(\sqrt{2}, \dfrac{3\pi}{4}\right)$; $\left(10, \pi + \tan^{-1}\dfrac{4}{3}\right)$; $\left(2, -\dfrac{\pi}{3}\right)$.

9 (i) $r = 2$; (ii) $r = 4 \sin \theta$; (iii) $r^2 \sin 2\theta = 2c^2$; (iv) $r^2 (\cos^2 \theta + 2 \sin^2 \theta) = 4$; (v) $r^2 \cos 2\theta = a^2$; (vi) $r^2 = a^2 \cos 2\theta$

10 (i) $x^2 + y^2 = 9$; (ii) $y = x$; (iii) $x = 2$; (iv) $(x^2 + y^2)^3 = a^4 y^2$; (v) $y^2 = 4(1 - x)$; (vi) $(x^2 + y^2)^3 = a^2 x^4$

EXAMPLES 9c (Page 131)

30 6π **31** $8\tfrac{8}{15}$ **32** $\dfrac{2\pi a^3}{3}(3 \ln 2 - 2)$ **33** $\dfrac{3\pi}{8}$

34 $\dfrac{16\sqrt{2}}{3}$ **35** $\tfrac{16}{3} a^2$ **36** $\tfrac{4}{3} \pi a b^2$

EXAMPLES 9d (Page 135)

13 $\dfrac{7 a^2 \pi^3}{384}$ **14** $\dfrac{a^2}{8}(9\pi + 16)$ **15** $\dfrac{1}{1 \cdot 88}$

17 $\dfrac{3\pi a^2}{2}$ **18** $\dfrac{a^2 \pi}{8}$ **19** a^2

MISCELLANEOUS EXAMPLES (Page 135)

1 Min. -1, Max. $\tfrac{1}{3}$ **2** $\pi, 3\pi$ **5** $\dfrac{1-\sqrt{2}}{2} \leqslant y \leqslant \dfrac{1+\sqrt{2}}{2}$

6 $\dfrac{\pi}{8}$ **10** $\dfrac{3}{7}$ **11** $\dfrac{3\pi a b}{8}$ **12** $3\sqrt{3} y = \pm 2x$

13 $\dfrac{a^2 \pi}{96}(\pi^2 - 6)$ **15** $3y + 2x = 2$; $(-\tfrac{5}{2}, \tfrac{7}{3})$ **16** $3\pi a^2$

17 $\dfrac{\pi a^2}{6}$ **18** $y \cos t + x \sin t = a \sin t \cos t$ **19** $\dfrac{4\sqrt{2}}{3}$

22 Max. $\left(1, \dfrac{1}{\sqrt{2}}\right)$, Min. $\left(1, -\dfrac{1}{\sqrt{2}}\right)$

23 Max. $\left(\dfrac{2\pi}{3}, \dfrac{3\sqrt{3}}{2}\right)$; Min. $\left(\dfrac{4\pi}{3}, -\dfrac{3\sqrt{3}}{2}\right)$

25 $(0, \tfrac{64}{99})$ **26** $\dfrac{2}{e}$ **27** $y = 3mx - 4am^3$

28 Max. $y = 0$, Min. $y = -\tfrac{4}{27}$; $(4, 4)$ **29** $y^2 = x^3(1-x)$ **30** $\tfrac{8}{3}a^2$

EXAMPLES 10a (Page 138)

1 $2y = 5x - 12$; $5y + 2x + 1 = 0$ **2** $79° 37'$ **3** $3y = 5x - 1$

4 $\dfrac{3}{\sqrt{5}}$ **5** $\dfrac{\sqrt{29}}{2}$; $5y = 2x + 9$

6 $y(\sqrt{5} - 1) + x(3 - \sqrt{5}) = 2\sqrt{5}$; $y(\sqrt{5} + 1) = x(\sqrt{5} + 3) + 2\sqrt{5}$

8 $y = x - 3$; $y + x + 1 = 0$ **9** $(\tfrac{4}{5}, \tfrac{19}{10})$ **10** 6 unit^2

11 $\sqrt{5}$ **12** $(4, 0)$; 12 unit^2 **13** $y(\sqrt{5} + 1) = x(\sqrt{5} + 3) - 4$

14 $(-\tfrac{15}{29}, \tfrac{6}{29})$ **15** ± 6 **16** (i) $32° 28'$; (ii) $\dfrac{7}{\sqrt{10}}$

17 $(\tfrac{13}{5}, \tfrac{11}{5})$ **18** $3y = x - 1$, $y + 3x + 7 = 0$ **19** $\dfrac{11}{2\sqrt{13}}$

20 Opposite

EXAMPLES 10b (Page 144)

1 (i) $y = \dfrac{2x}{5} + \dfrac{11}{5}$; (ii) $\dfrac{x}{-\dfrac{11}{2}} + \dfrac{y}{\dfrac{11}{5}} = 1$

3 (i) $x \cos(-60°) + y \sin(-60°) = 4$;

(ii) $x \cos 220° + y \sin 220° = 2$; (iii) $x \cos 53° 8' + y \sin 53° 8' = 2$;
(iv) $x \cos 143° 8' + y \sin 143° 8' = \tfrac{1}{5}$; (v) $x \cos 112° 37' + y \sin 112° 37' = 2$;
(vi) $x \cos(-35° 16') + y \sin(-35° 16') = 1$.

4 (i) $\dfrac{5}{\sqrt{2}} - 1$; (ii) $\dfrac{3\sqrt{3}}{2} - 2$; (iii) $\dfrac{33}{13}$; (iv) $2 \sin \alpha + 3 \cos \alpha - 1$ **5** Opposite

7 $11y - 7x + 18 = 0$ **8** $(-\tfrac{7}{9}, \tfrac{4}{9})$; $(-23, -4)$ **9** $3y + 3x - 19 = 0$

10 $1:3$ **11** $4y + 5x = 6$ **12** $x \cos 53° 8' + y \sin 53° 8' = 5$

14 $2:5$ **15** $22y + 55x - 71 = 0$ **16** $\left(\dfrac{16}{15}, \dfrac{2}{3}\right)$ **17** $\pm\sqrt{\dfrac{18}{5}}$

18 $3x + 8y - 6 = 0$. **19** $\pm 2\sqrt{15}$. **20** $y = 2x - 3$; $4y = 4x - 5$; $(\frac{7}{4}, \frac{1}{2})$.
21 $133° 42'$ **22** $2y = 5x - 10$; $23° 12'$, $78° 7'$ **23** $19y = x$

MISCELLANEOUS EXAMPLES (Page 146)

1 (i) $x + 1 = 0$; $4x + 3y = 20$; $y = 2x$; (ii) $36° 52'$, $26° 34'$, $116° 34'$,
2 $3y = x - 11$, $y + 3x = 3$ **3** $3y + 2x = 28$, $4y = 5x - 1$; $\frac{40}{23}$
4 $2x + 3y = 5$, $15x - 10y + 8 = 0$, $49x + 28y = 12$, $4x - 7y + 3 = 0$, $x - 3y + 2 = 0$, $3x + y + 1 = 0$
5 $(11, -5)$; $y = 2x - 2$
6 $(1, 8)$ **7** $(\frac{7}{3}, 3)$ **8** $x(2\sqrt{5} - 3) - y(4 + \sqrt{5}) + 8 - 3\sqrt{5} = 0$
9 $90°, 45°, 45°$; $(3, 5)$ **11** Same angle **12** $\frac{60}{13}$
13 $6x + y + 4 = 0$ **14** $\left(-\frac{1}{8}, \frac{9\sqrt{10} - 19}{8}\right)$ **15** $\left(\frac{45}{77}, \frac{65}{77}\right)$
16 $x + 8y - 14 = 0$; Externally 1:6 **17** 18 unit^2 **18** $(-\frac{46}{13}, -\frac{151}{26})$
19 $(\frac{14}{11}, -\frac{28}{11})$ **20** $(-\frac{1}{2}, -\frac{7}{2}), (\frac{7}{2}, \frac{3}{2})$ **21** $26\frac{1}{2}$ unit^2; $(\frac{86}{53}, -\frac{52}{53})$
22 $(\frac{21}{22}, -\frac{3}{22})$ **23** $(-\frac{4}{3}, \frac{14}{3}), (-\frac{2}{3}, \frac{7}{3}), (-3, 6), (-\frac{7}{8}, \frac{11}{8})$.
24 $\lambda = 34$ **25** $y - b = m(x - a)$; $\left(1 - \frac{2}{m}, 2 - m\right)$
26 $x = 0$; $3ky + xh = hk$; $xh + hk = 3yk$; $\left(0, \frac{h}{3}\right)$
27 $(a^2 + b^2)x_2 = -x_1(a^2 - b^2) - 2aby_1 - 2ac$;
$(a^2 + b^2)y_2 = y_1(a^2 - b^2) - 2abx_1 - 2bc$
33 Inside

EXAMPLES 11a (Page 150)

1 $(1, 1)$; 2 **2** $(3, -2)$; $\sqrt{10}$ **3** $(4, 0)$; 4 **4** $(0, -5)$; $5\sqrt{2}$
5 $(2, -2)$; $\sqrt{\frac{13}{2}}$ **6** $\left(1, -\frac{3}{2}\right)$; $\sqrt{\frac{19}{12}}$ **7** $x^2 + y^2 = 25$
8 $x^2 + y^2 - 2x - 4y - 4 = 0$ **9** $x^2 + y^2 + 4x - 6y - 36 = 0$
10 $x^2 + y^2 - 6x + 14y + 22 = 0$ **11** $x^2 + y^2 - 8x - 6y + 16 = 0$
12 $x^2 + y^2 + 4x + 6y + 9 = 0$ **13** $x^2 + y^2 - 2ax - 2ay = 0$
14 $x^2 + y^2 - 2ax - 2(a + 2)y + a^2 + 2a + 3 = 0$ **15** $x^2 + y^2 - 2x - 2y - 3 = 0$
16 $x^2 + y^2 - 6x + 4y + 11 = 0$ **17** $x^2 + y^2 - 10y + 9 = 0$
18 $2y + x = 0$ **21** $2x + y = 5$; $(5, -5)$ **22** $2y + 3x = 0$
23 $3x - 4y = 25$ **24** $(2, \frac{5}{6})$; $5y + 12x = 0$ **25** $2\sqrt{5}$
26 $4y = 3x + 10$ **27** $\sqrt{23}$ **28** $4y + x = 0$
30 $(\frac{14}{5}, \frac{22}{5})$ **31** $x^2 + y^2 - 2x + 4y + 1 = 0$

EXAMPLES 11b (Page 154)

1. $(\frac{5}{2}, -\frac{1}{2}); \frac{\sqrt{38}}{2}$ 2. $(0, -4); 4$ 3. $(1, -\frac{3}{2}); \frac{\sqrt{23}}{2}$
4. $(\frac{1}{3}, -\frac{1}{2}); \frac{1}{6}$ 5. $(3, 4); \sqrt{2}$ 6. $(-\frac{3}{2}, 2); \frac{5}{2}$
7. $(0, -3); 3\sqrt{2}$ 8. $x^2 + y^2 + 2x - 4y - 13 = 0$ 9. $(2 \pm \sqrt{7}, 0)$
10. 4 11. $y - 2x = 5; x + 2y = 0$ 12. $3x - 4y = 25; 4x + 3y = 0$
13. $2y = 3; x = 0$ 14. $3x - 2y = 0; 2x + 3y = 0$
15. $8x + y + 10 = 0; x - 8y = 15$ 16. $11x + 4y = 33; 4x = 11y + 12$
17. $y - 3x + 1 = 0; x + 3y + 3 = 0$ 18. $\sqrt{91}$ 19. $\sqrt{10}$
20. $\sqrt{23}$ 21. $\frac{\sqrt{294}}{2}$ 22. $\frac{\sqrt{129}}{3}$ 24. $67°23'; (\frac{13}{4}, \frac{1}{2})$
25. $18°26', 161°34'$ 26. $(3, 1)$ 27. $3x - 4y = 25; x = \pm 5$
28. $3x + 4y + 17 = 0$ 30. $4y = 3x + 30; x^2 + y^2 - 2x - 4y - 20 = 0$

EXAMPLES 11c (Page 157)

1. $x^2 + y^2 - 5x - y + 4 = 0$. 2. $x^2 + y^2 - 3x + y - 6 = 0$.
3. $x^2 + y^2 - 6x - 2y + 5 = 0$. 4. $2x^2 + 2y^2 + x - 11y - 1 = 0$.
9. $3x^2 + 3y^2 - 20x = 0$.
10. $(-\frac{7}{2}, 0); \frac{\sqrt{97}}{2}$,
12. $x^2 + y^2 - 21x + 89 = 0; x^2 + y^2 + 4x - 36 = 0$.
13. $\frac{25 \pm \sqrt{85}}{18}; \left(\frac{46 \pm 4\sqrt{85}}{43 \pm \sqrt{85}}, 0\right)$ 15. $x^2 + y^2 - 6x - 8y = 0$.

MISCELLANEOUS EXAMPLES (Page 159)

1. $x^2 + y^2 - 2y - 7 = 0$. 2. $3x^2 + 3y^2 - 26x - 16y + 61 = 0$. 3. $y = \pm\sqrt{\frac{7}{2}}x$.
4. $x^2 + y^2 - 2x - 2y + 1 = 0$. 6. $x + 2y = 1; 2x = y - 8$. 7. $78° 28^1$
8. $x^2 + y^2 - 24x - 10y + 88 = 0$. 9. $x^2 + y^2 + 2x - 14y + 30 = 0; (-1, 7), \sqrt{20}$.
10. $y = 2x \pm 2\sqrt{5}$ 11. $x^2 + y^2 - 6x - 4y + 8 = 0$. 12. $(\frac{3}{2} \pm \sqrt{6}, \frac{1}{2} \pm \sqrt{6})$.
13. $y + x = \pm 5\sqrt{2}$ 14. $\sqrt{\frac{8}{3}}; 85° 36'$. 16. $\frac{5}{6}$.
17. $10x + 6y = 25$ 19. $(-2, 5); (-22, 145)$ 20. $\tan^{-1}\frac{2 \pm 3\sqrt{2}}{4}$
21. $x^2 + y^2 - 4x - 4y + 3 = 0; 1$ 22. $x^2 + y^2 - 2x + 4y - 8 = 0$
24. $5x^2 + 5y^2 + 2x + 16y - 26 = 0$
25. $\left(2 - \frac{6}{\sqrt{13}}, -3 + \frac{9}{\sqrt{13}}\right); \left(2 + \frac{6}{\sqrt{13}}, -3 - \frac{9}{\sqrt{13}}\right)$
26. $x^2 + y^2 - 6x - 6y + 9 = 0, x^2 + y^2 - 30x - 30y + 225 = 0; x = 6, 3x + 4y = 30$

28 $17(x^2+y^2)+200x+108y-593=0$ **30** $(\frac{9}{2},\frac{11}{2})$
31 $x-2y+5=0; 2\sqrt{5}$ **33** $x(x-p)+(y-1)(y-q)=0$
35 $13(x^2+y^2)-102x-88y+333=0.$
36 $2gg_1+2ff_1-2g^2-2f^2+c-c_1=0; x+2y-2=0$

EXAMPLES 12a (Page 164)

13 (i) $(-2,0)$; (ii) $(-\frac{1}{2},0)$; (iii) $x=-2$; (iv) $x=-3\cdot5$
14 (i) $(0,\frac{1}{3})$; (ii) $(0,-2\frac{2}{3})$; (iii) $y=\frac{1}{3}$; (iv) $y=3\frac{1}{3}$
15 (i) $(-\frac{1}{6},1)$; (ii) $(1\frac{1}{3},1)$; (iii) $x=-\frac{1}{6}$; (iv) $x=-\frac{5}{3}$
16 (i) $(2,-2)$; (ii) $(1\frac{1}{2},-2)$; (iii) $x=2$; (iv) $x=\frac{5}{2}$
17 (i) $(-\frac{1}{2},-\frac{3}{4})$; (ii) $(-\frac{1}{2},0)$; (iii) $y=-\frac{3}{4}$; (iv) $y=-\frac{3}{2}$
18 (i) $(-\frac{7}{6},\frac{3}{2})$; (ii) $(-\frac{19}{24},\frac{3}{2})$; (iii) $x=-\frac{7}{6}$; (iv) $x=-\frac{37}{24}$
19 $(8,0)$ **20** 30 m above **21** $y=x+3; y+x=9$
22 $y+x+2=0$ **23** $48\sqrt{5}$ m **24** $2y=x+3$
25 $y^2-2y-10x-14=0$ **26** $y^2-6y+8x-23=0$
27 $x^2-6x+4y+5=0$ **28** $x^2+2x+4y-19=0$
29 $x^2+y^2-2xy+8x+8y-16=0$ **30** $x^2-4xy+4y^2-30x-10y+50=0$

EXAMPLES 12b (Page 166)

5 $\left(\frac{5t^2}{2},5t\right)$ **6** $(6t^2,12t)$ **7** $\left(-\frac{t^2}{2},t\right)$ **8** $\left(3t,\frac{3t^2}{2}\right)$
9 $(1+2t^2,4t)$ **10** $(6t,1+3t^2)$
11 (i) $(0,4); y=-4$; (ii) $(0,3); y=-3$; (iii) $(0,0); x=-2$;
(iv) $(3,-2); x=-5$; (v) $(4,1); x=2$; (vi) $(1,0); x=5$
12 24 **13** 4 **14** $3x+y=9$ **15** $(16,-16)$ **16** $(17,-12)$

EXAMPLES 12c (Page 168)

1 $3y-x=36$ **2** $2y+x+12=0$ **3** $2y=x+3$
4 $x+2y+10=0$ **5** $3y=2x+18$ **6** $y=x+a$
7 $12; y=x+3$ **8** $2x+y=48$ **9** $2y=4x+3$
10 $(0,2)$ **11** $y+x=6$ **12** $y=x-9$
13 $y+8x=66$ **14** $y+tx=2t+t^3$ **15** $1; \sqrt{3}$
16 $(12,12)$ **17** $2y=2mx+\frac{7}{m}; \left(\frac{7}{2m^2},\frac{7}{m}\right)$
18 $y+x=0; y-x+8=0$ **19** $(6,14)$ **20** $T(-4,0); G(12,0)$
21 $(-5,-\frac{25}{6})$ **24** $y^2=12x; \left(\frac{3}{m^2},\frac{6}{m}\right)$ **25** $90°$
27 $my+x=2+4m^2; y-mx=6m+4m^3$

EXAMPLES 12e (Page 170)

1 $45°$ **2** $y+x+1=0, 3y=x+9$ **3** $5y=4x+8$
4 $y^2=x^2+18x+9$ **5** $y=m(x-5)$ **7** $3y^2-32y-16=0; (\frac{35}{9},\frac{16}{3})$

8 $\left(\dfrac{4-3m+2m^2}{m^2}, \dfrac{4}{m}\right)$; $y^2 - 3y - 4x + 8 = 0$ **9** $y = 3x - 7$

11 $2y + 4x + a = 0$; $\left(\dfrac{a}{4}, -a\right)$ **14** $y^2 + 20y - 10c = 0$; -10

22 $(13, 8\sqrt{3})$ **24** $9x^2 - 24xy + 16y^2 - 44x + 192y + 276 = 0$; $(\tfrac{66}{25}, -\tfrac{63}{25})$

26 $2y^2 = 27x$ **27** $y^2(x-4) + (x-3)^3 = 0$ **28** $y^2 = a(x-a)$

EXAMPLES 13a (Page 179)

7 (i) $\dfrac{1}{\sqrt{2}}$; (ii) $(\mp 1, 0)$; (iii) $x = \mp 2$.

8 (i) $\dfrac{1}{\sqrt{3}}$; (ii) $(\mp \sqrt{2}, 0)$; (iii) $x = \mp 3\sqrt{2}$

9 (i) $\dfrac{1}{\sqrt{2}}$ (ii) $(0, \mp 2)$; (iii) $y = \mp 4$

10 (i) $\tfrac{1}{3}\sqrt{5}$; (ii) $(\mp \tfrac{2}{3}\sqrt{5}, 0)$; (iii) $x = \mp \dfrac{6}{\sqrt{5}}$.

11 (i) $\tfrac{1}{4}\sqrt{15}$; (ii) $(\mp \tfrac{5}{4}\sqrt{15}, 0)$; (iii) $x = \mp \dfrac{20}{\sqrt{15}}$

12 (i) $\dfrac{\sqrt{3}}{2}$; (ii) $(0, \mp \sqrt{3})$; (iii) $y = \mp \dfrac{4}{\sqrt{3}}$

13 (i) $\tfrac{1}{3}\sqrt{5}$; (ii) $(\mp \sqrt{5}, -1)$; (iii) $x = \mp \dfrac{9}{\sqrt{5}}$

14 (i) $\dfrac{\sqrt{3}}{2}$; (ii) $(-2 \mp \sqrt{3}, 1)$; (iii) $x = -2 \mp \dfrac{4}{\sqrt{3}}$

15 $x + 2 = 0$ **16** $x + y = 5$ **17** $x - 2y = 4$

18 $9x + 2y + 20 = 0$ **19** $(\pm \tfrac{12}{5}, \pm \tfrac{12}{5})$ **20** $y = 3x - 3$

21 (i) 8; (ii) $2\sqrt{7}$ **22** $\left(\pm \dfrac{2}{\sqrt{3}}, \pm \sqrt{\dfrac{2}{3}}\right)$ **23** $(-\tfrac{4}{3}, \tfrac{1}{3})$

24 $6, 2\sqrt{5}$; $\dfrac{(x-4)^2}{9} + \dfrac{(y-1)^2}{5} = 1$

EXAMPLES 13b (Page 182)

6 (i) $\dfrac{2\sqrt{2}}{3}$; (ii) $(\mp 4\sqrt{2}, 0)$; (iii) $x = \mp \dfrac{9}{\sqrt{2}}$

7 (i) $\dfrac{\sqrt{7}}{4}$; (ii) $(\mp \sqrt{7}, 0)$; (iii) $x = \mp \dfrac{16}{\sqrt{7}}$

Answers | 357

8 (i) $\dfrac{1}{\sqrt{2}}$; (ii) ($\mp 1, 0$); (iii) $x = \mp 2$

9 (i) $\dfrac{\sqrt{5}}{3}$; (ii) (0, $\mp\sqrt{5}$); (iii) $y = \mp\dfrac{9}{\sqrt{5}}$

10 6π **11** $2\sqrt{15}$ **12** $x = 3\cos\theta, y = 2\sin\theta$
13 $x = 2\cos\theta, y = \sin\theta$ **14** $x = \tfrac{3}{2}\cos\theta, y = \sin\theta$
15 $x = 1 + 2\cos\theta, y = \sin\theta$ **16** $x = -2 + 4\cos\theta, y = 1 + 3\sin\theta$

17 $\dfrac{\pi}{3} \pm \pi;\ -\dfrac{1}{2\sqrt{3}}$ **19** $\dfrac{2\sqrt{14}}{9}$ **20** $\dfrac{64\sqrt{7}\pi}{9}$

21 $(a\cos\phi, a\sin\phi)$

EXAMPLES 13c (Page 185)

1 $x + 2y = 2\sqrt{2}$ **2** $2x + 3\sqrt{3}y = 12$ **3** $x - 2\sqrt{3}y = 8$
4 $\dfrac{x\cos\phi}{5} + \dfrac{y\sin\phi}{4} = 1$ **5** $\dfrac{x\cos\phi}{2} + y\sin\phi = 1$ **6** $y = 3x \pm \sqrt{38}$.
7 $y + x = \pm 5$ **8** $y = \dfrac{x}{2} \pm \sqrt{12}$ **9** $y = \dfrac{\sqrt{3}}{2}x \pm \tfrac{1}{2}\sqrt{85}$
10 $\pm\sqrt{5}$ **11** $y = x \pm \sqrt{13}, y + x = \pm\sqrt{13}$ **12** $\tfrac{1}{2}, -2$.
13 $y + x = \pm 5$ **14** $3\sqrt{3}x + 2y = 12; 4x - 6\sqrt{3}y + 5\sqrt{3} = 0$
16 $P\left(\dfrac{a^2 - b^2}{a}\cos\phi, 0\right), Q\left(0, \dfrac{b^2 - a^2}{b}\sin\phi\right)$ **17** $3y + x = \pm\sqrt{31}$
18 $7m^2 + 12m + 5 = 0$; outside **19** Inside **20** $35°\,10'$

EXAMPLES 13d (Page 188)

1 (i) $3y + 2x = 0$; (ii) $3y + x = 0$; (iii) $3y - 2x = 0$; (iv) $3my + 2x = 0$;
2 (i) $8y + x = 0$; (ii) $4y - x = 0$; (iii) $4my + x = 0$
3 (i) $(-\tfrac{4}{5}, \tfrac{1}{5})$; (ii) $(-\tfrac{8}{13}, \tfrac{6}{13})$; (iii) $(\tfrac{8}{41}, \tfrac{5}{41})$.
4 (i) $16y + 9x = 25$; (ii) $9x - 8y = 26$; (iii) $8y - 3x = 30$; (iv) $x + 1 = 0$
5 -1 **6** $\sqrt{5}y + \sqrt{3}x = 0$ **7** $\pm\sqrt{\dfrac{5}{8}}$ **8** $\sqrt{\dfrac{2}{3}}$
9 5.196 units **11** $(x + 2y)^2 + (2y - x)^2 = 4$

EXAMPLES 13f (Page 192)

1 $\tfrac{16}{7}$ **2** (i) $4x - 2\sqrt{3}y = 9\sqrt{3}$; (ii) 3·57
3 (i) $\tan^{-1}\tfrac{8}{3}, \pi + \tan^{-1}\tfrac{8}{3}$; (ii) $\tan^{-1}(-\tfrac{2}{3}), \pi + \tan^{-1}(-\tfrac{2}{3})$
4 $\sin^{-1}\dfrac{b^2}{a^2 - b^2}$ **5** $\dfrac{\pi}{4}, \dfrac{7\pi}{4}, \dfrac{3\pi}{4}, \dfrac{5\pi}{4}$

6 $2y = 3x \pm 4\sqrt{3}$; $(-\sqrt{3}, \tfrac{1}{2}\sqrt{3})$, $(\sqrt{3}, -\tfrac{1}{2}\sqrt{3})$

7 $(3\cos\phi, 4\sin\phi)$; $\dfrac{x^2}{9} + \dfrac{y^2}{16} = 1$ **8** $(-2\sin\phi, \sqrt{2}\cos\phi)$

9 $\left(\dfrac{12\cos\phi}{4\cos^2\phi + 9\sin^2\phi}, \dfrac{18\sin\phi}{4\cos^2\phi + 9\sin^2\phi}\right)$

12 $(x^2 + y^2 - 5)^2 = 4(x^2 + 4y^2 - 4)$

16 $(\tfrac{5}{2}\cos\phi - \sin\phi, \tfrac{3}{2}\cos\phi + \sin\phi)$; $\left(\dfrac{x}{5} + \dfrac{y}{3}\right)^2 + \left(\dfrac{y}{3} - \dfrac{x}{5}\right)^2 = 1$

22 $\dfrac{\sqrt{3}(1 - 4m^2)}{4m^2 + 1}, \dfrac{8\sqrt{3}m}{4m^2 + 1}$ **23** $\left(\dfrac{10x}{27}\right)^2 + (10y)^2 = 1$ **25** $\sqrt{\dfrac{a^2 - b^2}{a^2}}$

EXAMPLES 14a (Page 197)

1 (i) $\dfrac{\sqrt{5}}{2}$; (ii) $(\mp\sqrt{5}, 0)$; (iii) $x = \mp\dfrac{4}{\sqrt{5}}$; (iv) $y = \pm\dfrac{x}{2}$

2 (i) $\sqrt{2}$; (ii) $(\mp 2\sqrt{2}, 0)$; (iii) $x = \mp\sqrt{2}$; (iv) $y = \pm x$.

3 (i) $\dfrac{2}{\sqrt{3}}$; (ii) $(\mp 4, 0)$; (iii) $x = \mp 3$; $\sqrt{3}y = \pm x$.

4 (i) $\tfrac{13}{12}$; (ii) $(\mp 13, 0)$; (iii) $x = \mp\tfrac{144}{13}$; $y = \pm\tfrac{5}{12}x$

5 (i) $\dfrac{\sqrt{5}}{2}$; (ii) $(\mp 3\sqrt{5}, 0)$; (iii) $x = \mp\dfrac{12}{\sqrt{5}}$; (iv) $y = \pm\dfrac{x}{2}$

6 (i) $\sqrt{2}$; (ii) $\left(\mp\dfrac{3\sqrt{2}}{2}, 0\right)$; (iii) $x = \mp\dfrac{3}{2\sqrt{2}}$; (iv) $y = \pm x$

7 (i) $\dfrac{\sqrt{13}}{3}$; (ii) $(1 \mp \sqrt{13}, 0)$; (iii) $x = 1 \mp \dfrac{9}{\sqrt{13}}$; (iv) $y = \tfrac{2}{3}(x - 1)$.

8 (i) $\tfrac{5}{4}$; (ii) $(-1 \pm 10, 2)$; (iii) $x = -1 \mp \tfrac{32}{5}$; (iv) $y - 2 = \pm\tfrac{3}{4}(x + 1)$.

9 $70°\,32'$ **10** $y + 1 = \pm(x - 3)$ **11** $(-1, 2)$ **12** $\dfrac{x^2}{16} - \dfrac{y^2}{9} = 1$

13 $5x^2 - 16x - 4y^2 = 16$ **14** $7x^2 - 9y^2 - 78x + 135 = 0$

15 $8xy + 4x + 4y = 7$

EXAMPLES 14b (Page 199)

1 $x - y = 1$ **2** $(\tfrac{9}{7}, \tfrac{8}{7})$ **3** $4y = 3x \pm 1$

6 $\left(\pm\dfrac{27}{4\sqrt{5}}, \pm\dfrac{1}{4\sqrt{5}}\right)$ **8** $29°\,56'$ **11** $bx - ay\sin\phi = ab\cos\phi$

12 $bx(t^2 + 1) - ay(t^2 - 1) = 2abt$

EXAMPLES 14c (Page 202)

1. $\left(3t, \dfrac{3}{t}\right)$ 2. $\left(4t, \dfrac{4}{t}\right)$ 3. $\left(\dfrac{5t}{2}, \dfrac{5}{2t}\right)$ 4. $\left(\dfrac{t}{3}, \dfrac{1}{3t}\right)$

5. $\left(\sqrt{2}t, \dfrac{\sqrt{2}}{t}\right)$ 6. $\left(2t, -\dfrac{2}{t}\right)$ 7. $\left(1+t, \dfrac{1}{t}\right)$ 8. $\left(-2+3t, \dfrac{3}{t}\right)$

9. $\left(-1+5t, 3+\dfrac{5}{t}\right)$ 10. (i) $xy = 25$; (ii) $(5, 5), (-5, -5)$

11. (i) $xy = 9$; (ii) $(3, 3), (-3, -3)$ 12. (i) $xy = 36$; (ii) $(6, 6), (-6, -6)$
13. (i) $xy = -1$; (ii) $(1, -1), (-1, 1)$
14. (i) $y(x-1) = 4$; (ii) $(3, 2), (-1, -2)$.
15. (i) $x(y+1) = 16$; (ii) $(4, 3), (-4, -5)$
25. (i) 12; (ii) $(6, 6), (-6, -6)$.
26. (i) $4\sqrt{2}$; (ii) $(2\sqrt{2}, 2\sqrt{2}), (-2\sqrt{2}, -2\sqrt{2})$.
27. (i) $16\sqrt{2}$; (ii) $(8\sqrt{2}, 8\sqrt{2}), (-8\sqrt{2}, -8\sqrt{2})$
28. (i) 10; (ii) $(5, 5), (-5, -5)$ 29. (i) 4; (ii) $(3, 2), (-1, -2)$
30. (i) $4\sqrt{2}$; (ii) $(1+2\sqrt{2}, 1+2\sqrt{2}), (-2\sqrt{2}+1, -2\sqrt{2}+1)$
31. $x + y = \pm 4\sqrt{2}$ 32. $t^2 y + x = 8t$; $ty - t^3 x = 4(1 - t^4)$
33. $2\sqrt{17}$ 35. $y + 3x = \pm 6$; $\dfrac{12}{\sqrt{10}}$ 36. $(10, \tfrac{2}{5}), (-2, -2)$

37. $\dfrac{17\sqrt{17}}{2}$ 38. $9y + x = 18, y + x + 6 = 0$

EXAMPLES 14d (Page 204)

1. $8x - 2y = 15a$ 2. $\tfrac{16}{15}\sqrt{34}$ 3. $\tfrac{1}{2}, -4; (\tfrac{16}{7}, -\tfrac{8}{7})$

4. $\left(\dfrac{4}{m}, 4m\right)$ 5. $69°\,38'$ 6. $t_1^2 t_2^2 + 1 = t_1^2 \sim t_2^2$

7. $\tfrac{82}{9}\sqrt{82}$ 8. $\left(\dfrac{2m-3}{m}, 3-2m\right)$; $xy = 3x + 2y$

9. $(4, 1), (-4, -1)$ 10. $(1, 2); (5, 6), (-3, -2); x = 1, y = 2$
12. $x = -\tfrac{1}{2}, y = \tfrac{5}{3}$ 14. $(\tfrac{3}{2}, 2); 14y = 2x + 25, 2y + 14x = 25$

20. $\left(\dfrac{n}{2l}, \dfrac{2c^2 l}{n}\right)$ 23. $y - t^2 x = \pm 4t$ 28. $(xy - 4)(x^2 + y^2) = xy$

MISCELLANEOUS EXAMPLES (Page 205)

1. $a = b = 3; (-1, 1), (-\tfrac{7}{13}, \tfrac{17}{13})$ 3. $yt - x = at^2, y + xt = 2at + at^3; 2a$
4. Tangent at the vertex 5. $\tfrac{4}{5}$ 6. $\tfrac{3}{4}$ 7. $x(t^2+1) - y(t^2-1) = 2at$
12. $y(t_1 + t_2) - 2x = 2at_1 t_2; yk = x - h$.
14. $9x^2 + 16y^2 - 24xy - 24x - 18y - 9 = 0$;
    $9x^2 + 16y^2 - 24xy + 96x + 72y - 144 = 0$

15 $a\sqrt{1+t^2}$ **16** $2bx^2 = a^2(y+b)$; $(0, -b)$
22 $yy_1 = 2a(x+x_1)$; $2a(y-y_1) + y_1(x-x_1) = 0$ **24** $2ax = y^2 + 8a^2$
28 $2x^2 + y^2 - 2ax = 0$ **33** $y\sin\theta + x\cos\theta = r + a\cos\theta$
37 $y^2 + x^2 - 2xy - 2ay - 2ax + a^2 = 0$

EXAMPLES 15a (Page 213)

1 $\dfrac{\pi}{6}$ **2** $\dfrac{\pi}{4}$ **3** $\dfrac{\pi}{3}$ **4** $-\dfrac{\pi}{4}$

5 $\dfrac{\pi}{6}$ **6** $-\dfrac{\pi}{4}$ **7** $\dfrac{5\pi}{6}$ **8** $-\dfrac{\pi}{6}$

9 $\dfrac{\pi}{3}$ **10** $\dfrac{\pi}{6}$ **11** $-\dfrac{\pi}{4}$ **12** $\dfrac{\pi}{6}$

13 $\dfrac{\pi}{4}$ **14** $\dfrac{\pi}{4}$ **15** 1·1071 **17** $\dfrac{3\pi}{10}$

18 $\dfrac{1}{2}$ **20** $\dfrac{2}{\sqrt{1-4x^2}}$ **21** $\dfrac{2}{4+x^2}$ **22** $-\dfrac{3}{\sqrt{1-9x^2}}$

23 $\dfrac{-1}{1+x^2}$ **24** $\dfrac{-1}{x\sqrt{x^2-1}}$ **25** $\dfrac{1}{2\sqrt{x-x^2}}$ **26** $\dfrac{-1}{1+x^2}$

27 $\dfrac{-2x}{\sqrt{1-x^4}}$ **28** $\dfrac{-1}{(x+1)\sqrt{x^2+2x}}$ **29** $\dfrac{1}{1+x^2}$ **30** $\dfrac{-1}{(1+x)\sqrt{x}}$

31 $\dfrac{1}{x\sqrt{9x^2-1}}$ **32** $\dfrac{\cos x}{1+\sin^2 x}$ **33** -1 **34** $\dfrac{2\cos x}{\sqrt{1-4\sin^2 x}}$

35 $\sin^{-1}\dfrac{x}{2}$ **36** $\dfrac{\pi}{4}$ **37** $\tfrac{1}{3}\sin^{-1} 3x$ **38** $\tfrac{1}{4}\sin^{-1} 4x$

39 $\tfrac{1}{2}\tan^{-1}\dfrac{x}{2}$ **40** $\dfrac{n}{12}$ **41** $\tfrac{1}{6}\tan^{-1}\dfrac{3x}{2}$ **42** $\tfrac{1}{20}\tan^{-1}\dfrac{5x}{4}$

43 $\dfrac{\pi}{4\sqrt{2}}$ **44** $\tan^{-1} 2$ **45** $\sin^{-1}\dfrac{x-3}{4}$ **46** $\dfrac{1}{\sqrt{6}}\tan^{-1}\sqrt{\dfrac{2}{3}}(x-1)$

EXAMPLES 15b (Page 216)

1 1·1752 **2** 3·7622 **3** 0·9640 **4** $-3\cdot6269$
5 $-1\cdot313$ **6** 0·4251 **9** $2\cosh 2x$ **10** $3\sinh 3x$
11 $2\operatorname{sech}^2 2x$ **12** $\tfrac{1}{3}\cosh\dfrac{x}{3}$ **13** $-\operatorname{cosech}^2 x$ **14** $-\tanh x \operatorname{sech} x$
15 $-\coth x \operatorname{cosech} x$ **16** $2\sinh x \cosh x$ **17** $2\cosh x \sinh x$
18 $\coth x$ **19** $\tanh x$ **20** $\cosh x \, e^{\sinh x}$ **21** $\tfrac{1}{2}\sinh 2x$
22 $\tfrac{1}{3}\cosh 3x$ **23** $2\cosh\dfrac{x}{2}$ **24** $\tfrac{1}{3}\tanh 3x$ **25** $\dfrac{\cosh 2x}{4}$
26 $\tfrac{1}{4}\sinh 2x - \dfrac{x}{2}$ **27** 0·6321 **28** $4(\sinh 2x + \cosh 2x)$

EXAMPLES 15c (Page 219)

1. 1.4436 2. 1.3169 3. -0.8812 4. 1.7627
5. 0.4810 6. 2.0947 8. $\ln(2x + \sqrt{4x^2 + 1})$.
9. $\ln(x + 1 + \sqrt{x^2 + 2x})$. 10. $\ln(2x - 1 + \sqrt{4x^2 - 4x + 2})$
11. $\ln\dfrac{x + 1 + \sqrt{x^2 + 2x - 3}}{2}$ 12. $\ln\dfrac{x^2 + \sqrt{x^4 + 4}}{2}$
13. $\ln(x + \sqrt{x^2 + 1})$ 14. $\ln\dfrac{x + \sqrt{x^2 - 4}}{2}$

15. $\dfrac{2}{\sqrt{4x^2 + 1}}$ 16. $\dfrac{1}{\sqrt{x^2 + 9}}$ 17. $\dfrac{1}{\sqrt{x^2 - 4}}$ 18. $\dfrac{4}{\sqrt{16x^2 + 9}}$

19. $\dfrac{1}{\sqrt{x^2 - 2}}$ 20. $\dfrac{\sqrt{3}}{\sqrt{3x^2 - 1}}$ 21. $\dfrac{-3}{\sqrt{2 - 6x + 9x^2}}$ 22. $\dfrac{1}{\sqrt{x^2 - 2x - 4}}$

23. $\dfrac{1}{\sqrt{x^2 - 2\sqrt{2}x + 4}}$ 24. $\dfrac{1}{2\sqrt{x}\sqrt{x + 1}}$ 25. $\dfrac{-1}{x\sqrt{1 - x^2}}$ 26. $\sec x$

27. $\dfrac{e^x}{\sqrt{e^{2x} + 1}}$ 29. $\dfrac{1}{1 - x^2}$ 30. (i) $\sec x$; (ii) $\dfrac{1}{2x}$ 31. $\cosh^{-1}\dfrac{x}{2}$

32. $\sinh^{-1}\dfrac{x}{3}$ 33. 0.481 34. 0.503 35. $\tfrac{1}{2}\cosh^{-1} 2x$

36. 0.6255 37. $\dfrac{1}{\sqrt{3}}\sinh^{-1}\sqrt{3}x$ 38. 0.2667

39. $\dfrac{1}{\sqrt{3}}\sinh^{-1}\sqrt{\dfrac{3}{2}}x$ 40. $\sinh^{-1}\dfrac{x + 2}{2}$ 41. 0.3796 42. $\dfrac{1}{\sqrt{2}}\sinh^{-1}\sqrt{2}(x - 3)$

EXAMPLES 15d (Page 222)

1. $\dfrac{1}{\sqrt{3}}\tan^{-1}\dfrac{x + 2}{\sqrt{3}}$ 2. $\dfrac{1}{4}\tan^{-1}\dfrac{2x - 1}{2}$ 3. $\sin^{-1}\dfrac{x - 1}{2}$

4. $\sinh^{-1}\dfrac{x + 1}{\sqrt{6}}$ 5. $\dfrac{1}{2}\tan^{-1}\dfrac{x + 1}{2}$ 6. $\sin^{-1}\dfrac{2x + 1}{\sqrt{17}}$

7. $\dfrac{2}{\sqrt{7}}\tan^{-1}\dfrac{2x - 1}{\sqrt{7}}$ 8. $\cosh^{-1}(2x - 1)$ 9. $\cosh^{-1}(2x + 1)$.

10. $\dfrac{1}{3}\tan^{-1}\dfrac{2x + 1}{3}$ 11. $\dfrac{1}{\sqrt{2}}\sin^{-1}(4x - 1)$ 12. $\dfrac{1}{\sqrt{2}}\sin^{-1}\dfrac{4x - 1}{\sqrt{33}}$

13. $\dfrac{1}{\sqrt{3}}\sinh^{-1}\sqrt{\dfrac{3}{5}}(x - 1)$ 14. $\dfrac{1}{3}\tan^{-1}(3x + 2)$ 15. $\dfrac{2}{\sqrt{3}}\tan^{-1}\dfrac{2x + 1}{\sqrt{3}}$

16. $\sinh^{-1}\dfrac{2x - 1}{\sqrt{3}}$ 17. $\dfrac{1}{\sqrt{3}}\sin^{-1}\dfrac{3x - 1}{2}$ 18. $\dfrac{\tfrac{1}{2}(2x) + 1}{\sqrt{x^2 + 1}}$

362 | Answers

19 $\dfrac{-\frac{3}{2}(2-2x)+1}{\sqrt{3+2x-x^2}}$; $-3\sqrt{3+2x-x^2}+\sin^{-1}\dfrac{x-1}{2}$

20 $\dfrac{1(x^2+x+1)+\frac{1}{2}(2x+1)-\frac{3}{2}}{x^2+x+1}$; $1+\dfrac{1}{2}\ln 3-\dfrac{\sqrt{3}\pi}{6}$

21 $\ln(x^2+2x+3)-\dfrac{3}{\sqrt{2}}\tan^{-1}\dfrac{x+1}{\sqrt{2}}$ **22** $1-\dfrac{\sqrt{3}}{2}+\dfrac{\pi}{6}$

23 $2\sqrt{x^2-1}-\cosh^{-1}x$ **24** $3\sqrt{5}-6-2\ln\left(\dfrac{1+\sqrt{5}}{2}\right)$

25 $\dfrac{1}{2}\ln(x^2+x+1)-\dfrac{1}{\sqrt{3}}\tan^{-1}\dfrac{2x+1}{\sqrt{3}}$ **26** $\sqrt{x^2-2x-3}+\cosh^{-1}\dfrac{x-1}{2}$

27 $\dfrac{3}{2}\sqrt{2x^2-4x+1}-\dfrac{1}{\sqrt{2}}\cosh^{-1}\sqrt{2}(x-1)$.   **28** $2-\dfrac{\pi}{2}$

29 $x+\ln(x^2-x+1)+\dfrac{2}{\sqrt{3}}\tan^{-1}\dfrac{2x-1}{\sqrt{3}}$

MISCELLANEOUS EXAMPLES (Page 223)

1 $\sqrt{1-a^2}, 2a\sqrt{1-a^2}$

2 (i) $-\dfrac{1}{1+x^2}$; (ii) $\dfrac{1}{2\sqrt{x^2+x}}$; (iii) $\dfrac{2}{\sqrt{x}(1+4x)}$; (iv) $\dfrac{\cos^{-1}x-x\sqrt{1-x^2}}{(1-x^2)^{3/2}}$

3 (i) $\dfrac{-2x}{x^4+1}$; (ii) $\dfrac{2}{1+x^2}$; (iii) $\dfrac{-1}{1+x^2}$ **4** $\dfrac{1}{2}$ **5** $x=\dfrac{-2+\sqrt{6}}{2}$

6 $\sqrt{x^2-a^2}$; $\dfrac{1}{2}x\sqrt{x^2-4}-2\cosh^{-1}\dfrac{x}{2}$ **7** $\dfrac{9\pi}{4}$

9 $\dfrac{x+(1+x^2)\tan^{-1}x}{(1+x^2)\{1+(x\tan^{-1}x)^2\}}$

10 (i) $\dfrac{1}{4}(\pi+1)$; (ii) $\dfrac{1}{3}-\dfrac{1}{3}\ln\dfrac{28}{13}+\dfrac{2\sqrt{3}}{9}(\tan^{-1}3\sqrt{3}-\tan^{-1}2\sqrt{3})$;

     (iii) $\dfrac{1}{2}\left(\cosh^{-1}\dfrac{25}{7}-\cosh^{-1}\dfrac{9}{7}\right)$; (iv) $2-\dfrac{\pi}{2}$; (v) $\dfrac{3}{\sqrt{2}}\sinh^{-1}\dfrac{1}{\sqrt{5}}$; (vi) $\dfrac{\pi}{2}-1$

11 (i) $x=2$; (ii) $x=\dfrac{-4+\sqrt{19}}{3}$

14 (i) $\sin^{-1}x-\sqrt{1-x^2}$; (ii) $\sqrt{x^2+x}-\frac{1}{2}\cosh^{-1}(2x+1)$;
     (iii) $\sqrt{x^2-1}-\cosh^{-1}x$

16 $x-\frac{1}{3}x^3+\frac{1}{5}x^5-\frac{1}{7}x^7$

EXAMPLES 16a (Page 228)

1. $\dfrac{1}{3(1-x)} - \dfrac{2}{3(2+x)}$ 2. $\dfrac{4}{7(2x+1)} + \dfrac{5}{7(x-3)}$ 3. $\dfrac{2}{x-2} + \dfrac{1}{x+1}$

4. $\dfrac{1}{(x-1)^2} - \dfrac{1}{2(x-1)} + \dfrac{1}{2(x+1)}$ 5. $-\dfrac{4}{x} + \dfrac{7x}{x^2+1}$

6. $1 + \dfrac{5}{2(x-2)} + \dfrac{1}{2(x+2)}$ 7. $\dfrac{3}{2(x-2)^2} - \dfrac{3}{4(x-2)} + \dfrac{3}{4x}$ 8. $\dfrac{1}{4x} - \dfrac{x}{4(x^2+4)}$

9. $1 + \dfrac{1}{x-2} - \dfrac{1}{x+1}$ 10. $-\dfrac{3}{x^3} + \dfrac{7}{x^2} - \dfrac{7}{x} + \dfrac{7}{x+1}$ 11. $1 + \dfrac{2}{x-1} + \dfrac{1}{(x-1)^2}$

12. $-\dfrac{13}{25(x+2)} + \dfrac{13}{25(x-3)} + \dfrac{12}{5(x-3)^2}$ 13. $\dfrac{8}{7(x+2)} + \dfrac{13x-12}{7(x^2+3)}$

14. $\dfrac{1}{1-x} + \dfrac{2+x}{1+x+x^2}$ 15. $-\dfrac{1}{x} + \dfrac{1}{x+1} + \dfrac{1}{x-1}$

16. $\dfrac{1}{5(x-2)} + \dfrac{1}{2(x+1)} - \dfrac{7}{10(x+3)}$ 17. $\dfrac{1}{4x^2} - \dfrac{1}{x} + \dfrac{9}{16(x+2)} + \dfrac{7}{16(x-2)}$

18. $\dfrac{1}{2\sqrt{2}}\left(\dfrac{1}{x-\sqrt{2}} - \dfrac{1}{x+\sqrt{2}}\right)$ 19. $\dfrac{1}{x-3} + \dfrac{6}{(x-3)^2} + \dfrac{9}{(x-3)^3}$

20. $\dfrac{9}{2(x-3)^2} + \dfrac{15}{16(x-3)} + \dfrac{1}{16(x+5)}$ 21. $1 - \dfrac{2}{3x} + \dfrac{2-3\sqrt{3}}{6(x+\sqrt{3})} + \dfrac{2+3\sqrt{3}}{6(x-\sqrt{3})}$

22. $\dfrac{1}{x+1} - \dfrac{x}{x^2-x+1}$ 23. $\dfrac{1}{16(x+2)} + \dfrac{1}{16(x-2)} - \dfrac{x}{8(x^2+4)}$

24. $\dfrac{1}{1+x^2} + \dfrac{1}{(1-x)^2} - \dfrac{2}{1-x}$ 25. $x + \dfrac{1}{2x} - \dfrac{5x}{2(x^2+2)}$

26. $-\dfrac{3}{2(x+1)} + \dfrac{3(\sqrt{3}+1)}{4(\sqrt{3}+x)} + \dfrac{3(\sqrt{3}-1)}{4(\sqrt{3}-x)}$

27. $\dfrac{7}{6(x+1)} + \dfrac{1}{2(x-1)} - \dfrac{3}{2(x+2)} - \dfrac{1}{6(x-2)}$

28. $\dfrac{1}{x^3} + \dfrac{2}{x^2} + \dfrac{4}{x} + \dfrac{8}{1-2x}$ 29. $\dfrac{1}{2\sqrt{2}}\left(\dfrac{1}{x+1-\sqrt{2}} - \dfrac{1}{x+1+\sqrt{2}}\right)$

30. $\dfrac{1-4x}{5(x^2+4)} - \dfrac{1}{(x-1)^2} + \dfrac{4}{5(x-1)}$

31. $\dfrac{1}{x} - \dfrac{1}{4(x+1)^2} - \dfrac{1}{2(x+1)} + \dfrac{1}{4(x-1)^2} - \dfrac{1}{2(x-1)}$

32. $\dfrac{1}{16x} - \dfrac{x}{4(x^2+4)^2} - \dfrac{x}{16(x^2+4)}$

Answers

EXAMPLES 16b (Page 230)

1. $|x| < 2$; $-\dfrac{1}{4}\left(1 + \dfrac{x^2}{4} + \dfrac{x^4}{16}\right)$

2. $|x| < \dfrac{1}{2}$; $-2(2 + x + 5x^2)$

3. $|x| < \dfrac{1}{3}$; $x - 5x^2 + 19x^3$

4. $|x| < 2$; $-\dfrac{1}{6} + \dfrac{x}{36} + \dfrac{11x^2}{216}$

5. $|x| < 1$; $-\dfrac{1}{2} + \dfrac{x}{4} - \dfrac{3x^2}{8}$

6. $|x| < 1$; $\dfrac{2x}{3} + \dfrac{4}{9}x^2 + \dfrac{14x^3}{27}$

7. $|x| < 1$; $1 + 4x + 9x^2$

8. $|x| < 1$; $\dfrac{1}{3} - \dfrac{x}{9} + \dfrac{19x^2}{27}$

9. $|x| < 1$; $2(1 + x + x^4)$

10. $|x| < \dfrac{1}{\sqrt{2}}$; $-\dfrac{1}{2} + \dfrac{5x}{4} + \dfrac{3x^2}{8}$

11. $|x| < \dfrac{1}{2}$; $\dfrac{9x}{2} - \dfrac{81x^2}{4} + \dfrac{513x^3}{8}$

12. $|x| < 1$; $\dfrac{1}{2} - \dfrac{x}{2} + \dfrac{5x^2}{4}$

13. $1 + 3x + 7x^2$

14. $-1 + x - 4x^2$

15. $\dfrac{1}{4x^2} - \dfrac{1}{4x} - \dfrac{1}{16} + \dfrac{x}{16} + \dfrac{x^2}{64}$

16. a^n; $\tfrac{1}{3}((-1)^n 2^{n+1} + 1)$

18. $\tfrac{2}{5}[2^{2n+2} + (-1)^n]$; $|x| < \tfrac{1}{2}$

19. $\dfrac{1}{x}\left(1 - \dfrac{2}{x} + \dfrac{4}{x^2} - \dfrac{8}{x^3} + \ldots\right)$; $|x| > 2$

20. $|x| > 2$; $\dfrac{2}{x} + \dfrac{2}{x^2} + \dfrac{6}{x^3}$

EXAMPLES 16c (Page 233)

1. $\tfrac{1}{2} \ln \dfrac{x-1}{x+1}$

2. $\ln \dfrac{4}{3}$

3. $-\tfrac{1}{5} \ln (2-x)^2 (3+x)^3$

4. $\dfrac{1}{2x} + \tfrac{1}{4} \ln \dfrac{x-2}{x}$

5. $5 \ln 3 - 4 \ln 4$

6. $\tfrac{1}{2} \ln \dfrac{x^2}{x^2+1}$

7. 0

8. $\dfrac{1}{121}\left(21 \ln \dfrac{2x-1}{x+5} - \dfrac{11}{2x-1}\right)$

9. $\dfrac{x}{2} + \dfrac{1}{12} \ln (2x+1) - \dfrac{4}{3} \ln (x+2)$

10. $\dfrac{1}{6} \ln 3 - \dfrac{1}{\sqrt{3}} \tan^{-1} 3\sqrt{3} + \dfrac{1}{\sqrt{3}} \tan^{-1} \dfrac{5}{\sqrt{3}}$

11. $\dfrac{1}{2\sqrt{2}} \ln \dfrac{\sqrt{2}+x}{\sqrt{2}-x}$

12. $\dfrac{1}{\sqrt{2}} \ln \dfrac{x+1-\sqrt{2}}{x+1+\sqrt{2}}$

13. $\dfrac{3}{4} - \dfrac{3}{2\sqrt{2}}\left(\tan^{-1} \sqrt{2} - \tan^{-1} \dfrac{1}{\sqrt{2}}\right)$

14. $-\ln (5+x)^2 (2-x)$

15. $\tfrac{1}{6} \ln \tfrac{5}{4}$

16. $-\tfrac{3}{2} \ln (x-1) + 2 \ln (x-2) - \tfrac{1}{2} \ln (x-3)$

17. $\dfrac{4x+3}{2(x+1)^2} + \ln (x+1)$

18. $\ln \dfrac{x}{\sqrt{x^2+x+1}} - \dfrac{1}{\sqrt{3}} \tan^{-1} \dfrac{2x+1}{\sqrt{3}}$

19. $\dfrac{1}{4} - \dfrac{1}{4} \ln 2$

20. $-\dfrac{1}{24(2x-1)^2} - \dfrac{5}{36(2x-1)} + \dfrac{1}{27} \ln \dfrac{2x-1}{x+1}$

21 $\dfrac{5}{3}+4\ln\dfrac{2}{3}$

22 $\dfrac{1}{11}\ln(x-1)-\dfrac{1}{22}\ln(2x^2+4x+5)+\dfrac{7}{22}\sqrt{\dfrac{2}{3}}\tan^{-1}\sqrt{\dfrac{2}{3}}(x+1)$

23 $\dfrac{x^2}{2}+x+2\ln(x^2-x-3)+\dfrac{5}{\sqrt{13}}\ln\dfrac{2x-(\sqrt{13}+1)}{2x+(\sqrt{13}-1)}$

24 $4\ln(x+2)-\dfrac{1}{2}\ln(x^2+5)+\dfrac{2}{\sqrt{5}}\tan^{-1}\dfrac{x}{\sqrt{5}}$

MISCELLANEOUS EXAMPLES (Page 233)

1 $A=\dfrac{55}{64},\ B=\dfrac{25}{8},\ C=\dfrac{9}{64}$

2 $-\dfrac{2}{3(x-1)^2}+\dfrac{7}{9(x-1)}-\dfrac{7x+1}{9(x^2+2)}$

3 (i) $2-\ln 3$; (ii) $\ln\tfrac{9}{8}$

4 $1+7x+28x^2+94x^3$

5 (i) $\dfrac{2}{3(x^2+2)}-\dfrac{1}{6(x+1)}+\dfrac{1}{6(x-1)}$; (ii) $x+1+\dfrac{8}{5(x+2)}+\dfrac{27}{5(x-3)}$

6 (i) $\dfrac{1}{2}\ln(x^2+6x+8)+\dfrac{3}{2}\ln\dfrac{x+4}{x+2}$; (ii) $\dfrac{1}{2}\ln(x^2+6x+10)-3\tan^{-1}(x+3)$

7 $\dfrac{2}{x-1}+\dfrac{1}{x+2}$; $-\dfrac{12}{(x-1)^4}-\dfrac{6}{(x+2)^4}$

8 $|x|<1$.

9 $1+x+3x^2+5x^3;\ 1+\dfrac{x}{2}+\dfrac{19x^2}{8}+\dfrac{53x^3}{16}$

11 $-\dfrac{3}{(x+2)^2}-\dfrac{1}{x+2}+\dfrac{2(x+1)}{2x^2+1}$

12 $\dfrac{1}{5(2x-1)}+\dfrac{2}{5(x+2)};\ \dfrac{384}{5(2x-1)^5}+\dfrac{48}{5(x+2)^5}$

13 (i) $|x|<1$; (ii) $|x|>3$; $\tfrac{1}{2}(1-(\tfrac{1}{3})^{n+2})$, $\tfrac{1}{2}(3^{n-2}-1)$

14 $1-2x+x^2+x^3;\ 1$

15 (i) $\dfrac{1}{2\sqrt{2}}\ln\dfrac{1}{3}$; (ii) $\dfrac{7}{3}-2\ln 3$; (iii) $\dfrac{1}{5}\ln\dfrac{1}{4}$

16 $1+\dfrac{2}{3(x+1)^2}-\dfrac{4}{3(x+1)}+\dfrac{x-1}{3(x^2-x+1)}$

17 $\tfrac{1}{2}\ln\tfrac{5}{9}$

18 $\dfrac{1}{(1-x)^2}-\dfrac{2}{1-x}+\dfrac{1}{x^2+1};\ 2x^3+4x^4+4x^5$

19 $\dfrac{1}{4}+\dfrac{\pi}{8}$

20 $\dfrac{x}{4(x^2+2)}+\dfrac{1}{4\sqrt{2}}\tan^{-1}\dfrac{x}{\sqrt{2}}$

21 $\dfrac{(-1)^n\,3\,.\,n!}{10}\left(\dfrac{2^n}{(2x-1)^{n+1}}+\dfrac{3\,.\,4^n}{(4x+3)^{n+1}}\right)$

22 $\dfrac{1}{9(x-1)}+\dfrac{2x-4}{3(x^2+2)^2}-\dfrac{x-8}{9(x^2+2)}$

23 $\tfrac{1}{4}[1+(-1)^n(6n+7)]$

24 (i) $\dfrac{1}{2}\tan^{-1}x+\dfrac{1}{4}\ln\dfrac{1+x}{1-x}$; (ii) $-\dfrac{1}{2(x-1)}+\dfrac{1}{4}\ln\dfrac{x+1}{x-1}$

25 $\dfrac{(-1)^n n!}{9}\left(\dfrac{8}{(x+2)^{n+1}}+\dfrac{1}{(x-1)^{n+1}}-\dfrac{12(n+1)}{(x+2)^{n+2}}\right)$ **26** $\dfrac{509}{16}$

28 $\dfrac{1}{1+x}+\dfrac{2}{(1-x)^2}-\dfrac{2}{1-x}; |x|>1$ **29** $A=-C=\dfrac{1}{2\sqrt{2}}, B=D=\dfrac{1}{2}$

EXAMPLES 17a (Page 240)

1 $-\dfrac{1}{8(2x-1)^4}$ **2** $\sqrt{x^2-1}$ **3** $\dfrac{\sin^6 x}{6}$ **4** $\tfrac{1}{2}(\ln x)^2$

5 $2\sqrt{x}-4\ln(2+\sqrt{x})$ **6** $2\sqrt{x-2}+\sqrt{2}\tan^{-1}\sqrt{\dfrac{x-2}{2}}$

7 $\dfrac{(x+5)^2}{2}-15(x+5)+\dfrac{125}{x+5}+75\ln(x+5)$ **8** $-\sqrt{1-x^2}$

9 $\tfrac{1}{4}$ **10** $\dfrac{\pi}{12}$ **11** $8\tfrac{23}{24}$ **12** $x-\ln(1+e^x)$

13 $2\sin\left(\dfrac{1}{\sqrt{2}}\right)$ **14** $\sqrt{2}-1$ **15** $\tan^{-1}e-\dfrac{\pi}{4}$ **16** $\tfrac{1}{4}\ln 3$

17 π **18** $\tfrac{5}{24}$ **19** $\tfrac{1}{4}\ln\dfrac{32}{17}$ **20** $\ln\dfrac{1+\sqrt{x}}{1-\sqrt{x}}$

EXAMPLES 17b (Page 243)

1 $2\sqrt{x-2}+2\sqrt{2}\tan^{-1}\sqrt{\dfrac{x-2}{2}}$ **2** $-\dfrac{1}{x-1}-\dfrac{1}{(x-1)^2}-\dfrac{1}{3(x-1)^3}$

3 $2\left(\dfrac{x}{2}-2\sqrt{x}+2\ln(\sqrt{x}+1)\right)$ **4** $\tfrac{3}{20}(3+2x)^{\frac{5}{3}}-\tfrac{9}{8}(3+2x)^{\frac{2}{3}}$

5 $\tfrac{1}{4}(\sin^{-1}2x+2x\sqrt{1-4x^2})$ **6** $6\tan^{-1}\sqrt{2x-1}$

7 $\dfrac{1}{\sqrt{7}}\ln\dfrac{\sqrt{7}+\tan\dfrac{x}{2}}{\sqrt{7}-\tan\dfrac{x}{2}}$ **8** $2\sin^{-1}\dfrac{x}{2}-\dfrac{x}{2}\sqrt{4-x^2}$ **9** $\ln(e^x-1)-x$

10 $\dfrac{1}{6}\ln(1+2x^3)$ **11** $\sin x-\dfrac{\sin^3 x}{3}$ **12** $-\dfrac{2}{1+\tan\dfrac{x}{2}}$

13 $\ln\dfrac{1+\tan\dfrac{x}{2}}{1-\tan\dfrac{x}{2}}=\ln\tan\left(\dfrac{\pi}{4}+\dfrac{x}{2}\right)$ **14** $\dfrac{\sin^3 x}{3}-\dfrac{\sin^5 x}{5}$

15 $\tfrac{2}{3}(x+2)^{\frac{3}{2}}-4\sqrt{x+2}$ **16** $\tfrac{1}{12}(4x^2-1)^{\frac{3}{2}}$

17 $-\dfrac{1}{4}\left[\dfrac{1}{3(2x-1)^3}+\dfrac{1}{4(2x-1)^4}\right]$ **18** $\dfrac{5}{7}(x-3)^7+\dfrac{(x-3)^8}{8}$

19 $\dfrac{8\sqrt{2}}{3} - 2\sqrt{3}$. **20** $-\dfrac{1}{360}$. **21** 1. **22** $\dfrac{203}{480}$.

23 $\dfrac{\pi}{4}$ **24** $\tan^{-1} 2 - \tan^{-1} 1 = \tan^{-1} \tfrac{1}{3}$ **25** $\ln(2 + \sqrt{3})$

26 $\tfrac{5}{24}$ **27** 2 **28** 0·9194

29 $\dfrac{e^4}{4} + \dfrac{4e^3}{3} + 3e^2 + 4e - 7\tfrac{7}{12}$ **30** π **31** $\dfrac{1}{\sqrt{119}} \ln \dfrac{12 + \sqrt{119}}{5}$

32 $\dfrac{\pi}{6}$ **33** $\sinh^{-1} 1 - \sinh^{-1} \tfrac{1}{2} = 0·40$ **34** $\dfrac{3\pi}{16}$

35 $\dfrac{13}{3 \cdot 2^{11}}$ **36** $\dfrac{1}{4\sqrt{17}}$

37 (i) $\dfrac{1}{2}\tan\theta$; (ii) $\dfrac{1}{\sqrt{3}} \tan^{-1} \dfrac{2\tan\theta + 1}{\sqrt{3}}$ **38** (i) $\dfrac{\pi}{4\sqrt{2}}$; (ii) $\dfrac{1}{\sqrt{15}} \tan^{-1} \sqrt{\dfrac{5}{3}}$

39 $\dfrac{1}{n} \ln \dfrac{\sqrt{1 + x^n} - 1}{\sqrt{1 + x^n} + 1}$ **40** $\dfrac{\pi}{8}$

EXAMPLES 17c (Page 245)

1 $x \sin x + \cos x$ **2** $xe^x - e^x$ **3** $\tfrac{1}{4} \sin 2x - \tfrac{1}{2} x \cos 2x$

4 $\dfrac{x^2}{2} \ln x - \dfrac{x^2}{4}$ **5** $-xe^{-x} - e^{-x}$ **6** $\dfrac{x^3}{3} \ln x - \dfrac{x^3}{9}$

7 $\tfrac{1}{3} x \sin 3x + \tfrac{1}{9} \cos 3x$ **8** $-\dfrac{1}{x} \ln x - \dfrac{1}{x}$ **9** $x^2 e^x - 2xe^x + 2e^x$

10 $x^2 \sin x + 2x \cos x - 2 \sin x$ **11** $\tfrac{1}{4}(\tfrac{1}{2} \sin 2x - x \cos 2x)$

12 $\dfrac{e^{2x}}{4}(2x^2 - 2x + 1)$ **13** π **14** $\tfrac{3}{16} - \tfrac{1}{8} \ln 2$ **15** $\dfrac{1}{4} - \dfrac{3}{4e^2}$

16 $\dfrac{\pi^2}{4}$ **17** $8(\pi - 2)$ **18** $\dfrac{\pi}{4} - \dfrac{\sqrt{2}}{2}$

19 $\dfrac{e^x}{2}(\sin x - \cos x); \dfrac{e^x}{2}(\cos x + \sin x)$

20 (i) $\dfrac{e^{2x}}{5}(2 \sin x - \cos x)$; (ii) $\dfrac{e^{-x}}{2}(\sin x - \cos x)$; (iii) $\dfrac{e^x}{10}(\sin 3x - 3 \cos 3x)$;

(iv) $\dfrac{e^{-2x}}{4}(\sin 2x - \cos 2x)$

EXAMPLES 17d (Page 247)

1 $x \sin^{-1} x + \sqrt{1 - x^2}$ **2** $x \cos^{-1} x - \sqrt{1 - x^2}$ **3** $x \sinh^{-1} x - \sqrt{x^2 + 1}$

4 $\dfrac{x}{2} \sqrt{4 - x^2} + 2 \sin^{-1} \dfrac{x}{2}$ **5** $\dfrac{x}{2} \sqrt{9 + x^2} + \dfrac{9}{2} \sinh^{-1} \dfrac{x}{3}$

6 $\dfrac{x^2}{2} \tan^{-1} x - \dfrac{x}{2} + \dfrac{1}{2} \tan^{-1} x$ **7** $x(\lg x - \lg e)$

8 $\tfrac{1}{2} x^2 \sin^{-1} x^2 + \tfrac{1}{2} \sqrt{1-x^4}$ **9** $x \tan^{-1} \dfrac{1}{x} + \dfrac{1}{2} \ln(x^2+1)$

10 $\dfrac{e^{3x}}{13}(3 \sin 2x - 2 \cos 2x)$ **11** $x \cot^{-1} x + \tfrac{1}{2} \ln(x^2+1)$

12 $\tfrac{1}{2} x\sqrt{3+2x-x^2} - \tfrac{1}{2}\sqrt{3+2x-x^2} + 2 \sin^{-1} \dfrac{x-1}{2}$

13 $-\dfrac{1}{2} e^{-x} + \dfrac{e^{-x}}{10}(\cos 2x - 2 \sin 2x)$ **14** $\dfrac{x^2}{2} \sin^{-1} x - \tfrac{1}{4} \sin^{-1} x + \tfrac{1}{4} x\sqrt{1-x^2}$

15 $\dfrac{x}{2}\sqrt{3-4x^2} + \dfrac{3}{4} \sin^{-1} \dfrac{2x}{\sqrt{3}}$ **16** $e^{2x}(\tfrac{1}{2} x^3 - \tfrac{3}{4} x^2 + \tfrac{3}{4} x - \tfrac{3}{8})$

17 $x - \sqrt{1-x^2} \sin^{-1} x$ **18** $\sqrt{1+x^2} \tan^{-1} x - \sinh^{-1} x$ **19** $2(x \ln x - x)$

20 $\dfrac{x}{2}\sqrt{2x^2+3} + \dfrac{3}{2\sqrt{2}} \sinh^{-1} \sqrt{\dfrac{2}{3}} x$ **21** $\dfrac{x^7}{7} \ln \dfrac{1}{x} + \dfrac{x^7}{49}$ **22** $\dfrac{1}{2}\left(\dfrac{\pi}{2} - 1\right)$

23 $\dfrac{1}{8}\left(3 - \dfrac{19}{e^2}\right)$ **24** $\dfrac{1}{10}\left(2\sqrt{2} e^{\frac{3\pi}{4}} - 3\right)$ **25** $\dfrac{\pi}{4} - \dfrac{1}{2}$ **26** $\dfrac{1}{6}$

27 $\dfrac{3\sqrt{2}}{4} - \dfrac{1}{8} \ln(3+\sqrt{8})$ **28** $e^x \left(x \dfrac{\sqrt{2}}{2} \sin\left(x - \dfrac{\pi}{4}\right) + \dfrac{1}{2} \cos x\right)$

30 $e^x \{\tfrac{1}{74}(\sin 6x - 6 \cos 6x) + \tfrac{1}{10}(\sin 2x - 2 \cos 2x)\}$

EXAMPLES 17e (Page 249)

1 (i) $-\tfrac{1}{3} \cos x \sin^2 x - \tfrac{2}{3} \cos x$; (ii) $\dfrac{35\pi}{256}$; (iii) $\dfrac{5\pi}{16}$

2 (i) $\dfrac{\tan^3 \theta}{3} - \tan \theta + \theta$; (ii) $\dfrac{\tan^2 \theta}{2} + \ln \cos \theta$; (iii) $\dfrac{13}{15} - \dfrac{\pi}{4}$

3 (i) $\dfrac{x}{2(x^2+1)} + \dfrac{1}{2} \tan^{-1} x$; (ii) $\dfrac{5\pi}{64} + \dfrac{11}{48}$;

   (iii) $\dfrac{x+1}{4(x^2+2x+2)^2} + \dfrac{3(x+1)}{8(x^2+2x+2)} + \dfrac{3}{8} \tan^{-1}(x+1)$

4 (i) $-\tfrac{1}{6} \sin^3 x \cos^3 x - \tfrac{1}{8} \sin x \cos^3 x + \dfrac{\sin 2x}{32} + \dfrac{x}{16}$;

   (ii) $-\tfrac{1}{6} \sin^2 x \cos^4 x - \dfrac{\cos^4 x}{12}$; (iii) $\dfrac{5\pi}{256}$

5 (i) $\dfrac{1}{5}\left[x^5(\ln x)^2 - \dfrac{2}{5}\left(x^5 \ln x - \dfrac{x^5}{5}\right)\right]$; (ii) $\dfrac{x^3}{3}[(\ln x)^3 - (\ln x)^2 + \tfrac{2}{3} \ln x - \tfrac{2}{9}]$;

   (iii) $\dfrac{1}{4} - \dfrac{5}{4e^2}$

Answers | 369

6 (i) $\frac{1}{4}\cos^3 x \sin x + \frac{3}{16}\sin 2x + \frac{3x}{8}$;

(ii) $\frac{1}{5}\cos^4 x \sin x + \frac{4}{15}\cos^2 x + \frac{8}{15}\sin x$; (iii) $\frac{5\pi}{32}$

7 $I_n = -x^n e^{-x} + nI_{n-1}$; (i) $-e^{-x}(x^3 + 3x^2 + 6x + 6)$; (ii) $24 - \frac{65}{e}$

8 (i) $-\frac{\cot^2 x}{2} - \ln \sin x$; (ii) $-\frac{\cot^3 x}{3} + \cot x + x$; (iii) $\frac{1}{2}\ln 2 - \frac{1}{4}$

9 $u_n = x^n \sin x + nx^{n-1}\cos x - n(n-1)u_{n-2}$;

(i) $x^4 \sin x + 4x^3 \cos x - 12x^2 \sin x - 24x \cos x + 24 \sin x$;

(ii) $x^3 \sin x + 3x^2 \cos x - 6x \sin x - 6 \cos x$; (iii) $\frac{\pi^5}{32} - \frac{5\pi^3}{2} + 60\pi - 120$

10 (i) $\frac{16}{35}$; (ii) $\frac{63\pi}{512}$; (iii) $\frac{3\pi}{8}$

MISCELLANEOUS EXAMPLES (Page 250)

1 -4 **2** $-\frac{2}{3}(4-x)^{\frac{3}{2}}$ **3** $-\frac{1}{2}\sqrt{1-2x^2}$ **4** $\frac{1}{5}$

5 $\frac{1}{3}e^{3x-1}$ **6** $-e^{-2x}\left(\frac{x}{2} + \frac{1}{4}\right)$ **7** $\frac{3}{40}(3^{\frac{2}{3}} + 3)$ **8** $\frac{1}{2}\ln(x^2 + 1) + \tan^{-1} x$

9 $\frac{\pi}{12}$ **10** $\frac{e-1}{2(e+1)}$ **11** $x - 8\ln(x+2) - \frac{16}{x+2}$

12 $-\frac{1}{b}\ln(a + b\cos x)$ **13** $\frac{1}{18}\frac{6x-1}{(3x-1)^2}$ **14** $\frac{\pi}{16}$ **15** $-\frac{\cos 5x}{10} - \frac{\cos 3x}{6}$

16 $\frac{x^2}{2} - \frac{24x^{\frac{11}{6}}}{11} + \frac{12x^{\frac{5}{3}}}{5}$ **17** $\frac{1}{2}\ln\frac{5}{6}$ **18** $\frac{3\pi}{32} - \frac{1}{4}$

19 $\sinh^{-1} x + \sqrt{x^2 + 1}$ **20** $\frac{x^2 \sin 2x}{2} + \frac{x \cos 2x}{2} - \frac{\sin 2x}{4}$

21 $-\frac{2}{7}\ln(2x-1) + \frac{9}{7}\ln(x+3)$ **22** $\frac{1}{2}\ln 2$ **23** $-\frac{1}{3}\cot 3x$

24 $4\ln 2 - \frac{15}{16}$ **25** $\sin^{-1}(x-1)$ **26** $\frac{2}{3}(19 + \ln\frac{26}{7})$

27 $2\sqrt{1+2x} + \ln\frac{\sqrt{1+2x}-1}{\sqrt{1+2x}+1}$ **28** $-\frac{3}{2} - \ln 4$ **29** $\frac{\sqrt{3}}{8}$

30 $-\frac{e^{-x}}{2}(\cos x + \sin x)$ **31** $\frac{1}{\sqrt{2}}\sinh^{-1}\sqrt{\frac{2}{3}}x$ **32** $\frac{6}{5}$ **33** $\frac{x}{2} + \frac{\sin 6x}{12}$

34 $\ln 2$ **35** $\frac{x}{16} + \frac{1}{6(e^x+2)^3} + \frac{1}{8(e^x+2)^2} + \frac{1}{8(e^x+2)} - \frac{1}{16}\ln(e^x+2)$

36 $\frac{9}{128}$ **37** $\frac{3}{2}\ln\frac{3}{4} + \frac{4}{3}\ln 2 - \frac{1}{6}\ln\frac{5}{4}$ **38** $\frac{1}{2}\sin^{-1} 2x + \frac{1}{2}\sqrt{1-4x^2}$

39 $10^x \lg e$ **40** $\frac{2}{3}\tan^{-1}\frac{1}{3}$ **41** $x\cos^{-1}\frac{1}{x} - \cosh^{-1} x$

42 $-\cos(\ln x)$ 43 $\frac{1}{10}(\tan^{-1} 2 + \ln \frac{5}{4})$ 44 $2\sqrt{e^x+4} + 2\ln\dfrac{\sqrt{e^x+4}-2}{\sqrt{e^x+4}+2}$

45 $1 - \dfrac{\pi}{4}$ 46 $-\frac{2}{3}\sqrt{1-x}\,(2+x)$ 47 $\dfrac{1}{\ln 2}(x\ln x - x)$

48 $-e^{-2x}\left(\dfrac{x^2}{2} + \dfrac{x}{2} + \dfrac{1}{4}\right)$ 49 $2\ln 2 - \frac{3}{4}$ 50 $\dfrac{\sin 3x}{6} - \dfrac{\sin 7x}{14}$

51 $\dfrac{\tan^3 x}{3} + \tan x$ 52 $\dfrac{2}{1 - \tan\dfrac{x}{2}}$ 53 $\dfrac{2}{27}$ 54 $\dfrac{na^4}{16}$

55 $-\ln 2 - \frac{5}{6}$ 56 $\tan^{-1}(2x-1)$ 57 $\frac{1}{2}\ln(1+\sin^2 x)$ 58 $\frac{15}{4}$

59 $\dfrac{\pi}{6}$ 60 $\frac{1}{6}\ln\dfrac{1+x^3}{1-x^3}$ 61 $-\frac{4}{3}(\sqrt{2}-1)$ 62 $\frac{3}{40}e^\pi - \dfrac{13}{40}$

64 $\frac{2}{7}$ 66 $\dfrac{35\pi}{256}$ 69 0 70 $2\displaystyle\int_0^\pi \dfrac{\cos(n-1)x.\cos x\, dx}{5 - 4\cos x}$

EXAMPLES 18a (Page 255)

1 $\dfrac{1}{27}(22^{\frac{3}{2}} - 8)$ 3 $c(e^{a/c} - e^{-a/c})$ 4 $\dfrac{1}{54\sqrt{2}}[59^{\frac{3}{2}} - 41^{\frac{3}{2}}]$ 9 πrl

11 $\dfrac{k-1}{2k}$ 12 $\dfrac{12\pi a^2}{5}$ 13 $10\,\text{cm}$ 15 $\dfrac{\pi c^2}{2}[\sinh 2 + 2]$

EXAMPLES 18b (Page 258)

1 $\frac{1}{2}(17)^{\frac{3}{2}}$ 2 $\frac{1}{2}(3)^{\frac{3}{2}}$ 3 $2^{\frac{3}{2}}$ 4 $\frac{1}{9}(10)^{\frac{3}{2}}$ 5 $\frac{1}{6}(13)^{\frac{3}{2}}$

6 $\frac{1}{6}$ 7 $4(3)^{\frac{3}{2}}$ 8 $\frac{1}{4}(17)^{\frac{3}{2}}$ 9 8 10 $\frac{1}{2}$

11 $\frac{3}{5}$ 12 $\frac{1}{10}(19)^{\frac{3}{2}}$ 13 ∞ 14 $\frac{1}{2}$ 15 $\frac{1}{6}(2)^{\frac{3}{2}}$

19 $3a\sin\theta\cos\theta$ 20 $(1, 2)$ 21 $(0, 1), (\frac{3}{4}, \frac{47}{128})$; Inflexion points

EXAMPLES 18c (Page 262)

1 $(0, -1)$ 2 $(575, -\frac{961}{24})$ 3 $\left(\dfrac{\pi}{2}, 0\right)$ 4 $(5, -2)$

5 $(-\frac{11}{2}, \frac{16}{3})$ 6 $(\frac{1}{2}, -\frac{129}{32})$ 7 $(\frac{49}{4}, 19)$ 8 $(0, -\frac{16}{3})$

9 $(\frac{125}{16}, \frac{27}{16})$ 10 $(\frac{3}{2}, -1)$ 11 $(-3, 1)$ 12 $(2\frac{1}{2}, 2\frac{1}{2})$

13 $(0, 2c)$ 14 $\left(\dfrac{\pi}{2} + \dfrac{5}{2}, \dfrac{11}{4}\right)$ 15 2 16 1

17 $y^2 = -x(x+2);\ 1$ 18 $\frac{1}{4}$ 19 54

22 $\{a\cos\theta(\cos^2\theta + 3\sin^2\theta),\ a\sin\theta(\sin^2\theta + 3\cos^2\theta)\}$

Answers | 371

MISCELLANEOUS EXAMPLES (Page 263)

1 (i) $80\pi \text{ cm}^2$; (ii) $\dfrac{416\pi}{3} \text{ cm}^3$ **2** $0.268:1$ **6** $2a(t^2+1)^{\frac{3}{2}}$

9 $8a$ **13** $\dfrac{13^{\frac{3}{2}}}{32}; \dfrac{1}{11}$ **14** $\dfrac{a}{2}$ **17** $r = 4a\cos\theta \sin^2\theta$

19 $3a\cos\theta \sin\theta$ **21** $\tfrac{1}{2}.3^{\frac{3}{2}}; (\tfrac{1}{2}\ln\tfrac{1}{2} - \tfrac{3}{2}, 2\sqrt{2})$

EXAMPLES 19a (Page 267)

1 $\dfrac{dy}{dx} = 2$ **2** $y = x\dfrac{dy}{dx} + 3$ **3** $2y = x\dfrac{dy}{dx} - 2$ **4** $y = (x+1)\dfrac{dy}{dx}$

5 $\dfrac{dy}{dx} = 3y$ **6** $x\dfrac{dy}{dx} + \ln x - y - 1 = 0$ **7** $e^x\dfrac{dy}{dx} - \dfrac{dy}{dx} = y$

8 $x + y\dfrac{dy}{dx} = 0$ **9** $y\dfrac{dy}{dx} + \dfrac{1}{x^2} = 0$ **10** $\dfrac{dy}{dx}(x^2 - y^2) = 2xy$

11 $\dfrac{dy}{dx}\sin x + y\cos x = \sin 2x$ **12** $(x^2 - y^2 - 4)\dfrac{dy}{dx} = 2xy$

13 $\dfrac{d^2y}{dx^2} = 0$ **14** $x^2\dfrac{d^2y}{dx^2} - 3x\dfrac{dy}{dx} + 3y = 0$ **15** $x^2\dfrac{d^2y}{dx^2} + x\dfrac{dy}{dx} - y = 0$

16 $x\dfrac{d^2y}{dx^2} + \dfrac{dy}{dx} = 0$ **17** $\dfrac{d^2y}{dx^2} - y = 0$ **18** $\dfrac{d^2y}{dx^2} - 4\dfrac{dy}{dx} + 3y = 0$ **19** $\dfrac{d^2y}{dx^2} + y = 0$

20 $\dfrac{d^2y}{dx^2} - 4\dfrac{dy}{dx} + 4y = 0$ **21** $\dfrac{d^2y}{dx^2} + 2\dfrac{dy}{dx} + 5y = 0$ **22** $\dfrac{d^2y}{dx^2} + 4y = 0$

23 $\dfrac{d^2y}{dx^2} - 4\dfrac{dy}{dx} + 5y = 0$ **24** $\dfrac{dy}{dx} - y = (x+3)e^{2x}$ **26** $y = x\dfrac{dy}{dx} + \left(\dfrac{dy}{dx}\right)^3$

EXAMPLES 19b (Page 270)

1 $3x^2 = 2y^3 + C$ **2** $x^3 + 3x = 3y + C$ **3** $y = Cx$
4 $\sin y = \sin x + C$ **5** $y^2 + 2e^{-x} = C$ **6** $y^2 = C(x^2 + 1)$
7 $y = \tan^{-1}x + C$ **8** $y^2 = x^2 + 4x + C$ **9** $\tan^{-1}y = \ln x + C$
10 $2y^3 + 3y^2 = 2x^3 + 3x^2 + C$ **11** $1 + y = C(1 + x)$ **12** $y(x+1) = Cx$
13 $\dfrac{x^2}{2} - \ln x = \dfrac{y^3}{3} - \dfrac{y^2}{2} + C$ **14** $y(x+1) = Ce^x$ **15** $\tan y = \tan x + C$
16 $e^y + e^{-x} = C$ **17** $x\cos y = C$ **18** $y = \ln(\sin x + \cos x) + C$
19 $\sin x \cos y = C$ **20** $\dfrac{2}{1-y} = \ln x + C$ **21** $\ln x\sqrt{y^2 + 1} + \tan^{-1}y = C$
22 $\tan^{-1}y - \tan^{-1}x + \tan^{-1}\tfrac{1}{3} = 0$ or $3(x - y) = 1 + xy$

372 | **Answers**

23 $y^2(x+1) = 3(x-1)$ **24** $x \ln(1-y) = 1$

25 $y^2 - \ln(y^2+1) + 2\tan^{-1} y = x^2 + C$ **26** $\sqrt{2} \ln \dfrac{\sqrt{2}+\sin y}{\sqrt{2}-\sin y} = x^4 + C$

27 $y^2 = 2ax + C$ **28** $y = 2e^{2x}$ **29** $xy = C$

30 $\tan^{-1}(x+y) = x + C$ **31** $\sec(x-y) + \tan(x-y) = Ce^x$

EXAMPLES 19c (Page 272)

1 $x = Ce^{y/x}$ **2** $x^3 - 3yx^2 = C$ **3** $\dfrac{2}{\sqrt{3}} \tan^{-1} \dfrac{2y-x}{\sqrt{3}x} = \ln x + C$

4 $\dfrac{1}{2\sqrt{2}} \ln \dfrac{(\sqrt{2}+1)x+y}{(\sqrt{2}-1)x-y} - \tfrac{1}{2} \ln x = C$ **5** $y^2 = 2x^2 \ln x + Cx^2$

6 $\ln x + \tfrac{1}{6} \ln \left(1 - \dfrac{2y^3}{x^3}\right) = C$ **7** $\sin^{-1} \dfrac{y}{x} = \ln x + C$

8 $\tan^{-1} \dfrac{y}{x} = \tfrac{1}{2} \ln(x^2 + y^2) + C$ **9** $\ln(y-x) - \dfrac{x}{y-x} = C$

10 $\ln xy + \dfrac{y^2}{2x^2} = C$ **11** $2x = (x-y)(\ln x + C)$ **12** $2y^2 \ln y - x^2 = Cy^2$

13 $2(y-x) - \ln(2x+2y+3) = C$ **14** $\dfrac{(y-x)^2}{2} + 5y - x = C$

15 $x - y = 4 \ln(2x - y + 7) + C$

EXAMPLES 19d (Page 274)

1 x^2 **2** $\sin x$ **3** $\sec^2 x$ **4** x^2 **5** $\sqrt{\cos x}$

6 $\dfrac{1}{x^3}$ **7** 2^x **8** $\sec x$ **9** $\sqrt[4]{\sin 2x}$ **10** $\sqrt{x^2+1}$

11 $\dfrac{1}{\sqrt{1-x^2}}$ **12** $\sqrt[3]{1+x^3}$ **13** $y = x - 1 + Ce^{-x}$ **14** $y = Ce^{\frac{x^2}{2}}$

15 $xy = x + C$ **16** $y = x^2 + Cx$ **17** $ye^x = x + C$ **18** $4y + 2x + 1 = Ce^{2x}$

19 $4y \sin x + \cos 2x = C$ **20** $y = x^3 + Cx^2$ **21** $3y \cos x = 3 \sin x - \sin^3 x + C$

22 $2y(x^2-1) = x^2 + C$ **23** $y\sqrt{1-x^2} = \sin^{-1} x + C$ **24** $2y = x^7 + Cx^5$

25 $2y = \sin x - \cos x + Ce^{-x}$ **26** $5ye^{3x} = e^{5x} + C$ **27** $y = (x+C)(x+1)^3$

28 $y(1 + \sin x) = \ln \sec x + C$ **29** $x = y^2 - 2y + 2 + Ce^{-y}$

30 $2x = y(y^2 + C)$ **31** $2 \tan y = x^2 - 1 + Ce^{-x^2}$

32 $\dfrac{dz}{dx} - \dfrac{z}{x} = -1; \ 1 + xy \ln x = Cxy$ **33** $xy^2 = 2y^5 + C$

34 $x \sin y = \ln x + C$

Answers | 373

EXAMPLES 19e (Page 278)

1 $y = \dfrac{x^4}{4} + Ax + B$ 2 $y + \ln x = Ax + B$ 3 $y = \ln \sec x + Ax + B$

4 $x = -\tfrac{1}{2}gt^2 + At + B$ 5 $x = -\dfrac{f}{n^2} \sin nt + At + B$ 6 $y = x \ln x + Ax + B$

7 $y = Ae^{2x} + B$ 8 $y = Ax^2 + B$ 9 $V = A \ln r + B$

10 $y = A \tan^{-1} x + B$ 11 $y = A \sin(x + B)$ 12 $\dfrac{1}{y} + Ax = B$

13 $\tfrac{1}{2} \ln \dfrac{3(y-1)}{y+1} = \pm \left(\dfrac{x}{\sqrt{2}} - 1 \right)$ 14 $y = -2 \ln \left(1 \pm \dfrac{x}{2} \right)$

15 $2(1 - x^{-\frac{1}{2}}) = \pm \sqrt{\dfrac{2k}{3}} \cdot t$

16 $-\sqrt{h}\sqrt{hx - x^2} + \tfrac{1}{2} h^{\frac{3}{2}} \sin^{-1} \dfrac{2x - h}{h} = \pm a\sqrt{2gt} + \tfrac{1}{4}\pi h^{\frac{3}{2}}$

17 $Cu = \cosh \sqrt{(k-1)} \, \theta$ where $k = \dfrac{\mu}{h^2}$

18 $ky = \dfrac{wx^4}{24} + \dfrac{Ax^3}{6} + \dfrac{Bx^2}{2} + Cx + D;\ B = D = 0,\ A = -\dfrac{wl}{2},\ C = \dfrac{wl^3}{24}$

20 $y = Ax^3 + \dfrac{B}{x^2}$

EXAMPLES 19f (Page 281)

1 $y = Ae^x + B$ 2 $y = Ae^{-\frac{x}{2}} + B$ 3 $y = Ae^{2x} + Be^{-2x}$

4 $y = Ae^{2x} + Be^{-x}$ 5 $y = e^x(Ax + B)$ 6 $y = Ae^{3x} + Be^{2x}$

7 $y = A \cos x + B \sin x$ 8 $y = Ae^{x/3} + Be^{-x}$ 9 $y = e^{-\frac{x}{2}}(Ax + B)$

10 $y = Ae^{5x} + Be^{-2x}$ 11 $y = Ae^{3x} + Be^{-3x}$ 12 $y = A \cos 3x + B \sin 3x$

13 $y = e^{-\frac{x}{2}} \left(A \cos \dfrac{\sqrt{3}x}{2} + B \sin \dfrac{\sqrt{3}x}{2} \right)$ 14 $y = e^{3x}(Ax + B)$

15 $y = e^{-x}(A \cos \sqrt{3}x + B \sin \sqrt{3}x)$ 16 $y = Ae^x + Be^{\frac{x}{3}}$

17 $y = e^{\frac{x}{a}}(Ax + B)$ 18 $y = e^{3x}(A \cos 2x + B \sin 2x)$

19 $y = Ae^{ax} + Be^{bx}$ 20 $y = A \cos mx + B \sin mx$

21 $x = \dfrac{2u}{k} e^{-\frac{kt}{2}} \sin \dfrac{kt}{2}$ 22 $x = a \cos \sqrt{\mu} t$ 23 $\theta = \alpha \sin \sqrt{\dfrac{g}{l}} \cdot t$

24 $s = e^{-2t} \left(a \cos 3t + \dfrac{u + 2a}{3} \sin 3t \right)$ 25 $y = e^{-ax}(a^2 x + a)$

26 $x = \tfrac{1}{2}(e^t + e^{-t}),\ y = \tfrac{1}{2}(e^{-t} - e^t)$

27 $x = 2e^{\frac{t}{2}} \sin \dfrac{\sqrt{3}t}{2},\ y = e^{\frac{t}{2}} \left(\sqrt{3} \cos \dfrac{\sqrt{3}t}{2} - \sin \dfrac{\sqrt{3}t}{2} \right)$ 28 $y = Ax^2 + Bx$

EXAMPLES 19g (Page 286)

1 $y = e^{\frac{3x}{2}}\left(A\cos\frac{\sqrt{3}}{2}x + B\sin\frac{\sqrt{3}}{2}x\right) + e^x$ **2** $y = Ae^{-2x} + Be^x + 2x + 1$

3 $y = A\cos 2x + B\sin 2x + 2\sin 3x$ **4** $y = e^{-2x}(Ax + B) + \frac{7}{3}$

5 $y = e^{-3x}(Ax + B) + 2e^{2x}$. **6** $y = A\cos 2x + B\sin 2x - \frac{1}{4}x\cos 2x$

7 $a = 2, b = 0$ **8** $a = -\frac{1}{4}, b = -2$ **9** $a = -2$.

10 $a = 0, b = \frac{2}{5}, p = 2$ **11** $a = -\frac{1}{5}, b = -4$ **12** $a = -\frac{1}{4}, b = 0, c = \frac{5}{8}$

13 $a = \frac{1}{4}, b = 0$ **14** $a = \frac{2}{3}, b = 1, c = 3$

15 $y = -\frac{2}{5}e^{2x}$; $y = Ae^{3x} + Be^{-3x} - \frac{2}{5}e^{2x}$ **16** $y = -2$; $y = Ae^x + Be^{-\frac{x}{4}} - 2$.

17 $y = -\cos 2x$; $y = A\cos x + B\sin x - \cos 2x$.

18 $y = -x^2 + 3x - 8$; $y = Ae^{\frac{x}{2}} + Be^{-\frac{x}{2}} - x^2 + 3x - 8$.

19 $y = -\frac{1}{9}e^{-x}$; $y = Ae^{8x} + Be^{-2x} - \frac{1}{9}e^{-x}$

20 $y = -\frac{1}{50}(3\cos 3x + 4\sin 3x)$; $y = e^{-x}(Ax + B) - \frac{1}{50}(3\cos 3x + 4\sin 3x)$.

21 $y = -\frac{1}{54}(9x^3 + 9x^2 + 42x + 26)$; $y = Ae^x + Be^{-\frac{3x}{2}} - \frac{1}{54}(9x^3 + 9x^2 + 42x + 26)$

22 $y = -\cos x - \sin x$; $y = e^{\frac{x}{2}}\left(A\cos\frac{\sqrt{3}}{2}x + B\sin\frac{\sqrt{3}}{2}x\right) - \cos x - \sin x$

23 $y = -\frac{1}{4}xe^{-2x}$; $y = Ae^{2x} + Be^{-2x} - \frac{1}{4}xe^{-2x}$

24 $y = 3$; $y = e^{-x}(A\cos 2x + B\sin 2x) + 3$

25 $y = e^x(A\cos 2x + B\sin 2x) + 10e^{3x}$ **26** $y = e^{-4x}(A\cos 3x + B\sin 3x) + 2$

27 $y = A\cos 3x + B\sin 3x + \frac{3}{5}\cos 2x - \frac{1}{5}\sin 2x$

28 $y = e^x(A\cos x + B\sin x) - \frac{1}{2}x^2 - \frac{1}{2}x + \frac{1}{2}$

29 $y = A\cos\frac{1}{2}x + B\sin\frac{1}{2}x + x + \frac{1}{5}e^x$ **30** $y = A + Be^{\frac{x}{2}} - 2x^2 - 9x$

31 $y = A\cos 4x + B\sin 4x - \frac{3}{8}x\cos 4x$ **32** $y = A\cos x + B\sin x + \frac{1}{2}\cosh x$

33 $\theta = \frac{\pi}{6}\cos 2t - \frac{1}{6}\sin 2t + \frac{1}{3}\sin t$ **34** $y = e^x + \frac{1}{4}e^{2x} + \frac{1}{2}x + \frac{11}{4}$

35 $y = -\frac{2}{27} + \frac{1}{135}e^{9x} - \frac{1}{45}\sin 3x + \frac{1}{15}\cos 3x$

MISCELLANEOUS EXAMPLES (Page 287)

1 $x = -\frac{1}{2}\tan^{-1}2$; Max. when $\sin 2x = \frac{2}{\sqrt{5}}$, $\cos 2x = -\frac{1}{\sqrt{5}}$; Min. when $\sin 2x = -\frac{2}{\sqrt{5}}$, $\cos 2x = \frac{1}{\sqrt{5}}$ **2** $\frac{dy}{dx} - y\tan x + 2\sin x = 0$

3 $x^4\frac{d^2y}{dx^2} + n^2y = 0$ **4** $y = \frac{1}{2}x^4 + Cx^2$

5 $\sec y + \tan y = C(\csc x + \cot x)$

6 $(x + y)^3 = C(x - y)$ **7** $e^y = e^x + C$ **8** $y = A\cos 2x + B\sin 2x$

9 $Cxy = y + \sqrt{y^2 - x^2}$ **10** $Cy = x(y + 1)$

Answers | 375

11 $y = e^{\frac{3x}{8}}\left(A\cos\frac{\sqrt{5}x}{8} + B\sin\frac{\sqrt{5}x}{8}\right)$

12 $y\sec^2 x = \sin x + C$ **13** $y = e^{-2x}(Ax + B)$

14 $\ln(y+1) = \tan^{-1}(\sin x) + C$ **15** $y = \sin x - 1 + Ce^{-\sin x}$

16 $Cx^2 = (y^2+1)(x^2+1)$ **17** $y = Ae^{4x} + Be^{-3x}$ **18** $y^2(1+x^2) = C$

19 $x + y = \tan(x+C)$ **20** $2y\cos x = \cos 2x + C$ **21** $\sin x \cos y = C$

22 $y = e^{-\frac{3x}{2}}\left(A\cos\frac{\sqrt{11}}{2}x + B\sin\frac{\sqrt{11}}{2}x\right)$ **23** $2y = x^2 - 1 + Ce^{-x^2}$

24 $y = Ae^{\sqrt{2}x} + Be^{-\sqrt{2}x}$ **25** $y(x-1) = x^4 - x^3 + Cx^2$ **26** $y(1 + \ln x) = 1$

27 $y = \frac{1}{6}W(l-x)^3 + \frac{Wl^2}{2}x - \frac{Wl^3}{6}$ **28** $y = x - 1 + e^{-x}$ **29** $y = \frac{1}{3}(4e^{-\frac{x}{2}} - e^{-2x})$

31 $x^4\frac{d^2y}{dx^2} + 2x^3\frac{dy}{dx} + 4y = 0$ **32** $y^2 = kx^2 + C$ **33** $2x = 1 + 3e^{-t^2}$

34 $2 + x^3y = Cxy;\ 2 + x^3y = 3xy$ **35** $y = \ln(e^x - A) - x + B$

36 $Q = e^{-Rt/2L}Q_0\left(\cos nt + \frac{R}{2Ln}\sin nt\right)$ where $n = \sqrt{\frac{1}{LC} - \frac{R^2}{4L^2}}$

37 $\frac{1}{x\ln z} - \frac{1}{2x^2} = C$ **38** $x^2 - y^2 = C$ **39** $z = \frac{x^4}{4} + \frac{Cx^2}{2} + Cx + C\ln(x-1) + D$

40 $x = \frac{k}{2}\left(1 - \frac{3}{5}e^{-\frac{6t}{11}} - \frac{2}{5}e^{-t}\right);\ y = \frac{1}{5}\left(e^{-t} - e^{-\frac{6t}{11}}\right)$ **41** $A + y = B(A-y)e^{2Ax}$

EXAMPLES 20a (Page 291)

1 (i) $2 + i$; (ii) $3 - i\sqrt{2}$; (iii) $1 + 3i$; (iv) $c + ki$; (v) $a(1 + i\sqrt{2})$

2 (i) -1; (ii) 1; (iii) $-i$; (iv) -1; (v) $-\frac{1}{2}i$

3 (i) $2 - 3i$; (ii) $3 + 4i$; (iii) $-2i$; (iv) $1 + i\sqrt{3}$; (v) $1 - i$; (vi) $\cos\theta - i\sin\theta$

5 (i) $\pm i$; (ii) $\pm 3i$; (iii) $\frac{1}{2}(1 \pm i\sqrt{3})$; (iv) $\frac{1}{4}(3 \pm i\sqrt{7})$; (v) $-4 \pm 3i$; (vi) $1 \pm 2i$; (vii) $\pm i,\ -1 \pm i$.

7 (i) $x = 0,\ y = 0$; (ii) $x = 2,\ y = -1$; (iii) $x = 2,\ y = -2$; (iv) $x = 1,\ y = 1$.

8 $3 - i$.   **9** 0.

EXAMPLES 20b (Page 293)

1 (i) $x = 2,\ y = 3$; (ii) $x = 0,\ y = \sqrt{3}$; (iii) $x = 1,\ y = -1$; (iv) $x = 1,\ y = -2$; (v) $x = 2,\ y = 3$.

2 (i) $4 - i,\ 2 - 3i$; (ii) $3,\ -1 - i2\sqrt{3}$; (iii) $4 + 2i,\ -4 + 4i$; (iv) $-4 + i,\ 8 - 5i$.

4 (i) $x = 3,\ y = 1$; (ii) $x = 1,\ y = -1$; (iii) $x = -1,\ y = 2$; (iv) $x = -\frac{2}{5},\ y = \frac{6}{5}$.

6 $u = \frac{x^2}{x^2 + y^2},\ v = \frac{y^2}{x^2 + y^2}$ **7** (i) $x = 1,\ y = -1$; (ii) $x = 2,\ y = 1$

9 $z = 2, a = -2$

10 $a = 3, b = 1$ or $a = 1, b = 3$; $a = -3, b = -1$ or $a = -1, b = -3$

11 (i) -1; (ii) 1; (iii) -1

EXAMPLES 20c (Page 295)

1 (i) $3 + i$; (ii) 13; (iii) $7 - 24i$; (iv) $\frac{1}{2}(1 + i)$; (v) $\frac{1}{13}(4 + 6i)$; (vi) $\frac{1}{5}(1 - 3i)$; (vii) $\frac{1}{17}(1 + 13i)$; (viii) $\frac{1}{10}(3 + i)$; (ix) $-2(1 + i)$; (x) $\frac{1}{25}(2 + 11i)$.

2 (i) $x = 5, y = -12$; (ii) $x = \frac{2}{29}, y = -\frac{5}{29}$; (iii) $x = -\frac{1}{2}, y = \frac{3}{2}$; (iv) $x = -\frac{2}{5}, y = -\frac{4}{5}$

3 (i) $z = \frac{1}{4}(3 - i\sqrt{7})$ **6** $z^2 - 4z + 13$ **7** $a = -12$

8 $\frac{1}{2}(1 + 3i)$ **9** $\sqrt{3}$ **10** $-1, \frac{1}{2}(1 \pm i\sqrt{3})$ **11** 0

12 $a = 2, b = 1; a = -2, b = -1; \pm(2 + i)$

13 $z = 4, 1 \pm i\sqrt{3}$ **14** $x = -2, y = 4$

15 $x = 3, y = -1$ or $x = -3, y = 1$

12 $(x - 1 + 2i)(x - 1 - 2i); (x + a + iy)(x + a - iy)$

17 $\dfrac{2x}{x^2 + a^2}$ **18** (i) 0; (ii) 0; (iii) 1

20 $a = -2, b = 4$ **21** $\cos\theta + i\sin\theta$

EXAMPLES 20d (Page 298)

1 (i) 5; (ii) $\sqrt{5}$; (iii) 4; (iv) $\sqrt{2}$; (v) $\sqrt{2}$; (vi) 13; (vii) 2; (viii) 5; (ix) 3; (x) $a\sqrt{2}$

2 (i) $\tan^{-1}\frac{4}{3}$; (ii) $\tan^{-1}2$; (iii) $\frac{1}{2}\pi$; (iv) $-\frac{1}{4}\pi$; (v) $\frac{3}{4}\pi$; (vi) $-(\pi - \tan^{-1}\frac{12}{5})$; (vii) $-\frac{1}{3}\pi$; (viii) $-(\pi - \tan^{-1}\frac{3}{4})$; (ix) $\tan^{-1}\frac{1}{2}\sqrt{5}$; (x) $\frac{1}{4}\pi$ if $a > 0$, $-\frac{3}{4}\pi$ if $a < 0$

3 (i) $5(\cos\theta + i\sin\theta)$ where $\theta = -\tan^{-1}\frac{3}{4}$; (ii) $\sqrt{2}(\cos\frac{1}{4}\pi + i\sin\frac{1}{4}\pi)$; (iii) $1(\cos 0 + i\sin 0)$; (iv) $1(\cos\pi + i\sin\pi)$; (v) $1(\cos\frac{1}{2}\pi + i\sin\frac{1}{2}\pi)$: (vi) $1[\cos(-\frac{1}{2}\pi) + i\sin(-\frac{1}{2}\pi)]$; (vii) $2(\cos\frac{2}{3}\pi + i\sin\frac{2}{3}\pi)$; (viii) $2[\cos(-\frac{1}{6}\pi) + i\sin(-\frac{1}{6}\pi)]$; (ix) $3(\cos\theta + i\sin\theta)$ where $\theta = -(\pi - \tan^{-1}\frac{1}{2}\sqrt{5})$; (x) $a\sqrt{2}(\cos\frac{1}{4}\pi + i\sin\frac{1}{4}\pi)$

4 $\cos\frac{2}{3}\pi + i\sin\frac{2}{3}\pi, \cos(-\frac{2}{3}\pi) + i\sin(-\frac{2}{3}\pi)$

5 $\frac{1}{2}(1 \pm i\sqrt{3}); \cos\frac{1}{3}\pi + i\sin\frac{1}{3}\pi, \cos(-\frac{1}{3}\pi) + i\sin(-\frac{1}{3}\pi)$

6 (i) $\sqrt{10}(\cos\theta + i\sin\theta)$ where $\theta = \tan^{-1}\frac{1}{3}$; (ii) $\frac{1}{5}\sqrt{10}(\cos\theta + i\sin\theta)$ where $\theta = \tan^{-1}\frac{1}{3}$

9 $2i\sin\theta$ **10** OP_1, OP_2, OP_3, where $OP_1P_3P_2$ is a parallelogram

12 Circle, centre O, radius 1 **15** -1

16 $\cos 4\theta = \cos^4\theta - 6\cos^2\theta\sin^2\theta + \sin^4\theta, \sin 4\theta = 4\cos^3\theta\sin\theta - 4\cos\theta\sin^3\theta$

MISCELLANEOUS EXAMPLES (Page 299)

1 $\frac{1}{2}(-1 + 7i)$ **2** -16 **3** $z = 3, \pm i\sqrt{3}$

5 $a = -1; z = -1, 1 - i$ **6** (i) $-3 + 2i$; (ii) $2 + 3i$

7 $a = -3, b = 5; z = -1$ **8** $2 + i$
9 $x = 16, y = 30$ **10** $(\cos\theta - i\sin\theta)^5$ or $\cos 5\theta - i\sin 5\theta$
11 (i) $\sqrt{13}$; (ii) $\sqrt{13}$; (iii) $\sqrt{2}$ **12** $u = x^2 - y^2, v = 2xy$
13 $\cos\frac{1}{3}\pi + i\sin\frac{1}{3}\pi, \cos\pi + i\sin\pi, \cos(-\frac{1}{3}\pi) + i\sin(-\frac{1}{3}\pi)$
14 $\dfrac{1 + \cos\theta - i\sin\theta}{2(1 + \cos\theta)}$ or $\frac{1}{2}(1 - i\tan\frac{1}{2}\theta)$
15 (i) $2\sqrt{2}, \frac{7}{12}\pi$; (ii) $8, \pi$ **17** $\dfrac{1}{\sqrt{2}}(\sqrt{3} + i)$
18 $z = \frac{1}{7}(3 \pm 2i\sqrt{3})$ **20** $\pm 2i, a = 5$
22 When the points z_1, z_2, z_3 are collinear   **24** $z = \pm 2 + i$

PAPER A (1) (Page 301)

1 $-\frac{10}{27}; x^2 - 10x + 9 = 0$
2 $x^6 - 6x^5 y + 15x^4 y^2 - 20x^3 y^3 + 15x^2 y^4 - 6xy^5 + y^6$; 59 348 000
3 $x < -4, 0 < x < 1$ **4** 11·54; 27·55
5 0, 0·17, 2·45, 4·20, 5·11   **6** $2\sqrt{13}$
7 $y + x(t+1) = a(t+1)(t^2 + 2t + 2)$
8 $18x(3x^2 + 7)^2$; $-\dfrac{2x}{(x^2 - 4)^2}$; $\frac{4}{5}x^5 - 4x^3 + 9x + c$; $\frac{11}{84}$
9 Maximum, $(4, \frac{1}{32})$, Minimum $(1\frac{1}{2}, 1\frac{1}{3})$ **10** $4\frac{1}{2}; 341·1\pi$

PAPER A (2) (Page 301)

1 (i) $x = 3, y = -5; x = -\frac{4}{17}, y = -\frac{5}{34}$; (ii) 95 433
2 (i) $-3, 9$; (ii) $n = 9$ **3** 174(·5) m
4 0   **5** 21·4 cm^2
6 4; 20; (11, 11), $(-1, -5)$
7 Centre (4, 8), Radius 5; $y = 13, 4y - 3x = 45; \frac{3}{4}$
8 (i) Minimum (0, 0), Maximum $(-1, -1)$; (ii) $2\frac{2}{3}$
9 (a) $3/(3-x)^2, \dfrac{\pi}{180}\cos x°$; (b) 1·9983
10 $s = 8t + 5t^2 - t^3; t = \frac{5}{3}s, s = 48$ m

PAPER A (3) (Page 302)

1 (a) $x = 3$; (b) $x = 2·54$ **2** $1 + 6x + 9x^2 - 20x^3 - 90x^4$; 1·1234
3 $A = 4, B = -1$; 398·3   **4** 78°54′; 6·354 m
5 (i) 0°, 45°, 135°, 180°, (ii) 35°15′, 144°45′; (iii) 90°
6 $(9, -10)$; Area 134   **7** Centroid $(\frac{7}{3}, \frac{8}{3})$; Circumcentre $(\frac{45}{22}, \frac{63}{22})$
8 (ii) $\dfrac{5}{4\pi}$ cm s^{-1}
9 (i) Maximum $\frac{4}{27}$, Minimum 0; (ii) $\frac{1}{4}\sin 2x + \frac{1}{2}x + c, \sin x - \frac{1}{3}\sin^3 x + c$
10 $6\frac{3}{4}$ unit^2

PAPER A (4) (Page 303)

1. (i) $x = 2, y = -1$; $x^2 - x(p^2 - 2q + 2) + p^2 + q^2 - 2q + 1 = 0$
3. $B = 38°41'$, $C = 95°19'$, $a = 2\cdot66$ m 4. $30°$ 6. $26/\sqrt{65}$
9. $6\frac{2}{3}, \frac{1}{2}(1 - 2/\pi)$; $\frac{1}{2}\sqrt{2x^2 - 3} + c$ 10. $6\pi^2 + 9\sqrt{3}\pi$

PAPER A (5) (Page 304)

1. (i) $x = 14$; (ii) $x = -2, y = 2$; $x = 4, y = -1$ 2. (ii) $0\cdot952996$
3. $-8(x - \frac{3}{4})^2 - \frac{1}{2}$ 5. $x = 0, 78°32', 127°40'$
6. $y - 3t^2 x + 2t^3 = 0$; $y = 28x^3$ 7. $44x + 117y = 0$
8. $\dfrac{4}{x+1} - \dfrac{1}{x-1}$; Minimum $\frac{9}{2}$, Maximum $\frac{1}{2}$
9. $4y = 11x - 20$; $(\frac{25}{16}, -\frac{45}{64})$ 10. (i) $0\cdot245, 0\cdot005$; (ii) $0\cdot959$

PAPER A (6) (Page 305)

1. (a) $\dfrac{p(p^2 - 3q)}{q}$; (b) $2x^2 - 6x + 9 = 0$. 2. (a) $40\,320$; (b) 4320
3. $37°27'$ 4. N $16°49'$E; $14°42'$. 5. $n\pi/3$ where $n = 1, 2 \ldots 5$.
6. $y = 2x + 1$ 7. $2x^2 + 2y^2 - 15x - 11y + 22 = 0$.
8. (i) $\dfrac{\cos\sqrt{x}}{2\sqrt{x}}$; (ii) $\dfrac{\cos x}{2\sqrt{\sin x}}$; (iii) $2\ln 10 \cdot 10^{2x}$
9. $y + 2t^3 x = 3at^2$; $9a^2 t^4 + \dfrac{9a^2}{4t^2}$; $\pm\dfrac{\sqrt{2}}{2}$. 10. (i) $\frac{1}{4}$; $\dfrac{5\sqrt{2}}{12}$; (ii) 6π.

PAPER A (7) (Page 306)

1. (a) $(3x + 2y - 4)(x - 4y + 3)$; (b) $x = -1, y = 2, z = 3$, 2. $n = 43$.
3. $2x - \dfrac{(2x)^2}{2} + \dfrac{(2x)^3}{3} - \dfrac{(2x)^4}{4}$; $-\frac{1}{2} < x \leq \frac{1}{2}$. 4. $7\cdot45$ cm.
5. $67°16', 0\cdot665$ 6. (a) Inside; (b) $(\frac{345}{16}, 10)$
8. (i) $\frac{1}{2}(ad - bc)(ax + b)^{-1/2}(cx + d)^{-3/2}$ 9. $\dfrac{av}{100h - a}$ m s^{-1}.
10. 18 unit^2; $\dfrac{256\pi}{15}$ unit^3

PAPER A (8) (Page 306)

1. $(\alpha + \beta)^3 - 3\alpha\beta(\alpha + \beta)$; $8x^2 - 43x - 343 = 0$
2. (i) $15\,600$; (ii) $17\,576$; (iii) $6\,300$
3. (a) $30°, 150°, 270°$; (b) $45°, 146°18', 225°, 326°18'$

Answers | 379

4 $r^2(\sqrt{3}-\pi/3)\,\text{cm}^2$ **5** $25\cos(x-73°44')$; (i) 25; (ii) $7°19'$, $140°9'$
6 $(\frac{33}{5}, \frac{26}{5})$; $19\frac{1}{2}$ unit^2.   **7** $(3, 2), (-5, -4)$
8 (i) $\dfrac{1}{2\sqrt{x}(1-x)^{3/2}}$ **9** $a = \frac{1}{2}, b = -1$.   **10** (a) $13\frac{41}{42}, \ln\frac{3}{2}$; (b) $1:3$

PAPER A (9) (Page 307)

1 (a) $\dfrac{n}{2}\{2\ln a + \overline{n-1}\ln b\}$ **2** 180; 60   **5** $\dfrac{b\sin\theta}{a}$

6 10, 20; $(8, -3), (13, 7), (-3, -5)$
7 $x^2 + y^2 - 4x - 2y - 4 = 0$; $3\frac{1}{2}$; $15y + 8x + 20 = 0$
9 (a) $6x^5 - 4x - 2/x^3$; $\dfrac{\cos x + \sin x}{\sin x - \cos x}$; (b) $(3\frac{4}{7}, 13)$

10 (a) $\dfrac{x}{2} - \dfrac{\sin 4x}{8} + c$; (b) $-2\sqrt{x} + \ln\dfrac{1+\sqrt{x}}{1-\sqrt{x}} + c$

PAPER A (10) (Page 308)

1 0·59, 2·0, 3·41; $0 < k < 1·45$
2 (a) $\frac{7}{40}$; (b) $4x + \dfrac{16x^3}{3} + \dfrac{64x^5}{5} + \dfrac{256x^7}{7}$; $-\frac{1}{2} < x < \frac{1}{2}$
3 $88°6'$; $62·52\,\text{cm}^2$ **4** 20 m   **6** Same side   **7** $(2, 1), (6, 13)$
8 $y = tx - t^3$; $(3t^2/4, -t^3/4)$ **9** $\dfrac{4\pi R^3}{3\sqrt{3}}$ **10** $2 - \ln 3$; $\pi(\frac{8}{3} - 2\ln 3)$

PAPER A (11) (Page 309)

1 (i) 1260; (ii) $\frac{2}{15}$ **2** (i) $-1\,399\,680$; (ii) $0·5004$.
3 $\dfrac{n}{3}\{4n^2 + 3n + 2\}$ **5** 93·1 m   **6** $(0, 0)$; 1:

7 Centre $(2, -1)$; Radius 5.
9 (i) $-\dfrac{1}{2x^{3/2}(1-x)^{1/2}}$; $xe^{-x}(2-x)$ (ii) $\dfrac{\lg e}{x}$; $0·0009$ **10** $0·0112$

PAPER A (12) (Page 310)

1 (a) $x = \pm 2\sqrt{2}$; $y = \pm\sqrt{2}$; $x = \pm\sqrt{3}, y = \pm\dfrac{4\sqrt{3}}{3}$. (b) $-2 < k < 1$
2 $n = 7, -4$; Coefficient = 56 or 45.   **3** 0·45, 1·40, 2·93(5).
5 (i) $0°, \pm 120°$; (ii) $-138°21', -97°39', 41°39', 82°21'$ **6** $1, \frac{1}{7}$; $3\sqrt{2}$
7 $3(x^2 + y^2) - 10x + 3 = 0$; Centre $(\frac{5}{3}, 0)$, Radius $\frac{4}{3}$; $3y + 4x = 0$

Answers

8 (i) $-\frac{105}{2}x^2(1-x)^{\frac{1}{2}}$; $x = 0, 1$; (ii) $-4e^{-2x}(x^2+4x+3)$; $x = -1, -3$
9 (i) $\frac{4}{3}x^3 - 8x^{3/2} + 9\ln x + c$; $\frac{1}{5}\cos^5 x - \frac{1}{3}\cos^3 x + c$; (ii) $14\frac{2}{15}$.
10 $1\frac{2}{3}$ unit^2; $\frac{14}{5}\pi$ unit^3

PAPER B (1) (Page 311)

1 (a) $\dfrac{n}{3}(4n^2 + 12n - 1)$; (b) $\dfrac{1}{3} - \dfrac{n+1}{(2n+1)(2n+3)}$

2 $\dfrac{1}{2(x-2)} + \dfrac{1}{x+2} + \dfrac{3x+2}{2(x^2+4)}$ 3 $a = 3$ 5 $\tan^{-1}(\sqrt{2}\cot\alpha)$

5 $1 - \dfrac{x^2}{2!} + \dfrac{x^4}{4!} - \dfrac{x^6}{6!}$; $\frac{3}{2}$ 6 (a) $y - a = mx$; $y = x + 6$ (b) $\frac{3}{2}$

8 Maximum $\frac{1}{2}$, Minimum $-\frac{1}{2}$; Inflexions $(0, 0)$, $\left(\pm\sqrt{3}, \pm\dfrac{\sqrt{3}}{4}\right)$

9 (i) $x + \frac{1}{3}x^3$ (ii) $16, \frac{1}{2}$ 10 $17\frac{1}{5}$ unit^2

PAPER B (2) (Page 312)

1 (i) $b^2 = 4ac$; (ii) $\dfrac{b^2}{ac} = \dfrac{b'^2}{a'c'}$ 2 $\dfrac{857}{1105}$

3 (a) $A = \frac{1}{2}$, $B = -\frac{3}{4}$, $C = \frac{1}{4}$, $D = 3$; (b) $x = 3$, $y = 5$, $z = 2$; $x = -1$, $y = -3$, $z = -\frac{2}{3}$.

5 0·49   7 $\dfrac{x^2}{25} + \dfrac{y^2}{9} = 1$; $20y + 9x = 75$; $(\frac{25}{4}, \frac{15}{16})$

10 (i) $-\frac{1}{3}$; (ii) $\dfrac{1}{e^x + 1} + x - \ln(e^x + 1) + c$

PAPER B (3) (Page 313)

2 (i) $-\frac{7}{18}$; (ii) $1 + 20x + 163x^2$.
3 (i) $p + q + r = pqr$; (ii) $pq + qr + rp = 1$; (iii) $p + q + pqr = r$
4 $1:2$; $60°\ 28'$.   5 (i) $45°, 60°, 135°, 225°, 300°, 315°$; (ii) $20°\ 16', 86°$.
6 (i) $y(bc' - b'c) + x(ac' - a'c) = 0$;
  (ii) $ya(ab' - a'b) - xb(ab' - a'b) + a(ac' - a'c) + b(bc' - b'c) = 0$.
7 $yt - x = at^2$ 8 (i) $8(3 - 4x - 4x^2)^{-\frac{3}{2}}$; $-2xe^{-x}\sin x$

9 $\dfrac{4}{x-1} - \dfrac{6}{2x-1}$ 10 $\frac{1}{8}$

PAPER B (4) (Page 314)

3 (a) Sum $= -(\alpha^2 + \beta^2)$, Product $= (\alpha - \beta)^2$; (b) $n = 12$, $k = 12$.
4 $a = 28\cdot24$ cm, $B = 101°\ 34'$, $C = 26°\ 52'$.

5 $\theta = 20°, 100°, 140°, 220°, 260°, 340°; x = 0.940, -0.174, -0.766$.

6 $x^2 + y^2 - 4x - 12y + 36 = 0$; $(0, 4)$.   **8** $\dfrac{5x}{\sqrt{5x^2-1}}, \dfrac{3x^2-3}{x^3-3x}$.

9 $\dfrac{3\pi}{8}$ **10** (a) $\ln\dfrac{1+x}{x} - \dfrac{1}{x} + c$, (b) $\tfrac{1}{4}\ln 3$.

PAPER B (5) (Page 315)

1 (a) $x = \tfrac{16}{9}, y = -\tfrac{3}{4}; x = 3, y = -\tfrac{5}{3}$ (b) $6p^2 \geqslant q^2$

2 (a) $1 - 2x - 2x^2 - 4x^3 - 10x^4$; (b) $a = \tfrac{4}{15}, b = \tfrac{3}{5}; c = \tfrac{8}{175}$

3 2·71   **5** $\tfrac{1}{2}(a-b)\cos 2\theta + \tfrac{1}{2}k\sin 2\theta + \tfrac{1}{2}(a-b)$

6 $(\tfrac{3}{5}, \tfrac{8}{5}), (-\tfrac{2}{5}, \tfrac{3}{5}), (\tfrac{7}{5}, -\tfrac{3}{5}), (\tfrac{12}{5}, \tfrac{2}{5})$; $y = x - 2, 2x + 3y = 1$

9 $(\ln\tfrac{1}{2}, 2); (\tfrac{1}{2}\ln\tfrac{1}{2}, \tfrac{4}{3}\sqrt{2})$ **10** $\tfrac{1}{2}$ unit^2; $\dfrac{19\pi}{80}$ unit^3

PAPER B (6) (Page 316)

1 270 725; 69 184; 6768; 76 145   **2** $1 + \dfrac{2}{3x} + \dfrac{5}{3(2x-3)}$

3 $a + b - 1, \tfrac{1}{2}(a^2 - b + 1), \tfrac{1}{6}(a^3 + 2b - 3); a = -1, b = 2$ **5** 34·31 m^2

6 $(0, 0)$ **8** $\dfrac{2\pi a^3}{3\sqrt{3}}$ **9** $x - \dfrac{x^2}{2} + \dfrac{x^3}{6}$ **10** $2\ln\tfrac{3}{2}; \dfrac{2\pi}{3}$

PAPER B (7) (Page 317)

2 (i) $\dfrac{12!(\cdot 03)^r}{(12-r)!\, r!}$ (ii) -3 **3** $15° \; 16°$

4 (i) Greatest $\sqrt{2}-1$, Least $-(\sqrt{2}+1)$ **6** $\dfrac{x\cos\theta}{a} + \dfrac{y\sin\theta}{b} = 1$

8 (i) $a = 4, y = \dfrac{x^2}{2} - \dfrac{4}{x} + 1$;   (ii) $a = \dfrac{u}{2p}, b = -\dfrac{u}{2p}$

9 $2\tan^{-1}\sqrt{x} + c; \tfrac{1}{3}\cos^3 x - \cos x + c$ **10** $2\sqrt{3\pi}$ unit^2

PAPER B (8) (Page 317)

1 2002; 252   **2** (a) $\dfrac{x+2}{x^2+1} + \dfrac{4}{x-2}$; (b) $\tfrac{1}{4}$

5 (a) $18°, 90°, 162°$; (b) $\cos\theta$

6 $AB \; y = 2x - 1, AD \; 2y + x = 3, CD \; y = 2x + 9, BC \; 2y + x = 18$

7 $y(t_1 + t_2) - 2x = 2at_1 t_2$ **8** 0·347, 0·414, 0·308

9 (i) $\dfrac{-4x}{\sqrt{3-4x^2}}$, (ii) $\dfrac{1}{\sqrt{1-x^2}} - \dfrac{x}{1-x^2}$, (ii) $3\cos 3x + 3\cos^2 x \sin x$ **10** 1·2

PAPER B (9) (Page 318)

1 $x = 1, y = 2, z = -3; x = -1, y = -2, z = 3$

2 (i) 105, 20. (ii) $1 + \dfrac{x}{2} + \dfrac{3x^2}{8} + \dfrac{5x^3}{16}$, 1·41421

4 (a) $0°, \pm 120°$; (b) $\pm \dfrac{x-y}{1+xy}, \pm \dfrac{1-xy}{x+y}$

6 $y\sqrt{3} + x = 2a, \ y - x\sqrt{3} + \dfrac{2a\sqrt{3}}{3} = 0$ **8** $\dfrac{3\sqrt{3}r^2}{8}, \dfrac{3\sqrt{3}r}{2}$

10 (i) $3\tfrac{1}{3}$, (ii) $\tfrac{1}{2}\tan^{-1}(2\tan^{-1}x)$

PAPER B (10) (Page 319)

2 1·43   **3** (i) $n\pi + \alpha$; (iii) $x = n180° + 10°18' + (-1)^n\,26°\,34'$

4 $41°\,39'$ **7** $\dfrac{xa}{\cos\phi} - \dfrac{yb}{\sin\phi} = a^2 - b^2$

8 (a) (i) $\dfrac{x^2 - 8x + 14}{(x-4)^2}$, (ii) $\dfrac{2}{\sqrt{x^2+1}}$, (iii) $e^{-5x}(3\cos 3x - 5\sin 3x)$

  (b) $-\dfrac{b}{a}\tan\theta, \ \dfrac{b}{3a^2\cos^4\theta\sin\theta}$

9 Curve $y = 2x^3 + 6x^2 - 6$, tangent $y + 6x + 8 = 0$

10 (i) $\dfrac{\pi}{8}$, (ii) $\tfrac{1}{4}\ln\tfrac{432}{125}$, (iii) $\dfrac{3\sqrt{3}}{8}$, (iv) $\dfrac{\pi}{8}$

PAPER B (11) (Page 320)

1 (a) 0·557 (b) Maximum $\tfrac{1}{25}$, Minimum 1.   **2** 31

5 $x^2 + y^2 - x(a+b) + ab = 0$ **9** $\bar{x} = \dfrac{4a}{3\pi}, \ \bar{y} = \dfrac{4b}{3\pi}$

10 (i) $\sin^{-1}(x-1) + c$; (ii) $\ln\dfrac{x-3}{\sqrt{x^2+1}} + c$; (iii) 1

PAPER B (12) (Page 321)

1 (i) $q = -3; x = -1, 2, 3$ (ii) $x = 2, 6$ **3** (a) $\tfrac{24}{25}, \tfrac{336}{625}$ (b) $32°\,46'$

5 $x^2 + y^2 - 6x - 6y + 10 = 0; \ y - x + 4 = 0, \ y = x + 4$

8 (i) $P(2, 12), Q(-1, -6)$; (ii) $\dfrac{3\sqrt{3}}{2}$

9 (i) $\ln\dfrac{x-1}{x} + \dfrac{1}{x} + c$; (ii) $\dfrac{\tan^3 x}{3} - \tan x + x + c; \ 4\tfrac{2}{5}$

10 (b) $\dfrac{32\sqrt{2}}{15} = 3\cdot 02 \text{ unit}^2$

Answers | 383

PAPER C (1) (Page 322)

1 $(-2, 4)$ Min., $(-\frac{2}{3}, -4)$ Max.; $x = -\frac{4}{3}$

2 $\ln(1+\frac{1}{2}) = \frac{1}{2} - \frac{1}{2}(\frac{1}{2})^2 + \frac{1}{3}(\frac{1}{2})^3 - \frac{1}{4}(\frac{1}{2})^4 + \frac{1}{5}(\frac{1}{2})^5 \ldots$;
$\ln(1-\frac{1}{2}) = -(\frac{1}{2} + \frac{1}{2}(\frac{1}{2})^2 + \frac{1}{3}(\frac{1}{2})^3 + \frac{1}{4}(\frac{1}{2})^4 + \frac{1}{5}(\frac{1}{2})^5 \ldots)$ **3** 0·45

5 $y^2 = 4ax$; 2 **7** (a) (i) $\tan^{-1} x$, (ii) $5e^{-x}(\cos 2x - 1)$; (b) $-8, -2$

8 $-\frac{3}{4}, \frac{75}{64}, \frac{5}{3}$ **9** (a) $-\frac{2}{3}\sqrt{1-x}(x+5) + c$, $\dfrac{\sin^4 x}{4} - \dfrac{\sin^6 x}{6} + c$ (b) $\frac{1}{3}$

PAPER C (2) (Page 323)

1 $\lambda = a + b \pm 2\sqrt{ab}$ **3** $120°, 153° 26', 240°, 333° 26'$ **4** $\dfrac{\sqrt{5}-1}{2}$

5 2·68; 1 **7** $3y - 4x + 2a = 0, \left(\dfrac{a}{3}, -\dfrac{2a}{9}\right)$ **8** (a) $y \pm 8x + 6 = 0$

9 $\dfrac{\pi}{12}$, $8\ln 2 - \frac{3}{4}$, $\frac{1}{2}\ln\frac{9}{8}$

PAPER C (3) (Page 324)

1 $\dfrac{n!}{\left(\dfrac{n}{2}!\right)^2}, \dfrac{n!}{\left(\dfrac{n+1}{2}\right)!\left(\dfrac{n-1}{2}\right)!}$; 45

3 $\dfrac{\tan A + \tan B + \tan C - \tan A \tan B \tan C}{1 - \tan B \tan C - \tan C \tan A - \tan A \tan B}$

5 $x = 0, y = \pm\left\{\dfrac{x(a-b)}{2\sqrt{ab}} + \sqrt{ab}\right\}$ **6** $\dfrac{x^2}{9} - \dfrac{y^2}{7} = 1$; $x^2 + y^2 = 2$

7 (a) $\dfrac{2}{x^2 + 4}, \dfrac{-x^2 e^{-x}}{(x-2)^2}$; (b) $x = \pm c$

8 (a) $x = 0$ (Min.); $x = \pm\dfrac{\pi}{2}$ (Max.) (b) (i) $-b$, (ii) $-\sqrt{a^2 + b^2}$

9 $\frac{1}{4}\ln\dfrac{x}{x+2} - \dfrac{1}{2x} + c$, $\dfrac{x^3}{3}\ln x - \dfrac{x^3}{9} + c$

10 (i) $\frac{1}{2}\ln\dfrac{1+y}{1-y} + \dfrac{1}{1+x} = A$; (ii) $y\sqrt{1+x^2} = \dfrac{x}{2}\sqrt{1+x^2} + \frac{1}{2}\sinh^{-1} x + A$

PAPER C (4) (Page 326)

1 $1 + \dfrac{x}{2} - \dfrac{x^2}{8} + \dfrac{x^3}{16} - \dfrac{5x^4}{128}$ **3** $a = \frac{1}{2}, b = 0, c = -\frac{1}{24}, d = 0$

4 $\dfrac{\pi}{5}, \dfrac{3\pi}{5}, \dfrac{7\pi}{5}, \dfrac{9\pi}{5}$ **7** $1 < x < 3, x > \frac{13}{4}$

384 | **Answers**

9 (i) 0; (ii) $\frac{1}{8}$; (iii) $\frac{1}{4}(e^2+1)$
10 (i) $xy = 2(x-1)$; (ii) $4y \sec^n x = \sin 2x + 2x$

PAPER C (5) (Page 327)

3 $x = 10°\,12', 109°\,48', 190°\,12', 289°\,48'$ **5** $y + tx = 2at + at^3$

7 0·41   **8** (i) $a^n \cos\left(ax + \frac{n\pi}{2}\right)$; (ii) $\frac{(-1)^{n-1}(n-1)!}{(1+x)^n}$

PAPER C (6) (Page 328)

1 $1 + nx + \frac{n(n+1)}{2!}x^2 + \frac{n(n+1)(n+2)}{3!}x^3 + \ldots,\ -1 < x < 1;\ 3(1-\frac{1}{2})^{-\frac{2}{3}} - 4$

3 (i) $n180° + (-1)^n 30°,\ n180° + (-1)^{n+1} 48°\,36'$; (ii) $(2n+1)\frac{\pi}{2},\ (4n+1)\frac{\pi}{10}$

7 Limits 0, 0

8 (i) $\tan\frac{x}{2} + x\tan\frac{x}{2} - 2\ln\sec\frac{x}{2} + c,\ \ln\frac{x}{\sqrt{x^2+1}} - \frac{1}{x} - \tan^{-1}x + c$; (ii) $\frac{27}{140}$

9 1·202

10 (i) $x^2(1-y) = A(1+y)$; (ii) $y = x\ln x + Ax$; (iii) $y = Ae^{2x} + Be^x$

PAPER C (7) (Page 329)

1 (a) $6n^2$; (b) $\frac{17}{36} - \frac{6n^2 + 21n + 17}{6(n+1)(n+2)(n+3)}$; (c) $\frac{x}{1-x} - \ln(1-x)$

3 $\theta - \frac{\theta^2}{3!} + \frac{\theta^5}{5!} - \ldots,\ 1 - \frac{\theta^2}{2!} + \frac{\theta^4}{4!} - \ldots$; Limit 2

4 $n360° + 53°\,8',\ n360° - 67°\,22';\ 65s^2 + 8s - 48 = 0$

5 $x^2 + y^2 - 2xp\cos\alpha - 2yp\sin\alpha + 2p^2 - r^2 = 0$ **6** $\frac{x\cos\theta}{a} + \frac{y\sin\theta}{b} = 1$

7 $\frac{(5+9x_1^4)^{\frac{3}{2}}}{30x_1},\ \left(\frac{1}{\sqrt{3}}, \frac{1}{3\sqrt{15}}\right),\ \left(-\frac{1}{\sqrt{3}}, -\frac{1}{3\sqrt{15}}\right)$ **8** $x^2 + y^2 = 1$

9 (a) (i) $\ln x$, (ii) $\ln(-x)$; $\frac{1}{3}\ln\frac{2}{5}$ (b) $\ln\tan\frac{x}{2} + c,$

$\frac{1}{4}\ln\frac{1+x}{1-x} - \frac{1}{2}\tan^{-1}x + c$

10 $\left(\frac{9a}{20}, \frac{9a}{10}\right)$

PAPER C (8) (Page 330)

2 $\frac{\pi}{12} \leqslant \alpha \leqslant \frac{7\pi}{12}, \frac{13\pi}{12} \leqslant \alpha \leqslant \frac{19\pi}{12}$ **3** $x - \frac{x^2}{2} + \frac{x^3}{3} - \frac{x^4}{4} + \ldots, -1 < x \leqslant 1$

4 (i) $n180° + 60°, n180° + 68°23', n180° + 141°37'$;
(ii) $n180° + (-1)^n 12°59', n360° \pm 90°$

5 $(at_1t_2, \overline{at_1 + t_2})$ **7** $-\frac{\pi}{8}, \frac{3\pi}{8}$ **8** $\frac{3}{2ku^2}$

9 (i) $\frac{1}{2}x^2(\ln x)^2 - \frac{1}{2}x^2 \ln x + \frac{1}{4}x^2 + c$; (ii) $\ln(\sec x + \tan x) - \cot \frac{1}{2}x + c$;

$\frac{a^4}{8}\left(\frac{\pi}{6} - \frac{\sqrt{3}}{8}\right)$ **10** (i) $y^2 = 2(x-1)$; (ii) $y^2 = \tan^2 x + 4\tan x - 1$

PAPER C (9) (Page 331)

1 $\lambda = -6, -1$; $5s \equiv 24(x+\frac{1}{2})^2 + (x-2)^2$, $5s' \equiv 4(x+\frac{1}{2})^2 + (x-2)^2$

2 (a) $\frac{1}{x-1} + \frac{1}{x+1} - \frac{x-1}{x^2+x+1} - \frac{x+1}{x^2-x+1}$;

(b) $\frac{1}{(1-x)^2}\left[1 + \frac{2x(1-x^n)}{1-x} - (n+1)^2 x^n + n^2 x^{n+1}\right], \frac{1+x}{(1-x)^3}$

4 $\frac{a(1 \sim \cot \alpha \cot \beta)}{\sqrt{(1-\cot\alpha)^2 + (1-\cot\beta)^2}}$ **5** $\alpha = 2\tan^{-1}\frac{a-1}{a+1}$

6 $y + tx = 2at + at^3$; fixed point $(-2a, 0)$ **7** $2y = tx - at^3$

8 $\frac{7^{\frac{3}{2}}}{4}$ **9** (i) $\frac{3}{4}$; (ii) $\frac{1}{\sqrt{2}}$; (iii) $-\frac{2}{3}$ **10** (i) $y\sin^2 x = \sin x - x\cos x$;

(ii) 495

PAPER C (10) (Page 332)

1 (b) $\frac{1}{x^2}\ln(1-x) - \frac{1}{x}\ln(1-x) + \frac{1}{x}$ **2** (i) $x = -1, 2 \pm \sqrt{3}$

4 $-\frac{1}{6}$ **5** a **6** $y = mx \pm \sqrt{a^2m^2 - b^2}, \left(\mp\frac{a^2m}{\sqrt{a^2m^2-b^2}}, \mp\frac{b^2}{\sqrt{a^2m^2-b^2}}\right)$

7 $-\frac{(x^2+y)}{(y^2+x)}; \frac{2xy}{(y^2+x)^3}; -\frac{(y^2+x)}{(x^2+y)}; \frac{2xy}{(x^2+y)^3}$ **8** $80\pi + 12\pi^2$ sq. units

9 $\frac{1}{2}\{\ln(x-1) - \frac{1}{2}\ln(x^2+1) + \tan^{-1} x + c\}$; $\ln\frac{e^x}{1+e^x} + c$;

$\frac{1}{2}e^x(\sin x + \cos x) + c$; $-\frac{\pi^2}{4}$ **10** $\sqrt{\frac{g}{k}}$

PAPER C (11) (Page 333)

2 (i) $\dfrac{1}{x-1} + \dfrac{1}{x-2} + \dfrac{1}{x-3}$; (ii) $a = -14$, $b = -44$, $c = -104$, $e = 14$, $f = -13$;

$\dfrac{1}{2(x+1)} + \dfrac{1}{54(x-1)} - \dfrac{2}{27}\left[\dfrac{7}{x+2} - \dfrac{6}{(x+2)^2} + \dfrac{36}{(x+2)^3}\right]$

3 (ii) $\dfrac{1}{3}, \dfrac{2}{3}$; (iii) $\dfrac{3\pi}{10}$

7 (a) $\dfrac{2}{1+x^2}$; $2n^2 x^{n-1} \dfrac{(x^n - 1)^{n-1}}{(x^n + 1)^{n+1}}$ (b) $n!(-1)^n \left[\dfrac{1}{(x+2)^{n+1}} + \dfrac{1}{(x-2)^{n+1}}\right]$

8 (i) $\dfrac{\pi}{4} - \tfrac{1}{2}\ln 2$; (ii) $\dfrac{\pi}{3} + \sqrt{3} - 2$; (iii) $\tfrac{1}{2}(1 + 2e^{-\pi} + e^{-2\pi})$

9 $6\sqrt{3}a$ **10** $y = e^{-a\cos \alpha x}\{A\cos(a\sin\alpha)x + B\sin(a\sin\alpha)x\}$

PAPER C (12) (Page 334)

2 $\cosh(x+y) = \cosh x \cosh y + \sinh x \sinh y$; $x = 2, \tfrac{1}{6}$.

4 $\dfrac{1}{2}\left[\dfrac{\sin(2n-1)\dfrac{x}{2}}{\sin\dfrac{x}{2}} - 1\right]$ **5** $P'Q' : PQ = 3 : 1$

8 (b) $\dfrac{\sqrt{3}\pi}{6} - \tfrac{1}{2}\ln 3$; $\tfrac{1}{6}\ln \tfrac{27}{5}$ **9** $\dfrac{2555a^2}{558}$ **10** $\dfrac{7\sqrt{2}}{16} + \tfrac{3}{16}\ln(1+\sqrt{2})$

INDEX

Angles with same sine, 94; same cosine, 94; same tangent, 95
Approximate methods, integration, 64; solution of equations, 45, 103
Arc length, 238
Area, ellipse, 172
Area of surface of revolution, 239
Area under a curve, 68
Argand diagram, 271
Arrangements, 2
Asymptotes, 119, 186
Auxiliary circle, 171; equation, 261

Binomial theorem, 51; proof for positive integral index, 52; greatest term or coefficient, 51

Centre of gravity, 75; of plane area, 76; of semicircle, 78; of hemisphere, 78
Chance, 16
Change of origin, 119
Circle, analytical geometry of, 142; general equation, 144; passing through three points, 148; passing through the intersection of two given circles, 148
Combinations, 1, 6
Complementary Function, 265
Complex numbers, 271; argument of, 278; conjugate numbers, 271; fundamental processes, 273; geometrical representation, 271; modulus of, 278
Conjugate diameters, ellipse 176; hyperbola, 188
Convergent series, 36
Cube roots of unity, 276
Curvature 241; radius of, 242; centre of, 244; at the origin, 245
Curve tracing: Cartesian coordinates, 120; parametric coordinates, 123; polar coordinates, 126

Diameters, of ellipse, 169 (see also conjugate diameters)
Differential equations: formation of, 249; first order and degree, 251; homogenous, 254; linear, 256; variables separable, 252; of form $a\dfrac{d^2y}{dx^2}+b\dfrac{dy}{dx}+cy=0$, 261; of form $a\dfrac{d^2y}{dx^2}+b\dfrac{dy}{dx}+cy=f(x)$, 264

Director circle of ellipse, 175
Directrix: ellipse, 168; hyperbola, 186; parabola, 154
Division of a line, 136

Eccentric angle, 171
Eccentricity, 168, 185
Ellipse, analytical geometry of, 167; area, 172; axes, 168; diameters, 169; equation, 167; geometrical properties, 178; parametric equations, 171
Equations, approximate solutions, 45, 103; differential, 249; graphical solution, 100; simultaneous linear, 90; simultaneous quadratic, 90; trigonometrical, 93
Equations of a curve, 114
Evolute, 244
Expansions, binomial, 51; exponential, 36; logarithmic, 39; trigonometrical, 44
Exponential functions, differentiation, 30; expansions, 36; graphs, 30; integration, 30

Factorial, 4
Factors, solution of trigonometrical equations by, 99
Focal chord, 157
Focus, parabola, 154; ellipse, 167; hyperbola, 185

Gradient form of straight line, 133
Gradient forms of tangents, parabola, 159; ellipse, 174; hyperbola, 187
Graphical solutions of equations, 100

Hemisphere, centre of gravity of, 78
Hyperbola, analytical geometry of, 185; asymptotes, 186; axes, 186; equation, 185; geometrical properties, 187; parametric equations, 204
Hyperbola, rectangular; 189; equations, 189, 190; parametric equations, 190
Hyperbolic functions, 204

Identities, hyperbolic functions, 204; inverse circular functions, 200
Inequalities, 106
Integration, approximate methods, 64; definite 62; indefinite, 62; methods